Studies in Logic

Volume 1

Proof-Theoretical Coherence

Texts in Computing Series Editor: Ian Mackie (ian@dcs.kcl.ac.uk)
Volume 1
Programming Languages and Operational Semantics
Maribel Fernandez

Volume 2
An Introduction to Lambda Calculi for Computer Scientists
Chris Hankin

Volume 3
Introduction to Logic
Rob Nederpelt and Fairouz Kamareddine

Volume 4
The Haskell Road to Logic, Maths and Programming
Kees Doets and Jan van Eijck

Volume 6
Bridges from Classical to Nonmonotonic Logic
David Makinson

Studies in Logic Series Editor: Dov Gabbay (dg@dcs.kcl.ac.uk)
Volume 1
Proof-Theoretic Coherence
Kosta Došen and Zoran Petrić

Proof-Theoretical Coherence

Kosta Došen and Zoran Petrić

ISBN 1-904-987-06-0
King's College Publications
Scientific Director: Dov Gabbay
Managing Director: Jane Spurr
Department of Computer Science
Strand, London WC2R 2LS, UK
kcp@dcs.kcl.ac.uk
www.dcs.kcl.ac.uk/kcl-publications/

Cover design by Richard Fraser, www.avalonarts.co.uk
Interior illustrations by Vincent F. Hendricks
Printed by Lightning Source, Milton Keynes, UK

EDITORIAL PREFACE

Logicians have been aware for quite some time that there is a serious problem with the proof theory of classical logic. All candidates for the formal representation of classical proofs, either in the lambda calculus or in category theory, seemed to lead to triviality—namely, all proofs with the same premises and conclusions would have to be equal. I remember receiving in 1969 a handwritten mimeographed manuscript from George Kreisel, raising and discussing the problem of identity of proofs. The idea did not catch on at the time but th eproblem remained in the minds of many logicians.[1] Since the 1990s several attempts have been made, more or less explicitly, to address this problem by finding a plausible notion of identity of proofs in classical propositional logic that would not be trivial. Research along these lines has been conducted mostly in the area of linear logic (Lafont, Danos, Joinet, Schellinx) and in the lambda-mu calculus (Parigot, Ong, Ritter, Hofmann, Streicher, Selinger, Herbelin, Ghilezan, Lescanne, Likavec). Sometimes this work is advertised as looking for a 'constructive' interpretation of classical logic, or as showing that classical logic is 'deterministic', or as showing that it is 'algorithmically consistent'. See also the chapter by S. Artemov and L. Beklemischev in the *Handbook of Philosophical Logic*, 2nd edition, Volume 13, pages 189–360. Kluwer, 2004.

That this problem is intensely studied these days is shown by the fact that the number of publications in the area seems to grow. Since the present book was written, an extended abstract of Fuhrmann's and Pym's work has appeared ('On the geometry of interaction for classical logic', in the Proceedings of the Nineteenth Annual IEEE Symposium LICS 2004, Turku, July 2004, pp. 211-220; see also by the same authors 'Order-enriched categorical models of the classical propositional sequent calculus ', draft, March 2004, which reports on work done within an EPSRC project also involving Hyland, Robinson and Urban). The common denominator of this parallel work and the present book is in realizing the importance for classical logic of dissociativity (alias linear distribution, originally introduced for linear

[1] I no longer have Kreisel's manuscript, but I remember one of his main examples: $p \wedge \neg p \vdash p \rightarrow p$. This consequence has at least two distinct proofs. Of course now I can offer a solution to the problem of identity of proofs using Labelled Deductive Systems, but at the time I did not know what to do with the problem.

logic by Cockett and Seely)—an idea that goes back to Dosen's programmatic paper 'Identity of proofs based on normalization and generality' (talk at the Logic Colloquium in Muenster, August 2002, http://wwwmath.uni-muenster.de/LC2002/proceedings.html, published in The Bulletin of Symbolic Logic vol. 9, 2003, pp. 477-503, section 6). The present book fulfills the program announced in that paper, and it is hoped that the book will prove an important contribution to several related currents of research in proof theory, category theory, theoretical computer science and the philosophy of logic.

Dov Gabbay
London, December 2004

Preface

This is a book in categorial (or categorical) proof theory, a field of general proof theory at the border between logic and category theory. In this field the language, more than the methods, of category theory is applied to proof-theoretical problems. Propositions are construed as objects in a category, proofs as arrows between these objects, and equations between arrows, i.e. commuting diagrams of arrows, are found to have proof-theoretical meaning. They provide a reasonable notion of identity of proofs by equating derivations that are reduced to each other in a cut-elimination or normalization procedure, or they may be involved in finding a unique normal form for derivations.

To enter into categorial proof theory one crosses what should be the watershed between proof theory and the rest of logic. We are not interested any more in provability only—namely, in the existence of proofs—which corresponds to a consequence *relation* between premises and conclusions. We have instead a consequence *graph*, where there may be more than one different proof with the same premise and the same conclusion. We describe these apparently different proofs, code them by terms for arrows, and find that some descriptions stand for the same proof, i.e. the same arrow, while others do not. Our consequence graph is a category, often of a kind that categorists have found important for their own reasons.

On the other hand, in categorial proof theory proof-theoretical, syntactical, methods are applied to problems of category theory. These are mainly methods of normalizing in the style of Gentzen or of the lambda calculus. (In this book, confluence techniques like those in the lambda calculus dominate in the first part, while cut elimination dominates in the second, bigger, part.) This syntactical standpoint is something that many categorists do not favour. Instead of dealing with language, they prefer to work as if they dealt with the things themselves. We find that for some problems of category theory, and in particular for so-called coherence problems, which make the subject matter of this book, paying attention to language is of great help.

The term "coherence" covers in category theory what from a logical point of view would be called problems of completeness, axiomatizability and decidability. Different authors put stress on different things. For our own purposes we will fix a particular notion of coherence, which agrees completely with Mac Lane's usage of the term in [93], the primordial paper on coherence.

In the 1960's, at the same time when coherence started being investigated in category theory, the connection between category theory and logic was established, mainly through Lawvere's ideas (see [88]). The roots of categorial proof theory date from the same years—they can be found in a series of papers by Lambek: [79], [80], [81] and [82]. Lambek introduced Gentzen's proof-theoretical methods in category theory, which Mac Lane and Kelly exploited in [76] to solve a major coherence problem (see also [95]).

There are not many books in categorial proof theory. The early attempt to present matters in [121] has shortcomings. Proofs are not systematically coded by terms for arrows; only the sources and targets of arrows are mentioned most of the time, and too much work is left to the reader. Some claims are excessively difficult to verify, and some are not correct (see [64], [9], Section 3, and [11], Section 1). Lambek's and Scott's book [85] is only partly about categorial proof theory and coherence, understood as a decidability problem for equality of arrows in cartesian closed categories. (Just a short chapter of [122], Chapter 8, touches upon this topic.) The only remaining book in categorial proof theory we know about, [34], is devoted to showing that cut elimination characterizes fundamental notions of category theory, in particular the notion of adjunction. Some parts of that book (Sections 4.10 and 5.9) are about coherence.

Papers in categorial logic often touch upon this or that point of categorial proof theory, but are not very often specifically within the field. And even when they are within this field, some authors prefer to advertise their work as "semantical". It should be clear, however, that this is not semantics in the established model-theoretical sense—the sense in which the word was used in logic in the twentieth century. We find this semantics of proofs more proof-theoretical than model-theoretical.

We will try to cover with the references in our book not the whole literature of categorial proof theory, but only papers relevant to the problems treated. To acknowledge more direct influences, we would like, however, to mention at the outset a few authors with whom we have been in contact, and whose ideas are more or less close to ours.

First, Jim Lambek's pioneering and more recent work has been for us, as for many others, a source of inspiration. Max Kelly's papers on coherence (see [72], [73], [49] and [75]) are less influenced by logic, though logical matters are implicit in them. Sergei Soloviev's contributions to categorial proof theory (see [112], [113] and [114]) and Djordje Čubrić's (see [25], [26] and [27]) are close to our general concerns, though they do not deal exactly with the subject matter of this book; the same applies to some work of Alex Simpson (in particular, [111]).

We extend Robert Seely's and Robin Cockett's categorial presentation of a fragment of linear logic, based upon what they call linear, *alias* weak, distribution (see [19]; other papers will be cited in the body of the book), which we call *dissociativity*. This is an associativity principle involving two operations, which in the context of lattices delivers distribution. While Cockett and Seely are concerned with dissociativity as it occurs in linear logic, and envisage also applications in the study of intuitionistic logic, we have been oriented towards the categorification of classical propositional logic. The subtitle of our book could be "General proof theory of classical propositional logic". We would have put this subtitle were it not that a great part of the book is about fragments of this proof theory, which are fragments of the proof theory of other logics too, and are also of an independent interest for category theory. Besides that, we are not sure our treatment of negation in the last chapter is as conclusive as what precedes it. (We also prefer a shorter and handier title.)

Proofs in the conjunctive-disjunctive fragment of logic, which is related to distributive lattices, may, but need not, be taken to be the same in classical and intuitionistic logic, and they are better not taken to be the same. Classical proof theory should be based on plural (multiple-conclusion) sequents, while intuitionistic proof theory, though it may be presented with such sequents, is more often, and more naturally, presented with singular (single-conclusion) sequents. By extending Cockett's and Seely's categorial treatment of dissociativity, we present in the central part of the book a categorification, i.e. a generalization in category theory, of the notion of distributive lattice, which gives a plausible notion of identity of proofs in classical conjunctive-disjunctive logic. This notion is related to Gentzen's cut-elimination procedure in a plural-sequent system. By building further on that, at the end of the book we provide a plausible categorification of the notion of Boolean algebra, which gives a nontrivial notion of identity of proofs for classical propositional logic, also related to Gentzen.

It is usually considered that it is hopeless to try to categorify the notion of Boolean algebra, because all plausible candidates based on the notion of bicartesian closed category (i.e. cartesian closed category with finite coproducts) led up to now to equating all proofs with the same premises and conclusions. In our Boolean categories, which are built on another base, this is not the case. The place where in our presentation of the matter classical and intuitionistic proof theory part ways is in understanding distribution. In intuitionistic proof theory distribution of conjunction over disjunction is an isomorphism, while distribution of disjunction over conjunction is not. This is how matters stand in bicartesian closed categories. We take that in classical proof theory neither of these distributions is an isomorphism, and restore symmetry, typical for Boolean notions.

We reach our notion of Boolean category very gradually. This gradual approach enables us to shorten calculations at latter stages. Moreover, along the way we prove coherence for various more general notions of category, entering into the notion of Boolean category or related to it. Coherence is understood in our book as the existence of a faithful structure-preserving functor from a freely generated category, built out of syntactical material, into the category whose arrows are relations between finite ordinals. This is a limited notion of coherence, and our goal is to explore the limits of this particular notion within the realm of classical propositional logic. We are aware that other notions of coherence exist, and that even our notion can be generalized by taking another category instead of the category whose arrows are relations between finite ordinals. These other notions and these generalizations are, however, outside the confines of our book, and we will mention them only occasionally (see, in particular, §12.5 and §14.3)

Mac Lane's primordial coherence results for monoidal and symmetric monoidal categories in [93] are perfectly covered by our notion of coherence. When the image of the faithful functor is a discrete subcategory of the category whose arrows are relations between finite ordinals, coherence amounts to showing that the syntactical category is a preordering relation, i.e. that "all diagrams commute". This is the case sometimes, but not always, and not in the most interesting cases. Mac Lane's coherence results are scrutinized in our book, and new aspects of the matter are made manifest. We also generalize previous results of [67] (Section 1) on strictification, i.e. on producing equivalent categories where some isomorphisms are turned into identity arrows. Our strictification is useful, because it facilitates the recording of lengthy calculations.

For categories with dissociativity, which cover proofs in the multiplicative conjunctive-disjunctive fragment of linear logic, and also proofs in the conjunctive-disjunctive fragment of classical logic, we provide new coherence results, and we prove coherence for our Boolean categories. These coherence theorems, which are the main results of the book, yield a simple decision procedure for the problem whether a diagram of canonical arrows commutes, i.e. for the problem whether two proofs are identical.

The most original contribution of our book may be that we take into account *union*, or *addition*, of proofs in classical logic. This operation on proofs with the same premise and same conclusion is related to the mix principle of linear logic. It plays an important role in our Boolean categories, and brings them close to linear algebra. Taking union of proofs into account saves Gentzen's cut-elimination procedure for classical logic from falling into triviality, as far as identity of proofs is concerned. This modified cut elimination is the cornerstone of the proof of our main coherence theorem for classical propositional logic.

We take into account also the notion of *zero proof*, a notion related to union of proofs—a kind of dual of it. With union of proofs hom-sets become semilattices with unit, but we envisage also that they be just commutative monoids, as in additive and abelian categories. Zero proofs, which are like a leap from any premise to any conclusion, are mapped into the empty relation in establishing coherence. Although they enable us to prove anything as far as provability is concerned, they are conservative with respect to the previously established identity of proofs in logic. We will show that envisaging zero proofs is useful. It brings logic closer to linear algebra, and facilitates calculations. We find also that the notion of zero proof may be present in logic even when we do not allow passing from any premise to any conclusion, but restrict ourselves to the types of the acceptable deductions connecting premises and conclusions, i.e. stick to provability in classical logic. Negation may be tied to such restricted zero proofs.

Zero proofs resemble what Hilbert called ideal mathematical objects, like imaginary numbers or points at infinity. If our concern is not with provability, but with proofs—namely, identity of proofs—zero proofs are useful and harmless. We don't think we have exhausted the advantages of taking them into account in general proof theory. We believe, however, we have fulfilled to a great extent the promises made in the programmatic survey [37] (summarized up to a point in the first chapter of the book), which provides further details about the context of our research.

We suppose our principal public should be a public of logicians, such as we are, but we would like no less to have categorists as readers. So we have strived to make our exposition self-contained, both on the logical and on the categorial side. This is why we go into details that logicians would take for granted, and into other details that categorists would take for granted. Only for the introductory first chapter, whose purpose is to give motivation, and for some asides, in particular at the very end, we rely on notions not defined in the book, but in the standard logical and categorial literature.

We suppose that the results of this book should be interesting not only for logic and category theory, but also for theoretical computer science. We do not control very well, however, the quickly growing literature in this field, and we will refrain from entering into it. We do not pretend to be experts in that area. Some of the investigations of proofs of classical logic that appeared since 1990 in connection with modal translations into linear logic or with the lambda-mu calculus, in which the motivation, the style and the jargon of computer science dominate, seem to be concerned with identity of proofs, but it is not clear to us how exactly these concerns are related to ours. We leave for others to judge.

This is more a research monograph than a textbook, but the text could serve nevertheless as the base for a graduate course in categorial proof theory. We provide after the final chapter a list of problems left open. To assist the reader, we also provide at the end of the book a list of axioms and definitions, and a list of categories treated in the book (which are quite numerous), together with charts for these categories indicating the subcategory relations established by our coherence results.

We would like to thank in particular Alex Simpson and Sergei Soloviev for encouraging and useful comments on the preprint of this book, which was distributed since May 2004. We would like to thank also other colleagues who read this preprint and gave compliments on it, or helped us in another manner.

Dov Gabbay was extremely kind to take care of the publishing of the book. We are very grateful to him and to Jane Spurr for their efforts and efficiency.

The results of this book were announced previously in a plenary lecture at the *Logic Colloquium* in Münster in August 2002, and in a talk at the *International Congress MASSEE* in Borovets in September 2003, with the support of the Alexander von Humboldt Foundation. We are indebted to Slobodan Vujošević and Milojica Jaćimović for the invitation to address the *Eleventh Congress of the Mathematicians of Serbia and Montenegro*, held in Petrovac in September 2004, with a talk introducing matters treated in the book. We had the occasion to give such introductory talks also at the Logic Seminar in Belgrade in the last two years and, thanks to Mariangiola Dezani-Ciancaglini and the Types project of the European Union, at the *Types* conference in Jouy-en-Josas in December 2004.

We would like to thank warmly the Mathematical Institute of the Serbian Academy of Sciences and Arts in Belgrade and the Faculty of Philosophy of the University of Belgrade for providing conditions in which we could write this book. Our work was generously supported by a project of the Ministry of Science of Serbia (1630: Representation of Proofs).

Belgrade, December 2004

CONTENTS

CHAPTER 1

INTRODUCTION

In this introductory chapter we provide in an informal manner motivation for the main themes of the book, without giving an exhaustive summary of its content (such summaries are provided at the beginning of every chapter). A great deal of the chapter (§§1.3-6) is based on the survey [37].

While in the body of the book, starting from the next chapter, our exposition, except for some asides, will be self-contained, both from a logical and from a categorial point of view, here we rely on some acquaintance with proof theory (which the reader may have acquired in classic texts like [55], [105] and [77], Chapter 15, or in the more recent textbook [122]), and on some notions of category theory (which may be found in [94] and [85]). Many, but not all, of the notions we need for this introduction will be defined later in the book.

To have read the present chapter is not essential for reading the rest of the book. A reader impatient for more precision can move to Chapter 2, where the book really starts, and return to this introduction later on.

§1.1. Coherence

It seems that what categorists call *coherence* logicians would, roughly speaking, call *completeness*. This is the question whether we have assumed for a particular brand of categories all the equations between arrows we should have assumed. Completeness need not be understood here as completeness with respect to models. We may have a syntactical notion of completeness—something akin to the Post completeness of the classical propositional calculus—but often some sort of model-theoretical completeness is implicit in coherence questions. Matters are made more complicated by the fact that categorists do not like to talk about syntax, and do not perceive the problem as being one of finding a match between syntax and semantics. They do not talk of formal systems, axioms and models.

Moreover, questions that logicians would consider to be questions of *decidability*, which is not the same as completeness, are involved in what categorists call coherence. A coherence problem often involves the question of deciding whether two terms designate the same arrow, i.e. whether a diagram of arrows commutes—we will call this the *commuting problem*—and sometimes it may involve the question of deciding whether there is in a category an arrow of a given type, i.e. with a given source and target—we will call this the *theoremhood problem* (cf. [34], Sections 0.2 and 4.6.1). Coherence is understood mostly as solving the commuting problem in [85] (see p. 117, which mentions [79] and [80] as the origin of this understanding). The commuting problem seems to be involved also in the understanding of coherence of [75] (Section 10).

Completeness and decidability, though distinct, are, of course, not foreign to each other. A completeness proof with respect to a manageable model may provide, more or less immediately, tools to solve decision problems. For example, the completeness proof for the classical propositional calculus with respect to the two-element Boolean algebra provides immediately a decision procedure for theoremhood.

The simplest coherence questions are those where it is intended that all arrows of the same type should be equal, i.e. where the category envisaged is a preorder. The oldest coherence problem is of that kind. This problem has to do with monoidal categories, and was solved by Mac Lane in [93] (where early related work by Stasheff and D.B.A. Epstein is mentioned; see [116] for historical notes, and also [117], Appendix B, co-authored with S. Shnider). The monoidal category freely generated by a set of objects is a preorder. So Mac Lane could claim that showing coherence is showing that "all diagrams commute". We provide in Chapter 4 a detailed analysis of Mac Lane's coherence result for monoidal categories.

In cases where coherence amounts to showing preorder, i.e. showing that from a given set of equations, assumed as axioms, we can derive all equations (provided the equated terms are of the same type), from a logical point of view we have to do with *axiomatizability*. We want to show that a decidable set of axioms (and we wish this set to be as simple as possible, preferably given by a finite number of axiom schemata) delivers all the intended equations. If preorder is intended, then all equations are intended. Axiomatizability is in general connected with logical questions of completeness, and a standard logical notion of completeness is completeness of a set of axioms. Where all diagrams should commute, coherence does not seem to be a question of model-theoretical completeness, but even in such cases it may be conceived that the model involved is a discrete category (cf. the end of §2.9).

Categorists are interested in axiomatizations that can be extended. These extensions are in a new language, with new axioms, and such extensions of the axioms of monoidal categories need not yield preorders any more. Categorists are also interested, when they look for axiomatizations, in finding the combinatorial building blocks of the matter. The axioms are such building blocks, as in knot theory the Reidemeister moves are the combinatorial building blocks of knot and link equivalence (see [91], Chapter 1, or any other textbook in knot theory).

In Mac Lane's second coherence result of [93], which has to do with symmetric monoidal categories, it is not intended that all equations between arrows of the same type should hold. What Mac Lane does can be described in logical terms in the following manner. On the one hand, he has an axiomatization, and, on the other hand, he has a model category where arrows are permutations; then he shows that his axiomatization is complete with respect to this model. It is no wonder that his coherence problem reduces to the completeness problem for the usual axiomatization of symmetric groups.

Algebraists do not speak of axiomatizations, but of *presentations by generators and relations*. All the axiomatizations in this book will be purely equational axiomatizations, as in algebraic varieties. Such were the axiomatizations of [93]. Categories are algebras with partial operations, and we are here interested in the equational theories of these algebras.

In Mac Lane's coherence results for monoidal and symmetric monoidal categories one has to deal only with natural isomorphisms. Coherence questions in the area of n-categories are usually restricted likewise to natural isomorphisms (see [90]). However, in the coherence result for symmetric monoidal closed categories of [76] there are already natural and dinatural transformations that are not isomorphisms.

A natural transformation is tied to a relation between the argument-places of the functor in the source and the argument-places of the functor in the target. This relation corresponds to a relation between finite ordinals, and in composing natural transformations we compose these relations (see §2.4 and §2.9). With dinatural transformations the matter is more complicated, and composition poses particular problems (see [103]). In this book we deal with natural transformations, and envisage only in some comments coherence for situations where we do not have natural transformations. Our general notion of coherence does not, however, presuppose naturality and dinaturality.

Our notion of a coherence result is one that covers Mac Lane's and Kelly's coherence results mentioned up to now, but it is more general. We call coherence a result that tells us that there is a faithful functor G from a category $\mathcal{S}$ freely generated in a certain class of categories to a "manageable" category $\mathcal{M}$. This calls for some explanation.

It is desirable, though perhaps not absolutely necessary, that the functor G be *structure-preserving*, which means that it preserves structure at least up to isomorphism (see §1.7 below, and, in particular, §2.8). In all coherence results we will consider, the functor G will preserve structure strictly, i.e. "on the nose". The categories $\mathcal{S}$ and $\mathcal{M}$ will be in the same class of categories, and G will be obtained by extending in a unique way a map from the generators of $\mathcal{S}$ into $\mathcal{M}$.

The category $\mathcal{M}$ is *manageable* when equations of arrows, i.e. commuting diagrams of arrows, are easier to consider in it than in $\mathcal{S}$. The best is if the commuting problem is obviously decidable in $\mathcal{M}$, while it was not obvious that it is such in $\mathcal{S}$.

With our approach to coherence we are oriented towards solving the commuting problem, and we are less interested in the theoremhood problem. In this book, we deal with the latter problem only occasionally, mostly when we need to solve it in order to deal with the commuting problem (see §4.2, §7.1, §§7.3-5, §§8.2-3 and §11.4). This should be stressed because other authors may give a more prominent place to the theoremhood problem. We find that the spirit of the theoremhood problem is not particularly categorial: this problem can be solved by considering only categories that are preorders. And ordinary, or perhaps less ordinary, logical methods for showing decidability of theoremhood are here more useful than categorial methods. For the categories in this book, the decidability of the theoremhood problem is shown by syntactical or semantical logical tools. Among the latter we also have sometimes simply truth tables. We have used on purpose the not very precise term "manageable" for the category $\mathcal{M}$ to leave room for modifications of our notion of coherence, which would be oriented towards solving another problem than the commuting problem. Besides the theoremhood problem, one may perhaps also envisage something else, but our official notion of coherence is oriented towards the commuting problem.

In this book, the manageable category $\mathcal{M}$ will be the category *Rel* with arrows being relations between finite ordinals, whose connection with natural transformations we have mentioned above. The commuting problem in *Rel* is obviously decidable. We do, however, consider briefly categories that may replace *Rel*—in particular, the category whose arrows are matrices (see §12.5).

The freely generated category $\mathcal{S}$ will be the monoidal category freely generated by a set of objects, or the symmetric monoidal category freely generated by a set of objects, and many others of that kind. The generating set of objects may be conceived as a discrete category. In our understanding of coherence, replacing this discrete generating category by an arbitrary category would prevent us to solve coherence—simply because the commuting problem in the arbitrary generating category may be undecidable. Far from having more general, stronger, results if the generating category is arbitrary, we may end up by having no result at all.

The categories $\mathcal{S}$ in this book are built ultimately out of *syntactic* material, as logical systems are built. Categorists are not inclined to formulate their coherence results in the way we do—in particular, they do not deal often with syntactically built categories (but cf. [125], which comes close to that). If, however, more involved and more abstract formulations of coherence that may be found in the literature (for early references on this matter see [74]) have practical consequences for solving the commuting problem, our way of formulating coherence has these consequences as well.

That there is a faithful structure-preserving functor G from the syntactical category $\mathcal{S}$ to the manageable category $\mathcal{M}$ means that for all arrows f and g of $\mathcal{S}$ with the same source and the same target we have

$$f = g \text{ in } \mathcal{S} \;\text{ iff }\; Gf = Gg \text{ in } \mathcal{M}.$$

The direction from left to right in this equivalence is contained in the functoriality of G, while the direction from right to left is faithfulness proper.

If $\mathcal{S}$ is conceived as a syntactical system, while $\mathcal{M}$ is a model, the faithfulness equivalence we have just stated is like a completeness result in logic. The left-to-right direction, i.e. functoriality, is soundness, while the right-to-left direction, i.e. faithfulness, is completeness proper.

In this book we will systematically separate coherence results involving special objects (such as unit objects, terminal objects and initial objects) from those not involving them. These objects tend to cause difficulties, and the statements and proofs of the coherence results gain by having these difficulties kept apart. When coherence is obtained both in the absence and in the presence of special objects, our results become sharper.

§1.2. Categorification

By *categorification* one can understand, very generally, presenting a mathematical notion in a categorial setting, which usually involves generalizing the notion and making finer distinctions. In this book, however, we have something more specific in mind. We say that we have a categorification of the notion of an algebraic structure in which there is a preordering, i.e. reflexive and transitive, relation R when we replace R with arrows in a category, and obtain thereby a more general categorial notion instead of the initial algebraic notion. If the initial algebraic structure is a completely free algebra of terms, like the algebra of formulae of a propositional language, the elements of the algebra just become objects in a free category in the class of categories resulting from the categorification. Otherwise, some splitting of the objects is involved in categorification.

Categorification is not a technical notion we will rely on later, and so we will not try to define it more precisely. What we have in mind should be clear from the following examples.

By categorifying the algebra of formulae of conjunctive logic with the constant true proposition, where the preordering relation R is induced by implication, we may end up with the notion of cartesian category. We may end up with the same notion by categorifying the notion of semilattice with unit, where the relation R is the partial ordering of the semilattice. A semilattice with unit is a cartesian category that is a partial order, i.e. in which whenever we have arrows from a to b and vice versa, then a and b are the same object. In the same sense, the notion of monoidal category is a categorification of the notion of monoid, and the notion of symmetric monoidal category is a categorification of the notion of commutative monoid, the preordering relation R in these two cases being equality.

There are other conceptions of categorification except that one. One may categorify an algebra by taking its objects to be arrows of a category. The notion of category is a categorification in this sense of the notion of monoid, monoids being categories with a single object. In that direction, one obtains more involved notions of categorification in the n-categorial setting (see [2] and [24]).

The motivation for categorification may be internal to category theory, but it may come from other areas of mathematics, like algebraic topology and mathematical physics—in particular, quantum field theory (many references are given in [2]). Our motivation comes from proof theory, as we will explain in latter sections of this introduction. We are replacing a consequence relation, which is a preordering relation, by a category, where arrows stand for proofs. In comparing our approach to others, note that the slogan "Replace equality by isomorphisms!", which is sometimes heard in connection with categorification, does not describe exactly what we are doing. Our slogan "Replace preorder by arrows!" implies, however, the other one, and so the same categorial notions, like, for example, the notion of monoidal category, may turn up under both slogans.

In this book one may find, in particular, categorifications, in our restricted sense, of the notions of distributive lattice and Boolean algebra. Alternatively, these may be taken as categorifications of conjunctive-disjunctive logic, or of the classical propositional calculus. Previously, a categorification of the notion of distributive lattice was obtained with so-called distributive categories, i.e. bicartesian categories with distribution arrows from $a \wedge (b \vee c)$ to $(a \wedge b) \vee (a \wedge c)$ that are isomorphisms (see [89], pp. 222-223 and Session 26, and [17]). Bicartesian closed categories, i.e. cartesian closed categories with finite coproducts (see [85], Section I.8), are distributive categories in this sense.

In our categorification of the notion of distributive lattice, distribution arrows of the type above need not be isomorphisms. This rejection of isomorphism is imposed by our wish to have coherence with respect to the category *Rel* of the preceding section, since the relation underlying the following diagram:

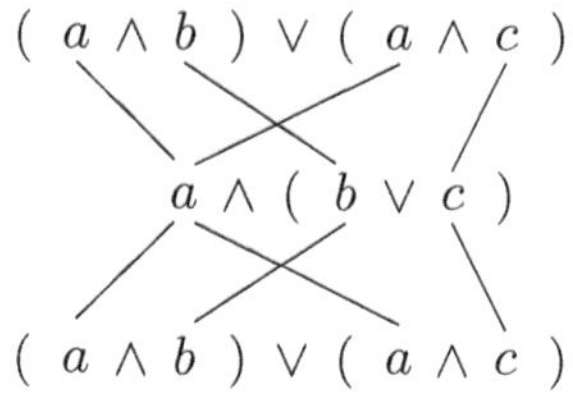

namely the relation underlying the diagram on the left-hand side below, is not the identity relation underlying the diagram on the right-hand side:

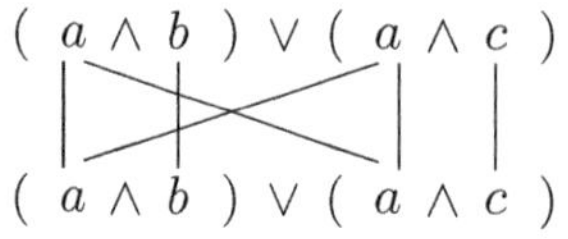

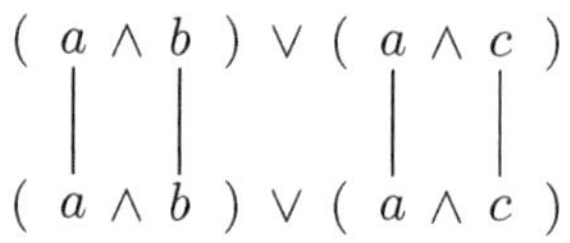

Our categorification of the notion of distributive lattice is based on arrows from $a \wedge (b \vee c)$ to $(a \wedge b) \vee c$, which Cockett and Seely studied in their categorial treatment of a fragment of linear logic (see [19]; further references are given in §7.1 and §7.9). At first, they called the principle underlying these arrows *weak* distribution, and then changed this to *linear* distribution in [22]. Since this is a principle that delivers ***dis***tribution in the context of lattices, but is in fact an as***sociativity*** involving two operations, we have coined the name *dissociativity* for it, to prevent confusion with what is usually called distribution. Cockett and Seely were concerned with establishing some sort of coherence for dissociativity with respect to proof nets.

Before appearing in proof nets and in categories, dissociativity was prefigured in universal algebra and logic (see §7.1 for references). Dissociativity is related to the modularity law of lattices (see §7.1), and we will see in §11.3 how in a context that is a categorification of the notion of lattice this two-sorted associativity delivers distribution arrows of the usual types, from $a \wedge (b \vee c)$ to $(a \wedge b) \vee (a \wedge c)$ and from $(a \vee b) \wedge (a \vee c)$ to $a \vee (b \wedge c)$ (the arrows of the converse types are there anyway), of which neither need to be an isomorphism. The arrows from $(a \vee b) \wedge (a \vee c)$ to $a \vee (b \wedge c)$ need not be isomorphisms in bicartesian closed categories too.

The categorification of the notion of Boolean algebra is usually deemed to be a hopeless task (see §14.3), because it is assumed this categorification should be based on the notion of bicartesian closed category. In that notion, as we said above, we have arrows corresponding to distribution of conjunction over disjunction that are isomorphisms. Natural assumptions in this context lead to triviality, i.e. to categories that are preorders. Our categorification of the notion of Boolean algebra is not trivial in this sense. It incorporates the notion of bicartesian category (i.e. category with finite products and coproducts), but does not admit cartesian closure. Its essential ingredient is our categorification of distributive lattices, in which the arrows corresponding to distribution of conjunction over disjunction are not isomorphisms.

We think it is a prejudice to assume that there must be an isomorphism behind distribution of conjunction over disjunction. It would likewise be a prejudice to assume that behind the idempotency law $a \wedge a = a$ or the absorption law $a \wedge (a \vee b) = a$ of lattices we must have isomorphisms. The categorification of the notion of lattice in bicartesian categories is not under the spell of the latter two assumptions, but the isomorphism corresponding to distribution of conjunction over disjunction is usually presupposed. This is presumably because in the category *Set* of sets with functions—the central category there is—distribution of cartesian product over disjoint union is an isomorphism. In the categorification of the notion of distributive lattice with distributive categories, where $a \wedge (b \vee c)$ is isomorphic to $(a \wedge b) \vee (a \wedge c)$, it is not required that $a \vee (b \wedge c)$ be isomorphic to $(a \vee b) \wedge (a \vee c)$, presumably because the latter isomorphism need not exist in *Set*. We assume neither of these isomorphisms in our categorification of the notion of distributive lattice.

§1.3. The Normalization Conjecture in general proof theory

Categorification is interesting for us because of its connection with general proof theory. The question "What is a proof?" was considered by Prawitz in [106] (Section I) to be the first question of general proof theory. To keep up with the tradition, we speak of "proof", though we could as well replace this term by the more precise term "deduction", since we have in mind deductive proofs from assumptions (including the empty collection of assumptions). Together with the question "What is a proof?", Prawitz envisaged the following as one of the first questions to be considered in general proof theory (see [106], p. 237):

> In the same way as one asks when two formulas define the same set or two sentences express the same proposition, one asks when two derivations represent the same proof; in other words, one asks for identity criteria for proofs or for a "synonymity" (or equivalence) relation between derivations.

An answer to the question of identity criteria for proofs might lead to an answer to the basic question "What is a proof?". A proof would be the equivalence class of a derivation. The related question "What is an algorithm?" could be answered by an analogous factoring through an equivalence relation on representations of algorithms. (Moschovakis stressed in [101], Section 8, the fundamental interest of identity criteria for algorithms.) Prawitz did not only formulate the question of identity criteria for proofs very clearly, but also proposed a precise mathematical answer to it.

Prawitz considered derivations in natural deduction systems and the equivalence relation between derivations that is the reflexive, transitive and symmetric closure of the immediate-reducibility relation between derivations. Of course, only derivations with the same premises and the same conclusion may be equivalent. Prawitz's immediate-reducibility relation is the one involved in reducing a derivation to normal form—a matter he studied previously in [105]. As it is well known, the idea of this reduction stems from Gentzen's thesis [55]. A derivation reduces immediately to another derivation (see [106], Section II.3.3) when the latter is obtained from the former either by removing a *maximum formula* (i.e. a formula with a connective α that is the conclusion of an introduction of α and the major premise of an elimination of α), or by performing one of the *permutative reductions* tied to the eliminations of disjunction and of the existential quantifier, which enables us to remove what Prawitz calls *maximum segments*. There are some further reductions, which Prawitz called *immediate simplifications*; they consist in removing eliminations of disjunction where no hypothesis is discharged, and there are similar immediate simplifications involving the existential quantifier, and "redundant" applications of the classical absurdity rule. Prawitz also envisaged reductions he called *immediate expansions*, which lead to the expanded normal form where all the *minimum formulae* are atomic (minimum formulae are those that are conclusions of eliminations and premises of introductions).

Prawitz formulates in [106] (Section II.3.5.6) the following conjecture, for which he gives credit (in Section II.5.2) to Martin-Löf, and acknowledges influence by ideas of Tait:

Conjecture. Two derivations represent the same proof if and only if they are equivalent.

We call this conjecture the *Normalization Conjecture.*

This conjecture, together with another conjecture, which will be considered in the next section, was examined in the survey [37]. The present section and the next three sections give an updated, somewhere shortened and somewhere expanded, variant of that survey, to which we refer for further, especially historical and philosophical, remarks. (Some other bits of that survey are in §14.3, where a mistaken statement is also corrected at the end of the section.)

The normalization underlying the Normalization Conjecture need not be understood always in the precise sense envisaged by Prawitz. For intuitionistic logic Prawitz's understanding of normalization, which is derived from Gentzen, is perhaps optimal. There are, however, other logics, and, in particular, there is classical logic, to which natural deduction is not so closely tied, and for which we may still have a notion of normalization, perhaps related Prawitz's, but different. What comes to mind immediately for classical logic is Gentzen's plural, i.e. multiple-conclusion, sequent systems (see below) and cut elimination for them.

Presumably, the notion of normalization we can envisage in the Normalization Conjecture cannot be based on an arbitrary notion of normal form. It is desirable that this normal form be unique, at least up to some superficial transformations (like alpha conversion in the lambda calculus). But uniqueness should not be enough. This normal form and the language for which it is formulated must be *significant*, where it is difficult to say what "significant" means exactly. The normal form and the language for which it is formulated should not be just a technical device, but they must be deeply tied to the logic, and exhibit its essential features. In the case of Prawitz's normal form for derivations in intuitionistic natural deduction, besides philosophical reasons having to do with the meaning of logical connectives, there are important ties with independently introduced mathematical structures.

The Normalization Conjecture was formulated by Prawitz at the time when the Curry-Howard correspondence between derivations in natural deduction and typed lambda terms started being recognized more and more (though the label "Curry-Howard" was not yet canonized). Prawitz's equivalence relation between derivations corresponds to beta-eta equality between typed lambda terms, if immediate expansions are taken into account, and to beta equality otherwise.

Besides derivations in natural deduction and typed lambda terms, where according to the Curry-Howard correspondence the latter can be conceived just as codes for the former, there are other, more remote, formal representations of proofs. There are first Gentzen's sequent systems, which are related to natural deduction, but are nevertheless different, and there are also representations of proofs as arrows in categories. The sources and targets of arrows are taken to be premises and conclusions respectively, and equality of arrows with the same source and target, i.e. commuting diagrams of arrows, should now correspond to identity of proofs via a conjecture analogous to the Normalization Conjecture.

The fact proved by Lambek (see [82] and [85], Part I; see also [36], [33] and [39]) that the category of typed lambda calculuses with functional types and finite product types, based on beta-eta equality, is equivalent to the category of cartesian closed categories, and that hence equality of typed lambda terms amounts to equality between arrows in cartesian closed categories, lends additional support to the Normalization Conjecture. Equality of arrows in bicartesian closed categories corresponds to equivalence of derivations in Prawitz's sense in full intuitionistic propositional logic (see [103], Section 3, for a detailed demonstration that the equations of bicartesian closed categories deliver cut elimination for intuitionistic propositional logic). The notion of bicartesian closed category is a categorification in the sense of the preceding section of the notion of Heyting algebra. The partial order of Heyting algebras is replaced by arrows in this categorification.

In category theory, the Normalization Conjecture is tied to Lawvere's characterization of the connectives of intuitionistic logic by adjoint situations. Prawitz's equivalence of derivations, in its beta-eta version, corresponds to equality of arrows in various adjunctions tied to logical connectives (see [88], [34], Section 0.3.3, [35] and [36]). Adjunction is the unifying concept for the reductions envisaged by Prawitz.

The fact that equality between lambda terms, as well as equality of arrows in cartesian closed categories, were first conceived for reasons independent of proofs is remarkable. This tells us that we are in the presence of a solid mathematical structure, which may be illuminated from many sides.

Prawitz formulated the Normalization Conjecture having in mind natural deduction, and so mainly intuitionistic logic. For classical logic we envisage something else. Our categorification of the notion of Boolean algebra, as the categorification of the notion of Heyting algebra with bicartesian closed categories, covers a notion of identity of proofs suggested by normalization via cut elimination in a plural-sequent system (see Chapters 11 and 14). This is in spite of the fact that for us distribution of conjunction over disjunction does not give rise to isomorphisms, as in bicartesian closed categories. This disagreement over the isomorphism of distribution may be explained as follows.

Classical and intuitionistic logic do not differ with respect to the consequence relation between formulae in the conjunction-disjunction fragment of propositional logic. In other words, they do not differ with respect to provable sequents of the form $A \vdash B$ where A and B are formulae of the conjunction-disjunction fragment. But, though these two logics do not differ with respect to provability, they may differ with respect to proofs. The standard sequent formulation of classical logic, the formulation that imposes itself by its symmetry and regularity, is based on *plural* sequents $\Gamma \vdash \Delta$, where Δ may be a collection with more than one formula, whereas the standard sequent formulation of intuitionistic logic is based on *singular*, i.e. single-conclusion, sequents $\Gamma \vdash \Delta$, where Δ cannot have more than one formula, while Γ can. There are presentations of intuitionistic logic with plural sequents (see [97] and [29], Section 5C4, with detailed historical remarks on pp. 249-250; cf. also [28], where the idea is already present), but they are not standard, and they do not correspond to natural deduction, as those with singular sequents do. Moreover, in these plural-sequent formulations of intuitionistic logic, a restriction based on singularity is kept for introduction of implication on the right-hand side, which corresponds to the deduction theorem. The deduction theorem enables the deductive metalogic to be mirrored with the help of implication in the object language, and when it comes to this mirroring, plural-sequent formulations of intuitionistic logic avow that their deductive metalogic is based on singular sequents.

The connection of intuitionistic logic with natural deduction, where there are possibly several premises, but never more than one conclusion, goes very deep. There are many reasons to hold that the meaning of intuitionistic connectives is explained in the framework of natural deduction, as suggested by Gentzen (see [55], Section II.5.13). Singular sequents are asymmetric, i.e. they have a plurality of premises versus a single conclusion. The asymmetries of intuitionistic logic, and, in particular, the asymmetry between conjunction and disjunction, can be explained by the asymmetry of singular sequents that underly this logic. One can suppose that the asymmetry of bicartesian closed categories, which consists in having $a \wedge (b \vee c)$ is isomorphic to $(a \wedge b) \vee (a \wedge c)$, without having $a \vee (b \wedge c)$ isomorphic to $(a \vee b) \wedge (a \vee c)$, has the same roots.

The dissociativity principle of the arrow that goes from $a \wedge (b \vee c)$ to $(a \wedge b) \vee c$ (see §1.2 and §7.1) delivers arrows that go from $a \wedge (b \vee c)$ to $(a \wedge b) \vee (a \wedge c)$ and from $(a \vee b) \wedge (a \vee c)$ to $a \vee (b \wedge c)$ (see §11.3; we have arrows of the converse types without assuming distribution), but neither of these arrows need to be isomorphisms. So symmetry, which is typical for Boolean notions, is restored. (Another possibility to restore symmetry would be to take that $a \wedge (b \vee c)$ is isomorphic to $(a \wedge b) \vee (a \wedge c)$ and $a \vee (b \wedge c)$ is isomorphic to $(a \vee b) \wedge (a \vee c)$, which is not the case in *Set*, but we will not explore that possibility in this book.)

The dissociativity principle, which is an essential ingredient of our categorification of the notions of distributive lattice and Boolean algebra, is built into the plural-sequent formulation of classical logic. It is tied to the cut rule of plural sequents (see §11.1, and also §7.7).

Prawitz envisaged the Normalization Conjecture for classical logic, but in a natural deduction formulation, i.e. with singular sequents. This is not the same as considering this conjecture with plural sequents.

§1.4. The Generality Conjecture

At the same time when Prawitz formulated the Normalization Conjecture, in a series of papers ([79], [80], [81] and [82]) Lambek was engaged in a project where arrows in various sorts of categories were construed as representing proofs. The source of an arrow corresponds to the premise, and the target to the conclusion. (Proofs where there is a finite number of premises different from one are represented by proofs with a single premise with the help of connectives like conjunction and the constant true proposition.) With this series of papers Lambek inaugurated the field of categorial proof theory.

The categories Lambek considered in [79] and [80] are first those that correspond to his substructural syntactic calculus of categorial grammar (these are monoidal categories where the functors $a \otimes \ldots$ and $\ldots \otimes a$ have right adjoints). Next, he considered monads, which besides being fundamental for category theory, cover proofs in modal logics of the S4 kind. In [81] and [82], Lambek dealt with cartesian closed categories, which cover proofs in the conjunction-implication fragment of intuitionistic logic. He also envisaged bicartesian closed categories, which cover the whole of intuitionistic propositional logic.

Lambek's insight is that equations between arrows in categories, i.e. commuting diagrams of arrows, guarantee cut elimination, i.e. composition elimination, in an appropriate language for naming arrows. (In [34] it is established that for some basic notions of category theory, and in particular for the notion of adjunction, the equations assumed are necessary and sufficient for composition elimination.) Since cut elimination is closely related to Prawitz's normalization of derivations, the equivalence relation envisaged by Lambek should be related to Prawitz's. (An early presentation of the connection between Prawitz and Lambek is in [99].)

The normalization of cut elimination does not involve only eliminating cuts, but also equations between cut-free terms for arrows, which may guarantee their uniqueness. (This is like adding the eta equations to the beta equations in the typed lambda calculus and natural deduction.)

Lambek's work is interesting not only because he worked with an equivalence relation between derivations amounting to Prawitz's, but also because he envisaged another kind of equivalence relation. Lambek's idea is best conveyed by considering the following example. In [81] (p. 65) he says that the first projection arrow $\hat{k}^1_{p,p}\colon p \wedge p \vdash p$ and the second projection arrow $\hat{k}^2_{p,p}\colon p \wedge p \vdash p$, which correspond to two derivations of conjunction elimination, have different *generality*, because they generalize to $\hat{k}^1_{p,q}\colon p \wedge q \vdash p$ and $\hat{k}^2_{p,q}\colon p \wedge q \vdash q$ respectively, and the latter two arrows do not have the same target; on the other hand, $\hat{k}^1_{p,q}\colon p \wedge q \vdash p$ and $\hat{k}^2_{q,p}\colon q \wedge p \vdash p$ do not have the same source. The idea of generality may be explained roughly as follows. We consider generalizations of derivations that diversify variables without changing the rules of inference. Two derivations have the same generality when every generalization of one of them leads to a generalization of the other, so that in the two generalizations we have the same premise and conclusion (see [79], p. 257). In the example above, this is not the case.

Generality induces an equivalence relation between derivations. Two derivations are equivalent if and only if they have the same generality. Lambek does not formulate so clearly as Prawitz a conjecture concerning identity criteria for proofs, but he suggests that two derivations represent the same proof if and only if they are equivalent in the new sense. We will call this conjecture the *Generality Conjecture*.

Lambek's own attempts at making the notion of generality precise (see [79], p. 316, where the term "scope" is used instead of "generality", and [80], pp. 89, 100) need not detain us here. In [81] (p. 65) he finds that these attempts were faulty.

The simplest way to understand generality is to use graphs whose vertices are occurrences of propositional letters in the premise and the conclusion of a derivation. We connect by an edge occurrences of letters that must remain occurrences of the same letter after generalizing, and do not connect those that may become occurrences of different letters. So for the first and second projection above we would have the two graphs

When the propositional letter p is replaced by an arbitrary formula A we have an edge for each occurrence of propositional letter in A.

The generality of a derivation is such a graph. According to the Generality Conjecture, the first and second projection derivations from $p \wedge p$ to p represent different proofs because their generalities differ.

One defines an associative composition of such graphs, and there is also an obvious identity graph with straight parallel edges, so that graphs make a category, which we call the *graphical category*. If on the other hand it is taken for granted that proofs also make a category, which we will call the *syntactical category*, with composition of arrows being composition of proofs, and identity arrows being identity proofs (an identity proof composed with any other proof, either on the side of the premise or on the side of the conclusion, is equal to this other proof), then the Generality Conjecture may be rephrased as the assertion that there is a faithful functor from the syntactical category to the graphical category. So the Generality Conjecture is analogous to a *coherence theorem* of category theory. The manageable category is a graphical category.

The coherence result of [76] proves the Generality Conjecture for the multiplicative conjunction-implication fragment of intuitionistic linear logic (modulo a condition concerning the multiplicative constant true proposition, i.e. the unit with respect to multiplicative conjunction), and, inspired by Lambek, it does so via a cut-elimination proof. The syntactical category in this case is a free symmetric monoidal closed category, and the graphical category is of a kind studied in [49]. The graphs of this graphical category are closely related to the *tangles* of knot theory. In tangles, as in braids, we distinguish between two kinds of crossings, but here we need just one kind, in which it is not distinguished which of the two crossed edges is above the other. (For categories of tangles see [128], [123] and [68], Chapter 12.) Tangles with this single kind of crossing are like graphs one encounters in Brauer algebras (see [12] and [126]). Here is an example of such a tangle:

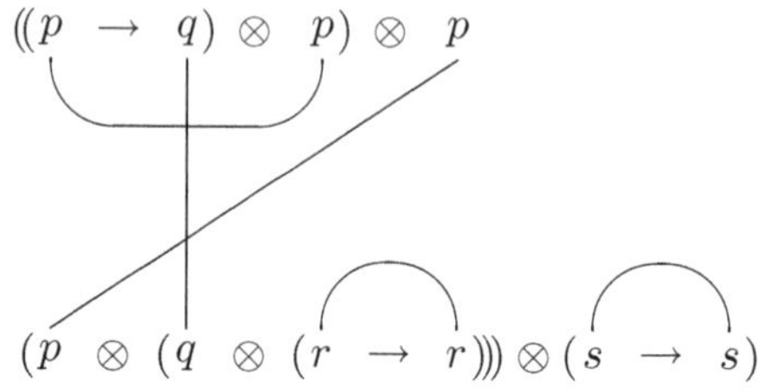

Tangles without crossings at all serve in [34] (Section 4.10; see also [38]) to obtain a coherence result for the general notion of adjunction, which according to Lawvere's Thesis underlies all the connectives of intuitionistic logic, as we mentioned in the preceding section. In terms of combinatorial low-dimensional topology, the mathematical content of the general notion of adjunction is caught by the Reidemeister moves of planar ambient isotopy. An analogous coherence result for self-adjunctions, where a single endofunctor is adjoint to itself, is proved in [45]. Through this latter result we reach the theory of Temperley-Lieb algebras, which play a prominent role in knot theory and low-dimensional topology, due to Jones' representation of Artin's braid groups in these algebras (see [69], [91], [104] and references therein).

In [45] one finds also coherence results for self-adjunctions where the graphical category is the category of matrices, i.e. the skeleton of the category of finite-dimensional vector spaces over a fixed field with linear transformations as arrows. Tangles without crossings may be faithfully represented in matrices by a representation derived from the orthogonal group case of Brauer's representation of Brauer algebras (see also [126], Section 3, and [65], Section 3). This representation is based on the fact that the Kronecker product of matrices gives rise to a self-adjoint functor in the category of matrices, and this self-adjointness is related to the fact that in this category, as well as in the category *Rel*, whose arrows are binary relations between finite ordinals, finite products and coproducts are isomorphic.

Graphs like graphs of the tangle type were tied to sequent derivations of classical logic in [15] and [16], but without referring to categories, coherence or the question of identity criteria for proofs.

In [102] there are several coherence results, which extend [93], for the multiplicative-conjunction fragments of substructural logics. But less us concentrate now on coherence results for classical and intuitionistic logic.

The Normalization Conjecture and the Generality Conjecture agree only for limited fragments of these two logics. They agree for purely conjunctive logic, with or without the constant true proposition $\top$ (see [42] and §9.1 and §9.2 below). Proofs in conjunctive logic are the same for classical and intuitionistic logic. Here the Normalization Conjecture is taken in its beta-eta version. By duality, the two conjectures agree for purely disjunctive logic, with or without the constant absurd proposition $\bot$. If we have both conjunction and disjunction, but do not yet have distribution, and have neither $\top$ nor $\bot$, then the two conjectures still agree for both logics, provided the graphical category is the category *Rel* whose arrows are relations between finite ordinals (see [44] and §9.4). And here it seems we have reached the limits of agreement as far as intuitionistic logic is concerned. With more sophisticated notions of graphs, matters may stand differently, and the area of agreement for the two conjectures may perhaps be wider, but it can be even narrower, as we will see below.

It may be questioned whether the intuitive idea of generality is caught by the category *Rel* in the case of conjunctive-disjunctive logic. The problem is that if $\hat{w}_p \colon p \vdash p \wedge p$ is a component of the diagonal natural transformation, and $\check{k}^1_{q,p} \colon q \vdash q \vee p$ is a first injection, then in categories with finite products and coproducts we have

$$(\mathbf{1}_q \vee \hat{w}_p) \circ \check{k}^1_{q,p} = \check{k}^1_{q,p \wedge p},$$

where the left-hand side cannot be further generalized, but the right-hand side can be generalized to $\check{k}^1_{q,p \wedge r}$. The intuitive idea of generality seems to require that in $\hat{w}_p \colon p \vdash p \wedge p$ we should not have only a relation between the domain and the codomain, as on the left-hand side below, but an equivalence relation on the union of the domain and the codomain, as on the right-hand side:

(see [46], and also [47]). With such equivalence relations, we can still get coherence for conjunctive logic, and for disjunctive logic, taken separately, but for conjunctive-disjunctive logic the left-to-right direction, i.e. the soundness part, of coherence would fail (see §14.3). So for conjunctive-disjunctive logic the idea of generality with which we have coherence is not quite the intuitive idea suggested by Lambek, but only something close to it, which involves the categorial notion of natural transformation (cf. the end of §14.3).

Even when we stay within the confines of the category *Rel*, our understanding of generality does not match exactly the intuitive notion of generality for conjunctive-disjunctive logic. Intuitively, the relations R of *Rel* corresponding to generality should satisfy *difunctionality* in the sense of [108]; namely, we should have $R \circ R^{-1} \circ R \subseteq R$. But this requirement is not satisfied for our images in *Rel* under G of proofs in conjunctive-disjunctive logic, even in the absence of distribution (see the end of §14.3). Generality is caught by *Rel* only for fragments of logic. Altogether, generality serves only as a loose motivation for taking *Rel* as our graphical category. Real grounds for *Rel* are in the notion of natural transformation, which has to do with permuting rules in derivations.

The Normalization Conjecture and the Generality Conjecture agree neither for the conjunction-disjunction fragment of intuitionistic logic with $\top$ and $\bot$ (see [43] and §9.6), nor for the conjunction-implication fragment of this logic. We do not have coherence for cartesian closed categories if the graphs in the graphical category are taken to be of the tangle type Kelly and Mac Lane had for symmetric monoidal closed categories combined with the graphs we have in *Rel* for cartesian categories—both the soundness part and the completeness part of coherence fail (for soundness see a counterexample in §14.3, with $\neg p \vee p$ replaced by $p \to p$, and for completeness see

[119]). The soundness part of coherence fails also for distributive bicartesian categories, and a fortiori for bicartesian closed categories. The problem is that in these categories distribution of conjunction over disjunction is taken to be an isomorphism, and *Rel* does not deliver that, as we have seen in §1.2.

The problem with the soundness part of coherence for cartesian closed categories may be illustrated with typed lambda terms in the following manner. By beta conversion and alpha conversion, we have the following equation:

$$\lambda_x\langle x, x\rangle\lambda_y y = \langle\lambda_y y, \lambda_z z\rangle$$

for y and z of type p, and x of type p^p (which corresponds to $p \rightarrow p$). The closed terms on the two sides of this equation are both of type $p^p \times p^p$. The type of the term on the left-hand side cannot be further generalized, but the type of the term $\langle\lambda_y y, \lambda_z z\rangle$, can be generalized to $p^p \times q^q$. The problem noted here does not depend essentially on the presence of surjective pairing $\langle \_, \_ \rangle$ and of product types; it arises also with purely functional types. This problem depends essentially on the multiple binding of variables, which we have in $\lambda_x\langle x, x\rangle$; that is, it depends on the structural rule of contraction. This throws some doubt on the right-to-left direction of the Normalization Conjecture, which Prawitz found relatively unproblematic. It might be considered strange that two derivations represent the same proof if, without changing inference rules, one can be generalized in a manner in which the other cannot be generalized.

The area of agreement between the Normalization Conjecture and the Generality Conjecture may be wider for classical logic, provided normalization is understood in the sense of cut elimination for plural sequent systems and generality is understood in the sense of the category *Rel*. It extends first to conjunctive-disjunctive logic without distribution with $\top$ and $\bot$ (see [43] and §9.4 below). Next, in conjunctive-disjunctive logic with distribution, with or without $\top$ and $\bot$, the agreement also holds (see Chapter 11). And it covers also the whole classical propositional calculus, with a particular way of understanding normalization involving zero proofs (see §1.6 and §§14.2-3). We do not pretend this particular way of understanding normalization in the presence of negation is the only possible one, but in the absence of negation we feel pretty secure, and the match between the two conjectures is indeed very good. Gentzen's cut elimination procedure for plural-sequent systems needs only to be modified in a natural way by admitting union of proofs, a rule that in this context amounts to the mix rule of linear logic (see Chapters 8 and 10). Admitting union of proofs saves Gentzen's cut-elimination procedure from falling into preorder and triviality. Our cut-elimination procedure differs also from Gentzen's in the way how it treats the structural rule of contraction, but in this respect it is more in the spirit of Gentzen. (We will point down at appropriate places in §§11.1-2 how our procedure is related to Gentzen's.)

Zero proofs (which were mentioned already in the preface) come up with negation. Their appearance is imposed by our wish to have coherence with respect to *Rel*. With other graphical categories they may disappear, but at the cost of many problems (which we discuss in §14.3). In particular, the match between the Normalization and the Generality Conjectures would be impaired (see §14.3). The price we have to pay with our categorification of the notion of Boolean algebra is that not all connectives will be tied to adjoint functors, as required by Lawvere. Conjunction and disjunction are tied to the usual adjunctions with the diagonal functor (the product bifunctor is right-adjoint to the diagonal functor, and the coproduct bifunctor is left-adjoint to the diagonal functor), but distribution is an additional matter, not delivered by these adjunctions, and classical negation and implication do not come with the usual adjunctions. (There are perhaps hidden adjunctions of some kind here.) Another price we have to pay with zero proofs is that all theorems, i.e. all propositions proved without hypotheses, will have zero proofs. So the theorems of classical propositional logic, in contradistinction to their intuitionistic counterparts, do not serve to encode the deductive metatheory of classical propositional logic. This metatheory exists, nevertheless, and its categorification is not given by categories that are preorders.

When we compare the two conjectures we should say something about their computational aspects. With the Normalization Conjecture we have to rely in intuitionistic logic on reduction to a unique normal form in the typed lambda calculus in order to check equivalence of derivations in the conjunction-implication fragment of intuitionistic propositional logic. Nothing more practical than that is known, and such syntactical methods may be tiresome. Outside of the conjunction-implication fragment, in the presence of disjunction and negation, such methods become uncertain.

Methods for checking equivalence of derivations in accordance with the Generality Conjecture, i.e. methods suggested by coherence results, often have a clear advantage. This is like the advantage truth tables have over syntactical methods of reduction to normal form in order to check tautologicality. However, the semantical methods delivered by coherence results have this advantage only if the graphical category is simple enough, as our category *Rel* is. When we enter into categories suggested by knot theory, this simplicity may be lost. Then, on the contrary, syntax may help us to decide equality in the graphical category. The Normalization Conjecture has made a foray in theoretical computer science, in the area of typed programming languages. It is not clear whether one could expect the Generality Conjecture to play a similar role.

The reflexive and transitive closure of the immediate-reducibility relation involved in normalization may be deemed more important than the equivalence relation engendered by immediate reducibility, which we have considered up to now. This matter leads outside our topic, which is *identity* of proofs, but it is worth mentioning. We may "categorify" the identity relation between proofs, and consider not only other relations between proofs, but maps between proofs. The proper framework for doing that seems to be the framework of weak 2-categories, where we have 2-arrows between arrows; or we could even go to n-categories, where we have $n+1$-arrows between n-arrows (one usually speaks of *cells* in this context). Composition of 1-arrows is associative only up to a 2-arrow isomorphism, and analogously for other equations between 1-arrows. Identity of 1-arrows is replaced by 2-arrows satisfying certain coherence conditions. In the context of the Generality Conjecture, we may also find it natural to consider 2-arrows instead of identity. The orientation would here be given by passing from a graph with various "detours" to a graph that is more "straight", which need not be taken any more as equal to the original graph.

With all this we would enter into a very lively field of category theory, interacting with other disciplines, mainly topology (see [90] and papers cited therein). The field looks very promising for general proof theory, both from Prawitz's and from Lambek's point of view, but, as far as we know, it has not yet yielded to proof theory much more than promises.

§1.5. Maximality

The fragments of logic mentioned in the preceding section where the Normalization Conjecture and the Generality Conjecture agree for intuitionistic logic all possess a property called *maximality*. Let us say a few words about this important property.

For the whole field of general proof theory to make sense, and in particular for considering the question of identity criteria for proofs, we should not have that any two derivations with the same premise and conclusion are equivalent. Otherwise, our field would be trivial.

Now, categories with finite nonempty products, cartesian categories and categories with finite nonempty products and coproducts have the following property. Take, for example, cartesian categories, and take any equation in the language of free cartesian categories that does not hold in free cartesian categories. If a cartesian category satisfies this equation, then this category is a preorder. We have an exactly analogous property with the other sorts of categories we mentioned (see §9.3 and §9.5). This property is a kind of syntactical completeness, analogous to the Post completeness of the usual axiomatizations of the classical propositional calculus. Any extension of the equations postulated leads to collapse.

Translated into logical language, this means that Prawitz's equivalence relation for derivations in conjunctive logic, disjunctive logic and conjunctive-disjunctive logic without distribution and without $\top$ and $\bot$, which in all these cases agrees with our equivalence relation defined via generality in the sense of *Rel*, is *maximal*. Any strengthening, any addition, would yield that any two derivations with the same premise and the same conclusion are equivalent.

If the right-to-left direction of the Normalization Conjecture holds, with maximality we can efficiently justify the left-to-right direction, which Prawitz found problematic in [106], and about which Kreisel was thinking in [78]. In the footnote on p. 165 of that paper Kreisel mentions that Barendregt suggested this justification via maximality. Suppose the right-to-left direction of the Normalization Conjecture holds, suppose that for some premise and conclusion there is more than one proof, and suppose the equivalence relation is maximal. Then if two derivations represent the same proof, they are equivalent. Because if they were not equivalent, we would never have more than one proof with a given premise and a given conclusion. Nothing can be missing from our equivalence relation, because whatever is missing, by maximality, leads to collapse on the side of the equivalence relation, and, by the right-to-left direction of the conjecture, it also leads to collapse on the side of identity of proofs.

Prawitz in [106] found it difficult to justify the left-to-right direction of the Normalization Conjecture, and Kreisel was looking for mathematical means that would provide this justification. Maximality is one such means.

Establishing the left-to-right direction of the Normalization Conjecture via maximality is like proving the completeness of the classical propositional calculus with respect to any kind of nontrivial model via Post completeness (which is proved syntactically by reduction to conjunctive normal form). Actually, the first proof of this completeness with respect to tautologies was given by Bernays and Hilbert exactly in this manner (see [129], Sections 2.4 and 2.5; see also [61], Section I.13, and §9.3 below).

Maximality for the sort of categories mentioned above is proved with the help of coherence in [42] and [44] (which is established proof-theoretically, by normalization, cut elimination and similar methods; see Chapter 9). Coherence is helpful in proving maximality, but maximality can also be proved by other means, as this is done for cartesian closed categories via a typed version of Böhm's theorem in [115], [111] and [41]. This justifies the left-to-right direction of the Normalization Conjecture also for the implicational and the conjunction-implication fragments of intuitionistic logic. The maximality of bicartesian closed categories, which would justify the left-to-right direction of the Normalization Conjecture for the whole of intuitionistic propositional logic is, as far as we know, an open problem. (A use for maximality similar to that propounded here and in [41] and [42] is envisaged in [127].)

In [34] (Section 4.11) it is proved that the general notion of adjunction is also maximal in some sense. The maximality we encountered above, which involves connectives tied to particular adjunctions, cannot be derived from the maximality of the general notion of adjunction, but these matters should not be foreign to each other.

Since we find maximality an interesting property, we pay attention to it in this book where we could establish it with the help of our coherence results, and where it is not a trivial property. Besides the maximality results from Chapter 9, mentioned above, there are analogous results in §12.3, §12.5 and §13.3. We also pay attention to maximality in cases where it cannot be established (see §10.3 and §11.5). In some cases where it does not hold, we still have relative maximality results (see §9.7, §11.5 and §12.5).

§1.6. Union of proofs and zero proofs

Gentzen's plural-sequent system for classical logic has implicitly a rule of *union*, or *addition*, of derivations, which is derived as follows:

$$\dfrac{\dfrac{\dfrac{f\colon A \vdash B}{\theta^R_C f\colon A \vdash B, C} \qquad \dfrac{g\colon A \vdash B}{\theta^L_C g\colon C, A \vdash B}}{\mathit{cut}(\theta^R_C f, \theta^L_C g)\colon A, A \vdash B, B}}{f \cup g\colon A \vdash B}\ \mathit{contractions}$$

Here $\theta^R_C f$ and $\theta^L_C g$ are obtained from f and g respectively by thinning on the right and thinning on the left, and $\mathit{cut}(\theta^R_C f, \theta^L_C g)$ may be conceived as obtained by applying to f and g a limit case of Gentzen's multiple-cut rule *mix*, where the collection of mix formulae is empty. A related principle was considered under the name *mix* in linear logic (see §8.1).

In a cut-elimination procedure like Gentzen's, $f \cup g$ is reduced either to f or to g (see [55], Sections III.3.113.1-2). If we have $f \cup g = f$ and $f \cup g = g$, then we get immediately $f = g$, that is collapse and triviality. In [59] (Appendix B.1 by Y. Lafont; see also [62], Section 1) this is taken as sufficient ground to conclude that cut elimination in the plural-sequent system for classical logic must lead to preorder and collapse. (In [59], the inevitability of this collapse is compared to the argument presented after Proposition 1 of §14.3, which shows that a plausible assumption about classical negation added to bicartesian closed categories leads to preorder, but these are different matters.) To evade collapse we may try keeping only one of the equations $f \cup g = f$ and $f \cup g = g$, and reject the other; then we must also reject the commutativity of $\cup$, but it seems such decisions would be arbitrary. (For similar reasons, even without assuming the commutativity of $\cup$, the assumptions of [121], p. 232, C.12, lead to preorder.) There is, however, a way to evade collapse here that is not arbitrary. The modification of Gentzen's cut-elimination procedure expounded in Chapter 11 (see also §12.5) and our coherence results (more precisely, the easy, soundness, i.e. functoriality, parts of these results) testify to that.

The Generality Conjecture tells us that we should have neither $f \cup g = f$ nor $f \cup g = g$. The union of two graphs may well produce a graph differing from each of the graphs entering into the union. It also tells us that union of proofs should be associative and commutative. The idempotency law $f \cup f = f$ is imposed by *Rel*, but it stands apart, and with another graphical category, we may do without it (see §12.5). Without idempotency, union of proofs is rather addition of proofs. Our way out of the problematic situation Gentzen found himself in is to take into account union or addition of proofs.

If we have union of proofs, it is natural to assume that we also have for every formula A and every formula B a *zero proof* $\mathbf{0}_{A,B} : A \vdash B$, with an empty graph, which with union of proofs makes at least a commutative monoid; with idempotency, it gives the unit of a semilattice. We may envisage having zero proofs $\mathbf{0}_{A,B} : A \vdash B$ only for those A and B where there is also a nonzero proof from A to B, as we do in our categorification of the notion of Boolean algebra, but the more sweeping assumption involving every A and every B makes sense too.

We should immediately face the complaint that with such zero proofs we have entered into inconsistency, since everything is provable. That is true, but not all proofs have been made identical, and we are here not interested in what is provable, but in what proofs are identical. If it happens—and with the Generality Conjecture it will happen indeed—that introducing zero proofs is conservative with respect to identity of proofs that do not involve zero proofs, then it is legitimate to introduce zero proofs, provided it is useful for some purpose. This is like extending our mathematical theories with what Hilbert called *ideal* objects; like extending the positive integers with zero, or like extending the reals with imaginary numbers.

The use of union of proofs is that it saves the agreement between the Normalization and Generality Conjectures in the presence of distribution, as we said in §1.4. The use of zero proofs is that it does the same in the presence of negation. The idempotency of union is essential in the absence of zero proofs, but not in their presence. Without idempotency our graphical category in the case of conjunctive-disjunctive logic turns up to be a category whose arrows are matrices, rather than the category *Rel*. Composition becomes matrix multiplication, and union is matrix addition. And in the presence of zero matrices, we obtain a unique normal form like in linear algebra: every matrix is the sum of matrices with a single 1 entry.

A number of logicians have sought a link between logic and linear algebra, and here is such a link. We have it not for an alternative logic, but for classical logic. We have it, however, not at the level of provability, but at the level of identity of proofs.

The unique normal form suggested by linear algebra is not unrelated to cut elimination. In the graphical category of matrices the result of cut elimination is obtained by multiplying matrices, and the equations of this category yield a cut-elimination procedure. They yield it even in the absence of zero proofs, provided we have $1+1 = 1$. Unlike Gentzen's cut-elimination procedures for classical logic, the new procedure admits a commutative addition or union of proofs without collapse. So, in classical logic, the Generality Conjecture is not foreign to cut elimination, and hence it is not foreign to the Normalization Conjecture, provided we understand the equivalence relation involved in this conjecture in a manner different from Prawitz's.

This need not exhaust the advantages of having zero proofs. They may be used also to analyze disjunction elimination. Without pursuing this topic very far, let us note that passing from $A \vee B$ to A involves a zero proof from B to A, and passing from $A \vee B$ to B involves a zero proof from A to B. If next we are able to reach C both from A and from B, we may add our two proofs from $A \vee B$ to C, and so to speak "cancel" the two zero proofs.

Logicians were, and still are, interested mostly in provability, and not in proofs. This is so even in proof theory. When we address the question of identity of proofs we have certainly left the realm of provability, and entered into the realm of proofs. This should become clear in particular when we introduce zero proofs.

§1.7. Strictification

Strictification is inverse to categorification. While categorification usually (but not always) involves splitting objects, strictification involves identifying objects. Factoring a set through an equivalence relation, i.e. replacing the objects of a set by equivalence classes of objects of this set, is a simple example of strictification. Logicians are very used to a kind of strictification that may be called "lindenbaumization", by which the algebra of formulae of conjunctive logic is replaced by a freely generated semilattice, or the algebra of formulae of intuitionistic propositional logic is replaced by a freely generated Heyting algebra, or the algebra of formulae of classical propositional logic is replaced by a freely generated Boolean algebra. The equivalence relation involved in these strictifications is mutual implication.

In this book we are, however, interested in strictification of categories. Precise notions of strictification, which we need for our work, will be introduced in Chapter 3. Let us say for the time being that the simpler of these notions is a kind of partial skeletization of a category. An equivalence relation, induced by a subcategory that is a groupoid and a preorder, is used to replace the objects of the category by equivalence classes of objects. In the other, more general and more involved notion, the partial skeletization is applied to a category generated out of a given category. (We are aware this preliminary rough description of the matter cannot be very informative.) After strictification, objects are replaced by equivalence classes, which may correspond to sequences, or multisets, or sets, or structures of that kind.

The idea is to obtain a strictified category equivalent to the initially given category in which computations are easier to record, because some arrows that were not identity arrows, like, typically, associativity isomorphisms, are replaced by identity arrows (see Chapters 5-8 and 11). This equivalence of categories is not meant to be any equivalence, but an equivalence via functors that preserve a particular categorial structure at least up to isomorphism. For that we will define precisely what it means for a functor to preserve a structure, such as interests us, up to isomorphism (see §2.8).

We were inspired by previous attempts to define this notion of functor for monoidal categories, and by the ensuing strictification results of Joyal and Street in [67] (Section 1) and of Mac Lane in [96] (Sections XI.2-3). We do not, however, find these definitions and results sufficient for our purposes, even when only the monoidal structure is strictified. We need something more general. We envisage strictifying structures other than just monoidal, and we will have occasion in this book to strictify also with respect to symmetry (see §6.5, §7.6 and §8.4). Another limitation of previous strictification results for monoidal categories is that they do not take into account that the monoidal structure may be just a part of a more complex ambient structure, and that the functors involved in equivalence should preserve this ambient, not strictified, structure up to isomorphism. To have just the monoidal structure preserved is rather useless from our point of view (see §3.1).

Our results on strictification will be much more general, but they are not such that they could not be further generalized. In particular, in defining the categorial structure preserved by our functors up to isomorphism we have presupposed that this structure is defined only with covariant binary endofunctors. A natural generalization is to take here into account also endofunctors of arbitrary arity, covariant in some argument-places and contravariant in others. We suppose that our results can be extended to cover such situations too. For the applications we need it was, however, enough to cover the simpler situation, excluding contravariance, and we did just that. We were afraid of complicating further a matter already full of details, to prove results for which we have no immediate application. (As Mac Lane says in [94], p. 103: "... good general theory does not search for the maximum generality, but for the right generality.")

So our notion of a logical system in the next chapter involves only conjunction and disjunction as binary connectives, together with the constants $\top$ and $\bot$. Implication is excluded, and negation is left for the end of the book. To cover these other connectives, we would need to extend our notion of logical structure to permit contravariance. We assume this can be done in a straightforward manner at the cost of complicating notation. We refrain, however, from doing so in this book, whose central piece is about conjunctive-disjunctive logic, and where negation appears only at the end. Anyway, as far as strictification goes, this limited notion of logical system is sufficient for our purposes.

Classical implication, defined in the usual way in terms of disjunction and negation, does not come out as a very important connective in our proof-theoretical perspective. It is not much of an implication, if the role of implication is to help in mirroring the deductive metatheory in the object language. Intuitionistic implication plays that role better.

Our results on strictification are still somewhat more general than what we strictly need. In strictifying a binary connective like conjunction, purely conjunctive formulae may be replaced by equivalence classes that correspond to sequences, or multisets, or sets, of the atomic formulae joined by conjunction. For our purposes, we could have stuck to the first two strictifications, but with our general treatment we cover also the third. With that, we stay within the limits of covariance.

Strictification, though an interesting topic on its own, is not absolutely essential for our main topic—coherence. It is for us just a tool, we could have dispensed with in principle. That would, however, be at the cost of making already pretty long records even longer. So strictification is for us a rather useful tool.

It is a tool more useful for recording computations than for discovering how they should be done. Blurring distinctions may sometimes hinder this discovery.

It is remarkable that the general notion of strictification may be found implicit in Gentzen's sequent systems, as we will try to explain in §11.1, in the central chapter of the book.

CHAPTER 2

SYNTACTICAL CATEGORIES

In this chapter, which is of a preliminary character, we define the notions of syntactical categories needed for our work. In particular, we introduce the notion of logical category (which should not be confused with the homonymous notion of [98], Section 3.4). Logical categories are obtained from logical systems in a propositional language by replacing derivations with equivalence classes of derivations. The equivalence producing these classes is of general mathematical interest, but it has also proof-theoretical meaning, so that the equivalence classes may be identified with proofs. This presupposes some notions of logic and category theory, which will be duly defined.

Many of these notions are quite standard, and we go over them just to fix terminology. Something less standard may be found in the section on definable connectives, where some intricacies inherent in this notion are made manifest. A new matter is also detailed definitions of notions of functors preserving the structure of a logical category. We are interested in particular in those of these functors that preserve the structure up to isomorphism. These definitions prepare the ground for Chapter 3. We treat these matters in generality greater than we strictly need after that chapter. It is not essential to master all the details we go into in order to follow the exposition later on.

After these syntactical matters, we introduce at the end of the chapter the semantics of our logical categories. This is a semantics of proofs, and not the usual kind of semantics of propositions. Our main semantical category is the category whose arrows are relations between finite ordinals. This category, which is tied to the notion of natural transformation, will serve for our coherence results. The interpretations that link our syntax to this semantics are functors that preserve the structure "on the nose", i.e. up to an isomorphism that is identity.

§2.1. Languages

A *language* is a set of *words*, each word being a finite (possibly empty) sequence of *symbols*. A symbol is a mathematical object of any kind. The *length* of a word is the number of occurrences of symbols in it, and this is the most standard measure of the complexity of a word. In particular cases, however, we may rely on various other measures of complexity, like, for example, the number of occurrences of some particular kind of symbol.

We introduce first several languages of the kind logicians call *propositional languages*. Such languages are generated from a set $\mathcal{P}$ of symbols called *letters*; logicians would call them *propositional letters* or *propositional variables*. Sometimes we require that $\mathcal{P}$ be infinite (see the end of §2.8), but $\mathcal{P}$ can also be finite, and even empty. Since nothing in particular is assumed about $\mathcal{P}$, the symbols of $\mathcal{P}$ can be arbitrary mathematical objects, and the definitions of notions built on $\mathcal{P}$ (such as that of logical system and logical category; see §2.6 and §2.7 below) do not depend on the particular $\mathcal{P}$ that was chosen.

Let π be a symbol of the kind called in logic *n-ary connective*, for $n \geq 0$. A 0-ary, i.e. *nullary*, connective is more commonly known as a *propositional constant*; 1-ary are *unary* connectives and 2-ary connectives are *binary* connectives. We assume, as usual, that $\mathcal{P}$ is disjoint from the set of connectives. Then a language $\mathcal{L}$ such as we need is built up with inductive clauses of the following kind:

($\mathcal{P}$) $\mathcal{P} \subseteq \mathcal{L}$,

(π) if $A_1, \ldots, A_n \in \mathcal{L}$, then $\pi A_1 \ldots A_n \in \mathcal{L}$.

It is assumed here that π is an n-ary connective. If $n = 0$, then $A_1 \ldots A_n$ is the empty sequence, and $\pi \in \mathcal{L}$. We have an analogous convention for all sorts of sequences that will appear in this work: if $n = 0$, then $x_1 \ldots x_n$ or $x_1, \ldots, x_n$ is the empty sequence, and $\{x_1, \ldots, x_n\}$ is the empty set $\emptyset$.

The elements of $\mathcal{L}$ are called *formulae*; logicians would say *propositional formulae*. We use p, $q, \ldots$, sometimes with indices, as variables for letters, i.e. elements of $\mathcal{P}$, and A, $B, \ldots$, sometimes with indices, as variables for formulae. The elements of $\mathcal{P}$ and nullary connectives are called *atomic formulae*. The *letter length* of a formula is the number of occurrences of letters in it.

We reserve ζ for nullary connectives and ξ for binary connectives. The formula $\xi\xi pqp$, which is in the Polish, *prefix*, notation, is more commonly written $((p \,\xi\, q) \,\xi\, p)$, and we will favour this common, *infix*, notation for binary connectives. Polish notation is handy for dealing with n-ary connectives where $n \geq 3$, but in the greatest part of this work we will have just nullary and binary connectives. A unary connective appears in Chapter 14. (Notation for unary connectives that would not be Polish, like Hilbert's negation $\bar{A}$, is uncommon in propositional logic; for nullary connectives there is no alternative.) We assume that we have as auxiliary symbols the right parenthesis) and the left parenthesis (, which are neither letters nor connectives, with whose help we formulate the clause

$$\text{if } A, B \in \mathcal{L}, \text{ then } (A \,\xi\, B) \in \mathcal{L}.$$

This clause replaces (π) for binary connectives. As usual, we take the outermost parentheses of formulae for granted, and omit them.

Consider a binary relation T on a set of elements called *nodes* such that when xTy we say x is the *predecessor* of y, or y is the *successor* of x. A *path* from a node x to a node y is a sequence $x_1 \ldots x_n$, with $n \geq 1$, such that x is x_1 and y is x_n, while for every $i \in \{1, \ldots, n-1\}$ we have $x_i T x_{i+1}$. A *root* is a node without predecessors, and a *leaf* a node without successors. We say that a node is of n-ary *branching*, with $n \geq 0$, when it has n successors. So leaves are of nullary branching.

A *finite tree* is such a relation T where the set of nodes is finite, there is exactly one root and every node except the root has exactly one predecessor. It is clear that in every finite tree there is exactly one path from the root to each node.

The *height* of a node in a finite tree is the number of nodes in the path from the root to this node. A finite tree is *planar* when all nodes of the same height $n \geq 1$ are linearly ordered by a relation $<_n$ such that if $x_1 T x_2$ and $y_1 T y_2$ and $x_1 <_n y_1$, then $x_2 <_{n+1} y_2$. When for two nodes x and y of the same height we have $x <_n y$ with this linear order, we say that *x is on the left-hand side of y*. Every formula of a language corresponds to a planar finite tree whose leaves are labelled with letters and nullary connectives, and whose remaining nodes are labelled with n-ary connectives for $n \geq 1$.

The language $\mathcal{L}_{\pi_1,\ldots,\pi_m}$ has exactly $\pi_1, \ldots, \pi_m$ as connectives. We are in particular interested in the languages where $\xi \in \{\wedge, \vee\}$ and $\varsigma \in \{\top, \bot\}$. These are the languages $\mathcal{L}_\wedge$, $\mathcal{L}_\vee$, $\mathcal{L}_{\wedge,\top}$, $\mathcal{L}_{\vee,\bot}$, $\mathcal{L}_{\wedge,\vee}$ and $\mathcal{L}_{\wedge,\vee,\top,\bot}$.

We use the word *subword* as usual: every word is a subword of itself, and if $w_1 w_2$ is a subword of a word w, then w_1 and w_2 are subwords of w. A *proper* subword of a word w is a subword of w different from w. A *subformula* of a formula is a subword that is a formula. The subformulae A and B are the *main conjuncts* of $A \wedge B$, and the *main disjuncts* of $A \vee B$.

Let $w(A)$ be the word obtained by deleting all parentheses in a formula A of a language $\mathcal{L}$. We say that the formulae A and B of $\mathcal{L}$ are *comparable* when $w(A)$ and $w(B)$ are the same word.

A *place* in A is a subword w' of $w(A)$. There is an obvious deleting map δ from subwords of A to places in A. We say that a subword v of A is *at a place* w' when $\delta(v) = w'$. (Note that different subwords of A can be at the same place.) For A and B comparable, a subword w_1 of A and a subword w_2 of B are *at the same place* when $\delta(w_1) = \delta(w_2)$.

We say, as usual, that an occurrence y of a symbol is within the *scope* of an occurrence x of an n-ary connective in a formula A when in A we have a subformula of the form $xA_1 \ldots A_n$ with y being in A_i for one $i \in \{1, \ldots, n\}$. We say that y is within the *immediate scope* of x when y is within the scope of x and there is no occurrence of a connective z within the scope of x such that y is within the scope of z.

§2.2. Syntactical systems

A *graph* is a pair of functions, called the *source* function and the *target* function, from a set of elements called *arrows* to a set of elements called *objects*. We use f, $g, \ldots$, sometimes with indices, as variables for arrows and a, $b, \ldots$, sometimes with indices, as variables for objects. In many cases in this work, objects will be formulae of a language such as we introduced in the preceding section, but we also need the more general notion.

The expression $f : a \vdash b$ means that the source function assigns a to f and the target function assigns b to f; we call a and b the *source* and *target* of f, respectively. In category theory, $\vdash$ is usually written $\rightarrow$, but we keep $\rightarrow$ for other purposes (for functions and implication), and we stress with the logical turnstile symbol $\vdash$ the proof-theoretical interpretation of our work. We call $a \vdash b$, which is just a peculiar notation for the ordered pair (a, b), the *type* of $f : a \vdash b$. A *hom-set* in a graph is the set of all arrows of the same type for a given type. A graph where for every $f, g : a \vdash b$ we have $f = g$, i.e. where hom-sets are either empty or singletons, amounts to a binary relation R on the set of its objects such that $(a, b) \in R$ iff there is an arrow of type $a \vdash b$ in the graph.

For a given graph $\mathcal{G}$, the *dual* graph $\mathcal{G}^{op}$ is defined by interchanging the source and target functions; namely, the source function of $\mathcal{G}^{op}$ is the target function of $\mathcal{G}$, and the target function of $\mathcal{G}^{op}$ is the source function of $\mathcal{G}$, while the sets of objects and arrows are the same. An object b in a graph $\mathcal{G}$ is *terminal* when for every object a in $\mathcal{G}$ there is a unique arrow of $\mathcal{G}$ of a type $a \vdash b$, and b is *initial* in $\mathcal{G}$ when it is terminal in $\mathcal{G}^{op}$.

A *deductive system* (in the sense of [85], Section I.1) is a graph that must have for every object a an *identity arrow* $\mathbf{1}_a : a \vdash a$, and whose arrows are closed under the partial operation of *composition*:

$$\frac{f : a \vdash b \qquad g : b \vdash c}{g \circ f : a \vdash c}$$

This fractional notation, taken over from the notation for rules in logic, conveys that if $f : a \vdash b$ and $g : b \vdash c$ are in the deductive system, then $g \circ f : a \vdash c$ is in the deductive system. We use an analogous notation in other cases.

A deductive system is *discrete* when all of its arrows are identity arrows. A deductive system is a *preorder* when for every $f, g : a \vdash b$ in this deductive system we have $f = g$. A deductive system that is a preorder amounts to a preordering, i.e. reflexive and transitive, relation on the set of its objects. A preorder is a *partial order* when the preordering relation is antisymmetric. Every discrete deductive system is a preorder, but not vice versa. In principle, one can envisage the empty deductive system, with an empty set of arrows and an empty set of objects, but we have no interest in it for our work, and we will exclude it.

The notion of a deductive system is a generalization of the notion of category. A *category* is a deductive system in which the following equations, called *categorial equations*, hold between arrows:

$$\begin{aligned} &(cat\ 1) \quad f \circ \mathbf{1}_a = \mathbf{1}_b \circ f = f : a \vdash b, \\ &(cat\ 2) \quad h \circ (g \circ f) = (h \circ g) \circ f. \end{aligned}$$

This notion of category covers only *small* categories, but in this work, where we have no foundational ambitions, we have no need for categories whose collections of objects or arrows are bigger than sets. When we speak occasionally of the category *Set* of sets with functions, we assume that the collection of objects of this category is the domain of a model of first-order axiomatic set theory, and hence it is a set. The functions between the elements of this domain also make a set. We make an analogous assumption for other categories mentioned in this book that seem not to be small.

A *syntactical system* is a particular kind of deductive system where arrows make an inductively defined language, whose members are called *arrow terms*. Arrow terms are words defined inductively out of *primitive arrow terms* with the help of symbols tied to partial or total finite *operations on arrow terms* and the auxiliary symbols of right and left parentheses. A *subterm* of a term is a subword that is a term.

Among the primitive arrow terms we must have the *identity arrow terms*, which make the identity arrows of the deductive system (so we must have them for every object), and among the symbols for operations on arrow terms we must have one tied to composition:

$$\frac{f : a \vdash b \qquad g : b \vdash c}{(g \circ f) : a \vdash c}$$

(As we said above, this is read: "If f of type $a \vdash b$ and g of type $b \vdash c$ are arrow terms, then the word $(g \circ f)$ is an arrow term of type $a \vdash c$.") So, officially, parentheses in arrow terms involving $\circ$ are compulsory; but, as usual, we will omit outermost parentheses, and other parentheses if this can be done without engendering ambiguity. Note that here $\circ$ is just a symbol.

The operation of composition tied to this symbol is the operation assigning to the pair of words (f, g), of the types $a \vdash b$ and $b \vdash c$ respectively, the word $(g \circ f)$, of type $a \vdash c$.

We say that a graph $\mathcal{G}_1$ is a *subgraph* of a graph $\mathcal{G}_2$ when the objects and arrows of $\mathcal{G}_1$ are included respectively in the objects and arrows of $\mathcal{G}_2$ and the arrows of $\mathcal{G}_1$ have in $\mathcal{G}_1$ the same source and target as in $\mathcal{G}_2$.

A deductive system $\mathcal{D}_1$ is a *subsystem* of a deductive system $\mathcal{D}_2$ when $\mathcal{D}_1$ is a subgraph of $\mathcal{D}_2$, the identity arrows of $\mathcal{D}_1$ are identity arrows in $\mathcal{D}_2$ and for every pair of arrows $(f : a \vdash b,\ g : b \vdash c)$ of $\mathcal{D}_1$ their composition in $\mathcal{D}_1$ is equal to their composition in $\mathcal{D}_2$. A *subcategory* is a subsystem of a category. A subcategory must be a category.

An arrow f in deductive system is *mono* when for every g and h the equation $f \circ g = f \circ h$ implies $g = h$, and f is *epi* when for every g and h the equation $g \circ f = h \circ f$ implies $g = h$.

An arrow $f : a \vdash b$ in a deductive system $\mathcal{D}$ is an *isomorphism* when there is an arrow $g : b \vdash a$ in $\mathcal{D}$ such that $g \circ f = \mathbf{1}_a$ and $f \circ g = \mathbf{1}_b$. The arrows f and g are here *inverses* of each other. Isomorphisms in categories are mono and epi. A category in which every arrow is an isomorphism is a *groupoid*. If there is an isomorphism of type $a \vdash b$, then a and b are said to be *isomorphic*.

A subsystem $\mathcal{D}_1$ of a deductive system $\mathcal{D}_2$ is *full* when for every arrow $f : a \vdash b$ of $\mathcal{D}_2$ if the objects a and b are in $\mathcal{D}_1$, then f is in $\mathcal{D}_1$. The *partial skeleton* $\mathcal{A}'$ of a category $\mathcal{A}$ is a full subcategory of $\mathcal{A}$ such that for every object a of $\mathcal{A}$ there is in $\mathcal{A}'$ an object a' isomorphic to a in $\mathcal{A}$. (So every category is a partial skeleton of itself.) If in this definition we require that the object a' be unique, then $\mathcal{A}'$ is called simply a *skeleton*. A skeleton is unique up to isomorphism of categories (see §2.4); so it is usual to speak about *the* skeleton of a category.

§2.3. Equational systems

An *equation in a syntactical system* $\mathcal{S}$ is a word $f = g$ where f and g are arrow terms of $\mathcal{S}$ of the same type. An *equational system* $\mathcal{E}$ in $\mathcal{S}$ is a set of equations in $\mathcal{S}$ such that the following conditions are satisfied:

(re) $f = f$ is in $\mathcal{E}$ for every arrow term f of $\mathcal{S}$;

(sy) if $f = g$ is in $\mathcal{E}$, then $g = f$ is in $\mathcal{E}$;

(tr) if $f = g$ and $g = h$ are in $\mathcal{E}$, then $f = h$ is in $\mathcal{E}$;

(co) if $f_1 = g_1, \dots, f_n = g_n$ for $n \geq 1$ are in $\mathcal{E}$, then $of_1 \dots f_n = og_1 \dots g_n$ is in $\mathcal{E}$, where $of_1 \dots f_n$ and $og_1 \dots g_n$ are arrow terms of $\mathcal{S}$ produced by an n-ary operation o on arrow terms.

For the congruence condition (co) to make sense, the operation o must be such that if f_i is of the same type as g_i, for $i \in \{1, \dots, n\}$, then $of_1 \dots f_n$ is of the same type as $og_1 \dots g_n$. We envisage only operations of this kind.

As in propositional languages above, when the arity of o is not greater than 2, we favour the infix notation with parentheses; so we write $f_1 o f_2$ (with outermost parentheses omitted) instead of $of_1 f_2$.

Consider the smallest equivalence relation $\equiv$ on the arrow terms of $\mathcal{S}$ that satisfies $f \equiv g$ iff $f = g$ is in $\mathcal{E}$. With the help of $\equiv$ we build a deductive system called $\mathcal{S}/\mathcal{E}$. The objects of $\mathcal{S}/\mathcal{E}$ are the objects of $\mathcal{S}$, and its arrows are equivalence classes $[f]$ of the arrow terms f of $\mathcal{S}$ with respect to $\equiv$. The identity arrows of $\mathcal{S}/\mathcal{E}$ are the equivalence classes of the identity arrow terms of $\mathcal{S}$, and for every n-ary operation o of $\mathcal{S}$, including in particular composition, we define an operation on equivalence classes by

$$o[f_1] \dots [f_n] =_{df} [of_1 \dots f_n].$$

The condition (co) above guarantees the correctness of this definition.

Most often, we do not write concrete equations, but equations with variables, like the categorial equations in the preceding section, where f, g and h are variables for arrows, while a and b are variables for objects. As usual, we call equations with variables simply equations. We say that such an equation *belongs* to an equational system $\mathcal{E}$ in $\mathcal{S}$ when every instance of it, with arrow terms of $\mathcal{S}$ substituted for variables for arrows and names of objects of $\mathcal{S}$ substituted for variables for objects, is an element of $\mathcal{E}$. In producing these instances we, of course, pay attention to types. For example, in instances of the categorial equation (cat 1) we have that a is the source of f, while b is its target, and in instances of (cat 2) we have that f, g and h have types that permit composition.

We say that an equation with variables holds in a graph $\mathcal{G}$ when *every* substitution instance of it holds in $\mathcal{G}$. (That such an instance holds in $\mathcal{G}$ means, of course, that the names on the two sides of the equation sign $=$ name the same thing.) It is quite common to understand holding of equations with variables in this universal manner, and that is how we will understand it, unless stated otherwise. That is how we understood holding for the categorial equations in the definition of category. Instead of saying that an equation *holds* in $\mathcal{G}$, we may say, synonymously, that it is *satisfied* in $\mathcal{G}$, or, simply, that we *have* it in $\mathcal{G}$.

To name the arrows of $\mathcal{S}/\mathcal{E}$, we use the arrow terms of $\mathcal{S}$, so that an arrow term names the equivalence class to which it belongs. Synonyms of *name* are *designate*, *denote* and *stand for*. Then every equation of $\mathcal{E}$ will hold in $\mathcal{S}/\mathcal{E}$. We say that the arrow terms f_1 and f_2 of $\mathcal{S}$ are *equal* in $\mathcal{S}/\mathcal{E}$ when $f_1 = f_2$ holds in $\mathcal{S}/\mathcal{E}$, which is equivalent with the equation $f_1 = f_2$ belonging to $\mathcal{E}$.

If the categorial equations belong to $\mathcal{E}$, then $\mathcal{S}/\mathcal{E}$ is a category, and, since such categories arise out of syntactical systems, we call them *syntactical categories*. We say that the category $\mathcal{S}/\mathcal{E}$ is *in* the system $\mathcal{S}$. If only instances of $f = f$ are in $\mathcal{E}$, then $\mathcal{S}/\mathcal{E}$ is $\mathcal{S}$ itself.

A set of *axioms* Ax of an equational system $\mathcal{E}$ is a proper subset of the set of equations $\mathcal{E}$ such that $\mathcal{E}$ may be generated from Ax by closing under the rules (sy), (tr) and (co). The set of axioms need not be finite, and it will usually be infinite in this work. Every equation of $\mathcal{E}$ is either an axiom in Ax or *derived* from previously obtained equations by applying one of the rules. More formally, a *derivation* is a finite tree of equations whose leaves are axioms, and where each node that is not a leaf is obtained from its successors, i.e. from nodes immediately above, by applying the rules. The root of the tree is the equation derived.

Instead of saying that an equation holds in $\mathcal{S}/\mathcal{E}$ because we can derive it in $\mathcal{E}$, we will sometimes say more simply that we can derive the equation *for* $\mathcal{S}/\mathcal{E}$. This way of speaking will often prove handier later in the book, and should not cause confusion.

§2.4. Functors and natural transformations

A *graph-morphism* F from a graph $\mathcal{G}_1$ to a graph $\mathcal{G}_2$ is a pair of maps, both denoted by F, from the objects of $\mathcal{G}_1$ to the objects of $\mathcal{G}_2$ and from the arrows of $\mathcal{G}_1$ to the arrows of $\mathcal{G}_2$, respectively, such that for every arrow $f: a \vdash b$ of $\mathcal{G}_1$ the type of the arrow Ff of $\mathcal{G}_2$ is $Fa \vdash Fb$.

A graph-morphism F from a graph $\mathcal{G}_1$ to a graph $\mathcal{G}_2$ is *faithful* when, for every pair (f, g) of arrows of $\mathcal{G}_1$ of the same type, if $Ff = Fg$ in $\mathcal{G}_2$, then $f = g$ in $\mathcal{G}_1$.

A *functor* F from a deductive system $\mathcal{D}_1$ to a deductive system $\mathcal{D}_2$ is a graph-morphism from $\mathcal{D}_1$ to $\mathcal{D}_2$ such that in $\mathcal{D}_2$ the following equations hold:

$$(\mathit{fun}\ 1) \quad F\mathbf{1}_a = \mathbf{1}_{Fa},$$
$$(\mathit{fun}\ 2) \quad F(g \circ f) = Fg \circ Ff.$$

Note that this definition of functor is more general than the usual one, which envisages only functors between categories. Otherwise, it is the same definition. We generalize similarly other notions introduced below.

The *product* $\mathcal{D}_1 \times \mathcal{D}_2$ of the deductive systems $\mathcal{D}_1$ and $\mathcal{D}_2$ is the deductive system whose objects are pairs (a_1, a_2) such that a_1 is an object of $\mathcal{D}_1$ and a_2 an object of $\mathcal{D}_2$, and analogously for arrows. The identity arrows of $\mathcal{D}_1 \times \mathcal{D}_2$ are of the form $(\mathbf{1}_{a_1}, \mathbf{1}_{a_2})$, and composition is defined by

$$(g_1, g_2) \circ (f_1, f_2) =_{df} (g_1 \circ f_1, g_2 \circ f_2).$$

A functor B from $\mathcal{D}_1 \times \mathcal{D}_2$ to $\mathcal{D}$ is called a *bifunctor*; for bifunctors, (*fun* 1) and (*fun* 2) amount to the following equations respectively:

$$(\mathit{bif}\ 1) \quad B(\mathbf{1}_a, \mathbf{1}_b) = \mathbf{1}_{B(a,b)},$$
$$(\mathit{bif}\ 2) \quad B(g_1 \circ f_1, g_2 \circ f_2) = B(g_1, g_2) \circ B(f_1, f_2),$$

which we call the *bifunctorial equations.*

Let $\mathcal{D}^0$ be the trivial deductive system with a single object $*$ and a single arrow $\mathbf{1}_* : * \vdash *$. (This deductive system is a category.) Let $\mathcal{D}^{n+1}$ be $\mathcal{D}^n \times \mathcal{D}$. It is clear that $\mathcal{D}^1$ is isomorphic to $\mathcal{D}$. A functor from $\mathcal{D}^n$ to $\mathcal{D}$ will be called an *n-endofunctor* in $\mathcal{D}$. An object of $\mathcal{D}$ may be identified with a 0-endofunctor.

The *identity functor* I of $\mathcal{D}$ is the 1-endofunctor in $\mathcal{D}$ for which we have $Ia = a$ and $If = f$. A functor from $\mathcal{D}_1$ to $\mathcal{D}_2^{op}$ is called a *contravariant* functor from $\mathcal{D}_1$ to $\mathcal{D}_2$.

A *natural transformation* from a functor F_1 from $\mathcal{D}_1$ to $\mathcal{D}_2$ to a functor F_2 from $\mathcal{D}_1$ to $\mathcal{D}_2$ is a family τ of arrows of $\mathcal{D}_2$ indexed by objects of $\mathcal{D}_1$ such that τ_a is of the type $F_1a \vdash F_2a$, and the following equations hold in $\mathcal{D}_2$ for $f\colon a \vdash b$ an arrow of $\mathcal{D}_1$:

$$F_2f \circ \tau_a = \tau_b \circ F_1f.$$

Consider now an m-endofunctor M and an n-endofunctor N in $\mathcal{D}$, and two functions $\mu\colon \{1,\ldots,m\} \to \{1,\ldots,k\}$ and $\nu\colon \{1,\ldots,n\} \to \{1,\ldots,k\}$ where $m, n \geq 0$ and $k \geq 0$ (if $m = 0$, then $\{1,\ldots,m\} = \emptyset$; if $k = 0$, then we must have $m = n = 0$). Then M^μ defined by

$$M^\mu(x_1,\ldots,x_k) =_{df} M(x_{\mu(1)},\ldots,x_{\mu(m)})$$

and N^ν defined analogously are k-endofunctors in $\mathcal{D}$. (If $m = 0$, then $M(f_1,\ldots,f_m)$ is $M(\mathbf{1}_*)$.)

A family α of arrows of $\mathcal{D}$ such that for every sequence $a_1,\ldots,a_k$ of objects of $\mathcal{D}$ there is an arrow $\alpha_{a_1,\ldots,a_k}$ of the type $M^\mu(a_1,\ldots,a_k) \vdash N^\nu(a_1,\ldots,a_k)$ is called a *transformation* of $\mathcal{D}$ of arity k. We say that the arrows $f_1,\ldots,f_k$ of $\mathcal{D}$, such that for $i \in \{1,\ldots,k\}$ the arrow f_i is of the type $a_i \vdash b_i$, *flow through* α *in* $\mathcal{D}$ when the following equation holds in $\mathcal{D}$:

$$(\alpha\ nat) \quad N^\nu(f_1,\ldots,f_k) \circ \alpha_{a_1,\ldots,a_k} = \alpha_{b_1,\ldots,b_k} \circ M^\mu(f_1,\ldots,f_k).$$

By the definition of natural transformation, a transformation α is a natural transformation from the k-endofunctor M^μ of $\mathcal{D}$ to the k-endofunctor N^ν of $\mathcal{D}$ when every k-tuple of arrows of $\mathcal{D}$ flows through α in $\mathcal{D}$. We say that $\alpha_{a_1,\ldots,a_k}$ is *natural in* $a_1,\ldots,a_k$ when it is a member of a natural transformation. The equations (α *nat*) will be called *naturality equations*.

A natural transformation τ in a deductive system $\mathcal{D}$ is a *natural isomorphism* when each member of the family τ is an isomorphism. Two functors are *naturally isomorphic* when there is a natural isomorphism from one to the other.

We say that the deductive systems $\mathcal{D}_1$ and $\mathcal{D}_2$ are *equivalent* via a functor F_2 from $\mathcal{D}_1$ to $\mathcal{D}_2$ and a functor F_1 from $\mathcal{D}_2$ to $\mathcal{D}_1$ when the composite functor F_1F_2 is naturally isomorphic to the identity endofunctor of $\mathcal{D}_1$ and the composite functor F_2F_1 is naturally isomorphic to the identity endofunctor of $\mathcal{D}_2$. It is easy to conclude that the functors via which two categories are equivalent are faithful functors.

The deductive systems $\mathcal{D}_1$ and $\mathcal{D}_2$ are *isomorphic* via a functor F_2 from $\mathcal{D}_1$ to $\mathcal{D}_2$ and a functor F_1 from $\mathcal{D}_2$ to $\mathcal{D}_1$ when the composite functor F_1F_2 is equal to the identity endofunctor of $\mathcal{D}_1$ and the composite functor F_2F_1 is equal to the identity endofunctor of $\mathcal{D}_2$. Two deductive systems are said to be *equivalent* when there is a pair of functors via which they are equivalent, and analogously for *isomorphic* deductive systems.

Suppose we have two syntactical systems $\mathcal{S}_i$, for $i \in \{1, 2\}$, together with the equational systems $\mathcal{E}_i$ in $\mathcal{S}_i$. A graph-morphism F from $\mathcal{S}_1$ to $\mathcal{S}_2$ *induces* an obvious graph-morphism from $\mathcal{S}_1/\mathcal{E}_1$ to $\mathcal{S}_2/\mathcal{E}_2$, such that $F[f] = [Ff]$, provided $f = g$ in $\mathcal{E}_1$ implies $Ff = Fg$ in $\mathcal{E}_2$. (We do not write $[f]$ usually, but use the arrow term f to designate $[f]$.) When F from $\mathcal{S}_1$ to $\mathcal{S}_2$ is a functor, then F is a functor from $\mathcal{S}_1/\mathcal{E}_1$ to $\mathcal{S}_2/\mathcal{E}_2$.

When we have the graph-morphisms F_1 from $\mathcal{S}_2$ to $\mathcal{S}_1$ and F_2 from $\mathcal{S}_1$ to $\mathcal{S}_2$ such that $\mathcal{S}_1/\mathcal{E}_1$ and $\mathcal{S}_2/\mathcal{E}_2$ are isomorphic deductive systems via functors induced by F_1 and F_2, we say that $\mathcal{S}_1$ and $\mathcal{S}_2$ are *synonymous* up to $\mathcal{E}_1$ and $\mathcal{E}_2$ via F_1 and F_2. A stronger notion of synonymity of syntactical systems, which we will usually encounter, is when F_1 and F_2 are functors between $\mathcal{S}_1$ and $\mathcal{S}_2$, and not any graph-morphisms.

§2.5. Definable connectives

Let $\mathcal{L}$ stand for one of the languages $\mathcal{L}_{\wedge}$, $\mathcal{L}_{\vee}$, $\mathcal{L}_{\wedge,\top}$, $\mathcal{L}_{\vee,\bot}$, $\mathcal{L}_{\wedge,\vee}$ and $\mathcal{L}_{\wedge,\vee,\top,\bot}$ of §2.1, generated by an arbitrary set of letters $\mathcal{P}$, and let $\mathcal{L}^{\Box}$ stand for the language with the same connectives as $\mathcal{L}$ generated by the set of letters $\{\Box\}$. We use M, $N, \ldots$, sometimes with indices, for elements of $\mathcal{L}^{\Box}$. Let $|M| = m \geq 0$ be the number of occurrences of $\Box$ in M, and let $w_1, \ldots, w_m$ be a sequence of m arbitrary words. Later on in this work, these words will denote either the objects or the arrows of a category. Then $M(w_1, \ldots, w_m)$ is the word obtained by putting w_i, where $i \in \{1, \ldots, m\}$, for the i-th occurrence of $\Box$ in M, counting from the left.

Let $\mathcal{L}^{con}$ be a set of pairs (M,μ), which we abbreviate by M^μ, where $M \in \mathcal{L}^\square$ and $|M| = m \geq 0$, while μ is a function from $\{1,\dots,m\}$ to $\{1,\dots,k\}$ for some $k \geq 0$. The *arity* of M^μ is k. We define $M^\mu(w_1,\dots,w_k)$ as $M(w_{\mu(1)},\dots,w_{\mu(m)})$ (cf. the definition of M^μ in the preceding section).

When $w_1,\dots,w_k$ are formulae of $\mathcal{L}$, the elements of $\mathcal{L}^{con}$ stand for the *definable connectives* of $\mathcal{L}$. Let us consider some examples of definable connectives. A primitive connective $\xi \in \{\wedge,\vee\}$ of $\mathcal{L}$ is represented in $\mathcal{L}^{con}$ by the definable connective $(\square\xi\square)^{\iota_{\{1,2\}}}$ where $\iota_{\{1,2\}}$ is the identity function on $\{1,2\}$, while $\varsigma \in \{\top,\bot\}$ is represented by $\varsigma^{\iota_\emptyset}$ where $\iota_\emptyset$ is the identity function on $\emptyset$, which is the empty function (the only possible function from $\emptyset$ to $\emptyset$). If $\iota_{\{1\}}$ is the identity function on $\{1\}$ (the only possible function from $\{1\}$ to $\{1\}$), then $\square^{\iota_{\{1\}}}$ is the identity unary connective, for which we have $\square^{\iota_{\{1\}}}(A) = A$. If μ is the function from $\{1,2\}$ to $\{1,2\}$ such that $\mu(x) = 3-x$, then for the definable connective $(\square\xi\square)^\mu$ we have $(\square\xi\square)^\mu(A,B) = (\square\xi\square)(B,A) = B\,\xi\,A$. If μ is the only possible function from $\{1,2\}$ to $\{1\}$, then for the definable connective $(\square\xi\square)^\mu$ we have $(\square\xi\square)^\mu(A) = (\square\xi\square)(A,A) = A\,\xi\,A$. If μ is the constant function with value 1 from $\{1\}$ to $\{1,2\}$, then for the definable connective $\square^\mu$ we have $\square^\mu(A,B) = \square(A) = A$.

For M^μ, $N_1^{\nu_1},\dots,N_k^{\nu_k}$ elements of $\mathcal{L}^{con}$, we want to define the element $M^\mu(N_1^{\nu_1},\dots,N_k^{\nu_k})$ of $\mathcal{L}^{con}$ resulting from the substitution of $N_1^{\nu_1},\dots,N_k^{\nu_k}$ within M^μ. In other words, we want to define generalized composition of elements of $\mathcal{L}^{con}$. This notion is rather simple when μ is an identity function, but in the general case we have the following, more involved, definition.

Let M^μ be of arity k such that $|M| = m$, and let $N_i^{\nu_i}$, for $i \in \{1,\dots,k\}$, be of arity l_i with $|N_i| = n_i$. To define the element $M^\mu(N_1^{\nu_1},\dots,N_k^{\nu_k})$ of $\mathcal{L}^{con}$ of arity

$$\sum_{1\leq j\leq k} l_j$$

we must first define what it means to substitute the functions $\nu_1,\dots,\nu_k$ within the function μ.

Let

$$\pi(i) = \sum_{1\leq j\leq i} n_{\mu(j)}, \qquad \lambda(i) = \sum_{1\leq j\leq i} l_j,$$

and let $\beta\colon \{1,\ldots,\pi(m)\} \to \{1,\ldots,m\}$ be defined by

$$\beta(x) =_{df} \min\{i \mid x-\pi(i) < 0\}.$$

Next we define the function $\mu(\nu_1,\ldots,\nu_k) : \{1,\ldots,\pi(m)\} \to \{1,\ldots,\lambda(k)\}$ by

$$\mu(\nu_1,\ldots,\nu_k)(x) =_{df} \nu_{\mu(\beta(x))}(x-\pi(\beta(x)-1)) + \lambda(\mu(\beta(x))-1).$$

With the help of $\mu(\nu_1,\ldots,\nu_k)$ we define $M^\mu(N_1^{\nu_1},\ldots,N_k^{\nu_k})$ as $M^\mu(N_1,\ldots,N_k)^{\mu(\nu_1,\ldots,\nu_k)}$, which is equal to $M(N_{\mu(1)},\ldots,N_{\mu(m)})^{\mu(\nu_1,\ldots,\nu_k)}$.

The definition of $\mu(\nu_1,\ldots,\nu_k)$ is pretty opaque, and we must make a few comments on it. We will consider as an example a simple case of $\mu(\nu_1,\ldots,\nu_k)$, which covers most of our needs in this book.

Let the function $\nu_1+\nu_2\colon \{1,\ldots,n_1+n_2\} \to \{1,\ldots,l_1+l_2\}$ be defined by

$$(\nu_1+\nu_2)(x) =_{df} \begin{cases} \nu_1(x) & \text{if } x \le n_1 \\ \nu_2(x-n_1)+l_1 & \text{if } n_1 < x \end{cases}$$

Then one can check that when $m = k$ and μ is the identity function on $\{1,\ldots,m\}$ we have

$$\mu(\nu_1,\ldots,\nu_k) = \nu_1 + \ldots + \nu_k.$$

The complications of the general definition of $\mu(\nu_1,\ldots,\nu_k)$ above come from the fact that we want to substitute within the function μ the functions $\nu_1,\ldots,\nu_k$ so that for every ordered pair (x,y) in μ we have a copy of ν_y. These complications are not essential for many of the latter parts of our work. In many cases, we will have for $M^\mu \in \mathcal{L}^{con}$ that μ is the identity function $\iota_{\{1,\ldots,m\}}$ on $\{1,\ldots,m\}$. We introduce the convention that $M^{\iota_{\{1,\ldots,m\}}}$ is abbreviated by M. With that in mind, the reader can forget about the indices μ in M^μ in many places. We have preferred, however, to state our results later on in greater generality.

§2.6. Logical systems

We will consider in this work a particular kind of syntactical system called *logical system*. A logical system $\mathcal{C}$ has as objects the formulae of a language

$\mathcal{L}$, as in the preceding section. We say that such a logical system $\mathcal{C}$ *is in* $\mathcal{L}$. The primitive arrow terms of $\mathcal{C}$ come in families α. The members of α are indexed by sequences $A_1, \ldots, A_k$, with $k \geq 0$, of objects of $\mathcal{C}$. With every family α we associate two elements M and N of $\mathcal{L}^{\Box}$, such that $|M| = m$ and $|N| = n$, and two functions $\mu\colon \{1, \ldots, m\} \to \{1, \ldots, k\}$ and $\nu\colon \{1, \ldots, n\} \to \{1, \ldots, k\}$. The type of $\alpha_{A_1,\ldots,A_k}$ is $M^{\mu}(A_1, \ldots, A_k) \vdash N^{\nu}(A_1, \ldots, A_k)$. So an α is a transformation in $\mathcal{C}$.

In Table 1 we present most of the transformations α we need for our work. In this table, $\emptyset$ denotes the empty function from the empty set. The types of the members of α are as in Table 2. In the leftmost column of that table we write down the name of the union of the families α on the right-hand side. So the b family includes the families, i.e. transformations, $\hat{b}^{\to}$, $\hat{b}^{\leftarrow}$, $\check{b}^{\leftarrow}$ and $\check{b}^{\to}$. Within the family b we have the subfamily $\hat{b}$, which includes $\hat{b}^{\to}$ and $\hat{b}^{\leftarrow}$, and the subfamily $\check{b}$, which includes $\check{b}^{\leftarrow}$ and $\check{b}^{\to}$. We have, analogously, the subfamilies $\hat{\delta}$-$\hat{\sigma}$ and $\check{\delta}$-$\check{\sigma}$ of the δ-σ family, and analogously in other cases.

Of course, an α from $\mathcal{L}_{\wedge}$ may be found also in a wider language $\mathcal{L}_{\wedge,\vee}$. In practice, one of μ and ν will be the identity function, as in the transformations in Table 1, but we allow for greater generality. For the sake of uniformity, we decided to take always μ as the identity function in the transformations with $\mathcal{L}_{\wedge}$ and $\mathcal{L}_{\top}$, and ν as the identity function in the transformations with $\mathcal{L}_{\vee}$ and $\mathcal{L}_{\bot}$, but for $\hat{c}$ and $\check{c}$ we could have done otherwise. The difference in indexing $\hat{c}_{A,B}$ and $\check{c}_{B,A}$ sometimes requires additional care when passing from matters involving $\wedge$ to matters involving $\vee$, but it helps to enhance the duality underlying $\wedge$ and $\vee$.

The labels b, c, w and k are borrowed from the combinators B, C, W and K of combinatory logic, d comes from "dissociativity" (see §1.2) and m from "mix" (see §8.1). The Greek labels δ, σ and κ involve $\top$ and $\bot$.

As every syntactical system, a logical system $\mathcal{C}$ will have the family $\mathbf{1}$ from Tables 1 and 2, which delivers its identity arrow terms. If we work with a language in which we have $\xi \in \{\wedge, \vee\}$, then for building the arrow terms of $\mathcal{C}$ we have the following clause corresponding to a total operation on arrow terms:

$\mathcal{L}$	α	k	M	N	m	n	$\mu(x)$	$\nu(x)$
$\mathcal{L}_\wedge$	$\stackrel{\wedge}{b}^{\rightarrow}$	3	$\Box \wedge (\Box \wedge \Box)$	$(\Box \wedge \Box) \wedge \Box$	3	3	x	x
$\mathcal{L}_\wedge$	$\stackrel{\wedge}{b}^{\leftarrow}$	3	$(\Box \wedge \Box) \wedge \Box$	$\Box \wedge (\Box \wedge \Box)$	3	3	x	x
$\mathcal{L}_\vee$	$\stackrel{\vee}{b}^{\rightarrow}$	3	$\Box \vee (\Box \vee \Box)$	$(\Box \vee \Box) \vee \Box$	3	3	x	x
$\mathcal{L}_\vee$	$\stackrel{\vee}{b}^{\leftarrow}$	3	$(\Box \vee \Box) \vee \Box$	$\Box \vee (\Box \vee \Box)$	3	3	x	x
$\mathcal{L}_{\wedge,\top}$	$\stackrel{\wedge}{\delta}^{\rightarrow}$	1	$\Box \wedge \top$	$\Box$	1	1	x	x
$\mathcal{L}_{\wedge,\top}$	$\stackrel{\wedge}{\delta}^{\leftarrow}$	1	$\Box$	$\Box \wedge \top$	1	1	x	x
$\mathcal{L}_{\wedge,\top}$	$\stackrel{\wedge}{\sigma}^{\rightarrow}$	1	$\top \wedge \Box$	$\Box$	1	1	x	x
$\mathcal{L}_{\wedge,\top}$	$\stackrel{\wedge}{\sigma}^{\leftarrow}$	1	$\Box$	$\top \wedge \Box$	1	1	x	x
$\mathcal{L}_{\vee,\bot}$	$\stackrel{\vee}{\delta}^{\rightarrow}$	1	$\Box \vee \bot$	$\Box$	1	1	x	x
$\mathcal{L}_{\vee,\bot}$	$\stackrel{\vee}{\delta}^{\leftarrow}$	1	$\Box$	$\Box \vee \bot$	1	1	x	x
$\mathcal{L}_{\vee,\bot}$	$\stackrel{\vee}{\sigma}^{\rightarrow}$	1	$\bot \vee \Box$	$\Box$	1	1	x	x
$\mathcal{L}_{\vee,\bot}$	$\stackrel{\vee}{\sigma}^{\leftarrow}$	1	$\Box$	$\bot \vee \Box$	1	1	x	x
$\mathcal{L}_\wedge$	$\stackrel{\wedge}{c}$	2	$\Box \wedge \Box$	$\Box \wedge \Box$	2	2	x	$3-x$
$\mathcal{L}_\vee$	$\stackrel{\vee}{c}$	2	$\Box \vee \Box$	$\Box \vee \Box$	2	2	$3-x$	x
$\mathcal{L}_\wedge$	$\stackrel{\wedge}{w}$	1	$\Box$	$\Box \wedge \Box$	1	2	x	1
$\mathcal{L}_\vee$	$\stackrel{\vee}{w}$	1	$\Box \vee \Box$	$\Box$	2	1	1	x
$\mathcal{L}_\wedge$	$\stackrel{\wedge}{k}^1$	2	$\Box \wedge \Box$	$\Box$	2	1	x	1
$\mathcal{L}_\wedge$	$\stackrel{\wedge}{k}^2$	2	$\Box \wedge \Box$	$\Box$	2	1	x	2
$\mathcal{L}_\vee$	$\stackrel{\vee}{k}^1$	2	$\Box$	$\Box \vee \Box$	1	2	1	x
$\mathcal{L}_\vee$	$\stackrel{\vee}{k}^2$	2	$\Box$	$\Box \vee \Box$	1	2	2	x
$\mathcal{L}_\top$	$\stackrel{\wedge}{\kappa}$	1	$\Box$	$\top$	1	0	x	$\mu = \emptyset$
$\mathcal{L}_\bot$	$\stackrel{\vee}{\kappa}$	1	$\bot$	$\Box$	0	1	$\mu = \emptyset$	x
$\mathcal{L}_{\wedge,\vee}$	d^L	3	$\Box \wedge (\Box \vee \Box)$	$(\Box \wedge \Box) \vee \Box$	3	3	x	x
$\mathcal{L}_{\wedge,\vee}$	d^R	3	$(\Box \vee \Box) \wedge \Box$	$\Box \vee (\Box \wedge \Box)$	3	3	x	x
$\mathcal{L}_{\wedge,\vee}$	m	2	$\Box \wedge \Box$	$\Box \vee \Box$	2	2	x	x
$\mathcal{L}_{\wedge,\vee}$	m^{-1}	2	$\Box \vee \Box$	$\Box \wedge \Box$	2	2	x	x
any	$\mathbf{1}$	1	$\Box$	$\Box$	1	1	x	x

TABLE 1

b	$\hat{b}^{\rightarrow}_{A,B,C}\colon A\wedge(B\wedge C)\vdash(A\wedge B)\wedge C$	$\check{b}^{\rightarrow}_{A,B,C}\colon A\vee(B\vee C)\vdash(A\vee B)\vee C$
	$\hat{b}^{\leftarrow}_{A,B,C}\colon (A\wedge B)\wedge C\vdash A\wedge(B\wedge C)$	$\check{b}^{\leftarrow}_{A,B,C}\colon (A\vee B)\vee C\vdash A\vee(B\vee C)$
δ-σ	$\hat{\delta}^{\rightarrow}_{A}\colon A\wedge\top\vdash A$	$\check{\delta}^{\rightarrow}_{A}\colon A\vee\bot\vdash A$
	$\hat{\delta}^{\leftarrow}_{A}\colon A\vdash A\wedge\top$	$\check{\delta}^{\leftarrow}_{A}\colon A\vdash A\vee\bot$
	$\hat{\sigma}^{\rightarrow}_{A}\colon \top\wedge A\vdash A$	$\check{\sigma}^{\rightarrow}_{A}\colon \bot\vee A\vdash A$
	$\hat{\sigma}^{\leftarrow}_{A}\colon A\vdash \top\wedge A$	$\check{\sigma}^{\leftarrow}_{A}\colon A\vdash \bot\vee A$
c	$\hat{c}_{A,B}\colon A\wedge B\vdash B\wedge A$	$\check{c}_{B,A}\colon A\vee B\vdash B\vee A$
w-k	$\hat{w}_{A}\colon A\vdash A\wedge A$	$\check{w}_{A}\colon A\vee A\vdash A$
	$\hat{k}^{1}_{A,B}\colon A\wedge B\vdash A$	$\check{k}^{1}_{A,B}\colon A\vdash A\vee B$
	$\hat{k}^{2}_{A,B}\colon A\wedge B\vdash B$	$\check{k}^{2}_{A,B}\colon B\vdash A\vee B$
κ	$\hat{\kappa}_{A}\colon A\vdash\top$	$\check{\kappa}_{A}\colon \bot\vdash A$
d	$d^{L}_{A,B,C}\colon A\wedge(B\vee C)\vdash(A\wedge B)\vee C$	
	$d^{R}_{C,B,A}\colon (C\vee B)\wedge A\vdash C\vee(B\wedge A)$	
	$m_{A,B}\colon A\wedge B\vdash A\vee B$	
	$m^{-1}_{A,B}\colon A\vee B\vdash A\wedge B$	
	$\mathbf{1}_{A}\colon A\vdash A$	

TABLE 2

$$\frac{f\colon A \vdash D \qquad g\colon B \vdash E}{f \,\xi\, g\colon A \,\xi\, B \vdash D \,\xi\, E}$$

together with the clause corresponding to the partial operation of composition mentioned in §1.2, with a, b and c replaced by A, B and C. This concludes our definition of logical system.

If β is one of the families of primitive arrow terms we have introduced, except the family $\mathbf{1}$, then we call *β-terms* the set of arrow terms introduced inductively as follows: every member of β is a β-term; if f is a β-term, then for every A in $\mathcal{L}$ we have that $\mathbf{1}_A \,\xi\, f$ and $f \,\xi\, \mathbf{1}_A$ are β-terms.

In every β-term there is exactly one subterm that belongs to β, which is called the *head* of the β-term in question. For example, the head of the $\hat{c}$-term $\mathbf{1}_A \wedge (\hat{c}_{B,C} \vee \mathbf{1}_D)$ is $\hat{c}_{B,C}$. An analogous definition where β is $\mathbf{1}$, yields arrow terms called *complex identities* (which are headless). Every complex identity is equal to an identity arrow term in the presence of bifunctorial equations.

If we build a language $\mathcal{L}(\mathcal{B})$ with the same connectives as $\mathcal{L}$ but with the generating set $\mathcal{P}$ replaced by a set $\mathcal{B}$ of the same cardinality as $\mathcal{P}$, then we obtain an isomorphic copy of $\mathcal{L}$. If $\mathcal{B}$ is not of the same cardinality as $\mathcal{P}$, then $\mathcal{L}(\mathcal{B})$ and $\mathcal{L}$ are not isomorphic, but one can be isomorphically embedded into the other. So we have a function that assigns to $\mathcal{B}$ the language $\mathcal{L}(\mathcal{B})$, and we call $\mathcal{L}(\mathcal{P})$ simply $\mathcal{L}$.

Our notion of logical system is such that for a logical system $\mathcal{C}$ in $\mathcal{L}$ we have a logical system $\mathcal{C}(\mathcal{B})$ in $\mathcal{L}(\mathcal{B})$. The logical system $\mathcal{C}(\mathcal{B})$ will be isomorphic to $\mathcal{C}$ if $\mathcal{P}$ and $\mathcal{B}$ are of the same cardinality. The possibility to build $\mathcal{C}(\mathcal{B})$ is ensured by requiring that the transformations α be indexed by *all* k-tuples of objects of $\mathcal{C}$ or $\mathcal{C}(\mathcal{B})$.

So what we have really defined with $\mathcal{C}$ in $\mathcal{L}$ is not a single logical system, but a function that assigns to an arbitrary generating set $\mathcal{P}$ a logical system $\mathcal{C}(\mathcal{P})$ in $\mathcal{L}(\mathcal{P})$, which we have chosen to denote by $\mathcal{C}$ and $\mathcal{L}$, respectively, not mentioning $\mathcal{P}$. Applied to a different generating set $\mathcal{B}$ of letters this function gives the logical system $\mathcal{C}(\mathcal{B})$ in $\mathcal{L}(\mathcal{B})$.

§2.7. Logical categories

For an equational system $\mathcal{E}$ in a logical system $\mathcal{C}$ in $\mathcal{L}$, we assume whatever we have assumed for equational systems in a syntactical system, namely the conditions (re), (sy), (tr) and (co), plus an additional condition. For an arrow term $f\colon A \vdash B$ of $\mathcal{C}$, let $f^p_C\colon A^p_C \vdash B^p_C$ be the arrow term of $\mathcal{C}$ obtained by uniformly replacing every occurrence of a letter p of $\mathcal{P}$ in f and in its type $A \vdash B$ by the formula C of $\mathcal{L}$. Then we assume closure of $\mathcal{E}$ under substitution; namely,

$$(su) \quad \text{if } f = g \text{ is in } \mathcal{E}, \text{ then } f^p_C = g^p_C \text{ is in } \mathcal{E}.$$

Closure under (su) means that the letters of $\mathcal{L}$ behave like variables for objects.

The equations of $\mathcal{E}$ will be introduced by axiomatic equations with variables in which letters of $\mathcal{P}$ do not occur. So we can assume these equations for an arbitrary set $\mathcal{P}$. This will also guarantee that $\mathcal{E}$ is closed under (su).

When the categorial equations belong to the equational systems $\mathcal{E}$ in a logical system $\mathcal{C}$, so that $\mathcal{C}/\mathcal{E}$ is a category, and, moreover, for every $\xi \in \{\wedge, \vee\}$ in the language $\mathcal{L}$ of $\mathcal{C}$ we have in $\mathcal{E}$ the bifunctorial equations $(bif\ 1)$ and $(bif\ 2)$ of §2.4 with B instantiated by ξ; namely, the following equations:

$$(\xi\ 1) \quad \mathbf{1}_A \,\xi\, \mathbf{1}_B = \mathbf{1}_{A \,\xi\, B},$$
$$(\xi\ 2) \quad (g_1 \circ f_1) \,\xi\, (g_2 \circ f_2) = (g_1 \,\xi\, g_2) \circ (f_1 \,\xi\, f_2),$$

so that $\mathcal{C}/\mathcal{E}$ has the 2-endofunctor ξ, we call the syntactical category $\mathcal{C}/\mathcal{E}$ a *logical category*. We say that a logical category $\mathcal{C}/\mathcal{E}$ *is in* $\mathcal{L}$ when $\mathcal{C}$ is in the language $\mathcal{L}$, and we also say that the category $\mathcal{C}/\mathcal{E}$ *is in* the logical system $\mathcal{C}$.

If for the families α of $\mathcal{C}$ the naturality equations $(\alpha\ nat)$ of §2.4 with $a_1, \ldots a_k$ and $b_1, \ldots b_k$ replaced by $A_1, \ldots, A_k$ and $B_1, \ldots, B_k$, respectively, belong to $\mathcal{E}$, then in a logical category $\mathcal{C}/\mathcal{E}$ we will have the natural transformations α from the k-endofunctor M^μ to the k-endofunctor N^ν in $\mathcal{C}/\mathcal{E}$. That M^μ and N^ν are k-endofunctors in $\mathcal{C}/\mathcal{E}$ is guaranteed by the bifunctorial equations. When the naturality equations belong to $\mathcal{E}$ for every α of $\mathcal{C}$ and $\mathcal{C}/\mathcal{E}$ is a logical category we say that $\mathcal{C}/\mathcal{E}$ is a *natural* logical category.

We separate naturality from bifunctoriality in our definition of logical category because there are reasons to envisage logical categories that need not be natural (cf. §14.3), though in this book bifunctoriality and naturality will go hand in hand. (We do not envisage rejecting bifunctoriality for logical categories, as some authors do; see §14.3.)

Here are the naturality equations for the transformations α in the tables of the preceding section, with $f\colon A\vdash D$, $g\colon B\vdash E$ and $h\colon C\vdash F$:

$$\begin{array}{ll}
(\hat{b}^{\rightarrow}\ nat) & ((f\wedge g)\wedge h)\circ\hat{b}^{\rightarrow}_{A,B,C}=\hat{b}^{\rightarrow}_{D,E,F}\circ(f\wedge(g\wedge h)),\\
(\hat{b}^{\leftarrow}\ nat) & (f\wedge(g\wedge h))\circ\hat{b}^{\leftarrow}_{A,B,C}=\hat{b}^{\leftarrow}_{D,E,F}\circ((f\wedge g)\wedge h),\\
(\check{b}^{\rightarrow}\ nat) & ((f\vee g)\vee h)\circ\check{b}^{\rightarrow}_{A,B,C}=\check{b}^{\rightarrow}_{D,E,F}\circ(f\vee(g\vee h)),\\
(\check{b}^{\leftarrow}\ nat) & (f\vee(g\vee h))\circ\check{b}^{\leftarrow}_{A,B,C}=\check{b}^{\leftarrow}_{D,E,F}\circ((f\vee g)\vee h),
\end{array}$$

$$\begin{array}{llll}
(\hat{\delta}^{\rightarrow}\ nat) & f\circ\hat{\delta}^{\rightarrow}_{A}=\hat{\delta}^{\rightarrow}_{D}\circ(f\wedge\mathbf{1}_{\top}), & (\check{\delta}^{\rightarrow}\ nat) & f\circ\check{\delta}^{\rightarrow}_{A}=\check{\delta}^{\rightarrow}_{D}\circ(f\vee\mathbf{1}_{\bot}),\\
(\hat{\delta}^{\leftarrow}\ nat) & (f\wedge\mathbf{1}_{\top})\circ\hat{\delta}^{\leftarrow}_{A}=\hat{\delta}^{\leftarrow}_{D}\circ f, & (\check{\delta}^{\leftarrow}\ nat) & (f\vee\mathbf{1}_{\bot})\circ\check{\delta}^{\leftarrow}_{A}=\check{\delta}^{\leftarrow}_{D}\circ f,\\
(\hat{\sigma}^{\rightarrow}\ nat) & f\circ\hat{\sigma}^{\rightarrow}_{A}=\hat{\sigma}^{\rightarrow}_{D}\circ(\mathbf{1}_{\top}\wedge f), & (\check{\sigma}^{\rightarrow}\ nat) & f\circ\check{\sigma}^{\rightarrow}_{A}=\check{\sigma}^{\rightarrow}_{D}\circ(\mathbf{1}_{\bot}\vee f),\\
(\hat{\sigma}^{\leftarrow}\ nat) & (\mathbf{1}_{\top}\wedge f)\circ\hat{\sigma}^{\leftarrow}_{A}=\hat{\sigma}^{\leftarrow}_{D}\circ f, & (\check{\sigma}^{\leftarrow}\ nat) & (\mathbf{1}_{\bot}\vee f)\circ\check{\sigma}^{\leftarrow}_{A}=\check{\sigma}^{\leftarrow}_{D}\circ f,
\end{array}$$

$$\begin{array}{ll}
(\hat{c}\ nat) & (g\wedge f)\circ\hat{c}_{A,B}=\hat{c}_{D,E}\circ(f\wedge g),\\
(\check{c}\ nat) & (g\vee f)\circ\check{c}_{B,A}=\check{c}_{E,D}\circ(f\vee g),
\end{array}$$

$$\begin{array}{llll}
(\hat{w}\ nat) & (f\wedge f)\circ\hat{w}_{A}=\hat{w}_{D}\circ f, & (\check{w}\ nat) & f\circ\check{w}_{A}=\check{w}_{D}\circ(f\vee f),\\
(\hat{k}^{1}\ nat) & f\circ\hat{k}^{1}_{A,B}=\hat{k}^{1}_{D,E}\circ(f\wedge g), & (\check{k}^{1}\ nat) & (g\vee f)\circ\check{k}^{1}_{B,A}=\check{k}^{1}_{E,D}\circ g,\\
(\hat{k}^{2}\ nat) & g\circ\hat{k}^{2}_{A,B}=\hat{k}^{2}_{D,E}\circ(f\wedge g), & (\check{k}^{2}\ nat) & (g\vee f)\circ\check{k}^{2}_{B,A}=\check{k}^{2}_{E,D}\circ f,
\end{array}$$

$$(\hat{\kappa}\ nat)\quad \mathbf{1}_{\top}\circ\hat{\kappa}_{A}=\hat{\kappa}_{D}\circ f,\quad (\check{\kappa}\ nat)\quad f\circ\check{\kappa}_{A}=\check{\kappa}_{D}\circ\mathbf{1}_{\bot},$$

$$\begin{array}{ll}
(d^{L}\ nat) & ((f\wedge g)\vee h)\circ d^{L}_{A,B,C}=d^{L}_{D,E,F}\circ(f\wedge(g\vee h)),\\
(d^{R}\ nat) & (h\vee(g\wedge f))\circ d^{R}_{C,B,A}=d^{R}_{F,E,D}\circ((h\vee g)\wedge f),
\end{array}$$

$$\begin{array}{ll}
(m\ nat) & (f\vee g)\circ m_{A,B}=m_{D,E}\circ(f\wedge g),\\
(m^{-1}\ nat) & (f\wedge g)\circ m^{-1}_{A,B}=m^{-1}_{D,E}\circ(f\vee g),
\end{array}$$

$$(\mathbf{1}\ nat)\quad f\circ\mathbf{1}_{A}=\mathbf{1}_{D}\circ f.$$

One side of the equations ($\hat{\kappa}$ *nat*) and ($\check{\kappa}$ *nat*) can, of course, be shortened by using the categorial equations (*cat* 1), and ($\mathbf{1}$ *nat*) is contained in (*cat* 1).

An arrow term of the form $f_n \circ \ldots \circ f_1$, where $n \geq 1$, with parentheses tied to $\circ$ associated arbitrarily, such that for every $i \in \{1, \ldots, n\}$ we have that f_i is composition-free is called *factorized.* In a factorized arrow term $f_n \circ \ldots \circ f_1$ the arrow terms f_i are called *factors.* A factorized arrow term $f_n \circ \ldots \circ f_1$ is *developed* when f_1 is of the form $\mathbf{1}_A$ and for every $i \in \{2, \ldots, n\}$ we have that f_i is a β-term for some β.

Then by using the categorial and bifunctorial equations we can easily prove by induction on the length of f the following lemma for logical categories $\mathcal{C}/\mathcal{E}$.

DEVELOPMENT LEMMA. *For every arrow term f there is a developed arrow term f' such that $f = f'$.*

Note that for a logical category $\mathcal{C}/\mathcal{E}$ our way of introducing $\mathcal{E}$ by axiomatic equations with variables in which letters of the generating set $\mathcal{P}$ do not occur is such that when $\mathcal{P}$ is replaced by another generating set $\mathcal{B}$ we have instructions for building another logical category $\mathcal{C}/\mathcal{E}(\mathcal{B})$ in the logical system $\mathcal{C}(\mathcal{B})$ in the language $\mathcal{L}(\mathcal{B})$. This logical category $\mathcal{C}/\mathcal{E}(\mathcal{B})$ will be isomorphic to $\mathcal{C}/\mathcal{E}$ if $\mathcal{P}$ and $\mathcal{B}$ are of the same cardinality. The axiomatic equations with variables assumed for $\mathcal{E}$ are applied to the arrow terms of $\mathcal{C}(\mathcal{B})$. We have really defined a function that assigns to an arbitrary generating set $\mathcal{B}$ a logical category $\mathcal{C}/\mathcal{E}(\mathcal{B})$, the logical category $\mathcal{C}/\mathcal{E}(\mathcal{P})$ being denoted simply by $\mathcal{C}/\mathcal{E}$ (cf. the end of the preceding section).

When the equational system $\mathcal{E}$ of a logical category $\mathcal{C}/\mathcal{E}$ has as axioms the elements of a set Ax of equations, and we speak of derivations of equations of $\mathcal{E}$, we need not count (su) among the rules of derivation if Ax is closed under (su). When later we produce sets of axioms, we always assume that they are closed under (su), so that the rules of derivation are just (sy), (tr) and (co). Hence, in general, Ax will be an infinite set of equations, though these equations are instances of a finite number of equations with variables.

§2.8. $\mathcal{C}$-functors

Let $\mathcal{C}$ be a logical system in $\mathcal{L}$. Deductive systems that have an operation ξ on objects and an operation ξ on arrows for every ξ of $\mathcal{L}$ such that for $f\colon a \vdash d$ and $g\colon b \vdash e$ we have $f \,\xi\, g\colon a \,\xi\, b \vdash d \,\xi\, e$,

an object ζ for every ζ of $\mathcal{L}$, and

a transformation α for every α of $\mathcal{C}$

are called *deductive systems of the $\mathcal{C}$ kind.* A *bifunctorial category* of the $\mathcal{C}$ kind is a category of the $\mathcal{C}$ kind in which the bifunctorial equations hold for every ξ of $\mathcal{C}$.

Let $\mathcal{A}_1$ and $\mathcal{A}_2$ be bifunctorial categories of the $\mathcal{C}$ kind. The operations ξ, the objects ζ and the transformations α are indexed by ^{1} and ^{2} when they are in $\mathcal{A}_1$ and $\mathcal{A}_2$ respectively. The type of $\alpha^i_{a_1,\ldots,a_k}$, where $i \in \{1,2\}$, is $M^\mu_i(a_1,\ldots,a_k) \vdash N^\nu_i(a_1,\ldots,a_k)$.

A *$\mathcal{C}$-functor* from $\mathcal{A}_1$ to $\mathcal{A}_2$ is made of

a functor F from $\mathcal{A}_1$ to $\mathcal{A}_2$,

for every ξ of $\mathcal{C}$ a family $\psi^{\Box\xi\Box}$ of arrows of $\mathcal{A}_2$ indexed by objects of $\mathcal{A}_1$ whose members are

$$\psi^{\Box\xi\Box}_{a,b} : Fa \,\xi^2\, Fb \vdash F(a \,\xi^1\, b),$$

for every ζ of $\mathcal{C}$ an arrow $\psi^\zeta : \zeta^2 \vdash F\zeta^1$ of $\mathcal{A}_2$.

In practice, when we refer to a $\mathcal{C}$-functor, we mention only F, taking the families ψ for granted. They will be mentioned explicitly when this is required.

A *dual $\mathcal{C}$-functor* from $\mathcal{A}_1$ to $\mathcal{A}_2$ is obtained from the definition of $\mathcal{C}$-functor by replacing $\psi^{\Box\xi\Box}_{a,b}$ by

$$\bar{\psi}^{\Box\xi\Box}_{a,b} : F(a \,\xi^1\, b) \vdash Fa \,\xi^2\, Fb$$

and ψ^ζ by $\bar{\psi}^\zeta : F\zeta^1 \vdash \zeta^2$.

For every $\mathcal{C}$-functor from $\mathcal{A}_1$ to $\mathcal{A}_2$ and every $M^\mu \in \mathcal{L}^{con}$ (see §2.5), we define in $\mathcal{A}_2$ the family of arrows ψ^{M^μ} by induction, with the following clauses:

$$(\psi^{M^\mu}\ 1) \qquad \psi^{\Box}_a = \mathbf{1}_{Fa},$$

$$(\psi^{M^\mu}\ 2) \qquad \psi^{M\xi N}_{\vec{a},\vec{b}} = \psi^{\Box\xi\Box}_{M(\vec{a}),N(\vec{b})} \circ (\psi^M_{\vec{a}}\ \xi^2\ \psi^N_{\vec{b}}),$$

where $\vec{a}$ and $\vec{b}$ stand for $a_1, \ldots, a_m$ and $b_1, \ldots, b_n$ respectively,

$$(\psi^{M^\mu}\ 3) \qquad \psi^{M^\mu}_{a_1,\ldots,a_k} = \psi^M_{a_{\mu(1)},\ldots,a_{\mu(m)}}.$$

We have a dual definition of $\bar{\psi}^{M^\mu}$ for dual $\mathcal{C}$-functors, where the clauses $(\psi^{M^\mu}\ 1)$ and $(\psi^{M^\mu}\ 3)$ have just ψ replaced by $\bar{\psi}$, while the clause $(\psi^{M^\mu}\ 2)$ is replaced by

$$\bar{\psi}^{M\xi N}_{\vec{a},\vec{b}} = (\bar{\psi}^M_{\vec{a}}\ \xi^2\ \bar{\psi}^N_{\vec{b}}) \circ \bar{\psi}^{\Box\xi\Box}_{M(\vec{a}),N(\vec{b})}.$$

We say that the members of ψ^{M^μ} are *ψ-arrows*. From the inductive definition of ψ^{M^μ} we can deduce the following equation of $\mathcal{A}_2$ for every $\mathcal{C}$-functor:

$$(\psi^{M^\mu}) \quad \psi^{M^\mu(L_1^{\lambda_1},\ldots,L_k^{\lambda_k})}_{\vec{a_1},\ldots,\vec{a_k}} = \psi^{M^\mu}_{L_1^{\lambda_1}(\vec{a_1}),\ldots,L_k^{\lambda_k}(\vec{a_k})} \circ M_2^\mu(\psi^{L_1^{\lambda_1}}_{\vec{a_1}}, \ldots, \psi^{L_k^{\lambda_k}}_{\vec{a_k}})$$

where $\vec{a_i}$ stands for a sequence $b_1, \ldots, b_l$, with $l \geq 0$, of objects of $\mathcal{A}_2$ and the $L_i^{\lambda_i}$ from the indices of ψ^{M^μ} on the right-hand side are from $\mathcal{A}_1$. It is easier to derive (ψ^{M^μ}) when the functions μ, $\lambda_1, \ldots, \lambda_k$ are all identity functions, and this easier equation is then used in the derivation of the general (ψ^{M^μ}) equation.

Let M_i^μ be obtained from M^μ by replacing ξ and ζ by ξ^i and ζ^i respectively. From the bifunctoriality of ξ^i in $\mathcal{A}_i$ we can deduce that M_i^μ defines a k-endofunctor in $\mathcal{A}_i$. The word $\Box\xi^i\Box$ stands for the 2-endofunctor ξ^i, the word $\Box$ for the identity 1-endofunctor of $\mathcal{A}_i$, and the word ζ^i for the 0-endofunctor ζ^i of $\mathcal{A}_i$.

We say that a $\mathcal{C}$-functor from $\mathcal{A}_1$ to $\mathcal{A}_2$ *preserves* an M^μ of $\mathcal{C}$ (i.e. an M^μ of $\mathcal{L}^{con}$, for $\mathcal{C}$ in $\mathcal{L}$) when $\psi^{M^\mu}_{a_1,\ldots,a_k}$ is natural in $a_1, \ldots, a_k$ (see §2.4); this means that the following equation holds in $\mathcal{A}_2$:

$$(\psi\ nat) \quad FM_1^\mu(f_1, \ldots, f_k) \circ \psi^{M^\mu}_{a_1,\ldots,a_k} = \psi^{M^\mu}_{b_1,\ldots,b_k} \circ M_2^\mu(Ff_1, \ldots, Ff_k).$$

It can be checked that it is enough to assume the following instance of $(\psi\ nat)$ for every ξ in M^μ:

$$(\psi^{\Box\xi\Box}\ nat)\quad F(f_1\,\xi\,f_2)\circ\psi^{\Box\xi\Box}_{a_1,a_2} = \psi^{\Box\xi\Box}_{b_1,b_2}\circ(Ff_1\,\xi^2\,Ff_2)$$

to derive by induction that F preserves M^μ. The following instances of $(\psi\ nat)$:

$$F\mathbf{1}_{\varsigma^1}\circ\psi^\varsigma = \psi^\varsigma\circ\mathbf{1}_{\varsigma^2},$$
$$Ff\circ\psi^\Box_a = \psi^\Box_b\circ Ff$$

follow from the functoriality of F and categorial equations.

We say that a $\mathcal{C}$-functor from $\mathcal{A}_1$ to $\mathcal{A}_2$ *preserves* an α of $\mathcal{C}$ when the following equation holds in $\mathcal{A}_2$:

$$(\psi\alpha)\quad F\alpha^1_{\vec{a}}\circ\psi^{M^\mu}_{\vec{a}} = \psi^{N^\nu}_{\vec{a}}\circ\alpha^2_{\vec{Fa}}$$

where if $\vec{a}$ is $a_1,\ldots,a_k$, then $\vec{Fa}$ is $Fa_1,\ldots,Fa_k$. We apply an analogous convention concerning $\vec{a}$ and $\vec{Fa}$ also later.

We say that a $\mathcal{C}$-functor is *partial* when it preserves every M^μ and every α of $\mathcal{C}$.

We say that a $\mathcal{C}$-functor from $\mathcal{A}_1$ to $\mathcal{A}_2$ is *fluent* in an α of $\mathcal{C}$ when every k-tuple of ψ arrows flows through α in $\mathcal{A}_2$ (see §2.4).

For every $\mathcal{C}$-functor from $\mathcal{A}_1$ to $\mathcal{A}_2$ that preserves α and is fluent in α we have the following equation in $\mathcal{A}_2$:

$$(\psi\alpha L)\quad F\alpha^1_{L_1^{\lambda_1}(\vec{a_1}),\ldots,L_k^{\lambda_k}(\vec{a_k})}\circ\psi^{M^\mu(L_1^{\lambda_1},\ldots,L_k^{\lambda_k})}_{\vec{a_1},\ldots,\vec{a_k}} =$$
$$\psi^{N^\nu(L_1^{\lambda_1},\ldots,L_k^{\lambda_k})}_{\vec{a_1},\ldots,\vec{a_k}}\circ\alpha^2_{L_1^{\lambda_1}(\vec{Fa_1}),\ldots,L_k^{\lambda_k}(\vec{Fa_k})}$$

where the $L_i^{\lambda_i}$ in the indices of α^1 are from $\mathcal{A}_1$, while those in the indices of α^2 are from $\mathcal{A}_2$. To derive $(\psi\alpha L)$ we just apply (ψ^{M^μ}), $(\psi\alpha)$ and fluency in α. The equation $(\psi\alpha)$ is the instance of $(\psi\alpha L)$ where every $L_i^{\lambda_i}$ is $\Box$, that is $\Box^{\iota\{1\}}$.

We are aware that the condition $(\psi\alpha L)$, with its multiple indexing, may look forbidding. Fortunately, in cases we deal with in this book, it will be equivalent to the simpler fluency in α condition, as we will see in a moment.

A $\mathcal{C}$-functor is called *total* when it preserves every M^μ of $\mathcal{C}$ and $(\psi\alpha L)$ holds for every α of $\mathcal{C}$. Every total $\mathcal{C}$-functor is a partial $\mathcal{C}$-functor. It can be verified that the composition of two partial $\mathcal{C}$-functors is a partial $\mathcal{C}$-functor, and the composition of two total $\mathcal{C}$-functors is a total $\mathcal{C}$-functor. The arrow $\psi^{\Box\xi\Box}_{a,b}$ tied to the functor F_3F_2 is defined as $F_3\psi^{2}{}^{\Box\xi\Box}_{a,b}\circ\psi^{3}{}^{\Box\xi\Box}_{F_2a,F_2b}$, and the arrow ψ^{ς} tied to the functor F_3F_2 as $F_3\psi^{2\varsigma}\circ\psi^{3\varsigma}$.

We say that a $\mathcal{C}$-functor is *groupoidal* when it is a $\mathcal{C}$-functor and a dual $\mathcal{C}$-functor with $\psi^{M^\mu}_{\vec{a}}$ and $\bar{\psi}^{M^\mu}_{\vec{a}}$ being inverses of each other. For $\mathcal{C}$-functors where the ψ^{M^μ} arrows are mono, $(\psi\alpha L)$ implies fluency in α, so that for groupoidal partial $\mathcal{C}$-functors, $(\psi\alpha L)$ is equivalent to fluency in α. The composition of two groupoidal $\mathcal{C}$-functors is a groupoidal $\mathcal{C}$-functor.

Let us call maps from the set of letters $\mathcal{P}$ into the objects of a deductive system $\mathcal{D}_i$ of the $\mathcal{C}$ kind *valuations* into $\mathcal{D}_i$. A valuation v_i into $\mathcal{D}_i$ is extended to two maps, both called v_i, from the objects and arrow terms of $\mathcal{C}$ to the objects and arrows, respectively, of $\mathcal{D}_i$ with the obvious clauses

$$\begin{aligned}
&v_i(A\,\xi\,B)=v_i(A)\,\xi\,v_i(B),\\
&v_i(\varsigma)=\varsigma,\\
&v_i(\alpha_{A_1,\ldots,A_k})=\alpha_{v_i(A_1),\ldots,v_i(A_k)},\\
&v_i(f\,\xi\,g)=v_i(f)\,\xi\,v_i(g),\\
&v_i(g\circ f)=v_i(g)\circ v_i(f).
\end{aligned}$$

We can prove the following.

PROPOSITION. *Let $\mathcal{A}_1$ and $\mathcal{A}_2$ be bifunctorial categories of the $\mathcal{C}$ kind. If F is a total $\mathcal{C}$-functor from $\mathcal{A}_1$ to $\mathcal{A}_2$, then for every arrow term $f: M(p_1,\ldots,p_m)\vdash N(q_1,\ldots,q_n)$ of $\mathcal{C}$, for every valuation v_1 into $\mathcal{A}_1$ and for every valuation v_2 into $\mathcal{A}_2$ such that $v_2(p)=Fv_1(p)$, in $\mathcal{A}_2$ we have*

$$(\psi t)\quad Fv_1(f)\circ\psi^{M}_{v_1(p_1),\ldots,v_1(p_m)}=\psi^{N}_{v_1(q_1),\ldots,v_1(q_n)}\circ v_2(f).$$

PROOF. We proceed by induction on the length of f. If f is

$$\alpha_{L_1^{\lambda_1}(\vec{p_1}),\ldots,L_k^{\lambda_k}(\vec{p_k})},$$

then we use $(\psi\alpha L)$. If f is $f_1\,\xi\,f_2$, then we use $(\psi\ nat)$, the bifunctorial equations for ξ and the induction hypothesis. If f is $f_2\circ f_1$, then we use $(fun\ 2)$ for F and the induction hypothesis. ⊣

Only a $\mathcal{C}$-functor that satisfies (ψt) may be said to preserve the $\mathcal{C}$-structure properly, up to ψ. But not everything in the definition of total $\mathcal{C}$-functor is a consequence of (ψt). In particular, $(\psi\ nat)$ with $f_1, \ldots, f_k$ foreign to the $\mathcal{C}$-structure is not such a consequence.

A groupoidal total $\mathcal{C}$-functor is called *strong.* A $\mathcal{C}$-functor F from $\mathcal{A}_1$ to $\mathcal{A}_2$ is called *strict* when $Fa\,\xi^2\,Fb = F(a\,\xi^1\,b)$, $\varsigma^2 = F\varsigma^1$ and $\psi^{\Box\xi\Box}_{a,b}$ and ψ^{ς} are identity arrows of $\mathcal{A}_2$. Every strict $\mathcal{C}$-functor is, of course, strong.

For a strict $\mathcal{C}$-functor F the equations $(\psi\ nat)$ and (ψt) become

$$FM^{\mu}_1(f_1, \ldots, f_k) = M^{\mu}_2(Ff_1, \ldots, Ff_k),$$
$$Fv_1(f) = v_2(f), \text{ where } v_2(p) = Fv_1(p),$$

while $(\psi\alpha L)$ is an easy consequence of $(\psi\alpha)$. Strict $\mathcal{C}$-functors preserve the $\mathcal{C}$ structure *on the nose* . (The expression "on the nose" is used in other analogous situations, when a structure is homomorphically preserved in an obvious manner.)

According to what we have said above about the composition of total $\mathcal{C}$-functors and groupoidal $\mathcal{C}$-functors, the composition of two strong $\mathcal{C}$-functors is a strong $\mathcal{C}$-functor. The composition of two strict $\mathcal{C}$-functors is, of course, a strict $\mathcal{C}$-functor.

These definitions are on the lines of Mac Lane's definition of monoidal functor of [96]. For example, with α being $\hat{b}^{\wedge}_{\rightarrow}$, the equation $(\psi\alpha)$ and the inductive clauses for ψ^M yield the following equation:

$$F\hat{b}^{\wedge}_{\overrightarrow{a,b,c}} \circ \psi^{\Box\wedge\Box}_{a,b\wedge^1 c} \circ (\mathbf{1}_{Fa} \wedge^2 \psi^{\Box\wedge\Box}_{b,c}) = \psi^{\Box\wedge\Box}_{a\wedge^1 b,c} \circ (\psi^{\Box\wedge\Box}_{a,b} \wedge^2 \mathbf{1}_{Fc}) \circ \hat{b}^{\wedge}_{\overrightarrow{Fa,Fb,Fc}},$$

which is used in [96] (Section XI.2) to define monoidal functors. Fluency is, however, only implicit for Mac Lane. Besides that, our definition, which does not presuppose as Mac Lane's that monoidal functors are between monoidal categories only, will enable us later to define monoidal categories via monoidal functors (see §4.6). We have analogous definitions via strict $\mathcal{C}$-functors for other sorts of categories too.

When every valuation into a bifunctorial category $\mathcal{A}$ of the $\mathcal{C}$ kind can be extended to a strict $\mathcal{C}$-functor from a logical category $\mathcal{C}/\mathcal{E}$ to $\mathcal{A}$ we say that $\mathcal{A}$ is a *$\mathcal{C}/\mathcal{E}$-category relative to $\mathcal{P}$*. The logical system $\mathcal{C}$ is here a logical system in the language $\mathcal{L}$ generated by the set of letters $\mathcal{P}$.

If something is a $\mathcal{C}/\mathcal{E}$-category relative to an infinite set of letters $\mathcal{P}$, then it must be a $\mathcal{C}/\mathcal{E}(\mathcal{B})$-category relative to any other generating set $\mathcal{B}$. (Depending on the number of different variables for formulae in the axiomatic equations with variables assumed for $\mathcal{E}$, we could do here with a finite set $\mathcal{P}$ of at least a certain cardinality instead of an infinite set $\mathcal{P}$, but, assuming uniformly that $\mathcal{P}$ is infinite, we are on the safe side.) A $\mathcal{C}/\mathcal{E}$-category relative to an infinite set $\mathcal{P}$ is then called simply a *$\mathcal{C}/\mathcal{E}$-category*.

If $\mathcal{C}/\mathcal{E}$ is a natural logical category, then a $\mathcal{C}/\mathcal{E}$-category $\mathcal{A}$ is a *natural* $\mathcal{C}/\mathcal{E}$-category when for every α in $\mathcal{C}$ the naturality equations hold in $\mathcal{A}$. The bifunctorial equations for every ξ of $\mathcal{C}$ are guaranteed in every $\mathcal{C}/\mathcal{E}$-category, and fluency in every α of $\mathcal{C}$ is guaranteed for every $\mathcal{C}$-functor into a natural $\mathcal{C}/\mathcal{E}$-category.

When for f and g arrow terms of $\mathcal{C}$ of the same type we say that the equation $f = g$ holds in a deductive system $\mathcal{A}$ of the $\mathcal{C}$ kind, we understand the letters $p, q, \ldots$ of $\mathcal{L}$ as variables for objects. If $\mathcal{A}$ is a $\mathcal{C}/\mathcal{E}$-category, then that $f = g$ holds in $\mathcal{A}$ amounts to saying that $Ff = Fg$ holds for every strict $\mathcal{C}$-functor F from $\mathcal{C}/\mathcal{E}$ to $\mathcal{A}$. Every equation of $\mathcal{E}$ holds in this sense in a $\mathcal{C}/\mathcal{E}$-category $\mathcal{A}$, but additional equations $f = g$, not in $\mathcal{E}$, may hold in $\mathcal{A}$ too.

§2.9. The category *Rel* and coherence

The objects of the category *Rel* are finite ordinals (we have $0 = \emptyset$ and $n{+}1 = n \cup \{n\}$), and its arrows are relations between finite ordinals. We write either $(x, y) \in R$ or xRy, as usual. In this category, $\mathbf{1}_n \colon n \vdash n$ is the identity relation, i.e. identity function, on n. If $n = 0$, then $\mathbf{1}_\emptyset \colon \emptyset \vdash \emptyset$ is the empty relation $\emptyset$, with domain $\emptyset$ and codomain $\emptyset$.

For $R_1 \colon n \vdash m$ (that is, $R_1 \subseteq n \times m$) and $R_2 \colon m \vdash k$, the set of ordered pairs of the composition $R_2 \circ R_1 \colon n \vdash k$ is $\{(x, y) \mid \exists z(xR_1z \textit{ and } zR_2y)\}$. For $R_1 \colon n \vdash m$ and $R_2 \colon k \vdash l$, let the set of ordered pairs of $R_1{+}R_2 \colon n{+}k \vdash m{+}l$ be

$$R_1 \cup \{(x{+}n, y{+}m) \mid (x, y) \in R_2\}.$$

With addition on objects, this operation on arrows gives a 2-endofunctor in *Rel*.

The category *Rel* is a category of the $\mathcal{C}$ kind for every logical system $\mathcal{C}$ whose families α are from Tables 1 and 2 of §2.6, and, moreover, the appropriate bifunctorial and naturality equations will hold in *Rel*. The 2-endofunctor ξ for every $\xi \in \{\wedge, \vee\}$ is $+$, and the object ζ for every $\zeta \in \{\top, \bot\}$ is 0. The natural transformation α for every α included in the families b, δ-σ, d, m, m^{-1} or **1** is the family **1** of *Rel*. In other cases, we have the following:

$$
\begin{array}{llllll}
(x,y) \in \hat{c}_{n,m} & \text{iff} & (y,x) \in \check{c}_{n,m} & \text{iff} & (x+m = y \text{ or } x = y+n); \\
(x,y) \in \hat{w}_{n} & \text{iff} & (y,x) \in \check{w}_{n} & \text{iff} & x \equiv y \pmod{n}; \\
(x,y) \in \hat{k}^1_{n,m} & \text{iff} & (y,x) \in \check{k}^1_{n,m} & \text{iff} & x = y; \\
(x,y) \in \hat{k}^2_{n,m} & \text{iff} & (y,x) \in \check{k}^2_{n,m} & \text{iff} & x = y+n;
\end{array}
$$

the relations $\hat{\kappa}_n \colon n \vdash \emptyset$ and $\check{\kappa}_n \colon \emptyset \vdash n$ are the empty relations.

It is not difficult to check that all these families α in *Rel* are natural transformations. This is clear from the diagrammatical representation of relations in *Rel*. Here are a few examples of such diagrammatical representations, with domains written at the top and codomains at the bottom:

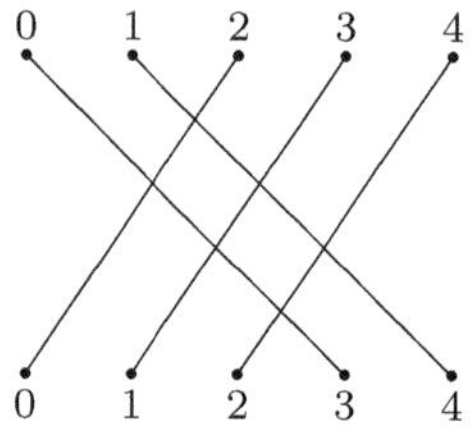

$\hat{c}_{2,3}$

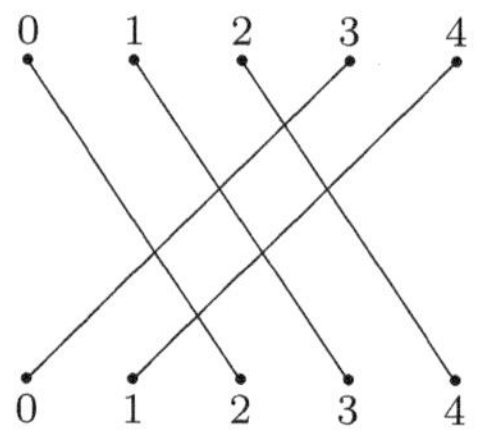

$\check{c}_{2,3}$

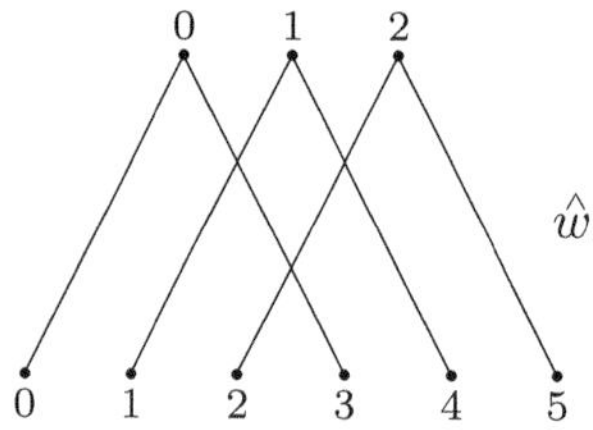

$\hat{w}_3$

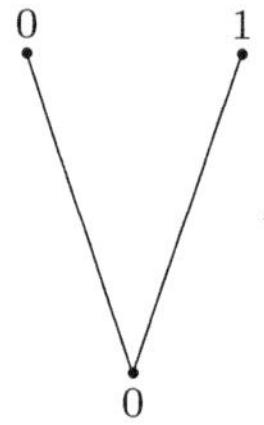

$\check{w}_1$

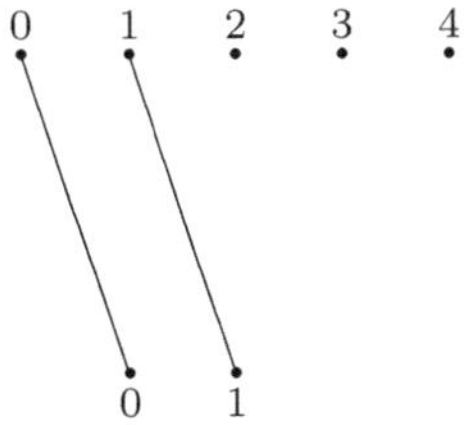
$\hat{k}^1_{2,3}$

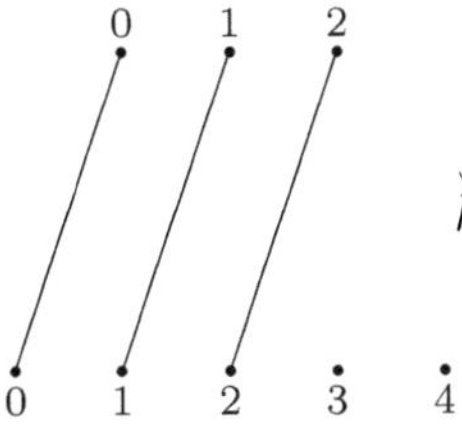
$\check{k}^1_{3,2}$

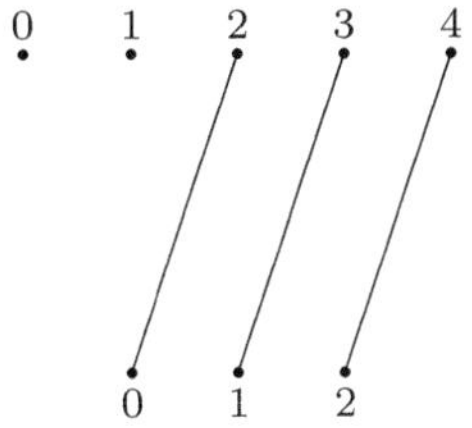
$\hat{k}^2_{2,3}$

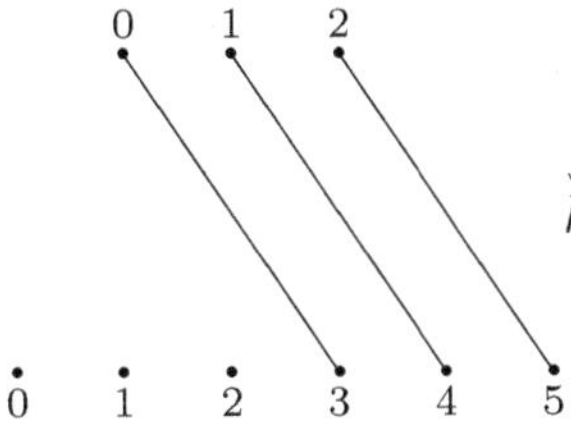
$\check{k}^2_{3,3}$

$\hat{\kappa}_3$

$\check{\kappa}_2$

For the identity relation, i.e. the identity function, $\mathbf{1}_n$ we have, of course,

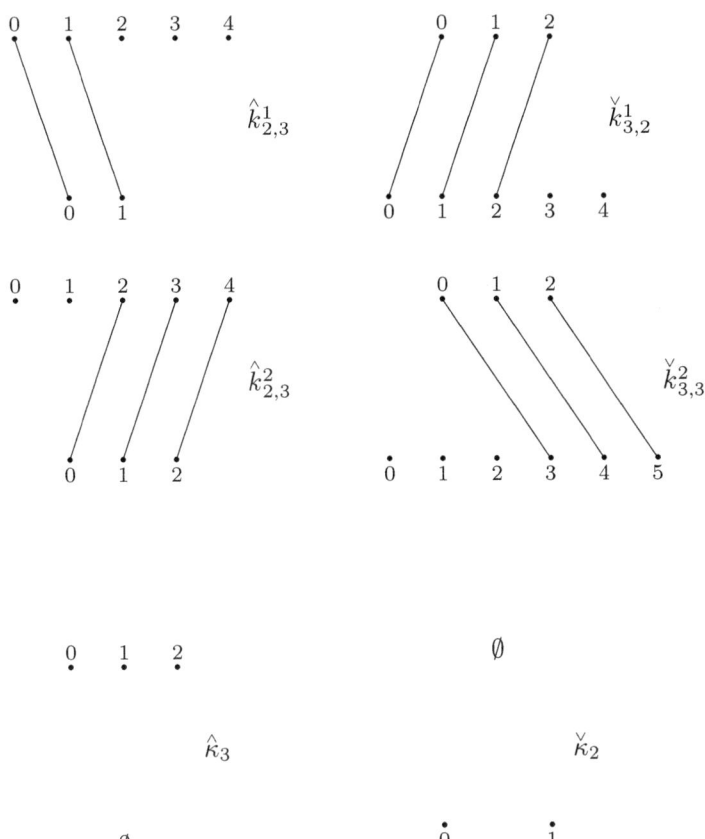

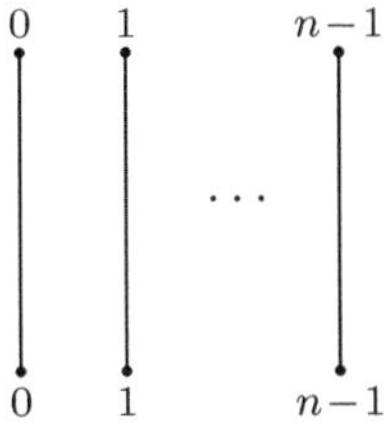

Such diagrams are composed in an obvious manner by putting them one below another; for example,

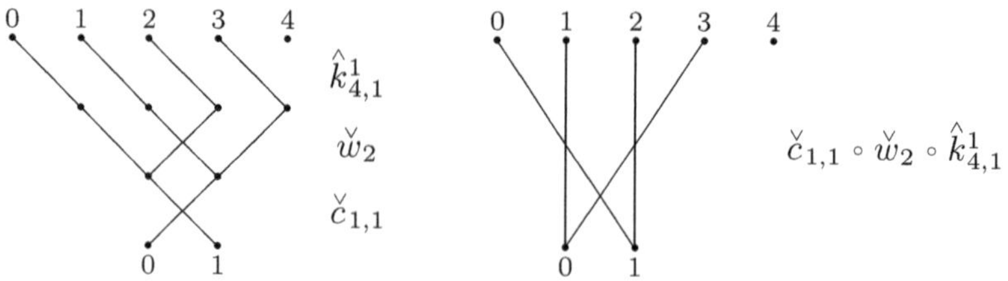

The equation

$$(\hat{w}\ nat) \quad (f \wedge f) \circ \hat{w}_n = \hat{w}_m \circ f,$$

which is an instance of $(\alpha\ nat)$, and which we take as an example, is justified in the following manner via diagrams:

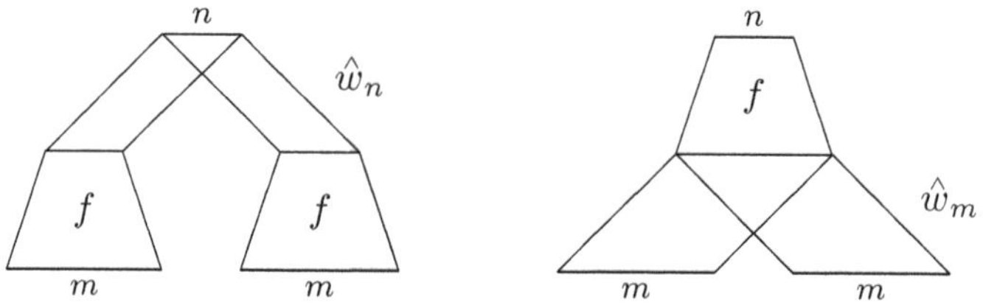

We can now define a function G from the objects of $\mathcal{C}$ to the objects of *Rel* such that for $p \in \mathcal{P}$, $\zeta \in \{\top, \bot\}$ and $\xi \in \{\wedge, \vee\}$ we have

$$\begin{aligned} & Gp = 1, \\ & G\zeta = 0, \\ & G(A\,\xi\,B) = GA + GB. \end{aligned}$$

Hence GA is just the letter length of A.

We can also define a function, called G too, from the arrow terms of $\mathcal{C}$ to the arrow terms of *Rel* such that

$$\begin{aligned} & G\alpha_{A_1,\dots,A_k} = \alpha_{GA_1,\dots,GA_k}, \\ & G(f\,\xi\,g) = Gf + Gg, \\ & G(g \circ f) = Gg \circ Gf. \end{aligned}$$

It is easy to check that for $f\colon A \vdash B$ we have that Gf is of type $GA \vdash GB$. It is also easy to check that if for $f\colon A \vdash B$ we have $(x, y) \in Gf$, then the $(x{+}1)$-th occurrence of letter in A, counting from the left, and the $(y{+}1)$-th occurrence of letter in B are occurrences of the same letter.

For many logical categories $\mathcal{C}/\mathcal{E}$ considered in this work, the two functions G we have just defined induce a strict $\mathcal{C}$-functor G from $\mathcal{C}/\mathcal{E}$ to *Rel* such that $G[f] = Gf$ (see the penultimate paragraph of §2.4). Whenever G is such a functor, it is straightforward to show that fact just by checking that *Rel* satisfies the equations of $\mathcal{E}$ (with A, $B, \ldots$ replaced by n,$m, \ldots$), and we will not dwell on the proof of that.

The greatest part of our work consists in demonstrating the faithfulness of such functors G. We call these faithfulness results *coherence theorems*, and say that $\mathcal{C}/\mathcal{E}$ is *coherent*.

If the image of $\mathcal{C}/\mathcal{E}$ by the functor G is a discrete subcategory of *Rel*, which is the case when we exclude c, w, k and κ, then $\mathcal{C}/\mathcal{E}$ is coherent iff $\mathcal{C}/\mathcal{E}$ is a preorder. So, in such cases, our coherence theorems will state that $\mathcal{C}/\mathcal{E}$ is a preorder (which is the narrow sense in which Mac Lane understood coherence originally in [93]).

It is clear that if $\mathcal{C}/\mathcal{E}$ is coherent in the sense just specified, then it is decidable whether arrow terms of $\mathcal{C}$ are equal in $\mathcal{C}/\mathcal{E}$. In logical terms, one would say that the coherence of $\mathcal{C}/\mathcal{E}$ implies the decidability of the equational system $\mathcal{E}$. This is because equality of arrows is clearly decidable in *Rel*. So, in the terminology of §1.1, coherence here implies a solution to the commuting problem.

CHAPTER 3

STRICTIFICATION

This chapter is devoted to strictification (a topic announced in §1.7). Our results are about categories that have as a subcategory a groupoid that is a preorder. For such a category we find an equivalent *strictified* category where the arrows of the groupoid are collapsed into identity arrows. The functors on which this equivalence of categories is based are functors that preserve structure up to isomorphism. The interest of strictification is that it shortens the coding of arrows, and facilitates the recording of computations.

§3.1. Strictification in general

We will prove a general theorem concerning the possibility of finding for a $\mathcal{C}/\mathcal{E}$-category $\mathcal{B}$ a $\mathcal{C}/\mathcal{E}$-category $\mathcal{B}^{\mathcal{G}}$ equivalent to $\mathcal{B}$ via strong $\mathcal{C}$-functors such that some isomorphisms of the $\mathcal{C}/\mathcal{E}$ structure of $\mathcal{B}$, which make a subcategory $\mathcal{G}$ of $\mathcal{C}/\mathcal{E}$, become identity arrows in $\mathcal{B}^{\mathcal{G}}$, and may hence be omitted according to the equation (*cat* 1) (see Chapter 11). So, instead of computing in $\mathcal{B}$ we can pass to $\mathcal{B}^{\mathcal{G}}$, where computations become shorter, and their recording is simplified. The category $\mathcal{B}^{\mathcal{G}}$ is here called *strict* with respect to the isomorphisms that have become identity arrows, and the procedure of passing from $\mathcal{B}$ to $\mathcal{B}^{\mathcal{G}}$ is called *strictification*.

Our theorem will generalize considerably analogous strictification results of Joyal and Street in [67] (Section 1) and of Mac Lane in [96] (Section XI.3). First, we strictify with respect to wider classes of isomorphisms $\mathcal{G}$, such as we will encounter in our work, and not only with respect to monoidal isomorphisms (for monoidal categories see §4.6). Secondly, even when we strictify only with respect to monoidal isomorphisms—i.e. when $\mathcal{G}$ is a free monoidal category—our $\mathcal{C}$-functors may preserve a wider structure, and not just the monoidal structure. They are not just monoidal functors.

As suggested by [67] and [96], strictification opens the way to alternative proofs of coherence results, in the sense of results about certain categories being preorders. (Some authors go so far as to call strictification results *coherence* results, but we believe this usage is confusing.) We will obtain such alternative proofs of coherence in some cases, but in general we favour the direct approach to proofs of coherence, in the style of [93] and [94] (Section VII.2), which is not more difficult. (The alternative proofs of coherence via strictification may look shorter when their presentation is sketchy.) The prime reason why we deal with strictification is not the production of alternative proofs of coherence—this is only an occasional byproduct. Our prime reason is a handy recording of lengthy computations, as mentioned above. Strictification enables us to have shorter records after coherence has been proved.

We formulate our strictification results with respect to a language $\mathcal{L}$ such as we specified at the beginning of §2.5, because it is mainly in this context that we mean to apply them. It will be clear, however, that analogous results hold also for contexts with richer languages.

Let $\mathcal{L}$ be a language such as in §2.5, and let the following condition be satisfied:

(I$\mathcal{B}$) $\mathcal{B}$ is a category that for every connective ξ and ζ of $\mathcal{L}$ has, respectively, a binary and a nullary operation on objects, denoted by ξ and ζ.

Let $\mathcal{L}(\mathcal{B}_{ob})$ be the language with the same connectives as $\mathcal{L}$ generated by the set $\mathcal{B}_{ob}$ of the objects of $\mathcal{B}$ instead of $\mathcal{P}$. The elements of $\mathcal{L}(\mathcal{B}_{ob})$ are also formulae. To distinguish the ξ and ζ of $\mathcal{L}(\mathcal{B}_{ob})$ from those of $\mathcal{B}_{ob}$ we write $\xi^{\mathcal{G}}$ and $\zeta^{\mathcal{G}}$ for the connectives ξ and ζ of $\mathcal{L}(\mathcal{B}_{ob})$. The connectives $\xi^{\mathcal{G}}$ and $\zeta^{\mathcal{G}}$ are new connectives, not to be confused with the operations ξ and ζ involved in the objects of $\mathcal{B}$.

Let $\equiv^{\mathcal{G}}$ be an equivalence relation on $\mathcal{L}(\mathcal{B}_{ob})$ such that if a and b are objects of $\mathcal{B}$ we have $a \equiv^{\mathcal{G}} b$ only if a is the same object as b. We call such an equivalence relation *generatively discrete*. Let $[A]$ be the equivalence class of the formula A of $\mathcal{L}(\mathcal{B}_{ob})$ with respect to $\equiv^{\mathcal{G}}$.

Out of $\mathcal{B}$ we build a category $\mathcal{B}^{\mathcal{G}}$ in the following manner. The objects of $\mathcal{B}^{\mathcal{G}}$ are all the classes $[A]$ for $A \in \mathcal{L}(\mathcal{B}_{ob})$. We use X, Y, $Z, \dots$, sometimes with indices, for the objects of $\mathcal{B}^{\mathcal{G}}$.

Consider the map E from $\mathcal{L}(\mathcal{B}_{ob})$ to $\mathcal{B}_{ob}$ defined inductively by

$$\begin{aligned} &Ea = a \quad \text{if } a \in \mathcal{B}_{ob},\\ &E\zeta^{\mathcal{G}} = \zeta,\\ &E(A\ \xi^{\mathcal{G}}\ B) = EA\ \xi\ EB, \end{aligned}$$

where ζ and ξ on the right-hand side are the ζ and ξ of $\mathcal{B}$. With the help of E we define a map F from the objects of $\mathcal{B}^{\mathcal{G}}$ to $\mathcal{B}_{ob}$ in the following manner. We choose first a fixed representative formula $A_F \in [A]$ so that if A is an object a of $\mathcal{B}$, then A_F is a. We have guaranteed above by generative discreteness that in $[a]$ we have no object of $\mathcal{B}$ different from a. Otherwise, the choice of the representative A_F is arbitrary. Then we define $F[A]$ as EA_F.

The arrows of $\mathcal{B}^{\mathcal{G}}$ are all the triples (f, X, Y) such that $f\colon FX \vdash FY$ is an arrow of $\mathcal{B}$. The arrow (f, X, Y) is of type $X \vdash Y$ in $\mathcal{B}^{\mathcal{G}}$. The arrow $(\mathbf{1}_{FX}, X, X)$ is an identity arrow of $\mathcal{B}^{\mathcal{G}}$, and for (f, X, Y) and (g, Y, Z) arrows of $\mathcal{B}^{\mathcal{G}}$ we define their composition in $\mathcal{B}^{\mathcal{G}}$ as $(g \circ f, X, Z)$. This defines the category $\mathcal{B}^{\mathcal{G}}$.

Out of the map on objects F we define a functor F from $\mathcal{B}^{\mathcal{G}}$ to $\mathcal{B}$ by setting $F(f, X, Y) = f$.

We define a functor $F^{\mathcal{G}}$ from $\mathcal{B}$ to $\mathcal{B}^{\mathcal{G}}$ by

$$\begin{aligned} &F^{\mathcal{G}}a = [a],\\ &F^{\mathcal{G}}f = (f, F^{\mathcal{G}}a, F^{\mathcal{G}}b), \quad \text{for } f\colon a \vdash b. \end{aligned}$$

For the definition of $F^{\mathcal{G}}f$ to be correct we must have that the type of f is $FF^{\mathcal{G}}a \vdash FF^{\mathcal{G}}b$ and this is guaranteed by $FF^{\mathcal{G}}c = F[c] = c$. It is trivial to check that F and $F^{\mathcal{G}}$ are indeed functors.

Let (I) be the collective name for (I$\mathcal{B}$) and the condition that $\equiv^{\mathcal{G}}$ is generatively discrete (i.e. the name for the conjunction of these two conditions). Then we can prove the following lemma.

LEMMA 1. *If* (I) *holds, then the categories* $\mathcal{B}$ *and* $\mathcal{B}^{\mathcal{G}}$ *are equivalent via the functors* $F^{\mathcal{G}}$ *and* F.

PROOF. We have $FF^{\mathcal{G}}a = a$, as we noted above, and we have also $FF^{\mathcal{G}}f = f$. On the other hand, $F^{\mathcal{G}}F[A] = F^{\mathcal{G}}EA_F = [EA_F]$. Note that EA_F is a generator in $\mathcal{L}(\mathcal{B}_{ob})$, and though $[A] = [A_F]$, the object $[A]$ may well differ from $[EA_F]$. However, we have in $\mathcal{B}^{\mathcal{G}}$ the natural isomorphism τ whose members are $\tau_X = (\mathbf{1}_{FX}, F^{\mathcal{G}}FX, X)$. ⊣

Note that to prove this lemma, we use just $f \circ \mathbf{1}_a = \mathbf{1}_b \circ f$ and $\mathbf{1}_a \circ \mathbf{1}_a = \mathbf{1}_a$, which are consequences of (*cat* 1), so that we could generalize the lemma to deductive systems $\mathcal{B}$ that are not categories.

Let the following conditions, called collectively (II), which strengthen (I$\mathcal{B}$), be satisfied:

(II$\mathcal{C}$) $\mathcal{C}$ is a logical system in $\mathcal{L}$;

(II$\mathcal{B}$) $\mathcal{B}$ is a bifunctorial category of the $\mathcal{C}$ kind;

(II$\mathcal{G}$) $\equiv^{\mathcal{G}}$ is a generatively discrete equivalence relation on $\mathcal{L}(\mathcal{B}_{ob})$.

We define, as before, the category $\mathcal{B}^{\mathcal{G}}$ and the functors F and $F^{\mathcal{G}}$ starting from the equivalence relation $\equiv^{\mathcal{G}}$ on $\mathcal{L}(\mathcal{B}_{ob})$. We ensure that $\mathcal{B}^{\mathcal{G}}$ is a bifunctorial category of the $\mathcal{C}$ kind with the following definitions:

$$\begin{aligned} X_1 \xi' X_2 &=_{df} F^{\mathcal{G}}(FX_1 \xi FX_2), \text{ where } \xi \text{ is the } \xi \text{ of } \mathcal{B}, \\ &= [FX_1 \xi FX_2], \end{aligned}$$

$$\begin{aligned} (f_1, X_1, Y_1)\, \xi' \,(f_2, X_2, Y_2) &=_{df} F^{\mathcal{G}}(F(f_1, X_1, Y_1)\, \xi\, F(f_2, X_2, Y_2)) \\ &= F^{\mathcal{G}}(f_1 \xi f_2), \text{ where } \xi \text{ is the } \xi \text{ of } \mathcal{B}, \\ &= (f_1 \xi f_2, X_1 \xi' X_2, Y_1 \xi' Y_2). \end{aligned}$$

$$\varsigma' =_{df} F^{\mathcal{G}}\varsigma = [\varsigma], \text{ where } \varsigma \text{ is the } \varsigma \text{ of } \mathcal{B}.$$

It is easy to check that the bifunctorial equations hold for ξ' in $\mathcal{B}^{\mathcal{G}}$, because they are inherited from $\mathcal{B}$. Then we define the k-endofunctor $(M')^{\mu}$ out of the k-endofunctor M^{μ}, where $M^{\mu} \in \mathcal{L}^{con}$, just by replacing ξ and ς everywhere by ξ' and ς' respectively, so that we have $(M')^{\mu}(X_1, \dots, X_k) = F^{\mathcal{G}}M^{\mu}(FX_1, \dots, FX_k)$. To complete the definition of the $\mathcal{C}$ structure we have

$$\alpha'_{X_1,\dots,X_k} =_{df} (\alpha_{FX_1,\dots,FX_k}, (M')^{\mu}(X_1, \dots, X_k), (N')^{\nu}(X_1, \dots, X_k)).$$

It is easy to obtain that $F(M')^{\mu}(X_1, \dots, X_k) = M^{\mu}(FX_1, \dots, FX_k)$, and analogously for N^{ν}.

Let $\mathcal{B}$ with the $\mathcal{C}$-structure be denoted by $\langle \mathcal{B}, M, \alpha \rangle$ and let $\mathcal{B}^{\mathcal{G}}$ with the $\mathcal{C}$-structure we have just defined be denoted by $\langle \mathcal{B}^{\mathcal{G}}, M', \alpha' \rangle$. Then we can check the following lemma in a straightforward manner.

LEMMA 2. *If* (II) *holds, then the functors* $F^{\mathcal{G}}$ *and* F *are strict* $\mathcal{C}$*-functors from* $\langle \mathcal{B}, M, \alpha \rangle$ *to* $\langle \mathcal{B}^{\mathcal{G}}, M', \alpha' \rangle$ *and vice versa, respectively.*

As a corollary of Lemma 2 we obtain that if $\langle \mathcal{B}, M, \alpha \rangle$ is a $\mathcal{C}/\mathcal{E}$-category for a logical category $\mathcal{C}/\mathcal{E}$, then $\langle \mathcal{B}^{\mathcal{G}}, M', \alpha' \rangle$ is a $\mathcal{C}/\mathcal{E}$-category too. A little bit of work is required only to demonstrate the bifunctorial equations for ξ', which follow from the definitions of $\mathcal{B}^{\mathcal{G}}$ and ξ', as we noted above.

It is easy to see that if (II) holds, and for an α of $\mathcal{C}$ we have that α is a natural transformation in $\mathcal{B}$, then α' is a natural transformation in $\mathcal{B}^{\mathcal{G}}$.

We call a logical system $\mathcal{C}$ *generatively discrete* when for every arrow term of $\mathcal{C}$ of type $p \vdash q$ we have that p is the same letter as q.

Let the following conditions, called collectively (III), which strengthen (II), be satisfied:

(III$\mathcal{C}$) $\mathcal{C}$ is a logical system in $\mathcal{L}$ and $\mathcal{C}'$ is a generatively discrete subsystem of $\mathcal{C}$ also in $\mathcal{L}$, so that $\mathcal{C}$ and $\mathcal{C}'$ have both as objects the formulae of $\mathcal{L}$; we abbreviate this condition involving $\mathcal{C}$, $\mathcal{C}'$ and $\mathcal{L}$ by $\mathcal{C}' \preceq_{\mathcal{L}} \mathcal{C}$;

(III$\mathcal{G}$) $\mathcal{G}$ is a logical category of the $\mathcal{C}'$ kind that is a groupoid;

(III$\mathcal{B}$) $\mathcal{B}$ is a $\mathcal{G}$-category and $\langle \mathcal{B}, M, \alpha \rangle$ is a category of the $\mathcal{C}$ kind.

Since $\mathcal{G}$ is a logical category, it is equal to $\mathcal{C}'/\mathcal{E}'$ for an equational system $\mathcal{E}'$ in $\mathcal{C}'$.

Let $\mathcal{G}(\mathcal{B})$ be defined as $\mathcal{G}$ save that the letters of $\mathcal{P}$ are replaced by the objects of $\mathcal{B}$ (see §2.7). More precisely, we proceed as follows. First, instead of $\mathcal{L}$ generated by $\mathcal{P}$ we have $\mathcal{L}(\mathcal{B}_{ob})$ generated by the set $\mathcal{B}_{ob}$ of objects of $\mathcal{B}$, as above, with new $\xi^{\mathcal{G}}$ and $\zeta^{\mathcal{G}}$. The set $\mathcal{L}(\mathcal{B}_{ob})$ is the set of objects of $\mathcal{G}(\mathcal{B})$. Next, we build the arrow terms of the logical system $\mathcal{C}'(\mathcal{B}_{ob})$ by indexing every α of $\mathcal{C}'$ with k-tuples of formulae of $\mathcal{L}(\mathcal{B}_{ob})$ instead of k-tuples of formulae of $\mathcal{L}$, and then closing under $\xi^{\mathcal{G}}$ for every ξ of $\mathcal{C}'$ and composition. Finally, the axiomatic equations with variables assumed for $\mathcal{E}'$ in $\mathcal{C}'$ are now interpreted in $\mathcal{C}'(\mathcal{B}_{ob})$. The variables for formulae range over the objects of $\mathcal{C}'(\mathcal{B}_{ob})$, and the connectives ξ and ζ in these equations now apply to $\xi^{\mathcal{G}}$ and $\zeta^{\mathcal{G}}$ of $\mathcal{L}(\mathcal{B}_{ob})$. The logical category $\mathcal{C}'/\mathcal{E}'(\mathcal{B}_{ob})$ is $\mathcal{G}(\mathcal{B})$ (see the end of §2.7).

Let $\equiv^{\mathcal{G}}$ be the binary relation on the objects of $\mathcal{G}(\mathcal{B})$ defined by $A \equiv^{\mathcal{G}} B$ iff there is an arrow of $\mathcal{G}(\mathcal{B})$ of type $A \vdash B$. Since $\mathcal{G}$, and hence also $\mathcal{G}(\mathcal{B})$, are groupoids, $\equiv^{\mathcal{G}}$ is an equivalence relation. Since $\mathcal{C}'$ is generatively discrete, the relation $\equiv^{\mathcal{G}}$ is generatively discrete; i.e., no two different objects of $\mathcal{B}$ are in the relation $\equiv^{\mathcal{G}}$. Let $[A]$ be, as before, the equivalence class of an object of $\mathcal{G}(\mathcal{B})$ with respect to $\equiv^{\mathcal{G}}$.

If (III) is fulfilled, then $\mathcal{B}$, which is a $\mathcal{G}$-category, is a $\mathcal{G}(\mathcal{B})$-category too (see the end of §2.8), and so the valuation that assigns to every generating object a of $\mathcal{G}(\mathcal{B})$ the object a of $\mathcal{B}$ itself can be extended to a strict $\mathcal{C}'$-functor E from $\mathcal{G}(\mathcal{B})$ to $\mathcal{B}$. Intuitively, E erases every superscript $^{\mathcal{G}}$ in $\xi^{\mathcal{G}}$ and $\zeta^{\mathcal{G}}$. We already introduced, when we defined the functor F from $\mathcal{B}$ to $\mathcal{B}^{\mathcal{G}}$, the function E on objects of the functor E.

For every object A of $\mathcal{G}(\mathcal{B})$ we have an isomorphism $\varphi_A \colon A_F \vdash A$ of $\mathcal{G}(\mathcal{B})$ where A_F is the chosen representative from $[A]$. The inverse of φ_A is the arrow $\varphi_A^{-1} : A \vdash A_F$ of $\mathcal{G}(\mathcal{B})$. A natural choice for $\varphi_{A_F} : A_F \vdash A_F$, and in particular for $\varphi_a : a \vdash a$, would be identity arrows, but this choice is not essential for the time being. It is also not essential to define φ_A by induction on the complexity of A. (Such an inductive definition of φ_A is possible if the representative A_F is chosen in a particular canonical way; cf. §4.5.) If, however, $\mathcal{G}$, and hence also $\mathcal{G}(\mathcal{B})$, are preorders, then φ_{A_F} must be $\mathbf{1}_{A_F}$. For every isomorphism φ_A of $\mathcal{G}(\mathcal{B})$ we have the isomorphism $E\varphi_A \colon EA_F \vdash EA$ in $\mathcal{B}$.

We can then define the following $\mathcal{C}$ structure in $\mathcal{B}^{\mathcal{G}}$, different from the $\langle \mathcal{B}^{\mathcal{G}}, M', \alpha' \rangle$ structure. On objects we have

$$[A]\, \xi''\, [B] =_{df} [A\, \xi^{\mathcal{G}}\, B] = [A_F\, \xi^{\mathcal{G}}\, B_F],$$
$$\zeta'' =_{df} [\zeta^{\mathcal{G}}].$$

That the definition of ξ'' is correct is guaranteed by the fact that if we have the isomorphisms $f \colon A \vdash A_1$ and $g \colon B \vdash B_1$ of $\mathcal{G}(\mathcal{B})$, then we have also the isomorphism $f\, \xi^{\mathcal{G}}\, g \colon A\, \xi^{\mathcal{G}}\, B \vdash A_1\, \xi^{\mathcal{G}}\, B_1$.

We define the following arrows of $\mathcal{B}^{\mathcal{G}}$:

$$\psi^{\square\xi\square}_{[A],[B]} =_{df} (E\varphi_{A_F\,\xi^{\mathcal{G}}\,B_F}, [A]\,\xi''\,[B], [A]\,\xi'\,[B]),$$
$$\psi^{\varsigma} =_{df} (E\varphi_{\varsigma^{\mathcal{G}}}, \varsigma'', \varsigma'),$$
$$\bar{\psi}^{\square\xi\square}_{[A],[B]} =_{df} (E\varphi^{-1}_{A_F\,\xi^{\mathcal{G}}\,B_F}, [A]\,\xi'\,[B], [A]\,\xi''\,[B]),$$
$$\bar{\psi}^{\varsigma} =_{df} (E\varphi^{-1}_{\varsigma^{\mathcal{G}}}, \varsigma', \varsigma'').$$

It is easy to check that the source of $E\varphi_{A_F\,\xi^{\mathcal{G}}\,B_F}$ is $F([A]\,\xi''\,[B])$, and its target $F([A]\,\xi'\,[B])$. The source of $E\varphi_{\varsigma^{\mathcal{G}}}$ is $F\varsigma''$ and its target is $F\varsigma'$ (which is equal to ς). Note that in the definition of $\psi^{\square\xi\square}$ and $\bar{\psi}^{\square\xi\square}$ we need the arrows φ_A and φ^{-1}_A only for A being of the form $A_F\,\xi^{\mathcal{G}}\,B_F$ and $\varsigma^{\mathcal{G}}$. We have no use for other φ and φ^{-1} arrows.

We define the following operation on the arrows of $\mathcal{B}^{\mathcal{G}}$:

$$(f_1, X_1, Y_1)\,\xi''\,(f_2, X_2, Y_2) =_{df} \bar{\psi}^{\square\xi\square}_{Y_1,Y_2} \circ ((f_1, X_1, Y_1)\,\xi'\,(f_2, X_2, Y_2)) \circ \psi^{\square\xi\square}_{X_1,X_2}.$$

It is clear that, since $E\varphi_A$ is an isomorphism, the bifunctorial equations hold for ξ'' in $\mathcal{B}^{\mathcal{G}}$.

For every $M^\mu \in \mathcal{L}^{con}$, let $(M'')^\mu$ be obtained by replacing ξ and ς with ξ'' and ς''. Starting from $\psi^{\square\xi\square}_{X_1,X_2}$ and ψ^{ς}, with the help of ξ'', we define $\psi^{M^\mu}_{X_1,\dots,X_k} : (M'')^\mu(X_1,\dots,X_k) \vdash (M')^\mu(X_1,\dots,X_k)$ by the inductive clauses $\psi^{\square}_X = \mathbf{1}_X = (\mathbf{1}_{FX}, X, X)$, which replaces ($\psi^{M^\mu}$ 1), and the clauses (ψ^{M^μ} 2), with ξ^2 replaced by ξ'', and (ψ^{M^μ} 3) of §2.8. We define analogously with the help of ξ'' the inverse $\bar{\psi}^{M^\mu}_{X_1,\dots,X_k}$ of $\psi^{M^\mu}_{X_1,\dots,X_k}$. Then we have the following definition in $\mathcal{B}^{\mathcal{G}}$:

$$\alpha''_{X_1,\dots,X_k} =_{df} \bar{\psi}^{N^\nu}_{X_1,\dots,X_k} \circ \alpha'_{X_1,\dots,X_k} \circ \psi^{M^\mu}_{X_1,\dots,X_k}.$$

This defines the $\mathcal{C}$ structure in $\mathcal{B}^{\mathcal{G}}$, and we can conclude that $\langle \mathcal{B}^{\mathcal{G}}, M'', \alpha'' \rangle$ is a bifunctorial category of the $\mathcal{C}$ kind.

It is straightforward to show by induction on $M \in \mathcal{L}^{\square}$ that in $\mathcal{B}^{\mathcal{G}}$ we have

$$(M'') \quad (M'')^\mu((f_1, X_1, Y_1), \dots, (f_k, X_k, Y_k)) = \bar{\psi}^{M^\mu}_{Y_1,\dots,Y_k} \circ (M')^\mu((f_1, X_1, Y_1), \dots, (f_k, X_k, Y_k)) \circ \psi^{M^\mu}_{X_1,\dots,X_k}.$$

We introduce the following definitions:

$$\psi'^{\Box\xi\Box}_{X_1,X_2} =_{df} \bar{\psi}^{\Box\xi\Box}_{X_1,X_2}, \qquad \psi'^{\varsigma} =_{df} \bar{\psi}^{\varsigma},$$
$$\bar{\psi}'^{\Box\xi\Box}_{X_1,X_2} =_{df} \psi^{\Box\xi\Box}_{X_1,X_2}, \qquad \bar{\psi}'^{\varsigma} =_{df} \psi^{\varsigma}.$$

Then starting from $\psi'^{\Box\xi\Box}_{X_1,X_2}$ and ψ'^{ς}, with the help of ξ', and not ξ'', we define

$$\psi'^{M^\mu}_{X_1,\ldots,X_k} : (M')^\mu(X_1,\ldots,X_k) \vdash (M'')^\mu(X_1,\ldots,X_k)$$

by the inductive clauses $\psi'^{\Box}_X = \mathbf{1}_X$, which replaces ($\psi^{M^\mu}$ 1), and the clauses (ψ^{M^μ} 2) and (ψ^{M^μ} 3). We define analogously with the help of ξ' the inverse $\bar{\psi}'^{M^\mu}_{X_1,\ldots,X_k}$ of $\psi'^{M^\mu}_{X_1,\ldots,X_k}$. Then we have the following in $\mathcal{B}^\mathcal{G}$:

$$\psi'^{M^\mu}_{X_1,\ldots,X_k} = \bar{\psi}^{M^\mu}_{X_1,\ldots,X_k},$$
$$\bar{\psi}'^{M^\mu}_{X_1,\ldots,X_k} = \psi^{M^\mu}_{X_1,\ldots,X_k}.$$

To show these equations, we use the equation (ψ^{M^μ}) of §2.8, its dual for $\bar{\psi}$, and the equation (M'') above.

It follows now immediately that in $\mathcal{B}^\mathcal{G}$ we have the following:

$$\alpha''_{X_1,\ldots,X_k} = \psi'^{N^\nu}_{X_1,\ldots,X_k} \circ \alpha'_{X_1,\ldots,X_k} \circ \bar{\psi}'^{M^\mu}_{X_1,\ldots,X_k}.$$

Since the ψ' arrows are defined in terms of ξ', which is easier to handle than ξ'', we will have occasion to apply later this equation, which we call the *alternative definition* of α''. We can prove the following lemma.

LEMMA 3. *If* (III) *holds, then the identity functor $I_{\mathcal{B}^\mathcal{G}}$ of $\mathcal{B}^\mathcal{G}$ with $\psi^{\Box\xi\Box}$ and ψ^{ς} is a groupoidal partial $\mathcal{C}$-functor from $\langle\mathcal{B}^\mathcal{G}, M', \alpha'\rangle$ to $\langle\mathcal{B}^\mathcal{G}, M'', \alpha''\rangle$, and $I_{\mathcal{B}^\mathcal{G}}$ with $\bar{\psi}^{\Box\xi\Box}$ and $\bar{\psi}^{\varsigma}$ is a groupoidal partial $\mathcal{C}$-functor from $\langle\mathcal{B}^\mathcal{G}, M'', \alpha''\rangle$ to $\langle\mathcal{B}^\mathcal{G}, M', \alpha'\rangle$.*

PROOF. The equations (ψ *nat*) follow from (M''), while the equations ($\psi\alpha$) follow immediately from the definition of α'', or the alternative definition of α''. ⊣

Note that with Lemma 3 we have asserted that ξ' and ξ'' are naturally isomorphic 2-endofunctors of $\mathcal{B}^\mathcal{G}$. Showing that $\psi^{\Box\xi\Box}$ and $\bar{\psi}^{\Box\xi\Box}$ involved in this isomorphism are natural transformations does not presuppose that the φ and φ^{-1} arrows in terms of which we define $\psi^{\Box\xi\Box}$ and $\bar{\psi}^{\Box\xi\Box}$ are members of natural transformations. We just assume that φ and φ^{-1} arrows are isomorphisms of $\mathcal{G}(\mathcal{B})$.

Suppose $\mathcal{C}' \preceq_{\mathcal{L}} \mathcal{C}$, as in (III$\mathcal{C}$). For every logical category $\mathcal{C}/\mathcal{E}$ we can determine out of $\mathcal{C}/\mathcal{E}$ and $\mathcal{C}'$ a logical subcategory $\mathcal{C}'/\mathcal{E}'$ of $\mathcal{C}/\mathcal{E}$ by restricting the equations of $\mathcal{E}$ to the transformations α of $\mathcal{C}'$.

We say that $\mathcal{C}'/\mathcal{E}'$ *flows* through $\mathcal{C}/\mathcal{E}$ when for every α of $\mathcal{C}$ every k-tuple of arrows of $\mathcal{C}'/\mathcal{E}'$ flows through α in $\mathcal{C}/\mathcal{E}$. If $\mathcal{C}/\mathcal{E}$ is a natural logical category, then for every subsystem $\mathcal{C}'$ of $\mathcal{C}$ we have that $\mathcal{C}'/\mathcal{E}'$ flows through $\mathcal{C}/\mathcal{E}$.

Suppose $\mathcal{C}' \preceq_{\mathcal{L}} \mathcal{C}$. For any deductive system $\mathcal{B}$ of the $\mathcal{C}$ kind there is a least subsystem of $\mathcal{B}$ of the $\mathcal{C}'$ kind, which we call the $\mathcal{C}'$*-core* of $\mathcal{B}$. We build the $\mathcal{C}'$-core of $\mathcal{B}$ by taking for objects all the objects of $\mathcal{B}$; for arrows we take the members of the transformations α of $\mathcal{B}$ for every α of $\mathcal{C}'$ with the same sources and targets as in $\mathcal{B}$, and then close under composition and the operations on arrows ξ of $\mathcal{B}$ for every ξ of $\mathcal{C}'$, i.e. of $\mathcal{L}$. Note that the $\mathcal{C}'$-core of $\mathcal{B}$ need not be a syntactical system: it inherits the equations between arrows of $\mathcal{B}$. As a limit case, we can determine also the $\mathcal{C}$-core of $\mathcal{B}$, which need not coincide with $\mathcal{B}$.

If $\mathcal{B}$ is a bifunctorial category of the $\mathcal{C}$-kind, then its $\mathcal{C}'$-core is a bifunctorial category of the $\mathcal{C}'$-kind; if $\mathcal{B}$ is a $\mathcal{C}/\mathcal{E}$-category, then its $\mathcal{C}'$-core is a $\mathcal{C}'/\mathcal{E}'$-category; and if $\mathcal{B}$ is a natural $\mathcal{C}/\mathcal{E}$-category, then its $\mathcal{C}'$-core is a natural $\mathcal{C}'/\mathcal{E}'$-category. The $\mathcal{C}'$-core of $\mathcal{C}/\mathcal{E}$ is $\mathcal{C}'/\mathcal{E}'$.

Consider now the following conditions, called collectively (IV), which strengthen (III):

(IV$\mathcal{C}$) $\mathcal{C}' \preceq_{\mathcal{L}} \mathcal{C}$, as in (III$\mathcal{C}$), and $\mathcal{C}/\mathcal{E}$ is a logical category in $\mathcal{L}$;

(IV$\mathcal{G}$) the $\mathcal{C}'$-core $\mathcal{C}'/\mathcal{E}' = \mathcal{G}$ of $\mathcal{C}/\mathcal{E}$ is a groupoid and $\mathcal{G}$ flows through $\mathcal{C}/\mathcal{E}$;

(IV$\mathcal{B}$) $\langle \mathcal{B}, M, \alpha \rangle$ is a $\mathcal{C}/\mathcal{E}$-category.

Lemma 3 holds if (III) is replaced by (IV) and "groupoidal partial $\mathcal{C}$-functor" is replaced by "strong $\mathcal{C}$-functor". Here is this new version of Lemma 3.

LEMMA 3(IV). *If* (IV) *holds, then the identity functor* $I_{\mathcal{B}^{\mathcal{G}}}$ *of* $\mathcal{B}^{\mathcal{G}}$ *with* $\psi^{\Box\xi\Box}$ *and* ψ^{ς} *is a strong* $\mathcal{C}$*-functor from* $\langle \mathcal{B}^{\mathcal{G}}, M', \alpha' \rangle$ *to* $\langle \mathcal{B}^{\mathcal{G}}, M'', \alpha'' \rangle$*, and* $I_{\mathcal{B}^{\mathcal{G}}}$ *with* $\bar{\psi}^{\Box\xi\Box}$ *and* $\bar{\psi}^{\varsigma}$ *is a strong* $\mathcal{C}$*-functor from* $\langle \mathcal{B}^{\mathcal{G}}, M'', \alpha'' \rangle$ *to* $\langle \mathcal{B}^{\mathcal{G}}, M', \alpha' \rangle$.

PROOF. We appeal to Lemma 3, and check, moreover, the fluency of the identity functor, with the arrows $\psi^{M^{\mu}}$ or $\bar{\psi}^{M^{\mu}}$, in every α, or we check directly $(\psi\alpha L)$ (which is slightly more complicated). ⊣

As a corollary, we obtain the following lemma.

LEMMA 4. *If* (IV) *holds, then the functor* $F^{\mathcal{G}}$*, with* $\psi^{\Box\xi\Box}$ *and* ψ^{ς}*, and the functor* F*, with* $F\bar{\psi}^{\Box\xi\Box}$ *and* $F\bar{\psi}^{\varsigma}$*, are strong* $\mathcal{C}$*-functors from* $\langle \mathcal{B}, M, \alpha \rangle$ *to* $\langle \mathcal{B}^{\mathcal{G}}, M'', \alpha'' \rangle$ *and vice versa, respectively.*

PROOF. As we noted in §2.8, the composition of two strong $\mathcal{C}$-functors is a strong $\mathcal{C}$-functor. Since strict $\mathcal{C}$-functors are strong, Lemmata 2 and 3(IV) deliver our lemma. ⊣

If in Lemma 4 the condition (IV) is replaced by (III), then we can only affirm that the functors in question are groupoidal partial $\mathcal{C}$-functors. If (IV) holds, then from Lemma 1 we obtain that $\langle \mathcal{B}, M, \alpha \rangle$ and $\langle \mathcal{B}^{\mathcal{G}}, M'', \alpha'' \rangle$ are equivalent categories via the two strong $\mathcal{C}$-functors of Lemma 4. We can prove the following lemma.

LEMMA 5. *If* (IV) *holds, then* $\langle \mathcal{B}^{\mathcal{G}}, M'', \alpha'' \rangle$ *is a* $\mathcal{C}/\mathcal{E}$*-category.*

PROOF. Note first that the bifunctorial equations hold for ξ'' in $\mathcal{B}^{\mathcal{G}}$, as noted after the definition of ξ'' on arrows. So $\langle \mathcal{B}^{\mathcal{G}}, M'', \alpha'' \rangle$ is a bifunctorial category of the $\mathcal{C}$ kind.

Then take a valuation v that maps the letters of $\mathcal{P}$ into the objects of $\mathcal{B}^{\mathcal{G}}$. As mentioned after Lemma 2, the valuation v can be extended to a strict $\mathcal{C}$-functor v' from $\mathcal{C}/\mathcal{E}$ to $\langle \mathcal{B}^{\mathcal{G}}, M', \alpha' \rangle$ such that $v'(p) = v(p)$.

Suppose that the equation $f = g$ for $f, g\colon M(p_1, \ldots, p_m) \vdash N(q_1, \ldots, q_n)$ belongs to $\mathcal{E}$. Then we know that $v'(f) = v'(g)$ holds in $\mathcal{B}^{\mathcal{G}}$, and hence we have in $\mathcal{B}^{\mathcal{G}}$ the following equation too:

$$\bar{\psi}^{N}_{v'(q_1),\ldots,v'(q_n)} \circ v'(f) \circ \psi^{M}_{v'(p_1),\ldots,v'(p_m)} = \bar{\psi}^{N}_{v'(q_1),\ldots,v'(q_n)} \circ v'(g) \circ \psi^{M}_{v'(p_1),\ldots,v'(p_m)}.$$

By Lemma 3(IV) and the Proposition of §2.8, we conclude that for the maps v'' from the objects and arrow terms of $\mathcal{C}$ to the objects and arrows of $\langle \mathcal{B}^{\mathcal{G}}, M'', \alpha'' \rangle$ defined by $v''(p) = v(p)$ and the inductive clauses as for v_i, which we gave just before that Proposition, we have $v''(f) = v''(g)$. So the valuation v is extended to a strict $\mathcal{C}$-functor v'' from $\mathcal{C}/\mathcal{E}$ to $\langle \mathcal{B}^{\mathcal{G}}, M'', \alpha'' \rangle$, which proves the lemma. ⊣

It is easy to see that if (III) holds, and for an α of $\mathcal{C}$ we have that α' is a natural transformation in $\mathcal{B}^{\mathcal{G}}$, then α'' is a natural transformation in $\mathcal{B}^{\mathcal{G}}$. So, together with the comment we made before introducing (III), we can conclude that if (III) holds, and for an α in $\mathcal{L}$ we have that α is a natural transformation in $\mathcal{B}$, then α'' is a natural transformation in $\mathcal{B}^{\mathcal{G}}$.

A deductive system $\mathcal{A}$ of the $\mathcal{C}$ kind is called *$\mathcal{C}$-strict* when for every α of $\mathcal{C}$ we have that the members of the transformation α in $\mathcal{A}$ are identity arrows. This presupposes that the objects $M^{\mu}(a_1, \ldots, a_k)$ and $N^{\nu}(a_1, \ldots, a_k)$ are equal in a $\mathcal{C}$-strict deductive system, though in $\mathcal{C}$ the formulae $M^{\mu}(A_1, \ldots, A_k)$ and $N^{\nu}(A_1, \ldots, A_k)$ need not be equal. A bifunctorial category $\mathcal{A}$ of the $\mathcal{C}$-kind is $\mathcal{C}$-strict iff its $\mathcal{C}$-core is discrete. We can now prove the following lemma.

LEMMA 6. *If* (IV) *holds and $\mathcal{G}$ is a preorder, then $\langle \mathcal{B}^{\mathcal{G}}, M'', \alpha'' \rangle$ is $\mathcal{C}'$-strict.*

PROOF. Suppose $\mathcal{G}$ is a preorder and take an α of $\mathcal{C}'$. Then in $\mathcal{G}(\mathcal{B})$ we have an arrow $\alpha_{A_1,\ldots,A_k} : (M^{\mathcal{G}})^{\mu}(A_1, \ldots, A_k) \vdash (N^{\mathcal{G}})^{\nu}(A_1, \ldots, A_k)$ where $(M^{\mathcal{G}})^{\mu}$ and $(N^{\mathcal{G}})^{\nu}$ are obtained from $M^{\mu}, N^{\nu} \in \mathcal{L}^{con}$ by replacing ξ and ζ by $\xi^{\mathcal{G}}$ and $\zeta^{\mathcal{G}}$ respectively. Hence $[(M^{\mathcal{G}})^{\mu}(A_1, \ldots, A_k)] = [(N^{\mathcal{G}})^{\nu}(A_1, \ldots, A_k)]$ and hence $(M'')^{\mu}([A_1], \ldots, [A_k]) = (N'')^{\nu}([A_1], \ldots, [A_k])$.

The arrow $\alpha''_{[A_1],\ldots,[A_k]}$ is so of the type $Y \vdash Y$ for some object Y of $\mathcal{B}^{\mathcal{G}}$. So the arrow $F\alpha''_{[A_1],\ldots,[A_k]}$ of $\mathcal{B}$ is of the type $FY \vdash FY$, and since this arrow belongs to the $\mathcal{C}'$-core of $\mathcal{B}$, with $\mathcal{G}$ being a preorder we obtain $F\alpha''_{[A_1],\ldots,[A_k]} = \mathbf{1}_{FY}$ in $\mathcal{B}$. From that we obtain that $\alpha''_{[A_1],\ldots,[A_k]}$ is $(\mathbf{1}_{FY}, Y, Y)$, which is an identity arrow in $\mathcal{B}^{\mathcal{G}}$. ⊣

Lemmata 1, 4, 5 and 6 yield the following theorem.

STRICTIFICATION THEOREM. *If* (IV) *holds and* $\mathcal{G}$ *is a preorder, then* $\langle \mathcal{B}, M, \alpha \rangle$ *is equivalent to the* $\mathcal{C}'$*-strict* $\mathcal{C}/\mathcal{E}$*-category* $\langle \mathcal{B}^{\mathcal{G}}, M'', \alpha'' \rangle$ *via the strong* $\mathcal{C}$*-functors* $F^{\mathcal{G}}$ *and* F.

As a corollary we obtain the following.

STRICTIFICATION COROLLARY. *If* $\mathcal{C}'/\mathcal{E}$ *is a natural logical category that is a groupoid and a preorder, then every* $\mathcal{C}'/\mathcal{E}$*-category* $\langle \mathcal{B}, M, \alpha \rangle$ *is equivalent to the* $\mathcal{C}'$*-strict* $\mathcal{C}'/\mathcal{E}$*-category* $\langle \mathcal{B}^{\mathcal{G}}, M'', \alpha'' \rangle$ *via the strong* $\mathcal{C}'$*-functors* $F^{\mathcal{G}}$ *and* F.

In [67] and [96] one finds the instance of this corollary where $\mathcal{C}'/\mathcal{E}$ is the free monoidal category generated by $\mathcal{P}$, i.e. our category $\hat{\mathbf{L}}_{\top}$ of §4.6. (From [67] and [96] one could get the wrong impression that something peculiar to monoidal categories has been discovered, while a more general result, stated in our Strictification Theorem and Strictification Corollary, looms behind.) We have no use, however, for the Strictification Corollary. Instead, we will rely on the stronger Strictification Theorem to record long computations concerning $\mathcal{B}$ with the help of $\mathcal{B}^{\mathcal{G}}$, as mentioned at the beginning of the section (cf. Chapter 11). The corollary is not sufficient for that, because it does not take into account the unstrictified $\mathcal{C}$ structure foreign to the strictified $\mathcal{C}'$ structure. This structure is not preserved by a functor that is just a $\mathcal{C}'$-functor and not also a $\mathcal{C}$-functor.

Suppose $\mathcal{C}' \preceq_{\mathcal{L}} \mathcal{C}$, as in (III$\mathcal{C}$). Then we say that a $\mathcal{C}/\mathcal{E}$-category $\mathcal{B}$ can be $(\mathcal{C}/\mathcal{E}, \mathcal{C}')$*-strictified* when there is a $\mathcal{C}'$-strict $\mathcal{C}/\mathcal{E}$-category $\mathcal{B}^*$ equivalent to $\mathcal{B}$ via two strong $\mathcal{C}$-functors. We can prove the following lemma.

LEMMA 7. *If* $\mathcal{C}' \preceq_{\mathcal{L}} \mathcal{C}$ *holds and every* $\mathcal{C}/\mathcal{E}$*-category can be* $(\mathcal{C}/\mathcal{E}, \mathcal{C}')$*-strictified, then for the* $\mathcal{C}'$*-core of* $\mathcal{C}/\mathcal{E}$ *we have that every arrow in it is an isomorphism of* $\mathcal{C}/\mathcal{E}$*, this* $\mathcal{C}'$*-core flows through* $\mathcal{C}/\mathcal{E}$ *and it is a preorder.*

PROOF. If every $\mathcal{C}/\mathcal{E}$-category can be $(\mathcal{C}/\mathcal{E}, \mathcal{C}')$-strictified, then the logical category $\mathcal{C}/\mathcal{E}$ itself can be so strictified. So there is a $\mathcal{C}'$-strict $\mathcal{C}/\mathcal{E}$-category $(\mathcal{C}/\mathcal{E})^*$ equivalent to $\mathcal{C}/\mathcal{E}$ via the strong $\mathcal{C}$-functors $\langle F, \psi^{\Box\xi\Box}, \psi^{\varsigma} \rangle$ from $(\mathcal{C}/\mathcal{E})^*$ to $\mathcal{C}/\mathcal{E}$ and $\langle F^*, \psi^{*\Box\xi\Box}, \psi^{*\varsigma} \rangle$ from $\mathcal{C}/\mathcal{E}$ to $(\mathcal{C}/\mathcal{E})^*$. Let σ be the natural isomorphism of $\mathcal{C}/\mathcal{E}$ whose members are $\sigma_A : FF^*A \vdash A$, and let σ_A^{-1} be the inverse of σ_A. Since $(\mathcal{C}/\mathcal{E})^*$ is a $\mathcal{C}/\mathcal{E}$-category, every valuation from $\mathcal{P}$ to the objects of $(\mathcal{C}/\mathcal{E})^*$ can be extended to a strict $\mathcal{C}$-functor. Let v^* be the strict $\mathcal{C}$-functor from $\mathcal{C}/\mathcal{E}$ to $(\mathcal{C}/\mathcal{E})^*$ such that $v^*(p) = F^*p$. (Note that the functors v^* and F^* need not coincide.)

Let $f : M(\vec{p}) \vdash N(\vec{q})$ be an arrow of the $\mathcal{C}'$-core of $\mathcal{C}/\mathcal{E}$. Then $v^*(f) : v^*M(\vec{p}) \vdash v^*N(\vec{q})$ is in the $\mathcal{C}'$-core of $(\mathcal{C}/\mathcal{E})^*$, and is hence an identity arrow of $(\mathcal{C}/\mathcal{E})^*$. By the Proposition of §2.8, we have in $(\mathcal{C}/\mathcal{E})^*$

$$F^*f = \psi^{*N}_{\vec{q}} \circ v^*(f) \circ \bar{\psi}^{*M}_{\vec{p}} = \psi^{*N}_{\vec{q}} \circ \bar{\psi}^{*M}_{\vec{p}}.$$

(The functor corresponding to v_1 in (ψt) of the Proposition of §2.8 is here the identity functor of $\mathcal{C}/\mathcal{E}$.) Hence,

$$f = \sigma^{-1}_{N(\vec{q})} \circ FF^*f \circ \sigma_{M(\vec{p})} = \sigma^{-1}_{N(\vec{q})} \circ F(\psi^{*N}_{\vec{q}} \circ \bar{\psi}^{*M}_{\vec{p}}) \circ \sigma_{M(\vec{p})},$$

which proves that f is an isomorphism of $\mathcal{C}/\mathcal{E}$.

For $f_1, \ldots, f_k$ in the $\mathcal{C}'$-core of $\mathcal{C}/\mathcal{E}$, with f_i of type $K_i(\vec{p_i}) \vdash L_i(\vec{q_i})$ for $i \in \{1, \ldots, k\}$, by using the Proposition of §2.8 and $(\psi\ nat)$, we have in $(\mathcal{C}/\mathcal{E})^*$

$$\begin{aligned}
F^*(N^\nu(f_1, \ldots, f_k) &\circ \alpha_{K_1(\vec{p_1}),\ldots,K_k(\vec{p_k})}) \\
&= \psi^{*N^\nu(L_1,\ldots,L_k)}_{\vec{q_1},\ldots,\vec{q_k}} \circ \bar{\psi}^{*N^\nu(K_1,\ldots,K_k)}_{\vec{p_1},\ldots,\vec{p_k}} \circ F^*\alpha_{K_1(\vec{p_1}),\ldots,K_k(\vec{p_k})} \\
&= \psi^{*N^\nu(L_1,\ldots,L_k)}_{\vec{q_1},\ldots,\vec{q_k}} \circ v^*(\alpha_{K_1(\vec{p_1}),\ldots,K_k(\vec{p_k})}) \circ \bar{\psi}^{*M^\mu(K_1,\ldots,K_k)}_{\vec{p_1},\ldots,\vec{p_k}} \\
&= F^*(\alpha_{K_1(\vec{p_1}),\ldots,K_k(\vec{p_k})} \circ M^\mu(f_1, \ldots, f_k)).
\end{aligned}$$

From that we easily infer, by applying F to both sides and composing with members of σ and σ^{-1}, that the $\mathcal{C}'$-core of $\mathcal{C}/\mathcal{E}$ flows through $\mathcal{C}/\mathcal{E}$.

Take now $f, g \colon M(\vec{p}) \vdash N(\vec{q})$ in the $\mathcal{C}'$-core of $\mathcal{C}/\mathcal{E}$. Since $v^*(f)$ and $v^*(g)$ are the same identity arrow of the $\mathcal{C}'$-core of $(\mathcal{C}/\mathcal{E})^*$, we obtain, as above,

$$F^*f = \psi^{*N}_{\vec{q}} \circ \bar{\psi}^{*M}_{\vec{p}} = F^*g,$$

from which $f = g$ follows. ⊣

It is easy to infer from our Strictification Theorem and Lemma 7 the following proposition.

STRICTIFICATION-COHERENCE EQUIVALENCE. *If* (IV) *holds, then every* $\mathcal{C}/\mathcal{E}$*-category can be* $(\mathcal{C}/\mathcal{E},\mathcal{C}')$*-strictified iff the* $\mathcal{C}'$*-core of* $\mathcal{C}/\mathcal{E}$ *is a preorder.*

As a corollary of this equivalence we have that if $\mathcal{C}$ is generatively discrete and $\mathcal{C}/\mathcal{E}$ is a natural logical category that is a groupoid, then every $\mathcal{C}/\mathcal{E}$-category can be $(\mathcal{C}/\mathcal{E},\mathcal{C})$-strictified iff $\mathcal{C}/\mathcal{E}$ is a preorder. Such a statement was suggested by [67] and [96] for the particular case when $\mathcal{C}/\mathcal{E}$ is the free monoidal category generated by $\mathcal{P}$.

A result analogous to Strictification-Coherence Equivalence is the following proposition:

If (IV) *holds, then* $\mathcal{C}/\mathcal{E}$ *can be* $(\mathcal{C}/\mathcal{E},\mathcal{C}')$*-strictified iff the* $\mathcal{C}'$*-core of* $\mathcal{C}/\mathcal{E}$ *is a preorder.*

We can also infer the following:

If (IV) *holds, then every* $\mathcal{C}/\mathcal{E}$*-category can be* $(\mathcal{C}/\mathcal{E},\mathcal{C}')$*-strictified iff* $\mathcal{C}/\mathcal{E}$ *can be* $(\mathcal{C}/\mathcal{E},\mathcal{C}')$*-strictified.*

If our goal is to use strictification to prove preorder, then appealing to the possibility of strictifying every $\mathcal{C}/\mathcal{E}$-category is irrelevant. The following statement, which is a corollary of Lemma 7, suffices.

STRICTIFICATION-COHERENCE IMPLICATION. *If* $\mathcal{C}$ *is generatively discrete,* $\mathcal{C}/\mathcal{E}$ *is a natural logical category that is a groupoid and* $\mathcal{C}/\mathcal{E}$ *can be* $(\mathcal{C}/\mathcal{E},\mathcal{C})$*-strictified, then* $\mathcal{C}/\mathcal{E}$ *is a preorder.*

We will rely on this implication to give alternative proofs of Associative Coherence and Monoidal Coherence in §4.5 ans §4.7.

§3.2. Direct strictification

The procedure of strictification of the preceding section can be simplified if the category $\mathcal{B}$ we want to strictify is a logical category. Then we can build a $\mathcal{C}'$-strict category simpler than $\mathcal{B}^{\mathcal{G}}$ equivalent to $\mathcal{B}$ via two strong $\mathcal{C}$-functors, one of which is even strict. This category replacing $\mathcal{B}^{\mathcal{G}}$, though not logical, will be a syntactical category, like the logical category $\mathcal{B}$.

Suppose the conditions (IV$\mathcal{C}$) and (IV$\mathcal{G}$) of the preceding section are fulfilled. Let $\equiv_{\mathcal{G}}$ be the binary relation on $\mathcal{L}$ defined by $A \equiv_{\mathcal{G}} B$ iff there is an arrow of type $A \vdash B$ in $\mathcal{G}$. Since $\mathcal{G}$ is a groupoid, $\equiv_{\mathcal{G}}$ is an equivalence relation. Since $\mathcal{C}'$ is generatively discrete, no two different letters of $\mathcal{P}$ are in the relation $\equiv_{\mathcal{G}}$.

The objects of the syntactical system $\mathcal{C}_{\mathcal{G}}$ are all the equivalence classes $[\![A]\!]$ with respect $\equiv_{\mathcal{G}}$ for A a formula of $\mathcal{L}$. We denote such classes by $X, Y, Z, \ldots$, sometimes with indices. On the objects of $\mathcal{C}_{\mathcal{G}}$ we define the operations ξ and ς by

$$[\![A]\!] \,\xi\, [\![B]\!] =_{df} [\![A \,\xi\, B]\!], \qquad\qquad \varsigma =_{df} [\![\varsigma]\!].$$

The definition of ξ is correct because $\mathcal{G}$ is a logical system in $\mathcal{L}$.

For every arrow term f of $\mathcal{C}$ let $f_{\mathcal{G}}$ be the arrow term obtained by replacing every letter p of $\mathcal{P}$ in the indices of f by $[\![p]\!]$. If the arrow term f of $\mathcal{C}$ is of type $A \vdash B$, then the arrow term $f_{\mathcal{G}}$ of $\mathcal{C}_{\mathcal{G}}$ is of type $[\![A]\!] \vdash [\![B]\!]$. The arrow terms of $\mathcal{C}_{\mathcal{G}}$ are the arrow terms $f_{\mathcal{G}}$ for every arrow term f of $\mathcal{C}$. As a subsystem of $\mathcal{C}_{\mathcal{G}}$ we have the syntactical system $\mathcal{C}'_{\mathcal{G}}$ obtained from $\mathcal{C}'$ as $\mathcal{C}_{\mathcal{G}}$ is obtained from $\mathcal{C}$. Since $\mathcal{G}$ is a groupoid, every arrow term of $\mathcal{C}'_{\mathcal{G}}$ is of the type $[\![A]\!] \vdash [\![A]\!]$ for some A in $\mathcal{L}$. The equational system $\mathcal{E}_{\mathcal{G}}$ in $\mathcal{C}_{\mathcal{G}}$ is obtained from the axiomatic equations with variables assumed for $\mathcal{E}$ (so that for every equation $f = g$ in $\mathcal{E}$ we have $f_{\mathcal{G}} = g_{\mathcal{G}}$ in $\mathcal{E}_{\mathcal{G}}$) by adding for every arrow term $f_{\mathcal{G}} : [\![A]\!] \vdash [\![A]\!]$ of $\mathcal{C}'_{\mathcal{G}}$ the equation $f_{\mathcal{G}} = \mathbf{1}_{[\![A]\!]}$, which we call a *strictifying equation*, and then closing under (sy), (tr) and (co) (see §2.3).

It is clear that the syntactical category $\mathcal{C}_{\mathcal{G}}/\mathcal{E}_{\mathcal{G}}$ is a $\mathcal{C}/\mathcal{E}$-category. If $\mathcal{C}/\mathcal{E}$ is a natural logical category, then $\mathcal{C}_{\mathcal{G}}/\mathcal{E}_{\mathcal{G}}$ is a natural $\mathcal{C}/\mathcal{E}$-category. Moreover, $\mathcal{C}_{\mathcal{G}}/\mathcal{E}_{\mathcal{G}}$ is $\mathcal{C}'$-strict. We will show that when $\mathcal{G}$ is a preorder the syntactical category $\mathcal{C}_{\mathcal{G}}/\mathcal{E}_{\mathcal{G}}$ is equivalent to $\mathcal{C}/\mathcal{E}$ via two strong $\mathcal{C}$-functors, one of which, going from $\mathcal{C}/\mathcal{E}$ to $\mathcal{C}_{\mathcal{G}}/\mathcal{E}_{\mathcal{G}}$, is even strict.

First we define as follows a graph-morphism $H_{\mathcal{G}}$ from $\mathcal{C}$ to $\mathcal{C}_{\mathcal{G}}$:

$$H_{\mathcal{G}}A =_{df} [\![A]\!],$$
$$H_{\mathcal{G}}f =_{df} f_{\mathcal{G}}.$$

This graph-morphism induces a functor $H_{\mathcal{G}}$ from $\mathcal{C}/\mathcal{E}$ to $\mathcal{C}_{\mathcal{G}}/\mathcal{E}_{\mathcal{G}}$ (see the penultimate paragraph of §2.4).

To define a functor from $\mathcal{C}_\mathcal{G}/\mathcal{E}_\mathcal{G}$ to $\mathcal{C}/\mathcal{E}$ we first choose in every $[\![A]\!]$ a fixed representative A_H, so that the representative of $[\![p]\!]$ is p for every p of $\mathcal{P}$. Because of the generative discreteness of $\mathcal{C}'$, this choice for $[\![p]\!]$ can be made unambiguously.

Next, we can choose for every A in $\mathcal{L}$ an isomorphism $\varphi_A\colon A_H \vdash A$ of $\mathcal{G}$, whose inverse is $\varphi_A^{-1}: A \vdash A_H$. A natural choice for $\varphi_{A_H}: A_H \vdash A_H$, and $\varphi_p : p \vdash p$ in particular, is identity arrows, but this choice is not essential for the time being. If, however, $\mathcal{G}$ is a preorder, then φ_{A_H} must be $\mathbf{1}_{A_H}$.

Then we define as follows a graph-morphism H from $\mathcal{C}_\mathcal{G}$ to $\mathcal{C}$:

$$\begin{aligned}
&H[\![A]\!] =_{df} A_H,\\
&H\alpha_{X_1,\ldots,X_k} =_{df} \varphi^{-1}_{N^\nu(HX_1,\ldots,HX_k)} \circ \alpha_{HX_1,\ldots,HX_k} \circ \varphi_{M^\mu(HX_1,\ldots,HX_k)},\\
&H(f^1_\mathcal{G} \xi f^2_\mathcal{G}) =_{df} \varphi^{-1}_{HY_1 \xi HY_2} \circ (Hf^1_\mathcal{G} \xi Hf^2_\mathcal{G}) \circ \varphi_{HX_1 \xi HX_2}, \text{ for } f^i_\mathcal{G}\colon X_i \vdash Y_i,\\
&H(g_\mathcal{G} \circ f_\mathcal{G}) =_{df} Hg_\mathcal{G} \circ Hf_\mathcal{G}.
\end{aligned}$$

It is clear that for $f_\mathcal{G}\colon [\![A]\!] \vdash [\![B]\!]$ we have that the type of $Hf_\mathcal{G}$ is $A_H \vdash B_H$, that is $H[\![A]\!] \vdash H[\![B]\!]$. We can prove the following lemma.

LEMMA 1. *If* (IV$\mathcal{C}$) *and* (IV$\mathcal{G}$) *hold and* $\mathcal{G}$ *is a preorder, then for every arrow term* $f\colon A \vdash B$ *of* $\mathcal{C}$*, in* $\mathcal{C}/\mathcal{E}$ *we have*

$$HH_\mathcal{G} f = \varphi_B^{-1} \circ f \circ \varphi_A.$$

PROOF. We proceed by induction on the length of f. If f is $\alpha_{A_1,\ldots,A_k}$, then

$$\begin{aligned}
HH_\mathcal{G}\alpha_{A_1,\ldots,A_k} &= H\alpha_{[\![A_1]\!],\ldots,[\![A_k]\!]}\\
&= \varphi^{-1}_{N^\nu(A_{1H},\ldots,A_{kH})} \circ \alpha_{A_{1H},\ldots,A_{kH}} \circ \varphi_{M^\mu(A_{1H},\ldots,A_{kH})}\\
&= \varphi^{-1}_{N^\nu(A_{1H},\ldots,A_{kH})} \circ N^\nu(\varphi^{-1}_{A_1},\ldots,\varphi^{-1}_{A_k}) \circ \alpha_{A_1,\ldots,A_k} \circ\\
&\qquad\qquad \circ M^\mu(\varphi_{A_1},\ldots,\varphi_{A_k}) \circ \varphi_{M^\mu(A_{1H},\ldots,A_{kH})},\\
&\qquad\qquad \text{since } \varphi \text{ is an isomorphism and } \mathcal{G} \text{ flows through } \mathcal{C}/\mathcal{E},\\
&= \varphi^{-1}_{N^\nu(A_1,\ldots,A_k)} \circ \alpha_{A_1,\ldots,A_k} \circ \varphi_{M^\mu(A_1,\ldots,A_k)}, \text{ by the preordering of } \mathcal{G}.
\end{aligned}$$

In the induction step, for $f^i\colon A^i \vdash B^i$ where $i \in \{1,2\}$, we have

$$HH_{\mathcal{G}}(f^1 \xi f^2) = \varphi^{-1}_{H[\![B^1]\!]\xi H[\![B^2]\!]} \circ (HH_{\mathcal{G}}f^1 \xi HH_{\mathcal{G}}f^2) \circ \varphi_{H[\![A^1]\!]\xi H[\![A^2]\!]}$$
$$= \varphi^{-1}_{B^1_H \xi B^2_H} \circ (\varphi^{-1}_{B^1} \xi \varphi^{-1}_{B^2}) \circ (f^1 \xi f^2) \circ (\varphi_{A^1} \xi \varphi_{A^2}) \circ \varphi_{A^1_H \xi A^2_H},$$
by the induction hypothesis and the bifunctoriality of ξ,
$$= \varphi^{-1}_{B^1 \xi B^2} \circ (f^1 \xi f^2) \circ \varphi_{A^1 \xi A^2},$$ by the preordering of $\mathcal{G}$;

$$HH_{\mathcal{G}}(g \circ f) = \varphi^{-1}_C \circ g \circ \varphi_B \circ \varphi^{-1}_B \circ f \circ \varphi_A,$$ by the induction hypothesis,
$$= \varphi^{-1}_C \circ g \circ f \circ \varphi_A,$$ since φ is an isomorphism. ⊣

We can then prove the following lemma.

LEMMA 2. *If $f_{\mathcal{G}} = g_{\mathcal{G}}$ in $\mathcal{E}_{\mathcal{G}}$, then $Hf_{\mathcal{G}} = Hg_{\mathcal{G}}$ in $\mathcal{C}/\mathcal{E}$.*

PROOF. We proceed by induction on the length of the derivation of $f_{\mathcal{G}} = g_{\mathcal{G}}$ in $\mathcal{E}_{\mathcal{G}}$. If $f_{\mathcal{G}} = g_{\mathcal{G}}$ is in $\mathcal{E}_{\mathcal{G}}$ because $f = g$ is in $\mathcal{E}$, then in $\mathcal{C}/\mathcal{E}$ we have

$$\varphi^{-1}_B \circ f \circ \varphi_A = \varphi^{-1}_B \circ g \circ \varphi_A,$$

and by Lemma 1 we obtain $Hf_{\mathcal{G}} = Hg_{\mathcal{G}}$. If $f_{\mathcal{G}} = \mathbf{1}_{[\![A]\!]}$ is a strictifying equation of $\mathcal{E}_{\mathcal{G}}$, then, by Lemma 1 and the preordering of $\mathcal{G}$, we have $Hf_{\mathcal{G}} = \mathbf{1}_{A_H}$. The induction step, where $f_{\mathcal{G}} = g_{\mathcal{G}}$ is obtained by (sy), (tr) or (co), is straightforward. ⊣

Lemma 2 guarantees that the graph-morphism H from $\mathcal{C}_{\mathcal{G}}$ to $\mathcal{C}$ induces a functor from $\mathcal{C}_{\mathcal{G}}/\mathcal{E}_{\mathcal{G}}$ to $\mathcal{C}/\mathcal{E}$.

Let $\psi^{\Box\xi\Box}_{X,Y}$ be the arrow $\varphi^{-1}_{HX\xi HY}$ of $\mathcal{G}$ and let ψ^{ς} be the arrow φ^{-1}_{ς} of $\mathcal{G}$. Then we have the following theorem.

DIRECT-STRICTIFICATION THEOREM. *If* (IV$\mathcal{C}$) *and* (IV$\mathcal{G}$) *hold and $\mathcal{G}$ is a preorder, then $\mathcal{C}/\mathcal{E}$ is equivalent to the $\mathcal{C}'$-strict $\mathcal{C}/\mathcal{E}$-category $\mathcal{C}_{\mathcal{G}}/\mathcal{E}_{\mathcal{G}}$ via the strict $\mathcal{C}$-functor $H_{\mathcal{G}}$ from $\mathcal{C}/\mathcal{E}$ to $\mathcal{C}_{\mathcal{G}}/\mathcal{E}_{\mathcal{G}}$ and the strong $\mathcal{C}$-functor $\langle H, \psi^{\Box\xi\Box}, \psi^{\varsigma}\rangle$ from $\mathcal{C}_{\mathcal{G}}/\mathcal{E}_{\mathcal{G}}$ to $\mathcal{C}/\mathcal{E}$.*

PROOF. We have $H_{\mathcal{G}}H[\![A]\!] = H_{\mathcal{G}}A_H = [\![A_H]\!] = [\![A]\!]$, and we also have

$$\begin{aligned} H_{\mathcal{G}}Hf_{\mathcal{G}} &= H_{\mathcal{G}}HH_{\mathcal{G}}f \\ &= H_{\mathcal{G}}(\varphi_B^{-1} \circ f \circ \varphi_A), \text{ by Lemma 1,} \\ &= H_{\mathcal{G}}\varphi_B^{-1} \circ H_{\mathcal{G}}f \circ H_{\mathcal{G}}\varphi_A, \\ &= f_{\mathcal{G}}, \text{ by the strictifying equations.} \end{aligned}$$

On the other hand, $HH_{\mathcal{G}}A = A_H$. We have in $\mathcal{C}/\mathcal{E}$, of which $\mathcal{G}$ is a subcategory, the isomorphism $\varphi_A : A_H \vdash A$. That φ is a natural transformation from $HH_{\mathcal{G}}$ to the identity functor follows immediately from Lemma 1. So $\mathcal{C}/\mathcal{E}$ and $\mathcal{C}_{\mathcal{G}}/\mathcal{E}_{\mathcal{G}}$ are equivalent via $H_{\mathcal{G}}$ and H.

It is clear that $H_{\mathcal{G}}$ is a strict $\mathcal{C}$-functor. It remains to show only that $\langle H, \psi^{\Box\xi\Box}, \psi^{\varsigma}\rangle$ is a strong $\mathcal{C}$-functor. That ($\psi^{\Box\xi\Box}$ *nat*) holds is built into the definition of $H(f^1_{\mathcal{G}} \,\xi\, f^2_{\mathcal{G}})$.

To prove ($\psi\alpha L$) for $\langle H, \psi^{\Box\xi\Box}, \psi^{\varsigma}\rangle$, note first that $HM^{\mu}(Z_1, \ldots, Z_k) = (M^{\mu}(HZ_1, \ldots, HZ_k))_H$. So $\psi^{M^{\mu}}_{Z_1,\ldots,Z_k} = \varphi^{-1}_{M^{\mu}(HZ_1,\ldots,HZ_k)}$ by the preordering of $\mathcal{G}$. Then ($\psi\alpha L$) follows by applying ($\psi^{M^{\mu}}$) and the facts that φ is an isomorphism and that $\mathcal{G}$ flows through $\mathcal{C}/\mathcal{E}$. Since it is clear that $\langle H, \psi^{\Box\xi\Box}, \psi^{\varsigma}\rangle$ is a groupoidal $\mathcal{C}$-functor, it follows that it is a strong $\mathcal{C}$-functor. ⊣

As a consequence of this theorem, we have that $f = g$ in $\mathcal{C}/\mathcal{E}$ iff $f_{\mathcal{G}} = g_{\mathcal{G}}$ in $\mathcal{C}_{\mathcal{G}}/\mathcal{E}_{\mathcal{G}}$. So instead of computing in $\mathcal{C}/\mathcal{E}$, we can pass to $\mathcal{C}_{\mathcal{G}}/\mathcal{E}_{\mathcal{G}}$, in which equations between arrow terms are easier to record. By omitting according to (*cat* 1) arrow terms equated with identity arrow terms, equations become shorter. We will avail ourselves of this opportunity provided by the Direct-Strictification Theorem in Chapters 5-8.

Consider the following subcategory $(\mathcal{C}/\mathcal{E})^{at}$ of $(\mathcal{C}/\mathcal{E})^{\mathcal{G}}$, where $(\mathcal{C}/\mathcal{E})^{\mathcal{G}}$ is defined as in the preceding section by taking that $\mathcal{B}$ is $\mathcal{C}/\mathcal{E}$. The objects of $(\mathcal{C}/\mathcal{E})^{at}$ are obtained from all the objects $[p]$ and ς'' of $(\mathcal{C}/\mathcal{E})^{\mathcal{G}}$ by closing under the operations ξ''. The category $(\mathcal{C}/\mathcal{E})^{at}$ is the full subcategory of $(\mathcal{C}/\mathcal{E})^{\mathcal{G}}$ with these objects.

The category $(\mathcal{C}/\mathcal{E})^{at}$ is isomorphic to $\mathcal{C}_{\mathcal{G}}/\mathcal{E}_{\mathcal{G}}$. It is easy to show that there is a bijection between the objects of $(\mathcal{C}/\mathcal{E})^{at}$ and $\mathcal{C}_{\mathcal{G}}/\mathcal{E}_{\mathcal{G}}$. For the arrows of these two categories we have the following bijections. To every arrow $(f, [A], [B])$ of $(\mathcal{C}/\mathcal{E})^{at}$, with $f : A_F \vdash B_F$ an arrow of $\mathcal{C}/\mathcal{E}$, we assign the arrow $H_{\mathcal{G}}f : [\![A_F]\!] \vdash [\![B_F]\!]$ of $\mathcal{C}_{\mathcal{G}}/\mathcal{E}_{\mathcal{G}}$; and to every arrow $f_{\mathcal{G}} : [\![A]\!] \vdash [\![B]\!]$ of

$\mathcal{C}_\mathcal{G}/\mathcal{E}_\mathcal{G}$ we assign the arrow $(\varphi_{B_F} \circ Hf_\mathcal{G} \circ \varphi^{-1}_{A_F} : A_F \vdash B_F, X, Y)$ of $(\mathcal{C}/\mathcal{E})^{at}$, where X and Y are obtained from $\llbracket A \rrbracket$ and $\llbracket B \rrbracket$, respectively, by replacing ξ and ζ in A and B by $\xi^\mathcal{G}$ and $\zeta^\mathcal{G}$, and the brackets $\rrbracket$ and $\llbracket$ by $]$ and $[$.

It seems rather natural to assume that a construction like our construction of $\mathcal{C}_\mathcal{G}/\mathcal{E}_\mathcal{G}$ out of $\mathcal{C}/\mathcal{E}$ will yield a category equivalent to $\mathcal{C}/\mathcal{E}$, and indeed this assumption may have been made tacitly by Mac Lane in [93] (proof of Theorem 4.2, p. 39) and [96] (Section XI.1, proof of Theorem 1, p. 254), where he establishes coherence for symmetric monoidal categories. A rather obvious interpretation of his text is that he constructs $\mathcal{C}_\mathcal{G}/\mathcal{E}_\mathcal{G}$ out of the symmetric monoidal category $\mathcal{C}/\mathcal{E}$ freely generated by $\mathcal{P}$. A more explicit assumption of a construction similar to our construction of $\mathcal{C}_\mathcal{G}/\mathcal{E}_\mathcal{G}$ out of $\mathcal{C}/\mathcal{E}$ is in [110] (proof of Theorem 6.1, p. 98; no details are given, and no justification that the construction will yield an equivalent category).

Though the assumption that $\mathcal{C}/\mathcal{E}$ and $\mathcal{C}_\mathcal{G}/\mathcal{E}_\mathcal{G}$ are equivalent is natural, this assumption is not warranted without assumptions concerning $\mathcal{C}/\mathcal{E}$, like our assumptions (IV$\mathcal{C}$), (IV$\mathcal{G}$) and the condition that $\mathcal{G}$ is a preorder. The main assumption here is that $\mathcal{C}/\mathcal{E}$ is freely generated, out of a generating set $\mathcal{P}$. This set can be conceived as a discrete category, and one may envisage free generation also out of other categories (which does not differ significantly from what we have been doing). Free generation is, however, essential.

Without our assumptions, or assumptions of the same kind, a construction analogous to our construction of $\mathcal{C}_\mathcal{G}/\mathcal{E}_\mathcal{G}$ out of $\mathcal{C}/\mathcal{E}$ need not yield an equivalent category, as it is shown by the following counterexample, which stems from Isbell (see [94], Section VII.1, p. 160).

Consider the logical system $\mathcal{C}$ in $\mathcal{L}_\wedge$ whose primitive arrow terms besides identity arrow terms are from the $\hat{b}$ and $\hat{w}$-$\hat{k}$ families, and consider the equational system $\mathcal{E}$ in $\mathcal{C}$ for which we assume all the equations that hold in the category $\hat{\mathbf{A}}$ (see §4.3; these are equations that hold in monoidal categories), the equations ($\hat{k}^1$ *nat*) and ($\hat{k}^2$ *nat*) (see §2.7) and, for $i \in \{1, 2\}$, the following additional equations (which hold in cartesian categories, and which we will encounter in §9.1):

$$(\hat{w}\hat{k}) \quad \hat{k}^i_{A,A} \circ \hat{w}_A = \mathbf{1}_A.$$

We build now a syntactical system $\mathcal{S}$ of the $\mathcal{C}$-kind that has a single object N. The operation $\wedge$ on the objects of $\mathcal{S}$ satisfies $N \wedge N = N$. The primitive arrow terms of $\mathcal{S}$ are $\mathbf{1}_N$, $\hat{b}^{\rightarrow}_{N,N,N}$, $\hat{b}^{\leftarrow}_{N,N,N}$, $\hat{k}^1_{N,N}$, $\hat{k}^2_{N,N}$ and $\hat{w}_N$, all of type $N \vdash N$. With the help of the equational system $\mathcal{E}$ above we obtain the syntactical category $\mathcal{S}/\mathcal{E}$, which is a $\mathcal{C}/\mathcal{E}$-category.

We want to show first that $\mathcal{S}/\mathcal{E}$ is not a preorder. Consider the skeleton *Card* of the category *Set* of sets with functions such that the set of natural numbers $\boldsymbol{N}$ is in *Card*. The category *Card* is equivalent to *Set*.

Consider then the functor S that maps $\mathcal{S}/\mathcal{E}$ into *Card* such that SN is $\boldsymbol{N}$, while for $\iota\colon \boldsymbol{N} \vdash \boldsymbol{N} \times \boldsymbol{N}$ being a chosen isomorphism between $\boldsymbol{N}$ and $\boldsymbol{N} \times \boldsymbol{N}$ we have $S\hat{k}^i_{N,N} = \hat{k}^i_{\boldsymbol{N},\boldsymbol{N}} \circ \iota$, where $\hat{k}^1_{\boldsymbol{N},\boldsymbol{N}}$ and $\hat{k}^2_{\boldsymbol{N},\boldsymbol{N}}$ are respectively the first and second projection in *Set*, and $S\hat{w}_N = \iota^{-1} \circ \hat{w}_{\boldsymbol{N}}$, where $\hat{w}_{\boldsymbol{N}}$ is the diagonal map of *Set* for which $\hat{w}_{\boldsymbol{N}}(n) = (n, n)$. The image of $\mathcal{S}/\mathcal{E}$ in *Card* under S is not a preorder, and hence $\mathcal{S}/\mathcal{E}$ is not a preorder.

For $\mathcal{C}'$ being the logical subsystem of $\mathcal{C}$ with the $\hat{b}$ arrow terms, let $\mathcal{G} = \mathcal{C}'/\mathcal{E}'$ be the $\mathcal{C}'$-core of $\mathcal{C}/\mathcal{E}$. Let us build out of the equations of $\mathcal{E}$ a category $\mathcal{S}_{\mathcal{G}}/\mathcal{E}_{\mathcal{G}}$, analogously to what we did to obtain $\mathcal{C}_{\mathcal{G}}/\mathcal{E}_{\mathcal{G}}$ out of $\mathcal{C}/\mathcal{E}$, which boils down to adding the equations $\hat{b}^{\rightarrow}_{N,N,N} = \hat{b}^{\leftarrow}_{N,N,N} = \mathbf{1}_N$ to $\mathcal{E}$ to obtain $\mathcal{E}_{\mathcal{G}}$. The unique object of $\mathcal{S}_{\mathcal{G}}/\mathcal{E}_{\mathcal{G}}$ may be identified with N. We have in $\mathcal{S}_{\mathcal{G}}/\mathcal{E}_{\mathcal{G}}$

$$\begin{aligned}
&(f \wedge g) \wedge h = f \wedge (g \wedge h), \;\; \text{by } (\hat{b}\ \mathit{nat}) \text{ and } (\mathit{cat}\ 1),\\
&(f \wedge g) \circ \hat{k}^1_{N,N} = f \circ \hat{k}^1_{N,N}, \;\; \text{by } (\hat{k}^1\ \mathit{nat}),\\
&f \wedge g = f, \;\; \text{by } (\hat{w}\hat{k}).
\end{aligned}$$

We derive analogously $f \wedge g = g$, starting from $h \wedge (f \wedge g) = (h \wedge f) \wedge g$ and by using $(\hat{k}^2\ \mathit{nat})$. So $f = g$, and $\mathcal{S}_{\mathcal{G}}/\mathcal{E}_{\mathcal{G}}$ is a preorder. Hence $\mathcal{S}_{\mathcal{G}}/\mathcal{E}_{\mathcal{G}}$ is not equivalent to $\mathcal{S}/\mathcal{E}$. This shows that direct strictification is not always innocuous.

Note that $\mathcal{S}/\mathcal{E}$ is not freely generated out of a set $\mathcal{P}$, and does not satisfy our assumption (IV$\mathcal{C}$), since it is not a logical category. The strictification of $\mathcal{S}/\mathcal{E}$ as a $\mathcal{C}/\mathcal{E}$-category, in the sense of the Strictification Theorem of the preceding section, is however allowed. The category $(\mathcal{S}/\mathcal{E})^{\mathcal{G}}$ is not a preorder, and it is equivalent to $\mathcal{S}/\mathcal{E}$. For $(\mathcal{S}/\mathcal{E})^{\mathcal{G}}$ we cannot, however, find its $(\mathcal{S}/\mathcal{E})^{at}$ subcategory.

§3.3. Strictification and diversification

For A and B formulae of a language $\mathcal{L}$, let the type $A \vdash B$ be called *balanced* when there is a bijection between the occurrences of letters in A and the occurrences of letters in B that maps the occurrence of a letter to an occurrence of the same letter. Let $\mathcal{C}$ be a logical system in $\mathcal{L}$ such that for each transformation α of $\mathcal{C}$ the type of the arrow term $\alpha_{A_1,\ldots,A_k}$: $M^\mu(A_1,\ldots,A_k) \vdash N^\nu(A_1,\ldots,A_k)$ is balanced. This is guaranteed if μ and ν are bijections. It is easy to show by induction that the type of every arrow term of $\mathcal{C}$ is balanced.

Let a formula A of $\mathcal{L}$ be called *diversified* when every letter occurs in A at most once. It is clear that for a balanced type $A \vdash B$ we have that A is diversified iff B is diversified. A type $A \vdash B$ is called *diversified* when A and B are diversified, and an arrow term is *diversified* when its type is diversified.

Let the conditions (IV$\mathcal{C}$) and (IV$\mathcal{G}$) of §3.1 be satisfied. Let $\mathcal{E}^{pr}$ be an equational system that is an extension of the equational system $\mathcal{E}$ such that $\mathcal{C}'/\mathcal{E}^{pr\prime}$ is a preorder and for every equation $f = g$ in $\mathcal{E}^{pr}$ that is not in $\mathcal{E}$ we have that the type of f and g is not diversified.

Let $(\mathcal{C}/\mathcal{E})^{div}$ be the full subcategory of $\mathcal{C}/\mathcal{E}$ whose objects are the diversified formulae of $\mathcal{L}$, and let $(\mathcal{C}/\mathcal{E}^{pr})^{div}$ be the analogous full subcategory of $\mathcal{C}/\mathcal{E}^{pr}$ whose objects are all the diversified formulae of $\mathcal{L}$.

Then it is straightforward to show that $(\mathcal{C}/\mathcal{E})^{div}$ and $(\mathcal{C}/\mathcal{E}^{pr})^{div}$ are isomorphic categories. On objects, this isomorphism is just identity, and the identity map on the arrow terms of $\mathcal{C}$ gives rise to a functor from $(\mathcal{C}/\mathcal{E})^{div}$ to $(\mathcal{C}/\mathcal{E}^{pr})^{div}$ and to a functor from $(\mathcal{C}/\mathcal{E}^{pr})^{div}$ to $(\mathcal{C}/\mathcal{E})^{div}$. To show the latter, it is enough to appeal to the fact that if $f = g$ in $\mathcal{E}^{pr}$ for f and g diversified, then $f = g$ in $\mathcal{E}$.

Then we can check that (IV$\mathcal{C}$) and (IV$\mathcal{G}$) hold when $\mathcal{C}/\mathcal{E}$ is replaced by $\mathcal{C}/\mathcal{E}^{pr}$. Now the $\mathcal{C}'$-core $\mathcal{G}$ of $\mathcal{C}/\mathcal{E}^{pr}$ is a preorder. So we can apply the Direct-Strictification Theorem of the preceding section to obtain the $\mathcal{C}'$-strict $\mathcal{C}/\mathcal{E}^{pr}$-category $\mathcal{C}_\mathcal{G}/\mathcal{E}_\mathcal{G}{}^{pr}$ equivalent to $\mathcal{C}/\mathcal{E}^{pr}$.

So, for diversified arrow terms f and g of $\mathcal{C}$ of the same type, we have $f = g$ in $\mathcal{C}/\mathcal{E}$ iff $f = g$ in $(\mathcal{C}/\mathcal{E})^{div}$, iff $f = g$ in $(\mathcal{C}/\mathcal{E}^{pr})^{div}$, iff $f = g$ in $\mathcal{C}/\mathcal{E}^{pr}$, iff $f_\mathcal{G} = g_\mathcal{G}$ in $\mathcal{C}_\mathcal{G}/\mathcal{E}_\mathcal{G}{}^{pr}$. And it is easier to compute in $\mathcal{C}_\mathcal{G}/\mathcal{E}_\mathcal{G}{}^{pr}$, as explained after the Direct-Strictification Theorem. We will take advantage of that in §§7.6-8 and §8.4.

CHAPTER 4

ASSOCIATIVE CATEGORIES

In this chapter we scrutinize Mac Lane's proof of coherence, in the sense of preordering, for monoidal categories (see [93] and [94], Section VII.2), and present a differently organized proof, making finer distinctions. We separate from this proof a proof of coherence for categories like monoidal categories that lack the unit object, but, in this respect, we do not differ from Mac Lane, who did the same in [93]. Throughout the book, it will be our policy to proceed in this manner, by separating coherence results with and without special objects such as unit objects. Besides obtaining sharper results in situations where we have coherence both with and without the special objects, this policy allows us to obtain coherence without the special objects in cases where adding the special objects causes difficulties.

As a new result, we have a proof of coherence for subcategories of monoidal categories where associativity arrows are not isomorphisms. Associativity goes just in one direction, and for coherence we need just naturality and Mac Lane's pentagonal coherence condition. The proof of Mac Lane's monoidal coherence may be built on this more basic coherence result, but it has also a shorter proof, such as Mac Lane's. Associativity that is not an isomorphism is interesting because of its relationship with dissociativity investigated from Chapter 7 on. Dissociativity is an associativity principle involving two operations, which is not an isomorphism.

We also explain in this chapter the effect of strictifying the monoidal structure of a category (which may have extra structure besides this monoidal structure), in accordance with the results of the preceding chapter. The methods of this chapter are based, as Mac Lane's, on confluence techniques, like those that may be found in the lambda calculus.

§4.1. The logical categories $\mathcal{K}$

For $\mathcal{C}$ a logical system in $\mathcal{L}$, let $\mathcal{E}_{\mathcal{C}}^{nat}$ be the least set of equations we must have in every equational system in $\mathcal{C}$ to make $\mathcal{C}/\mathcal{E}_{\mathcal{C}}^{nat}$ a natural logical category (see §2.7). So $\mathcal{E}_{\mathcal{C}}^{nat}$ has as axioms (re) (see §2.3), the categorial equations, the bifunctorial equations for every ξ of $\mathcal{L}$ and the naturality equations for every α of $\mathcal{C}$.

We will consider in this work a number of natural logical categories $\mathcal{K}$. Every such $\mathcal{K}$ will be $\mathcal{C}(\mathcal{K})/\mathcal{E}(\mathcal{K})$ for a logical system $\mathcal{C}(\mathcal{K})$ and an equational system $\mathcal{E}(\mathcal{K})$. To determine $\mathcal{C}(\mathcal{K})$ we will have to specify only the language $\mathcal{L}$ of $\mathcal{C}(\mathcal{K})$ and the transformations α of $\mathcal{C}(\mathcal{K})$. To determine $\mathcal{E}(\mathcal{K})$ it is enough to specify what equations besides those in $\mathcal{E}_{\mathcal{C}(\mathcal{K})}^{nat}$ have to be assumed as axioms. We call these equations *specific equations.* We always take for granted closure of the arrow terms of $\mathcal{C}(\mathcal{K})$ under the operations ξ and composition, the presence in $\mathcal{E}(\mathcal{K})$ of (re) and of bifunctorial and naturality equations, and the closure of $\mathcal{E}(\mathcal{K})$ under (sy), (tr), (co) and (su) of §2.3 and §2.7.

We will first deal with a number of categories $\mathcal{K}$ in the language $\mathcal{L}_{\wedge}$. These will make a hierarchy by having transformations α included in subfamilies involving $\wedge$ of the families specified below. The label $\mathcal{K}$ in such cases is as in the following table, sometimes with additional indices:

$\mathcal{K}$	**I**	**A**	**S**	**L**
α	**1**	**1**, b	**1**, b, c	**1**, b, c, w-k

When we come to natural logical categories $\mathcal{K}$ in $\mathcal{L}_{\wedge,\vee}$, we take the whole families mentioned. The label **I** is derived from "identity", **A** from "associativity", **S** from "symmetry", and **L** from "lattice".

If $\mathcal{K}$ is one of our logical categories in a language $\mathcal{L}$ without $\top$ and $\bot$, then $\mathcal{K}_{\top}$, $\mathcal{K}_{\bot}$ or $\mathcal{K}_{\top,\bot}$ will be obtained from $\mathcal{K}$ by adding to $\mathcal{L}$ either $\top$ or $\bot$, or both, and by adding to $\mathcal{C}(\mathcal{K})$ transformations α included in δ-σ, as appropriate. To obtain $\mathcal{E}(\mathcal{K}_{\top})$, $\mathcal{E}(\mathcal{K}_{\bot})$ or $\mathcal{E}(\mathcal{K}_{\top,\bot})$, we enlarge $\mathcal{E}(\mathcal{K})$. Our categories $\mathcal{K}$ where $\mathcal{C}(\mathcal{K})$ has transformations included in the family d will be named by prefixing **D** to **I**, **A**, **S** and **L**; in those where $\mathcal{C}(\mathcal{K})$ has m we prefix **M**, which comes from "mix", and where $\mathcal{C}(\mathcal{K})$ has m^{-1} we prefix **Z**, which comes from "zero". The categories $\mathcal{K}$ we will deal with are presented in the List of Categories at the end of the book.

It is easy to see that if we have proved coherence for the category $\mathcal{K}$ generated by an infinite set of letters $\mathcal{P}$, then we have proved coherence also for $\mathcal{K}$ generated by any set of letters $\mathcal{P}$. (If not more than $n \geq 0$ different letters occur in $f = g$, then, by substituting, every derivation of $f = g$ can be transformed into one in which not more than n different letters occur.) So we assume, when this is needed to prove coherence, that $\mathcal{K}$ is generated by an infinite set of letters $\mathcal{P}$.

Our first logical category $\mathcal{K}$ will be called $\hat{\mathbf{I}}$. The logical system $\mathcal{C}(\hat{\mathbf{I}})$ is in the language $\mathcal{L}_\wedge$ and its only transformation α is $\mathbf{1}$. The equational system $\mathcal{E}(\hat{\mathbf{I}})$ is just $\mathcal{E}^{nat}_{\mathcal{C}}$ for $\mathcal{C}$ being $\mathcal{C}(\hat{\mathbf{I}})$, with no additional equations. (The naturality equations follow here from categorial equations.) It is trivial to show that $\hat{\mathbf{I}}$ is discrete, and hence a preorder. So $\hat{\mathbf{I}}$ is coherent (see the end of Chapter 2).

A logical category closely related to $\hat{\mathbf{I}}$ is $\mathbf{I}$, where $\mathcal{C}(\mathbf{I})$ is in the language $\mathcal{L}_{\wedge,\vee}$, and where the only transformation α is again $\mathbf{1}$. The equational system $\mathcal{E}(\mathbf{I})$ is just $\mathcal{E}^{nat}_{\mathcal{C}(\mathbf{I})}$, and, as trivially as for $\hat{\mathbf{I}}$, we show that $\mathbf{I}$ is discrete. So $\mathbf{I}$ is coherent too.

§4.2. Coherence of semiassociative categories

For the label $\mathcal{K}$ of the preceding section being $\hat{\mathbf{A}}^{\rightarrow}$, let the logical system $\mathcal{C}(\hat{\mathbf{A}}^{\rightarrow})$ be in $\mathcal{L}_\wedge$ with the transformations α being $\mathbf{1}$ and $\hat{b}^{\rightarrow}$. The specific equations of $\mathcal{E}(\hat{\mathbf{A}}^{\rightarrow})$ are the instances of

$$(\hat{b}5) \quad \hat{b}^{\rightarrow}_{A\wedge B,C,D} \circ \hat{b}^{\rightarrow}_{A,B,C\wedge D} = (\hat{b}^{\rightarrow}_{A,B,C} \wedge \mathbf{1}_D) \circ \hat{b}^{\rightarrow}_{A,B\wedge C,D} \circ (\mathbf{1}_A \wedge \hat{b}^{\rightarrow}_{B,C,D}).$$

This is Mac Lane's *pentagonal* equation of [93] (Section 3; see also [94], Section VII.1).

We call natural $\hat{\mathbf{A}}^{\rightarrow}$-categories (in the sense of §2.8) *semiassociative* categories. Semiassociative categories differ from Mac Lane's monoidal categories by lacking $\hat{b}^{\leftarrow}$ and $\top$. We are now going to prove coherence for the category $\hat{\mathbf{A}}^{\rightarrow}$.

For every formula A of $\mathcal{L}$ and for x and y two different occurrences of $\wedge$ in A we write $xR_A y$ when BxC is a subformula of A such that y occurs in C. If A and B are formulae of $\mathcal{L}_\wedge$ that may differ only with respect to parentheses, then A and B are comparable (see §2.1), and we may take that R_A and R_B are relations between the same sets of occurrences of $\wedge$, and compare these relations. Formally, we could proceed as follows. Let $w(A)$ be, as in §2.1, the word obtained from A by deleting all parentheses. Then R_A gives rise to a relation R_{w_A} between occurrences of $\wedge$ in $w(A)$ such that we have $x' R_{w_A} y'$ iff x' and y' are occurrences of $\wedge$ in $w(A)$ corresponding respectively to the occurrences x and y of $\wedge$ in A and $x R_A y$. Then we do not compare R_A and R_B, but R_{w_A} and R_{w_B}. However, to switch all the time from R_A to R_{w_A} and back would be tedious, and we will not mention R_{w_A}. It is easy to see that for every arrow term $f\colon A \vdash B$ of $\mathcal{C}(\hat{\mathbf{A}}^{\rightarrow})$ the formulae A and B are comparable (namely, $w(A)$ and $w(B)$ are the same word), and $R_B \subseteq R_A$ (which means, officially, $R_{w_B} \subseteq R_{w_A}$). Moreover, if $\hat{b}^{\rightarrow}$ occurs in f, then R_B must be a proper subset of R_A; otherwise, $R_A = R_B$.

Since R_A and R_B are conceived as defined on the same sets when A and B are comparable, we may denote with the same symbol x the occurrences of $\wedge$ or of a letter in A and B that are at the same place (for the notion of place see §2.1). We proceed analogously in other similar cases in the future (cf. §7.1, §7.3, §7.5 and §8.3). We can prove the following.

EXTRACTION LEMMA. *Let A be a formula of $\mathcal{L}_\wedge$ with a subword $A_1 \wedge (^m q$, where $(^m$ stands for a sequence of $m \geq 0$ left parentheses. Then there is an arrow term $g\colon A \vdash C$ of $\mathcal{C}(\hat{\mathbf{A}}^{\rightarrow})$ such that in C we have as a subword $A_1 \wedge q$ at the place where A has $A_1 \wedge (^m q$. In addition,*

(∗) *if x is not the occurrence of $\wedge$ in the two subwords above, then $x R_A y$ implies $x R_C y$;*

(∗∗) *the first index of every $\hat{b}^{\rightarrow}$ in g is A_1.*

PROOF. We proceed by induction on m. If $m = 0$, then g is $\mathbf{1}_A : A \vdash A$. Suppose now $m > 0$. Then in A we have a subword of the form $A_1 \wedge (A_2 \wedge A_3)$ where A_2 is either q or beginning with $(^{m-1} q$. Then there is a $\hat{b}^{\rightarrow}$-term $h\colon A \vdash A'$, whose head is $\hat{b}^{\rightarrow}_{A_1,A_2,A_3}$. In A' we have $(A_1 \wedge A_2) \wedge A_3$ at the place where A has $A_1 \wedge (A_2 \wedge A_3)$, and hence in A' we have $A_1 \wedge (^{m-1} q$ at the place where A has $A_1 \wedge (^m q$. We apply the induction hypothesis to

A', and obtain an arrow term $g'\colon A' \vdash C$ of $\mathcal{C}(\hat{\mathbf{A}}^{\rightarrow})$ such that in C we have $A_1 \wedge q$ at the place where A has $A_1 \wedge (^m q$, and if x is not our occurrence of $\wedge$, then $xR_{A'}y$ implies $xR_C y$.

Suppose now x is not our occurrence of $\wedge$, and suppose $xR_A y$. Then we can conclude that $xR_{A'}y$, and we take that g is $g' \circ h : A \vdash C$. From this the lemma follows. ⊣

THEOREMHOOD PROPOSITION. *There is an arrow term* $f\colon A \vdash B$ *of* $\mathcal{C}(\hat{\mathbf{A}}^{\rightarrow})$ *iff* A *and* B *are comparable formulae of* $\mathcal{L}_{\wedge}$ *and* $R_B \subseteq R_A$.

PROOF. The direction from left to right is easy, as we noted above. For the other direction, we proceed by induction on the letter length $n \geq 1$ of A, which is equal to the letter length of B, because A and B are comparable. If $n = 1$, then $R_A = R_B = \emptyset$, and f is $\mathbf{1}_p\colon p \vdash p$.

Let $n > 1$. So in B there must be a subword of the form $(p \wedge q)$. Then we show that $R_B \subseteq R_A$ implies that A must have at the same place a subword $p \wedge (^m q$ for $m \geq 0$. Otherwise, A would have a subword $\wedge p)^n) \wedge (^m q$ with $n \geq 0$ at the place where B has $\wedge(^l(p \wedge q)$ with $l \geq 0$. From that it would follow that $R_B \not\subseteq R_A$, since for x being the left $\wedge$ in $\wedge(^l(p \wedge q)$ and y the right $\wedge$ we have $xR_B y$, but we do not have $xR_A y$.

Then, by the Extraction Lemma, there is an arrow term $g : A \vdash C$ of $\mathcal{C}(\hat{\mathbf{A}}^{\rightarrow})$ such that in C we have as a subword $(p \wedge q)$ at the place where B has $(p \wedge q)$, and, by $(*)$, if x is not the occurrence of $\wedge$ in $(p \wedge q)$, then $xR_A y$ implies $xR_C y$. We replace $(p \wedge q)$ in B and C by r, and obtain respectively B^r and C^r. We have $R_{B^r} \subseteq R_{C^r}$, and, by the induction hypothesis, there is an arrow term $f' : C^r \vdash B^r$ of $\mathcal{C}(\hat{\mathbf{A}}^{\rightarrow})$. Let $f'' : C \vdash B$ be obtained from this arrow term by putting back $(p \wedge q)$ at the place of r. Then f is $f'' \circ g\colon A \vdash B$. ⊣

It is clear that there is an arrow term of a given type $A \vdash B$ in a logical system $\mathcal{C}$ iff there is an arrow of type $A \vdash B$ in $\mathcal{C}/\mathcal{E}$. So, in the terminology of §1.1, with the Theoremhood Proposition we obtain a solution to the theoremhood problem for the category $\hat{\mathbf{A}}^{\rightarrow}$. The questions whether A and B are comparable and whether $R_B \subseteq R_A$ are clearly decidable. This is not a very difficult theoremhood problem, and we deal with it not so much because of its intrinsic interest, but because we need it for the proof of Semiassociative Coherence below. Besides that, it is a good introduction to our analogous treatment of other theoremhood problems in §7.1, §§7.3-5 and §8.3, some of which are less trivial. As here, analogues of the Theoremhood Proposition will be applied in establishing coherence results.

Note that the existential quantifier in the Theoremhood Proposition, as well as the existential quantifier in the Extraction Lemma, is constructive; namely, when the conditions are satisfied, we can actually construct the arrow term of the required type. This applies also to latter versions of the Extraction Lemma and of the Theoremhood Proposition.

Let $d(A)$ be the cardinality of the set of ordered pairs R_A. If $f\colon A \vdash B$ of $\hat{\mathbf{A}}^{\rightarrow}$ is not equal to $\mathbf{1}_A : A \vdash A$, then R_B is a proper subset of R_A and $d(B) < d(A)$. We can then prove the following.

SEMIASSOCIATIVE COHERENCE. *The category $\hat{\mathbf{A}}^{\rightarrow}$ is a preorder.*

PROOF. Let $f, g : A \vdash B$ be arrow terms of $\mathcal{C}(\hat{\mathbf{A}}^{\rightarrow})$. We proceed by induction on $d(A) - d(B)$ to show that $f = g$ in $\hat{\mathbf{A}}^{\rightarrow}$. (Until the end of this proof, we assume that equality between arrow terms is equality in $\hat{\mathbf{A}}^{\rightarrow}$.) If $d(A) = d(B)$, then we conclude that A and B are the same formula, and $f = g = \mathbf{1}_A$.

Suppose $d(B) < d(A)$. By the Development Lemma (see §2.7) we have that $f = f_2 \circ f_1$ and $g = g_2 \circ g_1$ for some $\hat{b}^{\rightarrow}$-terms $f_1 : A \vdash C$ and $g_1 : A \vdash D$, and some arrow terms $f_2 : C \vdash B$ and $g_2 : D \vdash B$ of $\mathcal{C}(\hat{\mathbf{A}}^{\rightarrow})$. We have here $d(C), d(D) < d(A)$. Let the head of f_1 be $\hat{b}^{\rightarrow}_{E,F,G}$, and let the head of g_1 be $\hat{b}^{\rightarrow}_{H,I,J}$. The following cases may arise.

(1) The formulae $E \wedge (F \wedge G)$ and $H \wedge (I \wedge J)$ have no occurrences of letters in common within A. Then we use $(\wedge\ 2)$ of §2.7 to obtain two $\hat{b}^{\rightarrow}$-terms $f_2' : C \vdash B'$ and $g_2' : D \vdash B'$ such that $f_2' \circ f_1 = g_2' \circ g_1$. We infer that $R_C \cap R_D = R_{B'}$, from which it follows by the Theoremhood Proposition that $R_B \subseteq R_{B'}$. Hence, again by the Theoremhood Proposition, there is an arrow term $h\colon B' \vdash B$ of $\mathcal{C}(\hat{\mathbf{A}}^{\rightarrow})$. By applying the induction hypothesis, we obtain that $f_2 = h \circ f_2'$ and $g_2 = h \circ g_2'$, from which $f = g$ follows.

(2) Suppose $E \wedge (F \wedge G)$ is a subformula of H or I or J in A; or, conversely,

$H \wedge (I \wedge J)$ is a subformula of E or F or G in A. Then we proceed as in case (1) by using the equation ($\hat{b}^{\rightarrow}$ *nat*).

(3) The subformulae $E \wedge (F \wedge G)$ and $H \wedge (I \wedge J)$ coincide in A. Then C is D and $f_1 = g_1$. We then apply the induction hypothesis to $f_2, g_2 \colon C \vdash B$ and obtain $f = g$.

(4) The subformula $F \wedge G$ is $H \wedge (I \wedge J)$ or $I \wedge J$ is $E \wedge (F \wedge G)$. Then we proceed as in case (1) by using the equation ($\hat{b}5$). ⊣

The technique used in the proof above is related to the *Church-Rosser*, or *confluence*, property of reductions in the lambda calculus (see [4], Chapter 3). Analogous techniques will be exploited in §7.1, §7.3, §7.5 and §8.3, where one finds proofs of coherence analogous to our proof of Semiassociative Coherence.

It is not difficult to see that $R_A = R_B$ implies that A and B are the same formula of $\mathcal{L}_\wedge$. Because, if $R_A = R_B$, then, by the Theoremhood Proposition, there is an arrow term $f \colon A \vdash B$ of $\mathcal{C}(\hat{\mathbf{A}}^{\rightarrow})$, in which $\hat{b}^{\rightarrow}$ cannot occur, since R_B is not a proper subset of R_A. Hence f must stand for an identity arrow. So there is a bijection between the formulae A of $\mathcal{L}_\wedge$ and the relations R_A, that is R_{w_A}. (A relation is not just a set of ordered pairs, but its domain and codomain must be specified; so $w(A)$ will be mentioned in specifying R_{w_A}.) From Semiassociative Coherence, we can conclude that $\hat{\mathbf{A}}^{\rightarrow}$ is isomorphic to the category whose objects are the relations R_A, and where an arrow exists between R_A and R_B when $R_B \subseteq R_A$. Note that this means that the category $\hat{\mathbf{A}}^{\rightarrow}$ is not just a preorder, but a partial order.

§4.3. Coherence of associative categories

To obtain the natural logical category $\hat{\mathbf{A}}$, we have that the logical system $\mathcal{C}(\hat{\mathbf{A}})$ is in $\mathcal{L}_\wedge$, with the transformations α included in $\mathbf{1}$ and $\hat{b}$. The specific equations of $\mathcal{E}(\hat{\mathbf{A}})$ are those of $\mathcal{E}(\hat{\mathbf{A}}^{\rightarrow})$ plus

$$(\hat{b}\hat{b}) \qquad \hat{b}^{\leftarrow}_{A,B,C} \circ \hat{b}^{\rightarrow}_{A,B,C} = \mathbf{1}_{A\wedge(B\wedge C)}, \qquad \hat{b}^{\rightarrow}_{A,B,C} \circ \hat{b}^{\leftarrow}_{A,B,C} = \mathbf{1}_{(A\wedge B)\wedge C}.$$

Note that it is enough to assume one of the equations ($\hat{b}^{\rightarrow}$ *nat*) and ($\hat{b}^{\leftarrow}$ *nat*) to derive the other one with the help of ($\hat{b}\hat{b}$), and ($\hat{b}\hat{b}$) enables us also to derive an equation analogous to ($\hat{b}5$) involving $\hat{b}^{\leftarrow}$. The equations

$(\hat{b}\hat{b})$, together with $(\hat{b}^{\rightarrow}\ nat)$ and $(\hat{b}^{\leftarrow}\ nat)$, say that $\hat{b}^{\rightarrow}$ and $\hat{b}^{\leftarrow}$ are natural isomorphisms.

We call natural $\hat{\mathbf{A}}$-categories *associative* categories. Associative categories are not necessarily monoidal in the sense of [94] (Section VII.1), because they may lack the unit object (see §4.6). The objects of an associative category that is a partial order make a semigroup.

A formula A of $\mathcal{L}_{\wedge}$ is said to be in *normal form* when R_A, defined as in the preceding section, is empty; i.e., when $d(A) = 0$. Such an A is of the form $(\ldots((p_1 \wedge p_2) \wedge p_3) \wedge \ldots \wedge p_n)$.

We can show that $\hat{\mathbf{A}}$ is a preorder by relying on the proof of Semiassociative Coherence of the preceding section, based on the Theoremhood Proposition, but we can establish more easily another, weaker, lemma. To formulate this lemma, we say that the arrow terms of $\mathcal{C}(\hat{\mathbf{A}})$ that are also arrow terms of $\mathcal{C}(\hat{\mathbf{A}}^{\rightarrow})$ are $\rightarrow$*-directed.* (This terminology will be extended later to arrow terms other than those of $\mathcal{C}(\hat{\mathbf{A}})$; see §4.6, §6.1 and §14.1.) Then the following lemma holds in the category $\hat{\mathbf{A}}^{\rightarrow}$, and hence also in $\hat{\mathbf{A}}$.

DIRECTEDNESS LEMMA. *If $f, g \colon A \vdash B$ are $\rightarrow$-directed arrow terms and B is in normal form, then $f = g$.*

The proof of this lemma, which is due to Mac Lane (see [93], Section 3), is a simplification of our proof of Semiassociative Coherence in the preceding section. The simplification consists in not having to refer to the full force of the Theoremhood Proposition, but only to a trivial case of it where R_B is empty.

Then we prove the following result of [93] (Section 3).

ASSOCIATIVE COHERENCE. *The category $\hat{\mathbf{A}}$ is a preorder.*

PROOF. For $f : A \vdash B$ an arrow term of $\mathcal{C}(\hat{\mathbf{A}})$, there are two $\rightarrow$-directed arrow terms $g \colon A \vdash C$ and $h \colon B \vdash C$ such that C is in normal form (these arrow terms are not uniquely determined). By the Development Lemma (see §2.7), the arrow term f is equal to a developed arrow term $f_n \circ \ldots \circ f_1$. We proceed by induction on n to show that $f = h^{-1} \circ g$, where h^{-1} is obtained from the arrow term h by inverting order in composition, and by replacing $\hat{b}^{\rightarrow}$ by $\hat{b}^{\leftarrow}$ and vice versa.

If $n = 1$, then, since f_1 is $\mathbf{1}_A$, by the Directedness Lemma we have $g = h$, from which $f = h^{-1} \circ g$ follows.

For $n > 1$ and $f_n : B' \vdash B$, we have by the induction hypothesis that $f_{n-1} \circ \ldots \circ f_1 : A \vdash B'$ is equal to $(h')^{-1} \circ g$ for $g : A \vdash C$ and $h' : B' \vdash C$. If f_n is a $\hat{b}^{\rightarrow}$-term, then, for $h : B \vdash C$, by the Directedness Lemma we have $h \circ f_n = h'$, and $f = h^{-1} \circ g$ follows. If f_n is a $\hat{b}^{\leftarrow}$-term, then by the Directedness Lemma we have $h' \circ f_n^{-1} = h$, and $f = h^{-1} \circ g$ follows again.

For $f' : A \vdash B$ we obtain in the same manner $f' = h^{-1} \circ g$, and so $f = f'$. ⊣

One might suppose that Semiassociative Coherence can be inferred directly from Associative Coherence. This would be so if we could find an independent proof that $\hat{\mathbf{A}}^{\rightarrow}$ is isomorphic to a subcategory of $\hat{\mathbf{A}}$, a proof that would not rely on Semiassociative Coherence. In fact, we use Semiassociative Coherence to conclude that $\hat{\mathbf{A}}^{\rightarrow}$ is isomorphic to a subcategory of $\hat{\mathbf{A}}$. That $\hat{\mathbf{A}}^{\rightarrow}$ is isomorphic to a subcategory of $\hat{\mathbf{A}}$ amounts to showing that for f and g arrow terms of $\mathcal{C}(\hat{\mathbf{A}}^{\rightarrow})$ we have $f = g$ in $\hat{\mathbf{A}}^{\rightarrow}$ iff $f = g$ in $\hat{\mathbf{A}}$. That $f = g$ in $\hat{\mathbf{A}}^{\rightarrow}$ implies $f = g$ in $\hat{\mathbf{A}}$ is clear without appealing to coherence, but for the converse implication we use Semiassociative Coherence (cf. §14.4).

In the proof of Associative Coherence above, we rely essentially on the normal form of formulae, and use both $\hat{b}^{\rightarrow}$-terms and $\hat{b}^{\leftarrow}$-terms. This is why for the proof of Semiassociative Coherence we could not rely on the Directedness Lemma, but we needed the Theoremhood Proposition of the preceding section. The proof of Semiassociative Coherence is not very difficult, but it is more difficult than the proof of Associative Coherence based on the Directedness Lemma. The proof of Associative Coherence can be based on Semiassociative Coherence, but it has also this simpler proof.

§4.4. Associative normal form

Once we have proved Associative Coherence, we can ascertain that every arrow term is equal to an arrow term in a normal form, which we are going to define. This normal form is unique, in the sense that arrow terms in normal form are equal in $\hat{\mathbf{A}}$ (i.e., they stand for the same arrow of $\hat{\mathbf{A}}$) iff they are the same arrow term.

First we prove the following analogue of the Extraction Lemma of §4.2 (see §2.1 for the notion of scope).

EXTRACTION LEMMA. *If there is an occurrence* z *of* $\wedge$ *in a formula* A *of* $\mathcal{L}_\wedge$, *then there is a formula* $A_1 \, z \, A_2$ *of* $\mathcal{L}_\wedge$ *such that there is an arrow term* $g\colon A \vdash A_1 \, z \, A_2$ *of* $\mathcal{C}(\hat{\mathbf{A}})$. *In addition,*

(∗) *for all occurrences* x *and* y *of* $\wedge$ *in* A_i, *where* $i \in \{1,2\}$, *we have that* y *is in the scope of* x *in* A *iff* y *is in the scope of* x *in* A_i;

(∗∗) *every subterm of* g *of the form* $\hat{b}^{\rightarrow}_{D,E,F}$ *is of the type* $D \wedge (E \, z \, F) \vdash (D \wedge E) \, z \, F$, *and every subterm of* g *of the form* $\hat{b}^{\leftarrow}_{F,E,D}$ *is of the type* $(F \, z \, E) \wedge D \vdash F \, z \, (E \wedge D)$.

PROOF. We proceed by induction on the number $n \geq 0$ of occurrences of connectives in A. If $n = 0$, then the antecedent of the lemma is false, and the lemma is trivially true.

If $n > 0$, then A is $A' \, u \, A''$ with u an occurrence of $\wedge$. If u is z, then g is $\mathbf{1}_A$. So suppose u is not z, and suppose z is in A'. Then, by the induction hypothesis, we have an arrow term $g' : A' \vdash A'_1 \, z \, A'_2$ of $\mathcal{C}(\hat{\mathbf{A}})$ satisfying the primed version of (∗). The arrow term $g' \wedge \mathbf{1}_{A''}$ is of the type $A \vdash (A'_1 \, z \, A'_2) \, u \, A''$, and we have the arrow term $\hat{b}^{\leftarrow}_{A'_1,A'_2,A''} \circ (g' \wedge \mathbf{1}_{A''})\colon A \vdash A'_1 \, z \, (A'_2 \, u \, A'')$ of $\mathcal{C}(\hat{\mathbf{A}})$.

To verify (∗), suppose x and y are two occurrences of $\wedge$ in A'_1. It is clear that y is in the scope of x in A iff it is in the scope of x in A'. So, by (∗) of the induction hypothesis, we have that y is in the scope of x in A iff it is in the scope of x in A'_1. We settle easily in a similar manner cases where x and y are both in A'_2 or A''. If x is u, then (∗) follows easily again.

The case where z is in A'' is settled analogously by using the arrow term $\hat{b}^{\rightarrow}_{A',A''_1,A''_2}\colon A' \, u \, (A''_1 \, z \, A''_2) \vdash (A' \, u \, A''_1) \, z \, A''_2$. We easily check (∗∗) by going over the proof above. ⊣

The analogue for $\hat{\mathbf{A}}$ of the Theoremhood Proposition of §4.2 would state simply that there is an arrow term of $\mathcal{C}(\hat{\mathbf{A}})$ of type $A \vdash B$ iff the formulae A and B are comparable.

We do not need the assertion (∗) of the Extraction Lemma of this section for the proof of the Associative Normal-Form Proposition below. We stated this assertion, nevertheless, because it is analogous to the assertion (∗) of the Extraction Lemma of §4.2.

We need some preliminary notions to introduce our normal form. For every formula A of $\mathcal{L}_\wedge$ we assign to every subformula of A a natural number $n \geq 2$ in the following manner. We assign to every occurrence of a letter p in A a prime number $i_A(p) \geq 2$, each occurrence having a different number from all other occurrences. (Note that this assignment is not unique.) Next, for a subformula $B \wedge C$ of A, we have $i_A(B \wedge C) = i_A(B) \cdot i_A(C)$. For every subformula D of A, we define $I(D)$ as follows:

if D is p, then $I(D)$ is p;
if D is $B \wedge C$, then $I(D)$ is $I(B) \wedge_{i_A(B \wedge C)} I(C)$.

So $I(D)$ is like a formula, but with subscripted occurrences of $\wedge$.

Let A be a formula comparable with B, and let x be an occurrence of $\wedge$ in A. The formula B has an occurrence of $\wedge$ at the same place, and we call that occurrence of $\wedge$ also x. Let A^* be obtained from A by adding to every x of A the subscript that x has in $I(B)$. This subscripting gives rise to a formula C^* with subscripted occurrences of $\wedge$ for every subformula C of A. For every arrow term $f\colon A_1 \vdash A_2$ of $\mathcal{C}(\hat{\mathbf{A}})$ such that both A_1 and A_2 are comparable with B, we have an arrow term $f^* : A_1^* \vdash A_2^*$ obtained by replacing every index C of f by C^*. Then we have the following proposition.

Associative Normal-Form Proposition. *If A and B are comparable formulae, then there is an arrow term $f : A \vdash B$ of $\mathcal{C}(\hat{\mathbf{A}})$ such that every subterm of $f^* : A^* \vdash I(B)$ of the form $\hat{b}^{\rightarrow}_{D,E,F}$ is of the type $D \wedge_l (E \wedge_k F) \vdash (D \wedge_l E) \wedge_k F$ where for every $\wedge_n$ in $D \wedge_l E$ and F we have that n divides k; analogously, every subterm of $f^* : A^* \vdash I(B)$ of the form $\hat{b}^{\leftarrow}_{F,E,D}$ is of the type $(F \wedge_k E) \wedge_l D \vdash F \wedge_k (E \wedge_l D)$ where for every $\wedge_n$ in F and $E \wedge_l D$ we have that n divides k.*

Proof. We proceed by induction on the number m of occurrences of connectives in B. If $m = 0$, then f is $\mathbf{1}_p : p \vdash p$. If $m > 1$, then B is of the form $B_1 \, z \, B_2$ for z an occurrence of $\wedge$, and in $I(B)$ we have z_k with $k = l \cdot n$ for $l, n \geq 2$, where l is any other subscript of $\wedge$ in $I(B)$. Then, by the

Extraction Lemma of this section, there is an arrow term $g\colon A \vdash A_1 \mathbin{z} A_2$ of $\mathcal{C}(\hat{\mathbf{A}})$ such that (**) is satisfied. This guarantees that all the subterms of g from the family $\hat{b}$ are as required in the statement of the proposition. By the induction hypothesis, we have arrow terms $f_1 : A_1 \vdash B_1$ and $f_2\colon A_2 \vdash B_2$ that satisfy the conditions of the proposition, and $f\colon A \vdash B$ is $(f_1 \wedge f_2) \circ g\colon A \vdash B$. $\dashv$

The procedure of the proof of this proposition, which presupposes the Extraction Lemma of this section, gives rise to a unique arrow term, which we may consider to be in normal form. We may transform this arrow term into a developed arrow term by replacing $(f_1 \wedge f_2) \circ g$ in the proof above by $(f_1 \wedge \mathbf{1}_{B_2}) \circ (\mathbf{1}_{A_1} \wedge f_2) \circ g$, or by $(\mathbf{1}_{B_1} \wedge f_2) \circ (f_1 \wedge \mathbf{1}_{A_2}) \circ g$, when neither of f_1 and f_2 is an identity arrow term.

§4.5. Strictification of associative categories

According to our definition of §3.1, an $\hat{\mathbf{A}}$-category, and in particular an associative category, is $\mathcal{C}(\hat{\mathbf{A}})$-strict when for all objects a, b and c we have

$$a \wedge (b \wedge c) = (a \wedge b) \wedge c,$$
$$\hat{b}^{\rightarrow}_{a,b,c} = \hat{b}^{\leftarrow}_{a,b,c} = \mathbf{1}_{a\wedge(b\wedge c)}.$$

The category *Rel* of §2.9 with $\wedge$ being $+$ is a $\mathcal{C}(\hat{\mathbf{A}})$-strict associative category.

For $\mathcal{G}$ being $\hat{\mathbf{A}}$, our construction of $\mathcal{B}^{\mathcal{G}}$ in §3.1 covers a construction exposed in [96] (pp. 257ff) and [67] (pp. 29-30), which builds out of an associative category $\mathcal{B}$ a $\mathcal{C}(\hat{\mathbf{A}})$-strict associative category $\mathcal{B}^{\mathcal{G}}$ equivalent to $\mathcal{B}$ via two strong $\mathcal{C}(\hat{\mathbf{A}})$-functors. As a matter of fact, [96] and [67] are about monoidal categories, i.e. natural $\hat{\mathbf{A}}_{\top}$-categories (see the next section), with whose strictification we deal in §4.7. The result one can extract from [96] and [67] is that every $\hat{\mathbf{A}}$-category can be $(\hat{\mathbf{A}}, \mathcal{C}(\hat{\mathbf{A}}))$-strictified ($\hat{\mathbf{A}}$ is $\mathcal{C}(\hat{\mathbf{A}})/\mathcal{E}(\hat{\mathbf{A}})$). We have shown something more than that in §3.1. We have shown, namely, that if $\mathcal{B}$ is a $\mathcal{C}/\mathcal{E}$-category for a logical category $\mathcal{C}/\mathcal{E}$ where $\mathcal{C}(\hat{\mathbf{A}}) \preceq_{\mathcal{L}_\wedge} \mathcal{C}$ and the $\mathcal{C}(\hat{\mathbf{A}})$-core $\mathcal{G}$ of $\mathcal{C}/\mathcal{E}$, which flows through $\mathcal{C}/\mathcal{E}$ and is a preorder, is $\hat{\mathbf{A}}$, then $\mathcal{B}$ can be $(\mathcal{C}/\mathcal{E}, \mathcal{C}(\hat{\mathbf{A}}))$-strictified. One passes from our construction to that of [96] (pp. 257ff) and [67] (pp. 29-30) by realizing that for $\mathcal{G}$ being $\hat{\mathbf{A}}$, and A and B formulae of $\mathcal{G}(\mathcal{B})$ we have $A \equiv^{\mathcal{G}} B$ iff after deleting every $\wedge^{\mathcal{G}}$ in A and B we obtain the same finite nonempty sequence of objects of $\mathcal{B}$. So there is a one-to-one correspondence between the classes $[A]$ with respect to $\equiv^{\mathcal{G}}$ and finite nonempty sequences of objects of $\mathcal{B}$. We are passing from the free groupoid generated by the objects of $\mathcal{B}$ to the free semigroup generated by these objects. The objects of the free semigroup may be represented by nonempty words.

When $\mathcal{G}$ is $\hat{\mathbf{A}}$, and $[A]$ corresponds to the sequence $a_1 \ldots a_n$ of objects of $\mathcal{B}$ in the sense just specified, we can take as the representative A_F of $[A]$ when $n \geq 2$ the formula $(\ldots(a_1 \wedge^{\mathcal{G}} a_2) \ldots \wedge^{\mathcal{G}} a_n)$ (where parentheses are associated to the left). Then, instead of choosing the arrows $\varphi_A \colon A_F \vdash A$ and $\varphi_A^{-1} \colon A \vdash A_F$ of $\mathcal{G}(\mathcal{B})$ arbitrarily, as we did in §3.1, we can define them inductively in the following manner. First, we define by induction the arrows $\varphi_{A_F \wedge^{\mathcal{G}} B_F} \colon (A \wedge^{\mathcal{G}} B)_F \vdash A_F \wedge^{\mathcal{G}} B_F$ and $\varphi^{-1}_{A_F \wedge^{\mathcal{G}} B_F} \colon A_F \wedge^{\mathcal{G}} B_F \vdash (A \wedge^{\mathcal{G}} B)_F$ (note that $(A \wedge^{\mathcal{G}} B)_F = (A_F \wedge^{\mathcal{G}} B_F)_F$):

$$\varphi_{A_F \wedge^{\mathcal{G}} b} = \varphi^{-1}_{A_F \wedge^{\mathcal{G}} b} = \mathbf{1}_{A_F \wedge^{\mathcal{G}} b}, \text{ for } b \text{ an object of } \mathcal{B},$$
$$\varphi_{A_F \wedge^{\mathcal{G}} (C_F \wedge^{\mathcal{G}} b)} = \hat{b}^{\leftarrow}_{A_F, C_F, b} \circ (\varphi_{A_F \wedge^{\mathcal{G}} C_F} \wedge^{\mathcal{G}} \mathbf{1}_b),$$
$$\varphi^{-1}_{A_F \wedge^{\mathcal{G}} (C_F \wedge^{\mathcal{G}} b)} = (\varphi^{-1}_{A_F \wedge^{\mathcal{G}} C_F} \wedge^{\mathcal{G}} \mathbf{1}_b) \circ \hat{b}^{\rightarrow}_{A_F, C_F, b}.$$

We have no need for other φ and φ^{-1} arrows except these to define $\psi^{\square\xi\square}$ and $\psi^{-1\square\xi\square}$ (see §3.1), but, for the sake of completeness, we can define inductively as follows φ_A and φ_A^{-1} for every object A of $\mathcal{G}(\mathcal{B})$:

$$\varphi_a = \varphi_a^{-1} = \mathbf{1}_a, \text{ for } a \text{ an object of } \mathcal{B},$$
$$\varphi_{A \wedge^{\mathcal{G}} B} = (\varphi_A \wedge^{\mathcal{G}} \varphi_B) \circ \varphi_{A_F \wedge^{\mathcal{G}} B_F},$$
$$\varphi^{-1}_{A \wedge^{\mathcal{G}} B} = \varphi^{-1}_{A_F \wedge^{\mathcal{G}} B_F} \circ (\varphi_A^{-1} \wedge^{\mathcal{G}} \varphi_B^{-1}).$$

To check the correctness of these definitions it is enough to verify that $\varphi_{A_F} = \varphi^{-1}_{A_F} = \mathbf{1}_{A_F}$. Note that φ_A and φ_A^{-1} are defined with arrow terms that, after deleting identity arrow terms, are in the associative normal form of the preceding section.

In Lemmata 1-5 of §3.1 we did not appeal to the preordering of $\mathcal{G}$. In Lemma 6, we had this assumption. We can prove, however, the following corollary of this lemma without appealing to the preordering of $\hat{\mathbf{A}}$, i.e. to its coherence.

LEMMA. *For $\mathcal{G}$ being $\hat{\mathbf{A}}$, the category $\langle \hat{\mathbf{A}}^{\mathcal{G}}, M'', \alpha'' \rangle$ is $\mathcal{C}(\hat{\mathbf{A}})$-strict.*

PROOF. For X and Y objects of $\hat{\mathbf{A}}^{\mathcal{G}}$, let us write XY for $X \wedge'' Y$, since $\wedge''$ in $\hat{\mathbf{A}}^{\mathcal{G}}$ corresponds to concatenation of sequences of formulae of $\mathcal{L}_{\wedge}$. We have, of course, $X \wedge'' (Y \wedge'' Z) = (X \wedge'' Y) \wedge'' Z = XYZ$.

Let $\hat{\varphi}_{[A],[B]}$ stand for $E\varphi_{A_F \wedge^{\mathcal{G}} B_F} : E(A \wedge^{\mathcal{G}} B)_F \vdash E(A_F \wedge^{\mathcal{G}} B_F)$ (see §3.1 for the functor E). Since $E(A \wedge^{\mathcal{G}} B)_F$ is $F([A][B])$ and $E(A_F \wedge^{\mathcal{G}} B_F)$ is $F[A] \wedge F[B]$, the type of $\hat{\varphi}_{X,Y}$ is $F(XY) \vdash FX \wedge FY$. If $\hat{\varphi}^{-1}_{[A],[B]}$ stands for $E\varphi^{-1}_{A_F \wedge^{\mathcal{G}} B_F}$, then $\hat{\varphi}^{-1}_{X,Y}\colon FX \wedge FY \vdash F(XY)$ is the inverse of $\hat{\varphi}_{X,Y}$ in $\hat{\mathbf{A}}$.

To show that for α being $\hat{b}^{\rightarrow}$ we have $\alpha''_{X,Y,Z} = \mathbf{1}_{XYZ}$ in $\hat{\mathbf{A}}^{\mathcal{G}}$ we proceed as follows. We make an induction on the length of the sequence corresponding to Z, and we use the alternative definition of α'' from §3.1.

If Z is a, then by the definition of $\hat{\varphi}_{U,V}$ and the fact that it is an isomorphism, we have

$$\begin{aligned} &\hat{\varphi}^{-1}_{XY,a} \circ (\hat{\varphi}^{-1}_{X,Y} \wedge \mathbf{1}_a) \circ \hat{b}^{\rightarrow}_{FX,FY,a} \circ (\mathbf{1}_{FX} \wedge \hat{\varphi}_{Y,a}) \circ \hat{\varphi}_{X,Ya} \\ &\qquad = (\hat{\varphi}^{-1}_{X,Y} \wedge \mathbf{1}_a) \circ \hat{b}^{\rightarrow}_{FX,FY,a} \circ \hat{\varphi}_{X,Ya} \\ &\qquad = \mathbf{1}_{F(XYa)}. \end{aligned}$$

If Z is Ua, then we have

$$\begin{aligned} &\hat{\varphi}^{-1}_{XY,Ua} \circ (\hat{\varphi}^{-1}_{X,Y} \wedge \mathbf{1}_{F(Ua)}) \circ \hat{b}^{\rightarrow}_{FX,FY,F(Ua)} \circ (\mathbf{1}_{FX} \wedge \hat{\varphi}_{Y,Ua}) \circ \hat{\varphi}_{X,YUa} \\ &= (\hat{\varphi}^{-1}_{XY,U} \wedge \mathbf{1}_a) \circ ((\hat{\varphi}^{-1}_{X,Y} \wedge \mathbf{1}_{FU}) \wedge \mathbf{1}_a) \circ \hat{b}^{\rightarrow}_{FX \wedge FY,FU,a} \circ \hat{b}^{\rightarrow}_{FX,FY,FU \wedge a} \circ \\ &\quad \circ (\mathbf{1}_{FX} \wedge \hat{b}^{\leftarrow}_{FY,FU,a}) \circ \hat{b}^{\leftarrow}_{FX,FY \wedge FU,a} \circ ((\mathbf{1}_{FX} \wedge \hat{\varphi}_{Y,U}) \wedge \mathbf{1}_a) \circ (\hat{\varphi}_{X,YU} \wedge \mathbf{1}_a), \\ &\qquad\qquad \text{by definition, } (\hat{b}\hat{b}) \text{ and naturality equations,} \\ &= (\hat{\varphi}^{-1}_{XY,U} \wedge \mathbf{1}_a) \circ ((\hat{\varphi}^{-1}_{X,Y} \wedge \mathbf{1}_{FU}) \wedge \mathbf{1}_a) \circ (\hat{b}^{\rightarrow}_{FX,FY,FU} \wedge \mathbf{1}_a) \circ \\ &\quad \circ ((\mathbf{1}_{FX} \wedge \hat{\varphi}_{Y,U}) \wedge \mathbf{1}_a) \circ (\hat{\varphi}_{X,YU} \wedge \mathbf{1}_a), \text{ by } (\hat{b}\hat{b}) \text{ and } (\hat{b}\ 5), \\ &= \mathbf{1}_{F(XYUa)}, \text{ by bifunctoriality and the induction hypothesis.} \end{aligned}$$

We proceed analogously when α is $\hat{b}^{\leftarrow}$. ⊣

So $\hat{\mathbf{A}}$ can be $(\hat{\mathbf{A}}, \mathcal{C}(\hat{\mathbf{A}}))$-strictified, and by the Strictification-Coherence Implication of §3.1, we can conclude that $\hat{\mathbf{A}}$ is a preorder.

§4.6. Coherence of monoidal categories

To obtain the natural logical category $\hat{\mathbf{A}}_{\top}$, we have that the logical system $\mathcal{C}(\hat{\mathbf{A}}_{\top})$ is in $\mathcal{L}_{\wedge,\top}$, with the transformations α included in $\mathbf{1}$, $\hat{b}$ and $\hat{\delta}$-$\hat{\sigma}$.

The specific equations of $\mathcal{E}(\hat{\mathbf{A}}_\top)$ are those of $\mathcal{E}(\hat{\mathbf{A}})$ plus

$$(\hat{\delta}\,\hat{\delta})\quad \hat{\delta}^{\rightarrow}_{A} \circ \hat{\delta}^{\leftarrow}_{A} = \mathbf{1}_A, \qquad \hat{\delta}^{\leftarrow}_{A} \circ \hat{\delta}^{\rightarrow}_{A} = \mathbf{1}_{A\wedge\top},$$

$$(\hat{\sigma}\,\hat{\sigma})\quad \hat{\sigma}^{\rightarrow}_{A} \circ \hat{\sigma}^{\leftarrow}_{A} = \mathbf{1}_A, \qquad \hat{\sigma}^{\leftarrow}_{A} \circ \hat{\sigma}^{\rightarrow}_{A} = \mathbf{1}_{\top\wedge A},$$

$$(\hat{b}\,\hat{\delta}\,\hat{\sigma})\quad \hat{b}^{\rightarrow}_{A,\top,C} = (\hat{\delta}^{\leftarrow}_{A} \wedge \mathbf{1}_C) \circ (\mathbf{1}_A \wedge \hat{\sigma}^{\rightarrow}_{C}).$$

From these equations one infers

$$(\hat{b}\,\hat{\delta})\quad \hat{b}^{\rightarrow}_{A,B,\top} = \hat{\delta}^{\leftarrow}_{A\wedge B} \circ (\mathbf{1}_A \wedge \hat{\delta}^{\rightarrow}_{B}),$$

$$(\hat{b}\,\hat{\sigma})\quad \hat{b}^{\rightarrow}_{\top,B,C} = (\hat{\sigma}^{\leftarrow}_{B} \wedge \mathbf{1}_C) \circ \hat{\sigma}^{\rightarrow}_{B\wedge C},$$

$$(\hat{\delta}\,\hat{\sigma})\quad \hat{\delta}^{\rightarrow}_{\top} = \hat{\sigma}^{\rightarrow}_{\top}$$

(see [70], Theorems 6 and 7, and §9.1 below).

The specific equations of $\mathcal{E}(\hat{\mathbf{A}})$ are introduced, as all our equations for logical categories, by axiomatic equations with variables. These equations with variables are now assumed for $\mathcal{E}(\hat{\mathbf{A}}_\top)$ (see §2.3 and §2.7). The equational system $\mathcal{E}(\hat{\mathbf{A}}_\top)$ will be closed under (su) for formulae C of $\mathcal{L}_{\wedge,\top}$, and not only of $\mathcal{L}_\wedge$. We assume tacitly from now on that we proceed in an analogous manner whenever we pass from an equational system formulated originally with respect to a poorer language to an equational system formulated with respect to a richer language.

The equations $(\hat{\delta}\,\hat{\delta})$ and $(\hat{\sigma}\,\hat{\sigma})$ above, together with the naturality equations for arrow terms in the family $\hat{\delta}$-$\hat{\sigma}$, say that in subfamilies of this family we find natural isomorphisms.

The natural $\hat{\mathbf{A}}_\top$-categories are commonly called *monoidal categories* (see [94], Section VII.1), and sometimes *tensor categories* (as in [66] and [67]). The objects of a monoidal category that is a partial order make a monoid.

For every formula A of $\mathcal{L}_{\wedge,\top}$ let $k(A)$ be the number of occurrences of $\top$ in A as main conjuncts in subformulae of A (i.e. the visible occurrences of $\top$ in subformulae of A of the form $B \wedge \top$ or $\top \wedge B$). So in $\mathcal{L}_{\wedge,\top}$ we do not count $\top$ only if A itself is $\top$, but our definition of $k(A)$ is adapted to other languages $\mathcal{L}$ too (cf. the end of §6.1). We say that A is in normal form when $d(A) = 0$ and $k(A) = 0$ (where $d(A)$ is defined in §4.2). So, for example, $(p \wedge q) \wedge r$ and $\top$ are in normal form.

An arrow term of $\mathcal{C}(\hat{\mathbf{A}}_\top)$ is called $\rightarrow$*-directed* when neither of $\hat{b}^{\leftarrow}$, $\hat{\delta}^{\leftarrow}$ and $\hat{\sigma}^{\leftarrow}$ occurs in it. (This definition extends the definition of $\rightarrow$-directed arrow terms of $\mathcal{C}(\hat{\mathbf{A}})$ in §4.3.) Then, by extending the proof of the Directedness Lemma of §4.3, which is a simplification of our proof of Semiassociative Coherence of §4.2, we can prove the following.

DIRECTEDNESS LEMMA. *If $f, g : A \vdash B$ are $\rightarrow$-directed arrow terms of $\mathcal{C}(\hat{\mathbf{A}}_\top)$ and B is in normal form, then $f = g$ in $\hat{\mathbf{A}}_\top$.*

PROOF. We proceed by induction on $d(A)+k(A)$. In the induction step we have the following new cases for $f = f_2 \circ f_1$ and $g = g_2 \circ g_1$ for some $\rightarrow$-directed arrow terms $f_2 : C \vdash B$ and $g_2 : D \vdash B$:

(I) $f_1 : A \vdash C$ is a $\hat{b}^{\rightarrow}$-term and $g_1 : A \vdash D$ is a $\hat{\delta}^{\rightarrow}$-term,

(II) $f_1 : A \vdash C$ is a $\hat{b}^{\rightarrow}$-term and $g_1 : A \vdash D$ is a $\hat{\sigma}^{\rightarrow}$-term,

(III) $f_1 : A \vdash C$ and $g_1 : A \vdash D$ are $\hat{\delta}^{\rightarrow}$-terms,

(IV) $f_1 : A \vdash C$ and $g_1 : A \vdash D$ are $\hat{\sigma}^{\rightarrow}$-terms,

(V) $f_1 : A \vdash C$ is a $\hat{\delta}^{\rightarrow}$-term and $g_1 : A \vdash D$ is a $\hat{\sigma}^{\rightarrow}$-term,

With (I), the only interesting additional case is when the head of f_1 is $\hat{b}^{\rightarrow}_{E,F,\top}$ and the head of g_1 is $\hat{\delta}^{\rightarrow}_F$, where we apply $(\hat{b}\,\hat{\delta})$.

With (II), the only interesting additional cases are when the head of f_1 is $\hat{b}^{\rightarrow}_{E,\top,G}$ and the head of g_1 is $\hat{\sigma}^{\rightarrow}_G$, where we apply $(\hat{b}\,\hat{\delta}\,\hat{\sigma})$, and when the head of f_1 is $\hat{b}^{\rightarrow}_{\top,F,G}$ and the head of g_1 is $\hat{\sigma}^{\rightarrow}_{F\wedge G}$, where we apply $(\hat{b}\hat{\sigma})$.

In the other, uninteresting cases, of (I) and (II), which are in principle covered by what we had in the proof of Semiassociative Coherence in §4.2, we apply bifunctorial and naturality equations.

We apply these equations also in cases (III), (IV) and (V); for the last case we also need the equation $(\hat{\delta}\,\hat{\sigma})$. The remaining cases are as in §4.2. $\dashv$

From this Directedness Lemma we infer the following result of [93] and [94] (Section VII.2), whose proof is analogous to the proof of Associative Coherence in §4.3.

MONOIDAL COHERENCE. *The category $\hat{\mathbf{A}}_\top$ is a preorder.*

§4.7. Strictification of monoidal categories

In a $\mathcal{C}(\hat{\mathbf{A}}_\top)$-strict $\hat{\mathbf{A}}_\top$-category for every object a we have

$$a \wedge \top = \top \wedge a = a,$$
$$\hat{\delta}^{\rightarrow}_{a} = \hat{\delta}^{\leftarrow}_{a} = \hat{\sigma}^{\rightarrow}_{a} = \hat{\sigma}^{\leftarrow}_{a} = \mathbf{1}_a,$$

in addition to what was mentioned at the very beginning of §4.5. The category *Rel* of §2.9 with $\wedge$ being $+$ and $\top$ being 0 is a $\mathcal{C}(\hat{\mathbf{A}}_\top)$-strict monoidal category.

What we have said at the beginning of §4.5 concerning the strictification of associative categories and previous results of [96] and [67], applies mutatis mutandis to the present context. One has to replace "associative" by "monoidal" and $\hat{\mathbf{A}}$ by $\hat{\mathbf{A}}_\top$.

For $\mathcal{G}$ being $\hat{\mathbf{A}}_\top$, the objects $[A]$ of $\mathcal{B}^\mathcal{G}$ correspond bijectively now to arbitrary finite sequences of objects of $\mathcal{B}$, including the empty sequence. The class $[\top^\mathcal{G}]$ corresponds to the empty sequence, and we can take $\top^\mathcal{G}$ as the representative A_F of $[\top^\mathcal{G}]$.

The inductive definition of φ_A of §4.5 can now be extended with the following clauses:

$$\varphi_{\top^\mathcal{G}} = \varphi^{-1}_{\top^\mathcal{G}} = \mathbf{1}_{\top^\mathcal{G}},$$
$$\varphi_{A_F \wedge^\mathcal{G} \top^\mathcal{G}} = \hat{\delta}^{\leftarrow}_{A_F}, \qquad \varphi^{-1}_{A_F \wedge^\mathcal{G} \top^\mathcal{G}} = \hat{\delta}^{\rightarrow}_{A_F},$$
$$\varphi_{\top^\mathcal{G} \wedge^\mathcal{G} A_F} = \hat{\sigma}^{\leftarrow}_{A_F}, \qquad \varphi^{-1}_{\top^\mathcal{G} \wedge^\mathcal{G} A_F} = \hat{\sigma}^{\rightarrow}_{A_F}.$$

The correctness of these definitions for the case $\varphi_{\top^\mathcal{G} \wedge^\mathcal{G} \top^\mathcal{G}}$ and $\varphi^{-1}_{\top^\mathcal{G} \wedge^\mathcal{G} \top^\mathcal{G}}$ is guaranteed by $(\hat{\delta}\hat{\sigma})$.

We can prove as before the following analogue of the Lemma of §4.5, without presupposing the preordering of $\hat{\mathbf{A}}_\top$.

LEMMA. *For $\mathcal{G}$ being $\hat{\mathbf{A}}_\top$, the category $\langle \hat{\mathbf{A}}^\mathcal{G}_\top, M'', \alpha'' \rangle$ is $\mathcal{C}(\hat{\mathbf{A}}_\top)$-strict.*

PROOF. We proceed as for the proof of the Lemma of §4.5, with the following additions.

Let us write $\emptyset$ instead of $[\top^\mathcal{G}]$. To show that for α being $\hat{b}^{\rightarrow}$ the equation $\alpha''_{X,Y,Z} = \mathbf{1}_{XYZ}$ holds in $\hat{\mathbf{A}}^\mathcal{G}_\top$, we have to consider new cases when X, Y or Z are $\emptyset$.

I. If X is $\emptyset$, then

$$\begin{aligned}&\hat{\varphi}^{-1}_{Y,Z}\circ(\hat{\varphi}^{-1}_{\emptyset,Y}\wedge\mathbf{1}_{FZ})\circ\hat{b}^{\rightarrow}_{\top,FY,FZ}\circ(\mathbf{1}_{\top}\wedge\hat{\varphi}_{Y,Z})\circ\hat{\varphi}_{\emptyset,YZ}\\&\quad=\hat{\varphi}^{-1}_{Y,Z}\circ(\hat{\sigma}^{\rightarrow}_{FY}\wedge\mathbf{1}_{FZ})\circ\hat{b}^{\rightarrow}_{\top,FY,FZ}\circ(\mathbf{1}_{\top}\wedge\hat{\varphi}_{Y,Z})\circ\hat{\sigma}^{\leftarrow}_{F(YZ)}\\&\quad=\mathbf{1}_{F(YZ)},\text{ by }(\hat{b}\hat{\sigma}),\text{ naturality and isomorphisms.}\end{aligned}$$

II. If Y is $\emptyset$, then

$$\begin{aligned}&\hat{\varphi}^{-1}_{X,Z}\circ(\hat{\varphi}^{-1}_{X,\emptyset}\wedge\mathbf{1}_{FZ})\circ\hat{b}^{\rightarrow}_{FX,\top,FZ}\circ(\mathbf{1}_{FX}\wedge\hat{\varphi}_{\emptyset,Z})\circ\hat{\varphi}_{X,Z}\\&\quad=\hat{\varphi}^{-1}_{X,Z}\circ(\hat{\delta}^{\rightarrow}_{FX}\wedge\mathbf{1}_{FZ})\circ\hat{b}^{\rightarrow}_{FX,\top,FZ}\circ(\mathbf{1}_{FX}\wedge\hat{\sigma}^{\leftarrow}_{FZ})\circ\hat{\varphi}_{X,Z}\\&\quad=\mathbf{1}_{F(XZ)},\text{ by }(\hat{b}\hat{\delta}\hat{\sigma})\text{ and isomorphisms.}\end{aligned}$$

III. If Z is $\emptyset$, then

$$\begin{aligned}&\hat{\varphi}^{-1}_{XY,\emptyset}\circ(\hat{\varphi}^{-1}_{X,Y}\wedge\mathbf{1}_{\top})\circ\hat{b}^{\rightarrow}_{FX,FY,\top}\circ(\mathbf{1}_{FX}\wedge\hat{\varphi}_{Y,\emptyset})\circ\hat{\varphi}_{X,Y}\\&\quad=\hat{\delta}^{\rightarrow}_{F(XY)}\circ(\hat{\varphi}^{-1}_{X,Y}\wedge\mathbf{1}_{\top})\circ\hat{b}^{\rightarrow}_{FX,FY,\top}\circ(\mathbf{1}_{FX}\wedge\hat{\delta}^{\leftarrow}_{FY})\circ\hat{\varphi}_{X,Y}\\&\quad=\mathbf{1}_{F(XZ)},\text{ by }(\hat{b}\hat{\delta}),\text{ naturality and isomorphisms.}\end{aligned}$$

To show that for α being $\hat{\delta}^{\rightarrow}$ the equation $\alpha''_X=\mathbf{1}_X$ holds in $\hat{\mathbf{A}}^{\mathcal{G}}_{\top}$, we have $\hat{\delta}^{\rightarrow}_{FX}\circ\hat{\varphi}_{X,\emptyset}=\mathbf{1}_{FX}$, since $\hat{\delta}^{\rightarrow}$ is an isomorphism, and analogously for $\hat{\sigma}^{\rightarrow}$. ⊣

So $\hat{\mathbf{A}}_{\top}$ can be $(\hat{\mathbf{A}}_{\top},\mathcal{C}(\hat{\mathbf{A}}_{\top}))$-strictified, and we can conclude that $\hat{\mathbf{A}}_{\top}$ is a preorder by the Strictification-Coherence Implication of §3.1.

In the presence of the unit object $\top$, when we deal with monoidal categories, there is a $\mathcal{C}(\hat{\mathbf{A}}_{\top})$-strict monoidal category alternative to $\mathcal{B}^{\mathcal{G}}$, inspired by Cayley's representation of monoids (see [67], pp. 26-27, and [96], p. 260, Exercises 1-3). This is a functor category in which $\wedge$ on objects is composition of functors. It is not clear how to adapt this functor category to cases where we have two monoidal structures, while $\mathcal{B}^{\mathcal{G}}$ covers that and much more. Moreover, the proof of Proposition 1.3 of [67] (p. 27), which states the faithfulness of a functor into the functor category, seems to rely essentially on the presence of the unit object $\top$. On the other hand, the approach through the category $\mathcal{B}^{\mathcal{G}}$ is, of course, possible in situations without unit objects.

The first proof proposed for the $(\hat{\mathbf{A}}_\top, \mathcal{C}(\hat{\mathbf{A}}_\top))$-strictification of monoidal categories in [67], viz. the proof of Corollary 1.4 on p. 27, does not stand, since the full image of the functor L in the functor category is not closed under composition of functors. The other proof of Corollary 1.4 in [67], on p. 30, is closer to what we have been doing in this section and in §4.5. (In the presentation of [96], pp. 255ff, which is more accessible than that in [67], there is a lapsus on p. 259; one should have there $1 \Box G_2$ and not $G_2 \Box 1$.)

CHAPTER 5

SYMMETRIC ASSOCIATIVE CATEGORIES

We present in this chapter a proof of coherence, with and without unit objects, for symmetric monoidal categories—a proof more thorough than Mac Lane's proof (see [93] and [96], Section XI.1), from which it stems. We provide with it a proof of the completeness of the usual axiomatization of symmetric groups via the normal form that stems from Burnside. We also make explicit the strictification of the monoidal structure involved in the proof, on which Mac Lane presumably also relies (as we noted in §3.2). Mac Lane seems to presuppose that this strictification is allowed, while we justify it by the results of Chapter 3.

§5.1. Coherence of symmetric associative categories

To obtain the natural logical category $\hat{\mathbf{S}}$, we have that the logical system $\mathcal{C}(\hat{\mathbf{S}})$ is in $\mathcal{L}_\wedge$, with the transformations α included in $\mathbf{1}$, $\hat{b}$ and $\hat{c}$. The specific equations of $\mathcal{E}(\hat{\mathbf{S}})$ are those of $\mathcal{E}(\hat{\mathbf{A}})$ plus

$$(\hat{c}\hat{c})\quad \hat{c}_{B,A} \circ \hat{c}_{A,B} = \mathbf{1}_{A\wedge B},$$

$$(\hat{b}\hat{c})\quad \hat{c}_{A,B\wedge C} = \hat{b}^{\rightarrow}_{B,C,A} \circ (\mathbf{1}_B \wedge \hat{c}_{A,C}) \circ \hat{b}^{\leftarrow}_{B,A,C} \circ (\hat{c}_{A,B} \wedge \mathbf{1}_C) \circ \hat{b}^{\rightarrow}_{A,B,C}.$$

The equation $(\hat{c}\hat{c})$, together with $(\hat{c}\ nat)$, says that $\hat{c}$ is a natural isomorphism, while the equation $(\hat{b}\hat{c})$ amounts to Mac Lane's *hexagonal* equation of [93] (Section 4; see also [94], Section VII.7, and [96], Section XI.1).

An alternative way to obtain $\hat{\mathbf{S}}$ is to extend $\mathcal{C}(\hat{\mathbf{A}}^{\rightarrow})$ with the transformation $\hat{c}$, and assume the following definition:

$$\hat{b}^{\leftarrow}_{A,B,C} =_{df} \hat{c}_{B\wedge C,A} \circ \hat{b}^{\rightarrow}_{B,C,A} \circ \hat{c}_{C\wedge A,B} \circ \hat{b}^{\rightarrow}_{C,A,B} \circ \hat{c}_{A\wedge B,C},$$

together with the equations $\mathcal{E}(\hat{\mathbf{S}})$ of $\hat{\mathbf{S}}$. In that context the equations $(\hat{b}\hat{b})$ become

$$\hat{c}_{B\wedge C,A} \circ \hat{b}^{\rightarrow}_{B,C,A} \circ \hat{c}_{C\wedge A,B} \circ \hat{b}^{\rightarrow}_{C,A,B} \circ \hat{c}_{A\wedge B,C} \circ \hat{b}^{\rightarrow}_{A,B,C} = \mathbf{1}_{A\wedge(B\wedge C)},$$
$$\hat{b}^{\rightarrow}_{A,B,C} \circ \hat{c}_{B\wedge C,A} \circ \hat{b}^{\rightarrow}_{B,C,A} \circ \hat{c}_{C\wedge A,B} \circ \hat{b}^{\rightarrow}_{C,A,B} \circ \hat{c}_{A\wedge B,C} = \mathbf{1}_{(A\wedge B)\wedge C},$$

while the equation $(\hat{b}\,\hat{c})$ amounts to

$$\hat{b}^{\rightarrow}_{C,A,B} \circ \hat{c}_{A\wedge B,C} \circ \hat{b}^{\rightarrow}_{A,B,C} = (\hat{c}_{A,C} \wedge \mathbf{1}_B) \circ \hat{b}^{\rightarrow}_{A,C,B} \circ (\mathbf{1}_A \wedge \hat{c}_{B,C}).$$

We call natural $\hat{\mathbf{S}}$-categories *symmetric associative* categories. Symmetric associative categories differ from Mac Lane's symmetric monoidal categories, which we will consider in §5.3, by not necessarily having the unit object $\top$. The objects of a symmetric associative category that is a partial order make a commutative semigroup.

It is easy to check that the two maps G of §2.9, defined on objects and on arrows, give rise to a strict $\mathcal{C}(\hat{\mathbf{S}})$-functor from $\hat{\mathbf{S}}$ to *Rel*. For that, it is enough to check that for $f = g$ being one of the equations $(\hat{c}\ nat)$, $(\hat{c}\,\hat{c})$ and $(\hat{b}\,\hat{c})$ we have $Gf = Gg$ in *Rel* (for the remaining equations in the axiomatization of $\hat{\mathbf{S}}$ this is trivial). It is clear that Gf corresponds to a permutation of a finite nonempty domain. Our goal is to prove coherence for the category $\hat{\mathbf{S}}$ with respect to *Rel*; namely, we will prove the following result of [93] (Section 4).

SYMMETRIC ASSOCIATIVE COHERENCE. *The functor G from $\hat{\mathbf{S}}$ to Rel is faithful.*

Coherence here does not mean, as for $\hat{\mathbf{A}}^{\rightarrow}$, $\hat{\mathbf{A}}$ and $\hat{\mathbf{A}}_{\top}$, that $\hat{\mathbf{S}}$ is a preorder. We do not have $G\ \hat{c}_{p,p} = G\mathbf{1}_{p\wedge p}$, and hence, by the functoriality of G, we do not have $\hat{c}_{p,p} = \mathbf{1}_{p\wedge p}$ in $\hat{\mathbf{S}}$.

For $\mathcal{G}$ being $\hat{\mathbf{A}}$ and $\mathcal{C}/\mathcal{E}$ being $\hat{\mathbf{S}}$, we have that the conditions (IV$\mathcal{C}$) and (IV$\mathcal{G}$) of §3.1 are satisfied, and $\mathcal{G}$ is moreover a preorder. To verify that $\hat{\mathbf{A}}$ is generatively discrete, we appeal to the fact that for every arrow term $f : A \vdash B$ of $\mathcal{C}(\hat{\mathbf{A}})$ we have that Gf is a bijection whose ordered pairs correspond to occurrences of the same letter in A and B. We have analogous arguments to establish generative discreteness in other cases of strictification, which we will encounter later, and we will not dwell on this matter any more.

Then we can apply the Direct-Strictification Theorem of §3.2 to obtain a category $\mathcal{C}_{\mathcal{G}}/\mathcal{E}_{\mathcal{G}}$, which we will call $\hat{\mathbf{S}}^{st}$. We call $\mathcal{C}_{\mathcal{G}}$ here $\mathcal{C}(\hat{\mathbf{S}}^{st})$. The category $\hat{\mathbf{S}}^{st}$ is equivalent to $\hat{\mathbf{S}}$ via the strict $\mathcal{C}(\hat{\mathbf{S}})$-functor $H_{\mathcal{G}}$ from $\hat{\mathbf{S}}$ to $\hat{\mathbf{S}}^{st}$, and the strong $\mathcal{C}(\hat{\mathbf{S}})$-functor $\langle H, \psi^{\Box\xi\Box}\rangle$ from $\hat{\mathbf{S}}^{st}$ to $\hat{\mathbf{S}}$.

Consider the composite functor GH from $\hat{\mathbf{S}}^{st}$ to *Rel*. It is easy to see that $GH[\![A]\!] = GA$, since all the formulae in $[\![A]\!]$ have the same letter length, and we also have

$$GH\alpha_{[\![A_1]\!],\ldots,[\![A_k]\!]} = G\alpha_{A_1,\ldots,A_k}.$$

We can conclude that G is equal to the composite functor $GHH_{\mathcal{G}}$. Hence it is enough to establish that GH is faithful to conclude that G is faithful, because we know that $H_{\mathcal{G}}$ is faithful.

Note that since in $\hat{\mathbf{S}}^{st}$ the equation $(\hat{b}\,\hat{c})$ becomes

$$\hat{c}_{X,Y\wedge Z} = (\mathbf{1}_Y \wedge \hat{c}_{X,Z}) \circ (\hat{c}_{X,Y} \wedge \mathbf{1}_Z),$$

and since we also have

$$\hat{c}_{X\wedge Y,Z} = (\hat{c}_{X,Z} \wedge \mathbf{1}_Y) \circ (\mathbf{1}_X \wedge \hat{c}_{Y,Z}),$$

every arrow term of $\mathcal{C}(\hat{\mathbf{S}}^{st})$ will be equal to a developed arrow term in which every $\hat{c}$-term is of one of the following forms:

$$\hat{c}_{[\![p]\!],[\![q]\!]},\quad \hat{c}_{[\![p]\!],[\![q]\!]} \wedge \mathbf{1}_X,\quad \mathbf{1}_X \wedge \hat{c}_{[\![p]\!],[\![q]\!]},\quad (\mathbf{1}_X \wedge \hat{c}_{[\![p]\!],[\![q]\!]}) \wedge \mathbf{1}_Y.$$

The $\hat{c}$-terms of $\mathcal{C}(\hat{\mathbf{S}}^{st})$ and their heads are defined analogously to what we had in §2.6.

For the first two arrow terms in this list we use the abbreviation s_1, and for the third and fourth we use s_{i+1}, where $i = GHX$. So our developed arrow term may be written in the form $s_{i_1} \circ \ldots \circ s_{i_n} \circ \mathbf{1}_X$, where $n \geq 0$.

It is easy to check that in $\hat{\mathbf{S}}^{st}$ we have the equations

$(s1)$ $\quad s_i \circ s_i = \mathbf{1}$,

$(s2)$ $\quad s_{i+k} \circ s_i = s_i \circ s_{i+k}$, $\quad$ for $k \geq 2$,

$(s3)$ $\quad s_i \circ s_{i+1} \circ s_i = s_{i+1} \circ s_i \circ s_{i+1}$, where **1** stands for $\mathbf{1}_X$ for some X. The equation ($s3$) is derived with the help of ($\hat{b}\,\hat{c}$) and ($\hat{c}$ *nat*) (see [96], Section XI.1, p. 254).

It is well known that the equations ($s1$), ($s2$) and ($s3$), together with the equations corresponding to the categorial equations (*cat* 1) and (*cat* 2)—namely, the equations of monoids—axiomatize symmetric groups (i.e., give a presentation of these groups by generators and relations; see [23], Section 6.2). A reader with this knowledge may now conclude that the functor GH is faithful. However, to make the matter self-contained, we will justify this conclusion in the next section.

§5.2. The faithfulness of GH

Let $s_{[i,j]}$ be an abbreviation for $s_i \circ s_{i-1} \circ \ldots \circ s_{j+1} \circ s_j$ if $i > j$, while $s_{[i,i]}$ stands for s_i. For $n \geq 0$, and **1** standing for $\mathbf{1}_X$ for some X, we say that

$$s_{[i_1,j_1]} \circ \ldots \circ s_{[i_n,j_n]} \circ \mathbf{1}$$

is in *normal form* when $i_1 < i_2 < \ldots < i_n$ (this normal form is implicit in [14], Note C, pp. 464-465).

Then from (*cat* 1), (*cat* 2), ($s1$), ($s2$) and ($s3$) we can prove the following equations for $i \geq k$:

$$\begin{aligned}
s_{[i,j]} \circ s_{[k,l]} &= s_{[k,l]} \circ s_{[i,j]}, && \text{if } k{+}1 < j,\\
&= s_{[i,l]}, && \text{if } k{+}1 = j,\\
&= s_{[k-1,l]} \circ s_{[i,j+1]}, && \text{if } k = j,\ i > j \text{ and } k > l,\\
&= s_{[i,j+1]}, && \text{if } k = j,\ i > j \text{ and } k = l,\\
&= s_{[k-1,l]}, && \text{if } k = j,\ i = j \text{ and } k > l,\\
&= \mathbf{1}, && \text{if } k = j,\ i = j \text{ and } k = l,\\
&= s_{[k-1,l]} \circ s_{[i,j+1]}, && \text{if } k > j \geq l,\\
&= s_{[k-1,l-1]} \circ s_{[i,j]}, && \text{if } k > j < l.
\end{aligned}$$

(Note that ($s1$) is the sixth equation, ($s2$) is an instance of the first equation, and ($s3$) is an instance of the last equation.) From these equations we easily infer the following.

NORMAL-FORM LEMMA. *Every arrow term of* $\mathcal{C}(\hat{\mathbf{S}}^{st})$ *is equal in* $\hat{\mathbf{S}}^{st}$ *to an arrow term in normal form.*

We can also prove the following.

UNIQUENESS LEMMA. *If the arrow terms $f, g : X \vdash Y$ of $\mathcal{C}(\hat{\mathbf{S}}^{st})$ are in normal form and $GHf = GHg$, then f and g are the same arrow term.*

PROOF. Let f and g be $s_{[i_1,j_1]} \circ \ldots \circ s_{[i_n,j_n]} \circ \mathbf{1}$ and $s_{[k_1,l_1]} \circ \ldots \circ s_{[k_m,l_m]} \circ \mathbf{1}$ respectively. Note that $GHs_{[i,j]}$ corresponds to the following diagram:

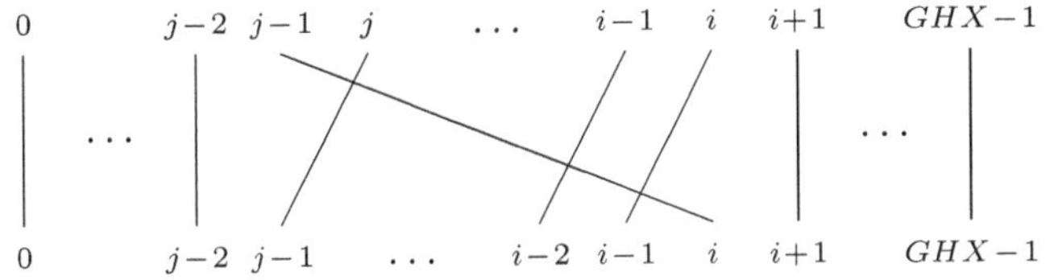

So, for $n > 0$, we have in GHf the ordered pair $(j_n{-}1, i_n)$, with $j_n{-}1 < i_n$, which we call the *last falling slope* of GHf. Note that for $l > i_n$ we have in GHf the ordered pairs (l, l).

Then we proceed by induction on n. If $n = 0$, then $m = 0$; otherwise, in GHf we would have only the ordered pairs (i, i), and in GHg we would have $(l_m{-}1, k_m)$ for $l_m{-}1 < k_m$.

If $n > 0$, then, as we have just shown, $m > 0$, while $i_n = k_m$ and $j_n = l_m$; otherwise, the last falling slopes of GHf and GHg would differ. Since for e being $s_{j_n} \circ s_{j_n+1} \circ \ldots \circ s_{i_n-1} \circ s_{i_n}$ we have

$$GHf \circ GHe = GHg \circ GHe,$$

we can conclude that for f' and g' being $s_{[i_1,j_1]} \circ \ldots \circ s_{[i_{n-1},j_{n-1}]} \circ \mathbf{1}$ and $s_{[k_1,l_1]} \circ \ldots \circ s_{[k_{m-1},l_{m-1}]} \circ \mathbf{1}$ respectively we have $GHf' = GHg'$, and we have by the induction hypothesis that f' and g' are the same arrow term. Hence f and g are the same arrow term. ⊣

As a matter of fact, it would be enough to prove instead of the Uniqueness Lemma that if f is in normal form and $GHf = GH\mathbf{1}$, then f is $\mathbf{1}$. (Altogether, this proof would not be shorter than the proof of the Uniqueness Lemma.)

From the Normal-Form Lemma, the functoriality of GH and the Uniqueness Lemma we infer easily that GH is faithful. An alternative proof of this faithfulness is obtained without the Uniqueness Lemma. Instead we establish that the number of different arrow terms $f: X \vdash Y$ in normal form is $n!$ for $n = GHX = GHY$, and that for every permutation π of an ordinal $n > 0$ (this permutation is an arrow of *Rel*) there is an arrow $f: X \vdash Y$ of $\hat{\mathbf{S}}^{st}$ such that $GHX = GHY = n$ and $GHf = \pi$. Then we use the fact that every onto function from $n!$ to $n!$ is also one-one. A proof in this alternative style is suggested by [14] (Note C, pp. 464-465). Our proof in this section is easily converted into a proof of completeness of the standard axiomatization of symmetric groups with respect to groups of permutations—a proof alternative to the proof in [14], mentioned above.

§5.3. Coherence of symmetric monoidal categories

To obtain the natural logical category $\hat{\mathbf{S}}_\top$, we have that the logical system $\mathcal{C}(\hat{\mathbf{S}}_\top)$ in $\mathcal{L}_{\wedge,\top}$, with the transformations α included in $\mathbf{1}$, $\hat{b}$, $\hat{c}$ and $\hat{\delta}$-$\hat{\sigma}$. The specific equations of $\mathcal{E}(\hat{\mathbf{S}}_\top)$ are obtained by taking the union of those of $\mathcal{E}(\hat{\mathbf{S}})$ and $\mathcal{E}(\hat{\mathbf{A}}_\top)$.

One can derive for $\hat{\mathbf{S}}_\top$ the following equation:

$$(\hat{c}\,\hat{\delta}\,\hat{\sigma}) \qquad \hat{c}_{A,\top} = \hat{\sigma}^{\leftarrow}_{A} \circ \hat{\delta}^{\rightarrow}_{A}$$

(see [70], Theorem 8). This equation says that one of $\hat{\delta}$ and $\hat{\sigma}$ is superfluous: it can be defined in terms of the other with the help of $\hat{c}$. Note that in the presence of $(\hat{c}\,\hat{\delta}\,\hat{\sigma})$, the instance of $(\hat{b}\,\hat{c})$ where A is $\top$, namely

$$\hat{c}_{\top,B\wedge C} = \hat{b}^{\rightarrow}_{B,C,\top} \circ (\mathbf{1}_B \wedge \hat{c}_{\top,C}) \circ \hat{b}^{\leftarrow}_{B,\top,C} \circ (\hat{c}_{\top,B} \wedge \mathbf{1}_C) \circ \hat{b}^{\rightarrow}_{\top,B,C},$$

is derivable without using $(\hat{b}\,\hat{c})$ (we apply $(\hat{b}\,\hat{\delta})$, $(\hat{b}\,\hat{\delta}\,\hat{\sigma})$ and $(\hat{b}\hat{\sigma})$), and analogously for the instances of $(\hat{b}\,\hat{c})$ where B or C is $\top$.

Natural $\hat{\mathbf{S}}_\top$-categories are usually called *symmetric monoidal* categories. The objects of a symmetric monoidal category that is a partial order make a commutative monoid.

The following result is from [93] (Section 5).

SYMMETRIC MONOIDAL COHERENCE. *The functor G from $\hat{\mathbf{S}}_\top$ to Rel is faithful.*

The proof of this faithfulness is easily obtained by extending our proof of Symmetric Associative Coherence in the two preceding sections. To obtain the category $\hat{\mathbf{S}}^{st}_{\top}$, we take $\hat{\mathbf{A}}_{\top}$ to be $\mathcal{G}$. Then in $\hat{\mathbf{S}}^{st}_{\top}$ we have $\hat{c}_{[\![A]\!],[\![\top]\!]} = \mathbf{1}_{[\![A]\!]}$.

Although the categories $\hat{\mathbf{S}}^{st}$ and $\hat{\mathbf{S}}^{st}_{\top}$ are not preorders, the categories $\hat{\mathbf{S}}^{div}$ and $\hat{\mathbf{S}}^{div}_{\top}$ are preorders (for the definition of these last two categories see §3.3). This follows from Symmetric Associative Coherence and Symmetric Monoidal Coherence. So extending $\mathcal{E}(\hat{\mathbf{S}}^{st})$ and $\mathcal{E}(\hat{\mathbf{S}}^{st}_{\top})$ with the equation $\hat{c}_{A,A} = \mathbf{1}_{A\wedge A}$, which yields preordering (see §6.5), does not add new equations to $\hat{\mathbf{S}}^{div}$ and $\hat{\mathbf{S}}^{div}_{\top}$. We will rely on that in §7.6.

CHAPTER 6

BIASSOCIATIVE CATEGORIES

In this chapter we prove coherence, in the sense of preordering, for categories that have two monoidal structures, with or without unit objects. We explain what are the effects of strictifying this double monoidal structure. With the help of that, we establish also coherence for categories with two symmetric monoidal structures. The proofs of the present chapter are based on the proofs of the preceding two chapters.

§6.1. Coherence of biassociative and bimonoidal categories

Let $\check{\mathbf{A}}$ be the natural logical category in $\mathcal{L}_\vee$ isomorphic to $\hat{\mathbf{A}}$ of §4.3. The only difference is that $\wedge$ is everywhere replaced by $\vee$. The primitive arrow terms of $\mathcal{C}(\check{\mathbf{A}})$ are included in $\mathbf{1}$ and $\check{b}$, while the equations of $\mathcal{E}(\check{\mathbf{A}})$ are obtained by replacing $\wedge$ by $\vee$ in those of $\mathcal{E}(\hat{\mathbf{A}})$ (see the List of Equations and the List of Categories at the end of the book).

To obtain the natural logical category $\mathbf{A}$, we have that the logical system $\mathcal{C}(\mathbf{A})$ is in $\mathcal{L}_{\wedge,\vee}$, with the transformations α included in the families $\mathbf{1}$ and b. The specific equations of $\mathcal{E}(\mathbf{A})$ are obtained by taking the union of those of $\mathcal{E}(\hat{\mathbf{A}})$ and $\mathcal{E}(\check{\mathbf{A}})$. We call natural $\mathbf{A}$-categories *biassociative* categories.

An arrow term of $\mathcal{C}(\mathbf{A})$ is called $\rightarrow$*-directed* when neither of $\hat{b}^{\leftarrow}$ and $\check{b}^{\leftarrow}$ occurs in it.

We define inductively as follows formulae of $\mathcal{L}_{\wedge,\vee}$ in *normal form*:

every letter is in normal form;

if A and B are in normal form and B is not of the form $B_1 \wedge B_2$, then $A \wedge B$ is in normal form;

> if A and B are in normal form and B is not of the form $B_1 \vee B_2$, then $A \vee B$ is in normal form.

So all parentheses within conjunctions and disjunctions are associated to the left as much as possible.

Then we can easily prove the Directedness Lemma of §4.3 for $\mathbf{A}$ by extending the proof in §4.3. From that we infer as before the following.

BIASSOCIATIVE COHERENCE. *The category* $\mathbf{A}$ *is a preorder.*

Let $\check{\mathbf{A}}_\bot$ be the natural logical category in $\mathcal{L}_{\vee,\bot}$ isomorphic to $\hat{\mathbf{A}}_\top$ of §4.6. The only difference is that $\wedge$ and $\top$ are everywhere replaced by $\vee$ and $\bot$ respectively. The primitive arrow terms of $\mathcal{C}(\check{\mathbf{A}}_\bot)$ are included in $\mathbf{1}$, $\check{b}$ and $\check{\delta}$-$\check{\sigma}$, while the equations of $\mathcal{E}(\check{\mathbf{A}}_\bot)$ are obtained by replacing $\wedge$ and $\top$ by $\vee$ and $\bot$ respectively in those of $\mathcal{E}(\hat{\mathbf{A}}_\top)$ (see the List of Equations and the List of Categories).

To obtain the natural logical category $\mathbf{A}_{\top,\bot}$, we have that the logical system $\mathcal{C}(\mathbf{A}_{\top,\bot})$ is in $\mathcal{L}_{\wedge,\vee,\top,\bot}$, with the transformations α included in the families $\mathbf{1}$, b and δ-σ. The specific equations of $\mathcal{E}(\mathbf{A}_{\top,\bot})$ are obtained by taking the union of those of $\mathcal{E}(\hat{\mathbf{A}}_\top)$ and $\mathcal{E}(\check{\mathbf{A}}_\bot)$. We call natural $\mathbf{A}_{\top,\bot}$-categories *bimonoidal* categories.

An arrow term of $\mathcal{C}(\mathbf{A}_{\top,\bot})$ is called $\rightarrow$*-directed* when neither of $\hat{b}^{\leftarrow}$, $\check{b}^{\leftarrow}$, $\hat{\delta}^{\leftarrow}$, $\check{\delta}^{\leftarrow}$, $\hat{\sigma}^{\leftarrow}$ and $\check{\sigma}^{\leftarrow}$ occurs in it.

We define inductively as follows formulae of $\mathcal{L}_{\wedge,\vee,\top,\bot}$ in *normal form*:

> every letter and the nullary connectives $\top$ and $\bot$ are in normal form;
>
> if A and B are in normal form, B is not of the form $B_1 \wedge B_2$ and neither A nor B is $\top$, then $A \wedge B$ is in normal form;
>
> if A and B are in normal form, B is not of the form $B_1 \vee B_2$ and neither A nor B is $\bot$, then $A \vee B$ is in normal form.

So, as for the normal form of formulae of $\mathcal{L}_{\wedge,\vee}$, all parentheses within conjunctions and disjunctions are associated to the left as much as possible, and, moreover, the conjuncts $\top$ and disjuncts $\bot$ are deleted.

Then we can prove the Directedness Lemma of §4.3 for $\mathbf{A}_{\top,\bot}$ by extending the proof in §4.3 and §4.6. From that we infer as before the following.

BIMONOIDAL COHERENCE. *The category* $\mathbf{A}_{\top,\bot}$ *is a preorder.*

§6.2. Form sequences

The classes $[A]$ involved in $(\mathcal{C}/\mathcal{E},\hat{\mathbf{A}})$-strictification correspond bijectively to finite nonempty sequences of objects (see §3.1 and §4.5). With $(\mathcal{C}/\mathcal{E},\mathbf{A})$-strictification the classes $[A]$ correspond analogously to a more complicated notion of sequence where we distinguish concatenation of the $\wedge$ kind from concatenation of the $\vee$ kind. To define this notion, let $\mathcal{X}$ be an arbitrary set, and let $\xi \in \{\wedge, \vee\}$. If ξ is $\wedge$, then ξ^c is $\vee$, and if ξ is $\vee$, then ξ^c is $\wedge$.

We define inductively as follows the notion of *form sequence of* $\mathcal{X}$ *of colour* ξ:

(1) every $x \in \mathcal{X}$ is a form sequence of $\mathcal{X}$ of colour ξ;

(2) if $X_1 \ldots X_n$, where $n \geq 2$, is a sequence of form sequences of colour ξ^c, then the ordered pair $(X_1 \ldots X_n,\ \xi)$ is a form sequence of $\mathcal{X}$ of colour ξ.

(*Finite nonempty form sequence* would be a more precise, but less concise, denomination for the notion of form sequence just introduced. We will introduce below a more general notion of form sequence that covers also empty form sequences of both colours.)

It is easy to see that every form sequence of $\mathcal{X}$ of colour ξ corresponds to a planar finite tree (see §2.1) with nodes of n-ary branching where $n \geq 2$, such that every leaf is labelled by an element of $\mathcal{X}$, every node that is not a leaf is labelled by $\wedge$ or $\vee$, for every node labelled $\beta \in \{\wedge, \vee\}$ its successor is labelled β^c, provided this successor is not a leaf, and the root is labelled ξ, provided this root is not a leaf.

We introduce now an alternative notation for form sequences, which is obtained by writing $(X_1\, \xi\, X_2\, \xi\, \ldots\, \xi\, X_n)$ for $(X_1 \ldots X_n,\ \xi)$ in clause (2) of the definition above. We call this notation, which we will need in latter sections, the *natural notation* for form sequences. The natural notation for $[A]$ may be conceived as obtained from any formula in $[A]$ by deleting parentheses corresponding to ξ in the immediate scope of ξ. For example, we replace $((p\wedge q)\wedge r)\vee s$ by $(p\wedge q\wedge r)\vee s$. Note that for form sequences in natural notation the variables for form sequences $X, Y, \ldots$ stand in different contexts for different syntactic objects. For example, if X is the form sequence $(p \wedge q \wedge r)$, then in $X \wedge s$, the variable X stands for $p \wedge q \wedge r$, while in $X \vee s$ it stands for $(p \wedge q \wedge r)$. As usually done, with other kinds of formulae and terms, we omit the outermost parentheses in natural notation.

In the context of $(\mathcal{C}/\mathcal{E},\mathbf{A})$-strictification, we may use as arrow terms for arrows of the directly strictified category, arrow terms in natural notation, i.e. arrow terms in which parentheses corresponding to ξ in the immediate scope of ξ are deleted, as above. For example, we replace $(((f_1 \wedge f_2) \wedge f_3) \vee f_4) \circ f_5$ by $((f_1 \wedge f_2 \wedge f_3) \vee f_4) \circ f_5$. Such arrow terms correspond to planar finite trees if $\circ$ does not occur in them.

If $\mathcal{B}_{ob}$ is the set of objects of a $\mathcal{C}/\mathcal{E}$ category $\mathcal{B}$, and if $\mathcal{G}$ is $\mathbf{A}$, and is related to $\mathcal{C}/\mathcal{E}$ as in (IV$\mathcal{C}$) and (IV$\mathcal{G}$) of §3.1, then the objects of $\mathcal{B}^{\mathcal{G}}$ correspond to form sequences of $\mathcal{B}_{ob}$ of both colours. If $X, X_1, \ldots, X_1', \ldots$ stand for form sequences of $\mathcal{B}_{ob}$ of colour ξ^c (if $X \in \mathcal{B}_{ob}$, then X is both of colour ξ^c and ξ), then we define the operations $\xi'' \in \{\wedge'', \vee''\}$ on the objects of $\mathcal{B}^{\mathcal{G}}$ in the following manner:

$$\begin{aligned}
&X_1\, \xi''\, X_2 =_{df} (X_1X_2,\ \xi),\\
&X\, \xi''\, (X_1 \ldots X_n,\ \xi) =_{df} (XX_1 \ldots X_n,\ \xi),\\
&(X_1 \ldots X_n,\ \xi)\, \xi''\, X =_{df} (X_1 \ldots X_nX,\ \xi),\\
&(X_1 \ldots X_n,\ \xi)\, \xi''\, (X_1' \ldots X_m',\ \xi) =_{df} (X_1 \ldots X_nX_1' \ldots X_m',\ \xi),
\end{aligned}$$

for $n, m \geq 2$. The operation ξ'' is, intuitively, concatenation of the ξ kind.

For the classes $[A]$ involved in $(\mathcal{C}/\mathcal{E},\mathbf{A}_{\top,\bot})$-strictification, we have to extend the notion of form sequence to take into account the empty sequences of colours $\wedge$ and $\vee$. For the formal definition that follows, let $\mathcal{X}$ be an arbitrary set, and let ξ^c for $\xi \in \{\wedge, \vee\}$ be as before.

A *form sequence of $\mathcal{X}$ of colour ξ* is defined inductively, as before, with the clauses (1), (2) and the following additional clause:

(0) if $\emptyset$ is the empty sequence of elements of $\mathcal{X}$, then $(\emptyset, \xi)$ is a form sequence of $\mathcal{X}$ of colour ξ.

When we need to distinguish the previous notion of form sequence from the new notion just introduced, we call the former notion *nonextended* and the latter one *extended.* Planar trees corresponding to form sequences in the extended sense have leaves labelled by elements of $\mathcal{X}$ or by $(\emptyset, \xi)$.

For $\mathcal{G}$ being $\mathbf{A}_{\top,\bot}$, we define the operations ξ'' and ζ'' on the objects of $\mathcal{B}^{\mathcal{G}}$ conceived as form sequences in the extended sense in the following manner. For $\top''$ we take $(\emptyset, \wedge)$, which corresponds to $[\top^{\mathcal{G}}]$, and for $\bot''$ we take $(\emptyset, \vee)$, which corresponds to $[\bot^{\mathcal{G}}]$. For ξ'' we enlarge the definition above with

$$Y\,\xi''\,(\emptyset, \xi) =_{df} Y,$$
$$(\emptyset, \xi)\,\xi''\,Y =_{df} Y,$$

for Y any form sequence of $\mathcal{B}_{ob}$.

§6.3. Coherence of symmetric biassociative categories

Let $\check{\mathbf{S}}$ be the natural logical category in $\mathcal{L}_{\vee}$ isomorphic to $\hat{\mathbf{S}}$ of §5.1. The difference is that $\wedge$ is everywhere replaced by $\vee$. The primitive arrow terms of $\mathcal{C}(\check{\mathbf{S}})$ are included in $\mathbf{1}$, $\check{b}$ and $\check{c}$, while the equations of $\mathcal{E}(\check{\mathbf{S}})$ are obtained by replacing $\wedge$ by $\vee$, and by permuting the indices of c in the equations of $\mathcal{E}(\hat{\mathbf{S}})$. So we obtain the equations $(\check{c}\,\check{c})$ and $(\check{b}\,\check{c})$ (see the List of Equations and the List of Categories).

To obtain the natural logical category $\mathbf{S}$, we have that the logical system $\mathcal{C}(\mathbf{S})$ is in $\mathcal{L}_{\wedge,\vee}$, with the transformations α included in the families $\mathbf{1}$, b and c. The specific equations of $\mathcal{E}(\mathbf{S})$ are obtained by taking the union of those of $\mathcal{E}(\hat{\mathbf{S}})$ and $\mathcal{E}(\check{\mathbf{S}})$. We call natural $\mathbf{S}$-categories *symmetric biassociative* categories.

For $\mathcal{G}$ being $\mathbf{A}$ and $\mathcal{C}/\mathcal{E}$ being $\mathbf{S}$, we have that (IV$\mathcal{C}$) and (IV$\mathcal{G}$) of §3.1 are satisfied, and $\mathcal{G}$ is moreover a preorder. Thus we can apply the Direct-Strictification Theorem of §3.2 to obtain a category $\mathcal{C}_{\mathcal{G}}/\mathcal{E}_{\mathcal{G}}$, which we will call $\mathbf{S}^{st}$. We call $\mathcal{C}_{\mathcal{G}}$ here $\mathcal{C}(\mathbf{S}^{st})$.

As in §3.2, we have the functor GH from $\mathbf{S}^{st}$ to *Rel*, and it is enough to show that this functor is faithful to conclude the following.

SYMMETRIC BIASSOCIATIVE COHERENCE. *The functor G from $\mathbf{S}$ to Rel is faithful.*

In the remainder of this section we prove that the functor GH from $\mathbf{S}^{st}$ to *Rel* is faithful.

For $\xi \in \{\wedge, \vee\}$, a $\overset{\xi}{c}$-term of $\mathcal{C}(\mathbf{S}^{st})$ is called *atomized* when, for every arrow term $\mathbf{1}_X$ occurring in it, X is $[\![p]\!]$ for some letter p of $\mathcal{P}$. Atomized $\overset{\xi}{c}$-terms may be designated by composition-free arrow terms in natural notation, as explained in the preceding section, and these arrow terms correspond to planar finite trees analogous to those that correspond to form sequences. To every atomized $\overset{\xi}{c}$-term we assign a planar finite tree with nodes of n-ary branching, where $n \geq 2$, such that exactly one leaf λ is labelled with the head $\overset{\xi}{c}_{X,Y}$ of our $\overset{\xi}{c}$-term, and all the other leaves are labelled with arrow terms of the form $\mathbf{1}_p$. Nodes that are not leaves are labelled with $\wedge$ or $\vee$, and for every node labelled $\beta \in \{\wedge, \vee\}$ its successor is labelled β^c, provided this successor is not a leaf.

Let ν^λ be either the leaf λ above when the predecessor of λ is labelled with ξ^c or λ is the root, or else let ν^λ be the predecessor of λ labelled with ξ. The *level* $l(f)$ of an atomized $\overset{\xi}{c}$-term f is the height of ν^λ (see §2.1 for this notion of height). The *span* $s(f)$ of an atomized $\overset{\xi}{c}$-term f is the number of nodes of the same height as ν^λ on the left-hand side of ν^λ.

It is easy to see that with the help of the bifunctorial and naturality equations every arrow term of $\mathcal{C}(\mathbf{S}^{st})$ is equal in $\mathbf{S}^{st}$ to a developed arrow term $f_n \circ \ldots \circ f_1 \circ \mathbf{1}_X$ where every factor f_i is an atomized $\overset{\xi}{c}$-term, and if $1 \leq i < j \leq n$, then $l(f_i) \leq l(f_j)$ and $s(f_i) \leq s(f_j)$. It is also easy to see that for every arrow term $f : X \vdash Y$ of $\mathcal{C}(\mathbf{S}^{st})$ there is an arrow term $f^{-1} : Y \vdash X$ of $\mathcal{C}(\mathbf{S}^{st})$ such that $f^{-1} \circ f = \mathbf{1}_X$ and $f \circ f^{-1} = \mathbf{1}_Y$ in $\mathbf{S}^{st}$. From that we conclude that to show the faithfulness of GH it is enough to prove for $f : X \vdash X$ that if $GHf = GH\mathbf{1}_X$, then $f = \mathbf{1}_X$ in $\mathbf{S}^{st}$.

Let $f : X \vdash X$ be the developed arrow term $f_n \circ \ldots \circ f_k \circ \ldots \circ f_1 \circ \mathbf{1}_X$ of the kind described above, such that $l(f_n) = \ldots = l(f_k)$, $s(f_n) = \ldots = s(f_k)$, and either $l(f_k) > l(f_{k-1})$ or $s(f_k) > s(f_{k-1})$. Suppose $GHf = GH\mathbf{1}_X$. If $GH(f_n \circ \ldots \circ f_k) = GH\mathbf{1}_X$, then by the faithfulness of GH proved in §5.2 we can conclude that $f_n \circ \ldots \circ f_k = \mathbf{1}_X$ in $\mathbf{S}^{st}$, and we must have $GH(f_n \circ \ldots \circ f_k) = GH\mathbf{1}_X$, because, otherwise, according to our conditions on l and s, we could not have $GHf = GH\mathbf{1}_X$. We repeat this reasoning with $f_{k-1} \circ \ldots \circ f_1 \circ \mathbf{1}_X$, until we obtain that $f = \mathbf{1}_X$ in $\mathbf{S}^{st}$. This concludes the demonstration of the faithfulness of GH, from which we infer Symmetric Biassociative Coherence.

§6.4. Coherence of symmetric bimonoidal categories

Let $\check{\mathbf{S}}_{\bot}$ be the natural logical category in $\mathcal{L}_{\vee,\bot}$ isomorphic to $\hat{\mathbf{S}}_{\top}$ of §5.3. The difference is that $\wedge$ and $\top$ are everywhere replaced by $\vee$ and $\bot$ respectively. The primitive arrow terms of $\mathcal{C}(\check{\mathbf{S}}_{\bot})$ are included in $\mathbf{1}$, $\check{b}$, $\check{c}$ and $\check{\delta}$-$\check{\sigma}$, while the equations of $\mathcal{E}(\check{\mathbf{S}}_{\bot})$ are obtained by replacing $\wedge$ and $\top$ by $\vee$ and $\bot$ respectively, and by permuting the indices of c in the equations of $\mathcal{E}(\hat{\mathbf{S}}_{\top})$.

To obtain the natural logical category $\mathbf{S}_{\top,\bot}$, we have that the logical system $\mathcal{C}(\mathbf{S}_{\top,\bot})$ is in $\mathcal{L}_{\wedge,\vee,\top,\bot}$, with the transformations α included in the families $\mathbf{1}$, b, c and δ-σ. The specific equations of $\mathcal{E}(\mathbf{S}_{\top,\bot})$ are obtained by taking the union of those of $\mathcal{E}(\hat{\mathbf{S}}_{\top})$ and $\mathcal{E}(\check{\mathbf{S}}_{\bot})$ plus

$$(\hat{c}\,\bot)\quad \hat{c}_{C,C} = \mathbf{1}_{C\wedge C},$$
$$(\check{c}\,\top)\quad \check{c}_{C,C} = \mathbf{1}_{C\vee C},$$

provided C is a letterless formula of $\mathcal{L}_{\wedge,\vee,\top,\bot}$.

From $(\hat{c}\,\bot)$ and $(\check{c}\,\top)$ we can derive the following equations. If $h\colon C \vdash D$ and $h^{-1} : D \vdash C$ are mutually inverse arrows of $\mathbf{S}_{\top,\bot}$, with C and D letterless formulae of $\mathcal{L}_{\wedge,\vee,\top,\bot}$, then in $\mathcal{E}(\mathbf{S}_{\top,\bot})$ we have

$$(\hat{c}\,h)\quad \hat{c}_{C,D} = (h \wedge \mathbf{1}_C)\circ(\mathbf{1}_C \wedge h^{-1}),$$
$$(\check{c}\,h)\quad \check{c}_{D,C} = (h \vee \mathbf{1}_C)\circ(\mathbf{1}_C \vee h^{-1}).$$

To derive $(\hat{c}\,h)$ we have

$$\begin{aligned}(h \wedge \mathbf{1}_C)\circ(\mathbf{1}_C \wedge h^{-1}) &= (h \wedge \mathbf{1}_C)\circ \hat{c}_{C,C}\circ(\mathbf{1}_C \wedge h^{-1}),\ \text{by } (\hat{c}\,\bot)\\ &= \hat{c}_{C,D},\end{aligned}$$

by naturality equations, bifunctorial equations, and by h being an isomorphism; we proceed analogously for $(\check{c}\,h)$. Conversely, we obtain $(\hat{c}\,\bot)$ and $(\check{c}\,\top)$ from $(\hat{c}\,h)$ and $(\check{c}\,h)$ by putting $\mathbf{1}_C\colon C \vdash C$ for h and h^{-1}.

We call natural $\mathbf{S}_{\top,\bot}$-categories *symmetric bimonoidal* categories.

The category *Set* of sets with functions is a bimonoidal category with $\wedge$ being cartesian product, $\vee$ being disjoint union, $\top$ being a singleton and $\bot$ being the empty set. But, although the instance $\hat{c}_{\bot,\bot} = \mathbf{1}_{\bot\wedge\bot}$ of $(\hat{c}\,\bot)$ holds in *Set*, the instance $\check{c}_{\top,\top} = \mathbf{1}_{\top\vee\top}$ of $(\check{c}\,\top)$ does not hold. So *Set* is not a symmetric bimonoidal category in the sense just specified, though it is a symmetric biassociative category.

Let the category $\mathbf{S}^{st}_{\top,\bot}$ be obtained as $\mathbf{S}^{st}$ in the preceding section by taking that $\mathcal{G}$ is $\mathbf{A}_{\top,\bot}$ instead of $\mathbf{A}$, and that $\mathcal{C}/\mathcal{E}$ is $\mathbf{S}_{\top,\bot}$. We call $\mathcal{C}_{\mathcal{G}}$ here $\mathcal{C}(\mathbf{S}^{st}_{\top,\bot})$.

A *constant object* of $\mathbf{S}^{st}_{\top,\bot}$ is $[\![A]\!]$ where A is a letterless formula of $\mathcal{L}_{\wedge,\vee,\top,\bot}$. The remaining objects of $\mathbf{S}^{st}_{\top,\bot}$ are called *variable objects.* The constant objects of $\mathbf{S}^{st}_{\top,\bot}$ are denumerable.

The arrow term $\overset{\xi}{c}_{X,Y}$ of $\mathcal{C}(\mathbf{S}^{st}_{\top,\bot})$, for $\xi \in \{\wedge,\vee\}$, is called *basic* when the following two conditions are satisfied: first, the form sequences (in the extended sense) corresponding to X and Y are of colour ξ^c, and, secondly, if the objects X and Y are both constant, then there is no arrow term of $\mathcal{C}(\mathbf{S}^{st}_{\top,\bot})$ of type $X \vdash Y$ (hence there is neither an arrow term of type $Y \vdash X$).

A developed arrow term $f_n \circ \ldots \circ f_1 \circ \mathbf{1}_Z$ of $\mathcal{C}(\mathbf{S}^{st}_{\top,\bot})$ such that for every $i \in \{1,\ldots,n\}$ the head of f_i is a basic arrow term $\overset{\xi}{c}_{X,Y}$ is called *basically developed.* We can prove the following.

BASIC-DEVELOPMENT LEMMA 1. *Every arrow term of* $\mathcal{C}(\mathbf{S}^{st}_{\top,\bot})$ *is equal in* $\mathbf{S}^{st}_{\top,\bot}$ *to a basically developed arrow term.*

PROOF. For $f\colon X \vdash Y$ an arrow term of $\mathcal{C}(\mathbf{S}^{st}_{\top,\bot})$, we proceed by induction on the number $n \geq 1$ of nodes in the planar finite tree corresponding to X (which must be equal to the number of nodes in the planar finite tree corresponding to Y). If $n = 1$, then $X = Y$, and f must be equal to $\mathbf{1}_{[\![p]\!]}$ or $\mathbf{1}_{(\emptyset,\xi)}$. For the induction step, we find first a developed arrow term equal to f, and then, by using the equations $(\overset{\xi}{b}\overset{\xi}{c})$, $(\overset{\xi}{c}\overset{\xi}{\delta}\overset{\xi}{\sigma})$ and $(\overset{\xi}{c}\, h)$, together with the induction hypothesis, we transform this arrow term into a basically developed one. ⊣

Analogously to what we had in the preceding section, we find for every $\overset{\xi}{c}$-term f its level $l(f)$. (Atomization is not here essential, since it leaves the level invariant.) Then we have the following.

BASIC-DEVELOPMENT LEMMA 2. *Every arrow term of* $\mathcal{C}(\mathbf{S}^{st}_{\top,\bot})$ *is equal in* $\mathbf{S}^{st}_{\top,\bot}$ *to a basically developed arrow term* $h_k \circ \ldots \circ h_1 \circ \mathbf{1}_X$ *such that* $k \geq 0$ *and, if* $k > 1$*, then for* $1 \leq i < j \leq k$ *we have* $l(h_i) \leq l(h_j)$.

To prove this lemma, we use Basic-Development Lemma 1 together with bifunctorial and naturality equations, which do not spoil basic development.

BASIC-DEVELOPMENT LEMMA 3. *Every arrow term $f : (X_1 \ldots X_n,\, \xi) \vdash (Y_1 \ldots Y_n,\, \xi)$ of $\mathcal{C}(\mathbf{S}^{st}_{\top,\bot})$, with $n \geq 2$, is equal in $\mathbf{S}^{st}_{\top,\bot}$ to an arrow term of the form*

$$(\ldots(f_1 \,\xi\, f_2)\,\xi\, \ldots \,\xi\, f_n) \circ h_k \circ \ldots \circ h_1 \circ \mathbf{1}_X$$

such that f_i, for $i \in \{1,\ldots,n\}$, is of type $X_{\pi(i)} \vdash Y_i$ for π a permutation of $\{1,\ldots,n\}$, while $h_k \circ \ldots \circ h_1 \circ \mathbf{1}_X$, with $k \geq 0$, is basically developed and $l(h_j) = 1$ for every $j \in \{1,\ldots,k\}$.

To prove this lemma we just apply Basic-Development Lemma 2 and bifunctorial equations. Now we can prove the following.

SYMMETRIC BIMONOIDAL COHERENCE. *The functor G from $\mathbf{S}_{\top,\bot}$ to Rel is faithful.*

PROOF. As before, it is enough to prove that the functor GH from $\mathbf{S}^{st}_{\top,\bot}$ to *Rel* is faithful. As in the preceding section, it is enough to show for $f\colon X \vdash X$ that if $GHf = GH\mathbf{1}_X$, then $f = \mathbf{1}_X$ in $\mathbf{S}^{st}_{\top,\bot}$.

We proceed by induction on the number $n \geq 1$ of nodes in the planar finite tree corresponding to X. If $n = 1$, then f is equal to $\mathbf{1}_{[\![p]\!]}$ or $\mathbf{1}_{(\emptyset,\xi)}$. For the induction step, suppose X corresponds to the form sequence $(X_1 \ldots X_n,\, \xi)$ with $n \geq 2$. Then, by Basic-Development Lemma 3, we have that f is equal in $\mathbf{S}^{st}_{\top,\bot}$ to $(\ldots(f_1 \,\xi\, f_2)\,\xi\, \ldots \,\xi\, f_n) \circ h$ where h, which is $h_k \circ \ldots \circ h_1 \circ \mathbf{1}_X$, is an instance of an arrow term

$$h'\colon [\![(\ldots(p_1 \,\xi\, p_2)\,\xi\, \ldots \,\xi\, p_n)]\!] \vdash [\![(\ldots(p_{\pi(1)} \,\xi\, p_{\pi(2)})\,\xi\, \ldots \,\xi\, p_{\pi(n)})]\!]$$

of $\mathcal{C}(\overset{\xi}{\mathbf{S}}{}^{st})$, and the type of f_i is $X_{\pi(i)} \vdash X_i$, for $i \in \{1,\ldots,n\}$.

For every $i \in \{1,\ldots,n\}$ we must have $\pi(i) = i$. If X_i is a variable object of $\mathbf{S}^{st}_{\top,\bot}$, then this follows from the fact that f is of type $X \vdash X$ and $GHf = GH\mathbf{1}_X$. If X_i is a constant object of $\mathbf{S}^{st}_{\top,\bot}$, then this follows from the fact that h is basically developed. Otherwise, there would be in h an arrow term $\overset{\xi}{c}_{X_j,X_{\pi(j)}}$ or $\overset{\xi}{c}_{X_{\pi(j)},X_j}$ for some $j \in \{1,\ldots,n\}$; but this is impossible since f_j is of type $X_{\pi(j)} \vdash X_j$. So $GHh' = GH\mathbf{1}_{[\![A]\!]}$ for A being

$(\ldots(p_1 \xi p_2) \xi \ldots \xi p_n)$. By the faithfulness of GH from $\overset{\xi}{\mathbf{S}}^{st}$ to *Rel*, we have $h' = \mathbf{1}_{[\![A]\!]}$, from which it follows that $h = \mathbf{1}_X$ in $\mathbf{S}^{st}_{\top,\bot}$. We have also that f_i is of type $X_i \vdash X_i$, and since $GHf_i = GH\mathbf{1}_{X_i}$, by applying the induction hypothesis we obtain $f_i = \mathbf{1}_{X_i}$. It follows that $f = \mathbf{1}_X$ in $\mathbf{S}^{st}_{\top,\bot}$. ⊣

In the induction step of this proof we deal with the *least* level, while in the induction step of the proof of the faithfulness of GH from $\mathbf{S}^{st}$ to *Rel* in the preceding section we dealt with the *greatest* level. Because of that, we had to introduce there the notion of span. The preceding proof could, however, be reworked in the style of the present section—with the least level.

§6.5. The category S′

To obtain the natural logical category $\mathbf{S}'$, we take the logical system $\mathcal{C}(\mathbf{S})$ in $\mathcal{L}_{\wedge,\vee}$ of §6.3. The specific equations of $\mathcal{E}(\mathbf{S}')$ are those of $\mathcal{E}(\mathbf{S})$ plus the equations

$$(\overset{\xi}{c}\mathbf{1}) \quad \overset{\xi}{c}_{A,A} = \mathbf{1}_{A\xi A}$$

for $\xi \in \{\wedge, \vee\}$. We call $\mathbf{S}'$ the natural logical category $\mathcal{C}(\mathbf{S})/\mathcal{E}(\mathbf{S}')$. The equations $(\overset{\xi}{c}\ h)$ of the preceding section hold in $\mathbf{S}'$ with A and B being any formulae of $\mathcal{L}_{\wedge,\vee}$ such that $h\colon A \vdash B$ and $h^{-1}\colon B \vdash A$ are mutually inverse arrows of $\mathbf{S}'$. Conversely, we obtain $(\overset{\xi}{c}\mathbf{1})$ from $(\overset{\xi}{c}\ h)$ by putting $\mathbf{1}_A\colon A \vdash A$ for h and h^{-1}.

To show that $\mathbf{S}'$ is a preorder we proceed analogously to what we had in the preceding section. Let $\mathbf{S}'^{st}$ be obtained as $\mathbf{S}^{st}$ by taking that $\mathcal{G}$ is $\mathbf{A}$ and $\mathcal{C}/\mathcal{E}$ is $\mathbf{S}'$. Here $\mathcal{C}_{\mathcal{G}}$ is $\mathcal{C}(\mathbf{S}^{st})$. Basic arrow terms $\overset{\xi}{c}_{X,Y}$ of $\mathcal{C}(\mathbf{S}^{st})$ are those where X and Y are of colour ξ^c and there is no arrow term of $\mathcal{C}(\mathbf{S}^{st})$ of type $X \vdash Y$. Basically developed arrow terms are then defined as in the preceding section, and we take over also the notion of level. We can then prove analogues of Basic-Development Lemmata 1-3 of the preceding section.

To conclude the proof that $\mathbf{S}'$ is a preorder, we prove that every arrow term $f\colon X \vdash X$ of $\mathcal{C}(\mathbf{S}^{st})$ is equal in $\mathbf{S}'^{st}$ to $\mathbf{1}_X$. We proceed by induction on the number $n \geq 1$ of nodes in the planar finite tree corresponding to X. This proof is analogous to the proof of Symmetric Bimonoidal Coherence in the preceding section. In the induction step, we reason as in the case where X_i is a constant object.

Note that, since $G\,\overset{\xi}{c}_{A,A} \neq G\mathbf{1}_{A\xi A}$, we have no functor G from $\mathbf{S}'$ to *Rel*. The fact that $\mathbf{S}'$ is a preorder cannot be reformulated as a coherence theorem stating that G is faithful.

The natural logical category $\hat{\mathbf{S}}'$, whose logical system is $\mathcal{C}(\hat{\mathbf{S}})$ in $\mathcal{L}_\wedge$, has the specific equations of $\mathcal{E}(\hat{\mathbf{S}})$ with $\hat{c}_{A,A} = \mathbf{1}_{A\wedge A}$ added. We can show that $\hat{\mathbf{S}}'$ is a preorder by simplifying the argument above.

CHAPTER 7

DISSOCIATIVE CATEGORIES

In this chapter we prove coherence, in the sense of preordering, for categories with a double monoidal structure without unit objects and with the linear distribution arrows of [19]. Linear distribution is an associativity principle involving two binary operations, and we have coined for it the name *dissociativity*. This principle will yield arrows based on the usual distribution principle in Chapter 11, where the two monoidal structures are made of a product and a coproduct.

We prove beforehand coherence for categories such as those mentioned above that lack the ordinary associativity arrows. We also prove coherence in cases where dissociativity is allowed only on one side. Our method in these proofs is based on confluence techniques, like those that may be found in the lambda calculus.

Next we prove coherence for the case where the two monoidal structures with dissociativity are symmetric, and we still lack the unit objects. Here the method of proof is more involved. It is based on a cut-elimination procedure in a sequent system strictified in the symmetric monoidal structure. We justify this strictification by the results of Chapters 3-6.

We are here at the watershed as far as method is concerned. Up to this chapter, confluence techniques predominated, while, from now on, cut elimination, or its simpler version, composition elimination, will take over the stage. (The two approaches are still mixed in the next chapter.) Cut elimination could have been employed in the first part of this chapter too. For the categories treated there, both approaches are available, and we opted for the first, the second being well illustrated in the second part of the chapter.

At the end of the chapter, we consider adding the unit objects, and we present the linearly distributive categories of [19], for which coherence in our sense does not obtain. Linearly distributive categories without unit objects, with which we deal in this chapter, do not seem to have been considered separately before.

§7.1. Coherence of dissociative categories

To obtain the natural logical category **DI**, we have that the logical system $\mathcal{C}(\mathbf{DI})$ is in $\mathcal{L}_{\wedge,\vee}$, with the transformations α included in $\mathbf{1}$ and d. The equations $\mathcal{E}(\mathbf{DI})$ are just those of $\mathcal{E}^{nat}_{\mathcal{C}(\mathbf{DI})}$ (see §4.1). We call natural **DI**-categories *dissociative* categories.

We have given in §1.2 our reasons for calling *dissociativity* the principle underlying the arrow terms in d. This principle may be found in [1] (Section 15.2), [60], [84] (Section 8) and [13]. In category theory it has been introduced by Cockett and Seely (see [18] and [19]; see also [63], Section 3.2). The dissociativity principle underlying the arrow terms in d resembles the modularity law for lattices:

$$\text{if } c \leq a, \text{ then } a \wedge (b \vee c) \leq (a \wedge b) \vee c$$

(see [6], Section I.7). The condition $a \wedge (b \vee c) \leq (a \wedge b) \vee c$, without the assumption $c \leq a$, has the same force as distribution in lattices (cf. §11.3).

For x and y occurrences of $\xi \in \{\wedge, \vee\}$ in a formula A of $\mathcal{L}_{\wedge,\vee}$ we define the relation S^{ξ}_{A} such that $xS^{\xi}_{A}y$ when x is in the scope of y in A. (For the notion of scope see §2.1.) Note that for $f : A \vdash B$ being a member of the family d we have that $S^{\wedge}_{B}$ is a proper subset of $S^{\wedge}_{A}$ and $S^{\vee}_{A}$ a proper subset of $S^{\vee}_{B}$. This holds also for f being any arrow term of $\mathcal{C}(\mathbf{DI})$ in which a member of the family d occurs; otherwise $S^{\xi}_{A} = S^{\xi}_{B}$. Here A and B are comparable formulae of $\mathcal{L}_{\wedge,\vee}$ (namely, formulae that yield the same word after deleting parentheses; see §2.1), and the relations S_A and S_B can be compared, as the relations R_A and R_B were compared in §4.2. It is clear that the following holds.

REMARK. *Let the formula A of $\mathcal{L}_{\wedge,\vee}$ be of the form $A_1 \, \xi \, A_2$ for $\xi \in \{\wedge, \vee\}$. Then, for x and y being occurrences of $\wedge$ or $\vee$ in A_i, where $i \in \{1, 2\}$, we have that x is in the scope of y in A iff x is in the scope of y in A_i.*

We have the following analogue of the Extraction Lemma of §4.4, which is proved by imitating the proof in §4.4.

EXTRACTION LEMMA. *If there is an occurrence z of $\vee$ in a formula A of $\mathcal{L}_{\wedge,\vee}$ such that there is no u with $zS^{\vee}_{A}u$, then there is a formula $A_1\, z\, A_2$ of $\mathcal{L}_{\wedge,\vee}$ such that there is an arrow term $g\colon A \vdash A_1\, z\, A_2$ of $\mathcal{C}(\mathbf{DI})$. In addition,*

(∗) *for all occurrences x and y of $\wedge$ or $\vee$ in A_i, where $i \in \{1,2\}$, we have that y is in the scope of x in A iff y is in the scope of x in A_i;*

(∗∗) *every subterm of g of the form $d^{L}_{D,E,F}$ is of the type $D\wedge(E\, z\, F) \vdash (D\wedge E)\, z\, F$, and every subterm of g of the form $d^{R}_{F,E,D}$ is of the type $(F\, z\, E)\wedge D \vdash F\, z\,(E\wedge D)$.*

We do not need (∗∗) for the proof of the Theoremhood Proposition below, but we stated this condition because it is analogous to (∗∗) of previous Extraction Lemmata in §4.2 and §4.4. The following lemma is analogous to the Theoremhood Proposition of §4.2.

THEOREMHOOD PROPOSITION. *There is an arrow term $f\colon A \vdash B$ of $\mathcal{C}(\mathbf{DI})$ iff A and B are comparable formulae of $\mathcal{L}_{\wedge,\vee}$, and we have $S^{\wedge}_{B} \subseteq S^{\wedge}_{A}$ and $S^{\vee}_{A} \subseteq S^{\vee}_{B}$.*

PROOF. The direction from left to right is easy. For the other direction, we proceed by induction on the letter length $n \geq 1$ of A. If $n = 1$, then $S^{\xi}_{A} = S^{\xi}_{B} = \emptyset$, and f is $\mathbf{1}_p\colon p \vdash p$.

If $n > 1$ and B is $B_1\, x\, B_2$ for x being an occurrence of $\wedge$, then from $S^{\wedge}_{B} \subseteq S^{\wedge}_{A}$ it follows that A is of the form $A_1\, x\, A_2$. Then, by the Remark, we have $S^{\wedge}_{B_i} \subseteq S^{\wedge}_{A_i}$ and $S^{\vee}_{A_i} \subseteq S^{\vee}_{B_i}$ for $i \in \{1,2\}$, and so, by the induction hypothesis, we have the arrow terms $f_i\colon A_i \vdash B_i$ of $\mathcal{C}(\mathbf{DI})$. The arrow term f is $f_1 \wedge f_2$.

If $n > 1$ and B is $B_1\, z\, B_2$ for z being an occurrence of $\vee$, then from $S^{\vee}_{A} \subseteq S^{\vee}_{B}$ we conclude that there is no u such $zS^{\vee}_{A}u$. So, by the Extraction Lemma of this section, there is an arrow term $g\colon A \vdash A_1\, z\, A_2$ of $\mathcal{C}(\mathbf{DI})$ such that (∗) of the Extraction Lemma holds. Since for x and y in A_i we have

$xS^{\xi}_{A_i}y$ iff $xS^{\xi}_{A}y$, by the induction hypothesis and the Remark we have the arrow terms $f_i \colon A_i \vdash B_i$ of $\mathcal{C}(\mathbf{DI})$, and f is $(f_1 \vee f_2) \circ g$. $\dashv$

As explained after the proof of the Theoremhood Proposition of §4.2, with the Theoremhood Proposition we have just proved we have solved the theoremhood problem for the category **DI**.

For a formula A let $d(A)$ be the cardinality of the set of ordered pairs $S^{\wedge}_{A}$. If $f : A \vdash B$ of **DI** is not equal to $\mathbf{1}_A : A \vdash A$, then $d(B) < d(A)$. We can prove the following.

DISSOCIATIVE COHERENCE. *The category* **DI** *is a preorder.*

PROOF. Let $f, g \colon A \vdash B$ be arrow terms of $\mathcal{C}(\mathbf{DI})$. We proceed by induction on $d(A) - d(B)$ to show that $f = g$ in **DI**. (Until the end of this proof, we assume that equality of arrow terms is equality in **DI**.) If $d(A) = d(B)$, then we conclude that $f = g = \mathbf{1}_A$.

Suppose $d(B) < d(A)$. By the Development Lemma of §2.7, we have that $f = f_2 \circ f_1$ and $g = g_2 \circ g_1$ for some d-terms $f_1 : A \vdash C$ and $g_1 : A \vdash D$, and some arrow terms $f_2 : C \vdash B$ and $g_2 : D \vdash B$ of $\mathcal{C}(\mathbf{DI})$. We have $d(C), d(D) < d(A)$. The following cases may arise.

(LL) The head of f_1 is $d^L_{E,F,G}$, and the head of g_1 is $d^L_{H,I,J}$. Under (LL), we have the following subcases.

$(LL1)$ The subformulae $E \wedge (F \vee G)$ and $H \wedge (I \vee J)$ have no occurrences of letters in common within A. Then we use bifunctorial equations to obtain two d-terms $f_2' : C \vdash B'$ and $g_2' : D \vdash B'$ such that $f_2' \circ f_1 = g_2' \circ g_1$. Then we can infer that $S^{\wedge}_{C} \cap S^{\wedge}_{D} = S^{\wedge}_{B'}$ and $S^{\vee}_{C} \cup S^{\vee}_{D} = S^{\vee}_{B'}$, from which it follows from the Theoremhood Proposition of this section that $S^{\wedge}_{B} \subseteq S^{\wedge}_{B'}$ and $S^{\vee}_{B'} \subseteq S^{\vee}_{B}$. Hence, again by the Theoremhood Proposition, there is an arrow term $h \colon B' \vdash B$ of $\mathcal{C}(\mathbf{DI})$. By applying the induction hypothesis, we obtain that $f_2 = h \circ f_2'$ and $g_2 = h \circ g_2'$, from which $f = g$ follows.

$(LL2)$ Suppose $E \wedge (F \vee G)$ is a subformula of H or of I or of J; or, conversely, suppose that $H \wedge (I \vee J)$ is a subformula of E or of F or of G. Then we proceed as in $(LL1)$ by using $(d^L\ \text{nat})$.

$(LL3)$ Suppose, finally, that E is H, F is I, and G is J. Then C is D, and $f_1 = g_1$. We then apply the induction hypothesis to $f_2, g_2 \colon C \vdash B$ and obtain $f = g$.

(LR) The head of f_1 is $d^L_{E,F,G}$, and the head of g_1 is $d^R_{J,I,H}$. Under (LR), we have two subcases that are settled analogously to $(LL1)$ and $(LL2)$. There are no remaining subcases under (LR). It might seem that E could be $J \vee I$, while $F \vee G$ is H; in other words, $E \wedge (F \vee G)$ and $(J \vee I) \wedge H$ would the same subformula of A of the form $(J\,x\,I) \wedge (F\,y\,G)$ for x and y occurrences of $\vee$. Then we would have $xS^\vee_C y$ and $yS^\vee_D x$, and, by the Theoremhood Proposition of this section, we would have both $xS^\vee_B y$ and $yS^\vee_B x$, which is a contradiction.

It remains to consider the following cases.

(RR) The head of f_1 is $d^R_{G,F,E}$, and the head of g_1 is $d^R_{J,I,H}$.

(RL) The head of f_1 is $d^R_{G,F,E}$, and the head of g_1 is $d^L_{H,I,J}$.

The case (RR) is settled analogously to (LL), while the case (RL) is the same as (LR). ⊣

It is not difficult to see that $S^\wedge_A = S^\wedge_B$ and $S^\vee_A = S^\vee_B$ implies that A and B are the same formula of $\mathcal{L}_{\wedge,\vee}$. Because, if $S^\wedge_A = S^\wedge_B$ and $S^\vee_A = S^\vee_B$, then, by the Theoremhood Proposition of this section, there is an arrow term $f \colon A \vdash B$ of $\mathcal{C}(\mathbf{DI})$, in which d^L and d^R cannot occur, because $S^\wedge_B$ is not a proper subset of $S^\wedge_A$ and $S^\vee_A$ is not a proper subset of $S^\vee_B$. Hence f must stand for an identity arrow. So there is a bijection between the objects A of $\mathbf{DI}$ and the pairs of relations $(S^\wedge_A, S^\vee_A)$. From Dissociative Coherence, we can conclude that $\mathbf{DI}$ is isomorphic to the category whose objects are such pairs, and where an arrow exists between $(S^\wedge_A, S^\vee_A)$ and $(S^\wedge_B, S^\vee_B)$ when $S^\wedge_B \subseteq S^\wedge_A$ and $S^\vee_A \subseteq S^\vee_B$. Note that, as the category $\hat{\mathbf{A}}^{\rightarrow}$ of §4.2, the category $\mathbf{DI}$ is not just a preorder, but a partial order.

§7.2. Net categories

To obtain the natural logical category $\mathbf{DA}$, we have that the logical system $\mathcal{C}(\mathbf{DA})$ is in $\mathcal{L}_{\wedge,\vee}$ with the transformations α included in $\mathbf{1}$, b and d. The specific equations of $\mathcal{E}(\mathbf{DA})$ are those of $\mathcal{E}(\mathbf{A})$ plus

$$
\begin{array}{ll}
(d^L\wedge) & d^L_{A\wedge B,C,D} = (\hat{b}^{\rightarrow}_{A,B,C} \vee \mathbf{1}_D) \circ d^L_{A,B\wedge C,D} \circ (\mathbf{1}_A \wedge d^L_{B,C,D}) \circ \hat{b}^{\leftarrow}_{A,B,C\vee D}, \\
(d^L\vee) & d^L_{D,C,B\vee A} = \check{b}^{\leftarrow}_{D\wedge C,B,A} \circ (d^L_{D,C,B} \vee \mathbf{1}_A) \circ d^L_{D,C\vee B,A} \circ (\mathbf{1}_D \wedge \check{b}^{\rightarrow}_{C,B,A}), \\
(d^R\wedge) & d^R_{D,C,B\wedge A} = (\mathbf{1}_D \vee \hat{b}^{\leftarrow}_{C,B,A}) \circ d^R_{D,C\wedge B,A} \circ (d^R_{D,C,B} \wedge \mathbf{1}_A) \circ \hat{b}^{\rightarrow}_{D\vee C,B,A}, \\
(d^R\vee) & d^R_{A\vee B,C,D} = \check{b}^{\rightarrow}_{A,B,C\wedge D} \circ (\mathbf{1}_A \vee d^R_{B,C,D}) \circ d^R_{A,B\vee C,D} \circ (\check{b}^{\leftarrow}_{A,B,C} \wedge \mathbf{1}_D), \\
(d\,\hat{b}) & d^R_{A\wedge B,C,D} \circ (d^L_{A,B,C} \wedge \mathbf{1}_D) = d^L_{A,B,C\wedge D} \circ (\mathbf{1}_A \wedge d^R_{B,C,D}) \circ \hat{b}^{\leftarrow}_{A,B\vee C,D}, \\
(d\,\check{b}) & (d^R_{A,B,C} \vee \mathbf{1}_D) \circ d^L_{A\vee B,C,D} = \check{b}^{\rightarrow}_{A,B\wedge C,D} \circ (\mathbf{1}_A \vee d^L_{B,C,D}) \circ d^R_{A,B,C\vee D}.
\end{array}
$$

Note that, after replacing $\vee$ by $\wedge$, the arrow term $d^L_{A,B,C}$ is of the same type as $\hat{b}^{\rightarrow}_{A,B,C}$, and, after replacing $\wedge$ by $\vee$, it is of the same type as $\check{b}^{\rightarrow}_{A,B,C}$. Dually, $d^R_{A,B,C}$ is of the type of $\overset{\xi}{b}{}^{\leftarrow}_{A,B,C}$ after these replacements. After such replacements, the equations (d^L *nat*) and (d^R *nat*) become the equations ($\hat{b}$ *nat*) and ($\check{b}$ *nat*) (see §2.7), while all the specific equations of $\mathcal{E}(\mathbf{DA})$ that are added to those of $\mathcal{E}(\mathbf{A})$ are related to the pentagonal equations ($\hat{b}\,5$) and ($\check{b}\,5$) (see §4.2 and the List of Equations at the end of the book). We may obtain all of these equations by starting from ($\hat{b}\,5$) and replacing one or two occurrences of $\wedge$ by $\vee$ in each of the types, at the same place. When only one occurrence is replaced, this forces three or four $\hat{b}$-terms to become d^L-terms or d^R-terms, and yields the equations ($d^L\wedge$), ($d^R\wedge$) and ($d\,\hat{b}$). The remaining three equations are obtained analogously from ($\check{b}\,5$) by replacing one occurrence of $\vee$ by $\wedge$. This covers all replacements of $\wedge$ by $\vee$ in ($\hat{b}\,5$), since the replacements in ($\check{b}\,5$) may be conceived as replacements of two occurrences of $\wedge$ by $\vee$ in ($\hat{b}\,5$). (When all the three occurrences of $\wedge$ are replaced by $\vee$ in ($\hat{b}\,5$), we obtain ($\check{b}\,5$).) There are many symmetries in these equations.

We call natural **DA**-categories *net* categories. Officially, in our nomenclature they would be called *dissociative biassociative* categories. A reason for switching to the handy denomination "net" is in the connection with the proof nets of linear logic (see [58] and [30]). The linearly distributive categories of [19] (the old denomination of these categories is "weakly distributive"; cf. [22] for the renaming) are net categories in the sense above, and all the specific equations of $\mathcal{E}(\mathbf{DA})$ may be found in [19] (Section 2.1;

see [18], Section 2.1, for an announcement). However, linearly distributive categories have also two objects $\top$ and $\bot$, with which one obtains a bimonoidal structure (see §7.9).

§7.3. Coherence of net categories

For $\mathcal{G}$ being **A** and $\mathcal{C}/\mathcal{E}$ being **DA**, we have that the conditions (IV$\mathcal{C}$) and (IV$\mathcal{G}$) of §3.1 are satisfied, and $\mathcal{G}$ is moreover a preorder. Thus we can apply the Direct-Strictification Theorem of §3.2 to obtain the category $\mathcal{C}_{\mathcal{G}}/\mathcal{E}_{\mathcal{G}}$, which we call $\mathbf{DA}^{st}$. We call $\mathcal{C}_{\mathcal{G}}$ here $\mathcal{C}(\mathbf{DA}^{st})$.

In order to prove that **DA** is a preorder, it is enough to prove that $\mathbf{DA}^{st}$ is a preorder. Our proof of the latter will be to a considerable extent analogous to the proofs of Semiassociative Coherence in §4.2 and Dissociative Coherence in §7.1.

We identify the objects of $\mathbf{DA}^{st}$ with form sequences of $\mathcal{P}$ (in the nonextended sense; see §6.2), which we call form sequences *of letters*, or, to simplify the exposition, simply form sequences. In this and in the next chapter, "form sequence" will mean "form sequence of letters". For these form sequences we use the variables X,Y,..., sometimes with indices. For every form sequence X in natural notation, we define a relation R_X between the set of occurrences of $\wedge$ in X and the set of occurrences of $\vee$ in X. For that we need some preliminary notions.

For every occurrence x of $\wedge$ in a form sequence X in natural notation, if y' is the rightmost occurrence of $\vee$ in X such that X has a subword $y'(X')$ with x in the form sequence X', then $l(x)$ is the leftmost occurrence of letter in X'; if there are no such occurrences of $\vee$ in X, then $l(x)$ is the leftmost occurrence of letter in X. Dually, if y'' is the leftmost occurrence of $\vee$ in X such that X has a subword $(X'')y''$ with x in the form sequence X'', then $r(x)$ is the rightmost occurrence of letter in X''; if there are no such occurrences of $\vee$ in X, then $r(x)$ is the rightmost occurrence of letter in X. For example, we have

$$
\begin{array}{ccccccccccccc}
p & \vee & ((\,q & \vee & r & \vee & s\,) & \wedge & ((\,t & \wedge & u\,) & \vee & v\,)) \\
 & y_1 & l(x_1) & y_2 & & y_3 & & x_2 & & x_1 & r(x_1) & y_4 & r(x_2) \\
 & & l(x_2) & & & & & & & & & &
\end{array}
$$

Then for an occurrence x of $\wedge$ in X and an occurrence y of $\vee$ in X we stipulate that xR_Xy when y is on the right-hand side of $l(x)$ and on the left-hand side of $r(x)$. If X is the form sequence in the example above, then we have $R_X = \{(x_1, y_2), (x_1, y_3), (x_2, y_2), (x_2, y_3), (x_2, y_4)\}$.

We can infer the following from the definitions of $l(x)$ and $r(x)$.

NONOVERLAPPING LEMMA. *A form sequence in which in natural notation an occurrence x_1 of $\wedge$ is on the left-hand side of an occurrence x_2 of $\wedge$ cannot have a subword of the form*

$$l(x_1)\, w_1\, l(x_2)\, w\, r(x_1)\, w_2\, r(x_2).$$

PROOF. By the definition of $r(x_1)$ and $l(x_2)$, we have a subword $(X_1)\vee$ with X_1 containing x_1 and ending in $r(x_1)$, and a subword $\vee(X_2)$ with X_2 containing x_2 and beginning with $l(x_2)$. Then either $(X_1)\vee$ is a proper subword of X_2, or $\vee(X_2)$ is a proper subword of X_1. This is because x_1 is on the left-hand side of x_2, and $l(x_2)$ is on the left-hand side of $r(x_1)$.

Suppose $\vee(X_2)$ is a proper subword of X_1. Then $r(x_2)$ must be in X_1, because $(X_1)\vee$ has x_2 as a subword. But $r(x_2)$ cannot be in X_1, because it is on the right-hand side of $r(x_1)$, which is the last occurrence of letter in X_1. We conclude analogously that $(X_1)\vee$ cannot be a proper subword of X_2. ⊣

Note that we cannot prove the Nonoverlapping Lemma without the assumption that x_1 is on the left-hand side of x_2. Here is a counterexample:

$$(\underset{l(x_1)}{p} \vee (\underset{l(x_2)}{q} \underset{x_2}{\wedge} r)) \wedge ((s \underset{x_1}{\wedge} \underset{r(x_1)}{t}) \vee \underset{r(x_2)}{v})$$

Note also that it is excluded that a form sequence in natural notation has a subword of the form

$$l(x_1)\, w_1\, r(x_1)\, w_2\, r(x_2)$$

with $l(x_2)$ being $r(x_1)$. Otherwise, for p being $r(x_1)$ and $l(x_2)$, we would have in our form sequence a word $({}^n p)^m$ with $n, m \geq 1$. (Here $({}^n$ is a sequence of n left parentheses, and $)^m$ a sequence of m right parentheses, as in the Extraction Lemma of §4.2.)

When X and Y are the same form sequence, or a pair of form sequences that in natural notation differ only with respect to parentheses, we say that

X and Y are *comparable* form sequences. For comparable form sequences X and Y, we may take that R_X and R_Y are relations between the same sets, and compare these relations (we did something analogous in §4.2). It is easy to see that for every arrow term $f : X \vdash Y$ of $\mathcal{C}(\mathbf{DA}^{st})$, the form sequences X and Y are comparable, and $R_Y \subseteq R_X$. Moreover, if d^L or d^R occurs in f, then R_Y is a proper subset of R_X; otherwise, $R_X = R_Y$. For example, with $d^L_{p,q,r} \wedge \mathbf{1}_s : p \wedge (q \vee r) \wedge s \vdash ((p \wedge q) \vee r) \wedge s$, if x_1 and x_2 are respectively the left and right $\wedge$, and y is the $\vee$, in $p \wedge q \vee r \wedge s$, then we have $R_{p\wedge(q\vee r)\wedge s} = \{(x_1, y), (x_2, y)\}$ and $R_{((p\wedge q)\vee r)\wedge s} = \{(x_2, y)\}$.

Two comparable form sequences X and Y in natural notation correspond to the same words $w(X)$ and $w(Y)$ written in letters, $\wedge$ and $\vee$, which are obtained from X and Y respectively by deleting all parentheses. A *place* in X is a subword w' of $w(X)$. There is an obvious deleting map δ from subwords of X to places in X. We say that a subword v of X is *at a place* w' when $\delta(v) = w'$. (Note that different subwords of X can be at the same place.) A subword x of X and a subword y of Y are *at the same place* when $\delta(x) = \delta(y)$. (These definitions are analogous to those we had in §2.1.) It is easy to see that the following holds.

REMARK. *Let X in natural notation be of the form $X_1 \,\xi\, X_2$ for $\xi \in \{\wedge, \vee\}$. Then, for x and y occurrences of $\wedge$ and $\vee$ respectively in X_i, for $i \in \{1, 2\}$, we have xR_Xy iff $xR_{X_i}y$.*

The following lemma is analogous to the Extraction Lemmata of §4.2, §4.4 and §7.1.

EXTRACTION LEMMA. *If there is an occurrence z of $\vee$ in the form sequence X, then there is a form sequence $X_1 \, z \, X_2$ in natural notation such that there is an arrow term $g : X \vdash X_1 \, z \, X_2$ of $\mathcal{C}(\mathbf{DA}^{st})$. In addition,*

(∗) *for every occurrence x of $\wedge$ in X_i and every occurrence y of $\vee$ in X_i, where $i \in \{1, 2\}$, if xR_Xy, then $xR_{X_i}y$;*

(∗∗) *every subterm of g of the form $d^L_{Y,Z,U}$ is of the type $Y \wedge (Z \, z \, U) \vdash (Y \wedge Z) \, z \, U$, and every subterm of g of the form $d^R_{U,Y,Z}$ is of the type $(U \, z \, Y) \wedge Z \vdash U \, z \, (Y \wedge Z)$.*

PROOF. We proceed by induction on the number $n \geq 1$ of occurrences of letters in X. If $n = 1$, then the antecedent of the lemma is false, and the lemma is trivially satisfied.

If $n > 1$, then X is $X' \xi X''$ for $\xi \in \{\wedge, \vee\}$. If ξ is z, then g is $\mathbf{1}_X$. Suppose ξ is not z, and suppose first ξ is $\wedge$ and z is in X'. Then, by the induction hypothesis, we have an arrow term $g' \colon X' \vdash X_1' \, z \, X_2'$ of $\mathcal{C}(\mathbf{DA}^{st})$ satisfying the primed version of $(*)$. The arrow term $g' \wedge \mathbf{1}_{X''}$ is of type $X \vdash (X_1' \, z \, X_2') \wedge X''$, and we have the arrow term $d^R_{X_1', X_2', X''} \circ (g' \wedge \mathbf{1}_{X''}) \colon X \vdash X_1' \, z \, (X_2' \wedge X'')$ of $\mathcal{C}(\mathbf{DA}^{st})$.

Suppose x is an occurrence of $\wedge$ and y an occurrence of $\vee$, and suppose $xR_X y$.

If x and y are both in X_1', then $xR_{X'}y$ by the Remark above, and hence, by the induction hypothesis, $(x, y) \in R_{X_1'}$. We settle easily in a similar manner, with the help of the Remark, cases where x and y are both in X_2' or both in X''.

If x is in X_2' and y is in X'', then $r(x)$ in $X_1' \, z \, (X_2' \wedge X'')$ is the rightmost occurrence of letter of X''. Otherwise, $r(x)$ would be a letter p in a subword $p)^l y'$ of X_2' such that y' is an occurrence of $\vee$. Then, since $xR_X y$ and y' is in between x and y, we must have $xR_X y'$, and, by the induction hypothesis and the Remark, we would have $(x, y') \in R_{X_1' \, z \, (X_2' \wedge X'')}$, which contradicts the fact that p in $p)^l y'$ is $r(x)$.

If x is in X'' and y is in X_2', then $l(x)$ in $X_2' \wedge X''$ is the leftmost occurrence of letter of X_2', and so $(x, y) \in R_{X_2' \wedge X''}$.

The case where z is in X'' is settled analogously by using $d^L_{X', X_1'', X_2''}$.

It remains to consider the case where ξ is $\vee$ but is not z. Suppose z is in X'. Then, by the induction hypothesis, we have an arrow term $g' \colon X' \vdash X_1' \, z \, X_2'$ of $\mathcal{C}(\mathbf{DA}^{st})$ satisfying the primed version of $(*)$. So we have the arrow term $g' \vee \mathbf{1}_{X''} \colon X' \xi X'' \vdash X_1' \, z \, X_2' \, \xi \, X''$ of $\mathcal{C}(\mathbf{DA}^{st})$.

Then we verify $(*)$ by the induction hypothesis and the Remark. The assertion $(**)$ is easily checked by going over the proof above. ⊣

Note that the implication converse to $(*)$ in the Extraction Lemma above holds trivially. We do not need $(**)$ for the proof of the Theoremhood Proposition below, but we stated this condition because it is analogous to $(**)$ of previous Extraction Lemmata. Here is the analogue of the Theoremhood Propositions of §4.2 and §7.1.

THEOREMHOOD PROPOSITION. *There is an arrow term* $f : X \vdash Y$ *of* $\mathcal{C}(\mathbf{DA}^{st})$ *iff* X *and* Y *are comparable form sequences and* $R_Y \subseteq R_X$.

PROOF. We have already verified above the easy direction from left to right. For the other direction, we proceed by induction on the number $n \geq 1$ of occurrences of letters in X. If $n = 1$, then $R_Y = R_X = \emptyset$, and f is $\mathbf{1}_p : p \vdash p$.

If $n > 1$ and Y is $Y_1\, x\, Y_2$ for x being an occurrence of $\wedge$, then, since for every occurrence y of $\vee$ in Y we have $xR_Y y$, we have $xR_X y$, which means that X is of the form $X_1\, x\, X_2$. Then, by the Remark, we have $R_{Y_i} \subseteq R_{X_i}$ for $i \in \{1,2\}$, and, by the induction hypothesis, we have the arrow terms $f_i : X_i \vdash Y_i$ of $\mathcal{C}(\mathbf{DA}^{st})$. The arrow term f is $f_1 \wedge f_2$.

If $n > 1$ and Y is $Y_1\, z\, Y_2$ for z being an occurrence of $\vee$, then, by the Extraction Lemma of this section, there is an arrow term $g : X \vdash X_1\, z\, X_2$ of $\mathcal{C}(\mathbf{DA}^{st})$ such that the assertion $(*)$ of the Extraction Lemma holds. If $xR_{Y_i} y$, then, since $R_Y \subseteq R_X$, by the Remark we have $xR_X y$. By $(*)$, we conclude that $xR_{X_i} y$. So, by the induction hypothesis, we have the arrow terms $f_i : X_i \vdash Y_i$ of $\mathcal{C}(\mathbf{DA}^{st})$, and f is $(f_1 \vee f_2) \circ g$. ⊣

As explained after the proof of the Theoremhood Proposition of §4.2, with the Theoremhood Proposition we have just proved we have solved the theoremhood problem for the category $\mathbf{DA}^{st}$. This yields also a solution of the theoremhood problem for the category $\mathbf{DA}$, but we will examine this latter problem separately in the next section.

For a form sequence X, let $d(X)$ be the cardinality of the set of ordered pairs R_X. If $f : X \vdash Y$ of $\mathbf{DA}^{st}$ is not equal to $\mathbf{1}_X : X \vdash X$, then R_Y is a proper subset of R_X and $d(Y) < d(X)$. We prove the following.

NET COHERENCE. *The category* $\mathbf{DA}$ *is a preorder.*

PROOF. It is enough to show that $\mathbf{DA}^{st}$ is a preorder. Let $f, g : X \vdash Y$ be arrow terms of $\mathcal{C}(\mathbf{DA}^{st})$. We proceed by induction on $d(X) - d(Y)$ to show that $f = g$ in $\mathbf{DA}^{st}$. (Until the end of this proof, we assume that equality of arrow terms is equality in $\mathbf{DA}^{st}$.) If $d(X) = d(Y)$, then we conclude that X is Y, and $f = g = \mathbf{1}_X$.

Suppose $d(Y) < d(X)$. By the Development Lemma of §2.7, we have that $f = f_2 \circ f_1$ and $g = g_2 \circ g_1$ for some d-terms $f_1 \colon X \vdash Z$ and $g_1 : X \vdash U$, and some arrow terms $f_2 \colon Z \vdash Y$ and $g_2 \colon U \vdash Y$ of $\mathcal{C}(\mathbf{DA}^{st})$. We have here $d(Z), d(U) < d(X)$. The following cases may arise.

(LL) The head of f_1 is $d^L_{E,F,G}$, and the head of g_1 is $d^L_{H,I,J}$. (Here E, F, G, H, I and J stand for form sequences.) Due to the presence of $(d^L\wedge)$ and $(d^L\vee)$, we can assume that E and H are not of the form $(X_1 \ldots X_n, \wedge)$ and G and J are not of the form $(X_1 \ldots X_n, \vee)$. Under (LL), we have the following subcases.

$(LL1)$ The form sequences $E\wedge(F\vee G)$ and $H\wedge(I\vee J)$ have no occurrences of letters in common within X. Then we use $(\wedge\ 2)$ and $(\vee\ 2)$ to obtain two d-terms $f_2' : Z \vdash Y'$ and $g_2' : U \vdash Y'$ such that $f_2' \circ f_1 = g_2' \circ g_1$. Then we can infer that $R_Z \cap R_U = R_{Y'}$, from which it follows by the Theoremhood Proposition of this section that $R_Y \subseteq R_{Y'}$. Hence, again by the Theoremhood Proposition, there is an arrow term $h \colon Y' \vdash Y$ of $\mathbf{DA}^{st}$. By applying the induction hypothesis, we obtain that $f_2 = h \circ f_2'$ and $g_2 = h \circ g_2'$, from which $f = g$ follows.

$(LL2)$ Suppose $E \wedge (F \vee G)$, in natural notation, is a subword of H or of I or of J; or, conversely, suppose that $H \wedge (I \vee J)$ is a subword of E or of F or of G. Then we proceed as in $(LL1)$ by using $(d^L\ nat)$.

$(LL3)$ Suppose, finally, that E is H and G is J. So F is I. (Due to our assumptions about E, H, G and J, there are no other remaining subcases under (LL).) Then Z is U, and $f_1 = g_1$. We then apply the induction hypothesis to $f_2, g_2 \colon Z \vdash B$, and obtain $f = g$.

(LR) The head of f_1 is $d^L_{E,F,G}$, and the head of g_1 is $d^R_{J,I,H}$. Due to the presence of $(d^L\wedge)$, $(d^L\vee)$, $(d^R\wedge)$ and $(d^R\vee)$, we can assume that E and H are not of the form $(X_1 \ldots X_n, \wedge)$ and G and J are not of the form $(X_1 \ldots X_n, \vee)$.

Under (LR), we have the following subcases. There are first two subcases that are settled analogously to $(LL1)$ and $(LL2)$. The remaining subcases are:

$$(LR4) \quad E \text{ is } J \vee I \ \text{ and } \ F \vee G \text{ is } H,$$

and when $F \vee G$ is $J \vee I$ we have the following two subcases:

$$\begin{array}{lll} (LR5) & F \text{ is } J & (\text{so } G \text{ is } I), \\ (LR6) & F \text{ is } J \vee F'' & (\text{so } I \text{ is } F'' \vee G). \end{array}$$

(There is no subcase named $(LR3)$, which would be analogous to $(LL3)$.)

$(LR4)$ Then by $(d\overset{\vee}{b})$ we have

$$(d^R_{J,I,F} \vee \mathbf{1}_G) \circ d^L_{E,F,G} = (\mathbf{1}_J \vee d^L_{I,F,G}) \circ d^R_{J,I,H}.$$

Let f'_2 and g'_2 be obtained from g_1 by replacing its head $d^R_{J,I,H}$ by $d^R_{J,I,F} \vee \mathbf{1}_G$ and $\mathbf{1}_J \vee d^L_{I,F,G}$ respectively. It is clear that $f'_2 \circ f_1 = g'_2 \circ g_1 : X \vdash Y'$. Then we infer that $R_Z \cap R_U = R_{Y'}$, and we continue reasoning as in $(LL1)$, by applying the Theoremhood Proposition.

$(LR5)$ Then by $(d\hat{b})$ we have

$$d^R_{E \wedge F,G,H} \circ (d^L_{E,F,G} \wedge \mathbf{1}_H) = d^L_{E,F,G \wedge H} \circ (\mathbf{1}_E \wedge d^R_{F,G,H}).$$

Let f'_2 and g'_2 be obtained from g_1 by replacing $\mathbf{1}_E \wedge d^R_{F,G,H}$ by $d^R_{E \wedge F,G,H}$ and $d^L_{E,F,G \wedge H}$ respectively. It is clear that $f'_2 \circ f_1 = g'_2 \circ g_1 : X \vdash Y'$. Then we infer that $R_Z \cap R_U = R_{Y'}$, and we continue reasoning as in $(LL1)$, by applying the Theoremhood Proposition.

$(LR6)$ We prove first that there is an occurrence z of $\vee$ in $J \vee F'' \vee G$ such that for every occurrence x of $\wedge$ in $E \wedge (J \vee F'' \vee G) \wedge H$ we do not have $x R_Y z$.

Let u be an occurrence of $\wedge$ in the word $E\wedge$. For every such u the occurrence of letter $r(u)$ in Y is either in J or in F''. Let p be the rightmost of these occurrences of letters.

If p is in J, then we take z to be the $\vee$ between J and F''. By the definition of $r(u)$, there is no occurrence u of $\wedge$ in $E\wedge$ such that $u R_Y z$. Since $R_Y \subseteq R_X$, there is no occurrence x of $\wedge$ in J such that $x R_Y z$, and, since $R_Y \subseteq R_U$, there is no occurrence x of $\wedge$ in the word $F'' \vee G) \wedge H$ such that $x R_Y z$.

If p is in F'', then we take z to be the occurrence of $\vee$ on the right-hand side of p nearest to p. This z is either in F'', or it is the $\vee$ between F'' and G. By the definition of $r(u)$, there is no occurrence u of $\wedge$ in $E\wedge$ such that

$uR_Y z$. Since $R_Y \subseteq R_X$, there is no occurrence x of $\wedge$ in J such that $xR_Y z$. If there is an occurrence x of $\wedge$ in $F'' \vee G) \wedge H$ such that $xR_Y z$, then in Y we have that $l(x)$ is on the left-hand side of p, which is $r(u)$ for some occurrence u of $\wedge$ in $E\wedge$. (As we said after the proof of the Nonoverlapping Lemma, it is excluded that $l(x)$ coincides with $r(u)$.) Since $R_Y \subseteq R_U$, we must have that $l(u)$ is on the left-hand side of $l(x)$, and, since $xR_Y z$, we must have that $r(x)$ is on the right-hand side of $r(u)$. Since x is on the right-hand side of u, all this contradicts the Nonoverlapping Lemma. Hence we do not have $xR_Y z$.

There are now two possibilities for the z we have found. Suppose first that $J \vee F'' \vee G$ is of the form KzL. Then we have three subcases:

(LR6.1) K is J (so L is $F'' \vee G$),
(LR6.2) L is G (so K is $J \vee F''$),
(LR6.3) K is $J \vee F_1''$ and L is $F_2'' \vee G$ (so F'' is $F_1'' \vee F_2''$).

(LR6.1) Then by $(d^L \vee)$ we have

$$d^L_{E,J,I} \wedge \mathbf{1}_H = ((d^L_{E,J,F''} \vee \mathbf{1}_G) \wedge \mathbf{1}_H) \circ (d^L_{E,F,G} \wedge \mathbf{1}_H).$$

Let f_1' and f_1'' be obtained from g_1 by replacing $\mathbf{1}_E \wedge d^R_{J,I,H}$ by $d^L_{E,J,I} \wedge \mathbf{1}_H$ and $(d^L_{E,J,F''} \vee \mathbf{1}_G) \wedge \mathbf{1}_H$ respectively. It is clear that $f_1' = f_1'' \circ f_1 \colon X \vdash Z'$. Suppose $xR_Y y$. If x is not in $E \wedge (J\, y\, F'' \vee G) \wedge H$, then it is easy to infer that $xR_{Z'} y$. If x is in $E \wedge (J\, y\, F'' \vee G) \wedge H$, then it is either in $E \wedge (J$ or in $F'' \vee G) \wedge H$. In the first case, y is on the left-hand side of z, and in the second case, it is on the right-hand side of z. In both cases, we get $xR_{Z'} y$. So $R_Y \subseteq R_{Z'}$, and, by the Theoremhood Proposition of this section, we obtain an arrow term $f_2' \colon Z' \vdash Y$ of $\mathcal{C}(\mathbf{DA}^{st})$. By the induction hypothesis, we have that $f_2 = f_2' \circ f_1''$, where $d(Z) < d(X)$. We continue reasoning as in subcase (LR5), starting from f_1' and g_1. There we apply $(d\hat{b})$.

The subcase (LR6.2) is settled analogously to (LR6.1) by using $(d^R \vee)$, and for subcase (LR6.3) we use both $(d^L \vee)$ and $(d^R \vee)$ to reduce it to (LR5), where we apply $(d\hat{b})$.

Suppose now that $J \vee F'' \vee G$ is not of the form $K\, z\, L$. So z is in F'', but F'' is not of the form $F_1''\, z\, F_2''$. Then, by the Extraction Lemma of this section, there is a form sequence F''' of the form $F_1'''\, z\, F_2'''$ in natural notation such that there is an arrow term $h\colon F'' \vdash F'''$ of $\mathcal{C}(\mathbf{DA}^{st})$ with $(*)$ being satisfied. Let f_1', g_1', h', f_2'' and g_2'' be obtained from g_1 by replacing $\mathbf{1}_E \wedge d^R_{J,I,H}$ respectively by

$$d^L_{E,J\vee F''',G} \wedge \mathbf{1}_H,$$
$$\mathbf{1}_E \wedge d^R_{J,F'''\vee G,H},$$
$$\mathbf{1}_E \wedge (\mathbf{1}_J \vee h \vee \mathbf{1}_G) \wedge \mathbf{1}_H,$$
$$((\mathbf{1}_E \wedge (\mathbf{1}_J \vee h)) \vee \mathbf{1}_G) \wedge \mathbf{1}_H,$$
$$\mathbf{1}_E \wedge (\mathbf{1}_J \vee ((h \vee \mathbf{1}_G) \wedge \mathbf{1}_H)).$$

Then, by $(d^L\ nat)$ and $(d^R\ nat)$, we have that $f_1' \circ h' = f_2'' \circ f_1$ and $g_1' \circ h' = g_2'' \circ g_1$. For h' being of the type $X \vdash X'$, we have that $R_Y \subseteq R_{X'}$, which follows easily from our assumption about z and from $(*)$ of the Extraction Lemma. For f_2'' being of the type $Z \vdash Z'$ and g_2'' of the type $U \vdash U'$, we infer that $R_{Z'} = R_Z \cap R_{X'}$ and $R_{U'} = R_U \cap R_{X'}$. So, by the Theoremhood Proposition of this section, we have the arrow terms $f_2' : Z' \vdash Y$ and $g_2' : U' \vdash Y$ of $\mathcal{C}(\mathbf{DA}^{st})$. Then we apply the induction hypothesis to $f_2, f_2' \circ f_2'' : Z \vdash Y$ and $g_2, g_2' \circ g_2'' : U \vdash Y$, and also to $f_2' \circ f_1', g_2', g_1' : X' \vdash Y$, where $d(Z), d(U), d(X') < d(X)$.

It remains to consider the following cases:

(*RR*) the head of f_1 is $d^R_{G,F,E}$, and the head of g_1 is $d^R_{J,I,H}$;

(*RL*) the head of f_1 is $d^R_{G,F,E}$, and the head of g_1 is $d^L_{H,I,J}$.

The case (*RR*) is settled analogously to (*LL*), while the case (*RL*) is the same as (*LR*). ⊣

It is not difficult to see that $R_X = R_Y$ implies that the form sequences X and Y coincide. Because, if $R_X = R_Y$, then, by the Theoremhood Proposition of this section, there is an arrow term $f\colon X \vdash Y$ of $\mathcal{C}(\mathbf{DA}^{st})$, in which d^L and d^R cannot occur, because R_Y is not a proper subset of R_X. Hence f must stand for an identity arrow. So there is a bijection between the objects X of $\mathbf{DA}^{st}$ and the relations R_X. From Net Coherence, we can conclude that $\mathbf{DA}^{st}$ is isomorphic to the category whose objects are the relations R_X, and where an arrow exists between R_X and R_Y when $R_Y \subseteq R_X$.

§7.4. Net normal form

In this section we will examine the theoremhood problem (in the sense of §1.1) for the category **DA**, and we will find a solution for it different from that suggested by the Theoremhood Proposition of the preceding section. This solution will also yield a unique normal form for arrow terms of $\mathcal{C}(\mathbf{DA})$, i.e. a normal form such that arrow terms of $\mathcal{C}(\mathbf{DA})$ in normal form are equal in **DA** iff they are the same arrow term.

Consider a formula B of $\mathcal{L}_{\wedge,\vee}$. Let $B^{\wedge}$ be obtained from B by replacing every $\vee$ by $\wedge$. Let $I(B)$ be obtained from $I(B^{\wedge})$ (see §4.4) by putting back the occurrences of $\vee$ where they were in B, while keeping the subscripts of $I(B^{\wedge})$.

Let A be a formula comparable with B (which means that A and B are the same after deleting parentheses). Next, let A^* be obtained from A by adding to every occurrence x of $\wedge$ or $\vee$ in A the subscript x has in $I(B)$. Then we have the following proposition.

THEOREMHOOD PROPOSITION. *There is an arrow term $f\colon A \vdash B$ of $\mathcal{C}(\mathbf{DA})$ iff A and B are comparable formulae of $\mathcal{L}_{\wedge,\vee}$ and*

(†) *in A^* defined with respect to $I(B)$, for every $n, m \geq 2$ there is no $\wedge_{nm}$ in the scope of $\vee_n$.*

PROOF. From left to right, suppose we have an arrow term $f : A \vdash B$ of $\mathcal{C}(\mathbf{DA})$ such that (†) fails in A^*. Then for $f_{\mathcal{G}}\colon X \vdash Y$ in $\mathbf{DA}^{st}$ we can find in X and Y an occurrence x of $\wedge$ corresponding to $\wedge_{nm}$ and an occurrence y of $\vee$ corresponding to $\vee_n$. Since (†) fails in A^*, we do not have xR_Xy, but the subscripts of $I(B)$ tell us that we have xR_Yy, which contradicts the easy, left-to-right, direction of the Theoremhood Proposition of §7.3.

For the other direction, we proceed as follows. By the Associative Normal-Form Proposition of §4.4, there is an arrow term $f^{\wedge}\colon A^{\wedge} \vdash B^{\wedge}$ of $\mathcal{C}(\hat{\mathbf{A}})$ such that in $(f^{\wedge})^*\colon (A^{\wedge})^* \vdash I(B^{\wedge})$ for every subterm of the form $\hat{b}^{\rightarrow}_{D,E,F}$ of type $D \wedge_l (E \wedge_k F) \vdash (D \wedge_l E) \wedge_k F$, and every subterm of the form $\hat{b}^{\leftarrow}_{F,E,D}$ of type $(F \wedge_k E) \wedge_l D \vdash F \wedge_k (E \wedge_l D)$, we have that l and every subscript in D, E and F divides k. We build out of $(f^{\wedge})^*$ an arrow term $f^*\colon A^* \vdash I(B)$ by putting back $\vee$ at some places, as required by A and B. In transforming

$(f^\wedge)^*$ into f^*, some subterms of $(f^\wedge)^*$ in the family $\hat{b}$ may remain in that family, and some may be transformed into arrow terms in the families $\check{b}$, d^L and d^R. It is excluded that the type of a subterm of $(f^\wedge)^*$ in the family $\hat{b}$ becomes $D \vee_l (E \wedge_k F) \vdash (D \vee_l E) \wedge_k F$ or $(F \wedge_k E) \vee_l D \vdash F \wedge_k (E \vee_l D)$, which would prevent its being transformed in an arrow term in the families $\hat{b}$, $\check{b}$, d^L or d^L. This is guaranteed by (†), and by the fact that for every subterm of $(f^\wedge)^*$ in the family $\hat{b}$ of a type $(G^\wedge)^* \vdash (H^\wedge)^*$ we have that H^* satisfies (†), as A^* does. We obtain $f \colon A \vdash B$ by deleting the subscripts of f^*. ⊣

The procedure of the proof of the right-to-left direction of this proposition, which presupposes the results of §4.4, gives rise to a unique arrow term, which we may consider to be in normal form.

We could imagine a proof of Net Coherence where instead of relying on the Theoremhood Proposition of the preceding section, we would rely on a strictified version of the Theoremhood Proposition of this section.

§7.5. Coherence of semidissociative biassociative categories

To obtain the natural logical category $\mathbf{D}^L\mathbf{A}$, we have that the logical system $\mathcal{C}(\mathbf{D}^L\mathbf{A})$ is in $\mathcal{L}_{\wedge,\vee}$, with the transformations α included in $\mathbf{1}$, b and d^L. So, in contradistinction to $\mathcal{C}(\mathbf{DA})$, we do not have d^R. The specific equations of $\mathcal{E}(\mathbf{D}^L\mathbf{A})$ are those of $\mathcal{E}(\mathbf{A})$ plus $(d^L\wedge)$ and $(d^L\vee)$ of §7.2. We call natural $\mathbf{D}^L\mathbf{A}$-categories *semidissociative biassociative* categories.

For $\mathcal{G}$ being $\mathbf{A}$ and $\mathcal{C}/\mathcal{E}$ being $\mathbf{D}^L\mathbf{A}$, we have that the conditions (IV$\mathcal{C}$) and (IV$\mathcal{G}$) of §3.1 are satisfied, and $\mathcal{G}$ is moreover a preorder. Thus we can apply the Direct-Strictification Theorem of §3.2 to obtain a category $\mathcal{C}_\mathcal{G}/\mathcal{E}_\mathcal{G}$, which we call $\mathbf{D}^L\mathbf{A}^{st}$. We call $\mathcal{C}_\mathcal{G}$ here $\mathcal{C}(\mathbf{D}^L\mathbf{A}^{st})$.

Our proof that $\mathbf{D}^L\mathbf{A}^{st}$ is a preorder is to a considerable extent analogous to the proof that $\mathbf{DA}^{st}$ is a preorder, and we assume the notions defined in §7.3. The new proof is somewhat more complicated as far as the definitions of the relation R_X is concerned.

For every object X of $\mathbf{D}^L\mathbf{A}^{st}$, i.e. for every form sequence X in the natural notation of §6.2, we define two relations R^l_X and R^r_X between the set of occurrences of $\wedge$ in X and the set of occurrences of $\vee$ in X. We have

$xR^l_X y$ when the occurrence y of $\vee$ is in between $l(x)$ and the occurrence x of $\wedge$, and we have $xR^r_X y$ when y is between x and $r(x)$. It is clear that R_X is the disjoint union of R^l_X and R^r_X.

It is easy to verify that, for every arrow term $f : X \vdash Y$ of $\mathcal{C}(\mathbf{D}^L\mathbf{A}^{st})$, we have $R^r_Y \subseteq R^r_X$ and $R^l_X = R^l_Y$. Moreover, if d^L occurs in f, then R^r_Y is a proper subset of R^r_X. It is also easy to verify that the Remark of §7.3 holds when we replace R by R^l and R^r. Then we can prove the following analogue of the Extraction Lemma of §7.3.

EXTRACTION LEMMA. *If there is an occurrence z of $\vee$ in the form sequence X such that there is no occurrence x of $\wedge$ in X with $xR^l z$, then there is a form sequence $X_1 \, z \, X_2$ in natural notation such that there is an arrow term $g\colon X \vdash X_1 \, z \, X_2$ of $\mathcal{C}(\mathbf{D}^L\mathbf{A}^{st})$. In addition,*

(∗) *for every occurrence x of $\wedge$ in X_i and every occurrence y of $\vee$ in X_i, where $i \in \{1, 2\}$, if $xR^r_X y$, then $xR^r_{X_i} y$; moreover, $R^l_X = R^l_{X_1 \, z \, X_2}$,*

(∗∗) *every subterm of g of the form $d^L_{Y,Z,U}$ is of the type $Y \wedge (Z \, z \, U) \vdash (Y \wedge Z) \, z \, U$.*

The proof is obtained by excluding the case where ξ is $\wedge$ and z is in X' in the proof of the Extraction Lemma of §7.3.

Next we state the analogue of the Theoremhood Proposition of §7.3.

THEOREMHOOD PROPOSITION. *There is an arrow term $f : X \vdash Y$ of $\mathcal{C}(\mathbf{D}^L\mathbf{A}^{st})$ iff X and Y are comparable form sequences, and we have $R^r_Y \subseteq R^r_X$ and $R^l_X = R^l_Y$.*

The proof is again a slight modification of the proof of the Theoremhood Proposition of §7.3.

The proof that the category $\mathbf{D}^L\mathbf{A}^{st}$ is a preorder is then obtained by proceeding as in the proof of Net Coherence of §7.3. We keep just the cases analogous to (LL) cases. So we have the following.

SEMIDISSOCIATIVE BIASSOCIATIVE COHERENCE. *The category $\mathbf{D}^L\mathbf{A}$ is a preorder.*

Analogously to what we had at the end of §4.2, §7.1 and §7.3, with the help of Semidissociative Biassociative Coherence, we obtain that $\mathbf{D}^L\mathbf{A}^{st}$ is isomorphic to a category whose objects are pairs of relations (R^r_X, R^l_X), and where an arrow exists between (R^r_X, R^l_X) and (R^r_Y, R^l_Y) when $R^r_Y \subseteq R^r_X$ and $R^l_X = R^l_Y$.

§7.6. Symmetric net categories

To obtain the natural logical category **DS**, we have that the logical system $\mathcal{C}(\mathbf{DS})$ is in $\mathcal{L}_{\wedge,\vee}$, with the transformations α included in $\mathbf{1}$, b, c and d. The specific equations of $\mathcal{E}(\mathbf{DS})$ are obtained by taking the union of those of $\mathcal{E}(\mathbf{DA})$ and $\mathcal{E}(\mathbf{S})$ plus

$$(d^R c) \quad d^R_{C,B,A} = \check{c}_{C,B\wedge A} \circ (\hat{c}_{A,B} \vee \mathbf{1}_C) \circ d^L_{A,B,C} \circ (\mathbf{1}_A \wedge \check{c}_{B,C}) \circ \hat{c}_{C\vee B,A}.$$

We call natural **DS**-categories *symmetric net* categories. In §12.4 we will give a concrete example of a symmetric net category in which $\wedge$ and $\vee$ are not isomorphic. (See §11.3 for the question whether the category *Set* of sets with functions is a symmetric net category.)

In the presence of $(d^R c)$, the equations $(d^R\ nat)$, $(d^R\wedge)$ and $(d^R\vee)$ become derivable from the remaining equations. Note that $(d^R c)$ may be conceived as a definition of d^R in terms of d^L, $\hat{c}$ and $\check{c}$. So we may as well assume that in $\mathcal{C}(\mathbf{DS})$ we do not have d^R, but only d^L, and that d^R is defined by $(d^R c)$. We make this assumption in §§7.6-8, and we write simply d for d^L, omitting the superscript ^{L}. This convention will be in force also later on whenever we have $(d^R c)$ (especially in Chapter 11).

To give some alternative axioms for $\mathcal{E}(\mathbf{DS})$ we introduce the following definitions:

$\hat{e}_{A,B,C,D} =_{df} d_{A,D,B\wedge C} \circ (\mathbf{1}_A \wedge \check{c}_{D,B\wedge C}) \circ (\mathbf{1}_A \wedge d_{B,C,D}) \circ \hat{b}^{\leftarrow}_{A,B,C\vee D}$
is of type $(A \wedge B) \wedge (C \vee D) \vdash (A \wedge D) \vee (B \wedge C)$;

$\hat{e}'_{A,B,C,D} =_{df} \hat{e}_{A,B,D,C} \circ (\mathbf{1}_{A\wedge B} \wedge \check{c}_{D,C})$
is of type $(A \wedge B) \wedge (C \vee D) \vdash (A \wedge C) \vee (B \wedge D)$.

Dually, we have that

$\check{e}_{D,C,B,A} =_{df} \check{b}^{\leftarrow}_{D\wedge C,B,A} \circ (d_{D,C,B} \vee \mathbf{1}_A) \circ (\hat{c}_{C\vee B,D} \vee \mathbf{1}_A) \circ d_{C\vee B,D,A}$
is of type $(C \vee B) \wedge (D \vee A) \vdash (D \wedge C) \vee (B \vee A)$;

$\check{e}'_{D,C,B,A} =_{df} (\hat{c}_{C,D} \vee \mathbf{1}_{B\vee A}) \circ \check{e}_{C,D,B,A}$
is of type $(D \vee B) \wedge (C \vee A) \vdash (D \wedge C) \vee (B \vee A)$.

Then we can state the following equations:

$$(\hat{e}) \quad \check{c}_{B\wedge C,A\wedge D} \circ \hat{e}_{A,B,C,D} = \hat{e}'_{B,A,C,D} \circ (\hat{c}_{A,B} \wedge \mathbf{1}_{C\vee D}),$$
$$(\check{e}) \quad (\mathbf{1}_{D\wedge C} \vee \check{c}_{B,A}) \circ \check{e}'_{D,C,A,B} = \check{e}_{D,C,B,A} \circ \hat{c}_{D\vee A,C\vee B}.$$

These two equations are mirror images of each other. The equation $(\hat{e})$ can replace $(d\hat{b})$, and the equation $(\check{e})$ can replace $(d\check{b})$, in our axiomatization of $\mathcal{E}(\mathbf{DS})$.

For every transformation α in the logical system $\mathcal{C}(\mathbf{DS})$ we have that in $\alpha_{A_1,\ldots,A_k} : M^\mu(A_1,\ldots,A_k) \vdash N^\nu(A_1,\ldots,A_k)$ the functions μ and ν are bijections, and hence the type $M^\mu(A_1,\ldots,A_k) \vdash N^\nu(A_1,\ldots,A_k)$ is balanced (see §3.3). Therefore, the type of every arrow term of $\mathcal{C}(\mathbf{DS})$ is balanced.

For $\mathcal{C}/\mathcal{E}$ being $\mathcal{C}(\mathbf{DS})/\mathcal{E}(\mathbf{DS})$, that is $\mathbf{DS}$, and $\mathcal{C}'$ being $\mathcal{C}(\mathbf{S})$ of §6.3, we have that the condition (IV$\mathcal{C}$) of §3.1 is satisfied. Next, let $\mathcal{G}$ be the $\mathcal{C}'$-core $\mathcal{C}'/\mathcal{E}'$ of $\mathcal{C}/\mathcal{E}$. By Symmetric Biassociative Coherence, and by the fact that if $f = g$ in $\mathcal{E}(\mathbf{DS})$, then $Gf = Gg$ in *Rel*, we can conclude that $\mathcal{G}$ is the natural logical category $\mathbf{S}$. The category $\mathcal{G}$ is a groupoid, and it flows through $\mathbf{DS}$, so that the condition (IV$\mathcal{G}$) of §3.1 is satisfied.

Let $\mathcal{E}^{pr}$ be the equational system obtained by extending $\mathcal{E}(\mathbf{DS})$ with the equation $(\overset{\xi}{c}\,\mathbf{1})$ of §6.5, namely $\overset{\xi}{c}_{A,A} = \mathbf{1}_{A\xi A}$ for $\xi \in \{\wedge, \vee\}$. We know that $\mathcal{C}'/\mathcal{E}^{pr\prime}$, that is $\mathcal{C}(\mathbf{S})/\mathcal{E}^{pr\prime}$, which is the category $\mathbf{S}'$ of §6.5, is a preorder. Next, for every equation $f = g$ in $\mathcal{E}^{pr}$ that is not in $\mathcal{E}(\mathbf{DS})$, we can show that the type of f and g is not diversified. We prove by induction on the length of derivation that if $f = g$ is in $\mathcal{E}^{pr}$ and the arrow terms f and g are diversified, then every derivation of $f = g$ is made of equations between diversified arrow terms. (The only problem is when in such a derivation we pass from $f_1 = f_2$ and $g_1 = g_2$ to $g_1 \circ f_1 = g_2 \circ f_2$, in which case we appeal to the fact that the types of arrow terms of $\mathcal{C}(\mathbf{DS})$ are always balanced.)

Then, as in §3.3, we have that (IV$\mathcal{C}$) and (IV$\mathcal{G}$) hold when $\mathcal{C}(\mathbf{DS})/\mathcal{E}(\mathbf{DS})$ is replaced by $\mathcal{C}(\mathbf{DS})/\mathcal{E}^{pr}$. Now the $\mathcal{C}'$-core $\mathcal{G}$ of $\mathcal{C}(\mathbf{DS})/\mathcal{E}^{pr}$ is a preorder.

By the Direct-Strictification Theorem of §3.2, we obtain the $\mathcal{C}(\mathbf{S})$-strict $\mathcal{C}(\mathbf{DS})/\mathcal{E}^{pr}$-category $\mathcal{C}(\mathbf{DS})_{\mathcal{G}}/\mathcal{E}^{pr}_{\mathcal{G}}$ equivalent to $\mathcal{C}(\mathbf{DS})/\mathcal{E}^{pr}$. As in §3.3, for diversified arrow terms f and g of $\mathcal{C}(\mathbf{DS})$ of the same type, we have $f = g$ in $\mathbf{DS}$ iff $f_{\mathcal{G}} = g_{\mathcal{G}}$ in $\mathcal{C}(\mathbf{DS})_{\mathcal{G}}/\mathcal{E}^{pr}_{\mathcal{G}}$.

Since the type of every arrow term of $\mathcal{C}(\mathbf{DS})$ is balanced, for every arrow term $f\colon A \vdash B$ of $\mathcal{C}(\mathbf{DS})$ there is a diversified arrow term $f^{div}\colon A^{div} \vdash B^{div}$ of $\mathcal{C}(\mathbf{DS})$ such that f is obtained by substituting uniformly letters for some letters in $f^{div} : A^{div} \vdash B^{div}$. Namely, f is a letter-for-letter substitution instance of f^{div}. Here we assume that the generating set $\mathcal{P}$ is infinite (see §4.1).

Our purpose is to show the following.

Symmetric Net Coherence. *The functor G from* $\mathbf{DS}$ *to Rel is faithful.*

According to what we said above, to prove this coherence we can proceed as follows. Suppose $Gf = Gg$ in *Rel* for the arrow terms f and g of $\mathcal{C}(\mathbf{DS})$ of the same type. Then we can find f^{div} and g^{div} of the same type, and we will prove

$$(div) \quad f^{div}_{\mathcal{G}} = g^{div}_{\mathcal{G}} \text{ in } \mathcal{C}(\mathbf{DS})_{\mathcal{G}}/\mathcal{E}^{pr}_{\mathcal{G}},$$

which implies $f^{div} = g^{div}$ in $\mathbf{DS}$, from which we can conclude, by applying (su) (see §2.7), that $f = g$ in $\mathbf{DS}$. So, to prove Symmetric Net Coherence, we have only to prove (div) under the assumption $Gf = Gg$.

We proceed with this proof in the next two sections. In §7.7, we prove a theorem that says that the equations of $\mathcal{C}(\mathbf{DS})_{\mathcal{G}}/\mathcal{E}^{pr}_{\mathcal{G}}$ cover a normalization procedure analogous to Gentzen's cut-elimination procedure of [55]. In §7.8, we prove additional results, which together with our cut elimination will yield (div) under the assumption $Gf = Gg$. In logic, these results correspond to inverting rules in derivations, i.e. passing from conclusions to premises. This invertibility is guaranteed by the possibility to permute rules, i.e. change their order in derivations, and we show for that permuting that it is covered by the equations that hold in $\mathcal{C}(\mathbf{DS})_{\mathcal{G}}/\mathcal{E}^{pr}_{\mathcal{G}}$. This means that the equations of $\mathbf{DS}$ also cover a cut elimination and invertibility, but this cut elimination and invertibility are more cumbersome to record within $\mathbf{DS}$ than within $\mathcal{C}(\mathbf{DS})_{\mathcal{G}}/\mathcal{E}^{pr}_{\mathcal{G}}$.

§7.7. Cut elimination in GDS

To formulate the cut-elimination result announced at the end of the preceding section, we need some preliminary notions. The objects of the category $\mathcal{C}(\mathbf{DS})_{\mathcal{G}}/\mathcal{E}^{pr}_{\mathcal{G}}$ (see the preceding section) correspond bijectively to something we will call *form multisets* of letters. We define this notion as follows.

We say that the form sequences of letters X and Y are *c-equivalent* when there is an arrow term of $\mathcal{C}(\mathbf{S})_{\mathcal{G}}$ of type $X \vdash Y$. It is clear that c-equivalence is an equivalence relation congruent with the operations ξ'' on form sequences of letters for $\xi \in \{\wedge, \vee\}$. A *form multiset* of letters is the equivalence class of a form sequence of letters with respect to c-equivalence. (We exclude here the empty form sequences $(\emptyset, \xi)$.) As before in this chapter, we presuppose that form multisets and form sequences are *of letters*, i.e. *of* $\mathcal{P}$, and omit mentioning that all the time. We can use form sequences, and, in particular, form sequences in natural notation, to designate form multisets. For example, $p \wedge q \wedge (p \vee r \vee p)$ in natural notation stands for the same form multiset as $q \wedge (r \vee p \vee p) \wedge p$.

For A a diversified formula of $\mathcal{L}_{\wedge,\vee}$ (see §3.3), the form multiset $[\![A]\!]$ is such that every letter of $\mathcal{P}$ occurs in it at most once. Such a form multiset is called a *form set*.

Let **GDS** be the full subcategory of $\mathcal{C}(\mathbf{DS})_{\mathcal{G}}/\mathcal{E}^{pr}_{\mathcal{G}}$ whose objects are all the objects that correspond to form sets of letters. We write **G** in the name of **GDS** because of the relationship we are going to establish between this category and Gentzen's sequent systems. The category **GDS** is a syntactical category in a syntactical system called $\mathcal{C}(\mathbf{GDS})$, which is a subsystem of $\mathcal{C}(\mathbf{DS})_{\mathcal{G}}$.

In this section, $X, Y, Z, \ldots, X_1, \ldots$ will be form sequences of letters that stand for form sets of letters, and the operations ξ'' on form sequences, for $\xi \in \{\wedge, \vee\}$, will be written simply ξ, with $''$ omitted.

We define by induction a set of terms for arrows of **GDS**, which we call *Gentzen terms*. First, we stipulate that for every letter p the term $\mathbf{1}_p \colon p \vdash p$, which denotes the arrow $\mathbf{1}_{[\![p]\!]}$ of **GDS**, is a Gentzen term. The remaining Gentzen terms are obtained by closing under the following operations on Gentzen terms, which we call *Gentzen operations*. We present these operations by inductive clauses in fractional notation, which are interpreted as saying that if the terms above the horizontal line are Gentzen terms, then the term below the horizontal line is a Gentzen term (cf. §2.2). The schema on the left-hand side of the $=_{dn}$ sign stands for the Gentzen term, while the schema on the right-hand side of $=_{dn}$ is the arrow denoted by this term. Our Gentzen operations correspond to Gentzen's rules for cut, introduction of conjunction on the right and introduction of disjunction on the left:

$$\frac{f\colon U \vdash X \vee Z \qquad g\colon X \wedge Y \vdash W}{cut_X(f,g) =_{dn} (g \vee \mathbf{1}_Z) \circ d_{Y,X,Z} \circ (f \wedge \mathbf{1}_Y)\colon U \wedge Y \vdash Z \vee W}$$

$$\frac{f\colon U \vdash X \qquad g\colon X \wedge Y \vdash W}{cut_X(f,g) =_{dn} g \circ (f \wedge \mathbf{1}_Y)\colon U \wedge Y \vdash W}$$

$$\frac{f\colon U \vdash X \vee Z \qquad g\colon X \vdash W}{cut_X(f,g) =_{dn} (g \vee \mathbf{1}_Z) \circ f\colon U \vdash W \vee Z}$$

$$\frac{f\colon U \vdash X \qquad g\colon X \vdash W}{cut_X(f,g) =_{dn} g \circ f\colon U \vdash W}$$

$$\frac{f_1\colon U_1 \vdash X_1 \vee Z_1 \qquad f_2\colon U_2 \vdash X_2 \vee Z_2}{\wedge_{X_1,X_2}(f_1,f_2)\colon U_1 \wedge U_2 \vdash (X_1 \wedge X_2) \vee Z_1 \vee Z_2}$$

where $\wedge_{X_1,X_2}(f_1,f_2) =_{dn} (d_{X_2,X_1,Z_1} \vee \mathbf{1}_{Z_2}) \circ d_{X_1 \vee Z_1,X_2,Z_2} \circ (f_1 \wedge f_2)$,

$$\frac{f_1\colon U_1 \vdash X_1 \vee Z_1 \qquad f_2\colon U_2 \vdash X_2}{\wedge_{X_1,X_2}(f_1,f_2) =_{dn} d_{X_2,X_1,Z_1} \circ (f_1 \wedge f_2)\colon U_1 \wedge U_2 \vdash (X_1 \wedge X_2) \vee Z_1}$$

$$\frac{f_1\colon U_1 \vdash X_1 \qquad f_2\colon U_2 \vdash X_2}{\wedge_{X_1,X_2}(f_1,f_2) =_{dn} f_1 \wedge f_2\colon U_1 \wedge U_2 \vdash X_1 \wedge X_2}$$

$$\frac{f_1 \colon X_1 \wedge Z_1 \vdash U_1 \qquad\qquad f_2 \colon X_2 \wedge Z_2 \vdash U_2}{\vee_{X_1,X_2}(f_1, f_2) \colon (X_1 \vee X_2) \wedge Z_1 \wedge Z_2 \vdash U_1 \vee U_2}$$

where $\vee_{X_1,X_2}(f_1, f_2) =_{dn} (f_1 \vee f_2) \circ d_{Z_2,X_2,X_1 \wedge Z_1} \circ (d_{Z_1,X_1,X_2} \wedge \mathbf{1}_{Z_2})$,

$$\frac{f_1 \colon X_1 \wedge Z_1 \vdash U_1 \qquad\qquad f_2 \colon X_2 \vdash U_2}{\vee_{X_1,X_2}(f_1, f_2) =_{dn} (f_1 \vee f_2) \circ d_{Z_1,X_1,X_2} \colon (X_1 \vee X_2) \wedge Z_1 \vdash U_1 \vee U_2}$$

$$\frac{f_1 \colon X_1 \vdash U_1 \qquad\qquad f_2 \colon X_2 \vdash U_2}{\vee_{X_1,X_2}(f_1, f_2) =_{dn} f_1 \vee f_2 \colon X_1 \vee X_2 \vdash U_1 \vee U_2}$$

Note that $\wedge_{X_1,X_2}(f_1, f_2) = \wedge_{X_2,X_1}(f_2, f_1)$ holds in **GDS**. (In case we have $f_1 : U_1 \vdash X_1 \vee Z_1$ and $f_2 : U_2 \vdash X_2 \vee Z_2$, we apply $(d\overset{\vee}{b})$ of §7.2.) We will consider the terms on the two sides of this equation as the same Gentzen term. Analogously, $\vee_{X_1,X_2}(f_1, f_2) = \vee_{X_2,X_1}(f_2, f_1)$ holds in **GDS**. (In case we have $f_1 \colon X_1 \wedge Z_1 \vdash U_1$ and $f_2 \colon X_2 \wedge Z_2 \vdash U_2$, we apply $(d\hat{b})$ of §7.2.) We will consider also the terms on the two sides of this equation as the same Gentzen term. We do something analogous for arrow terms of $\mathcal{C}(\mathbf{GDS})$ built with $\wedge$ and $\vee$. Namely, we may omit some parentheses without ambiguity, and order is irrelevant. For example, $f \wedge g \wedge h$ stands for $(f \wedge g) \wedge h$, or $g \wedge (f \wedge h)$, etc., because all these arrow terms are equal in **GDS**.

In all the inductive clauses of Gentzen operations above, the Gentzen terms defined must denote arrows of **GDS**. So, for example, for $f_1 : U_1 \vdash X_1$ and $f_2 \colon U_2 \vdash X_2$ in $\wedge_{X_1,X_2}(f_1, f_2) \colon U_1 \wedge U_2 \vdash X_1 \wedge X_2$, we must have that $U_1 \wedge U_2$ and $X_1 \wedge X_2$ correspond to form sets of letters, which means that U_1 and U_2 cannot have letters in common, and the same for X_1 and X_2. So all our Gentzen operations are partial operations.

We can then prove the following lemma.

GENTZENIZATION LEMMA. *Every arrow of* **GDS** *is denoted by a Gentzen term.*

PROOF. We show by induction on the number of letters in the form set X that $\mathbf{1}_X$ is denoted by a Gentzen term. For that, we rely on the following equations of **GDS**:

$$\mathbf{1}_{X_1 \xi X_2} = \mathbf{1}_{X_1} \, \xi \, \mathbf{1}_{X_2}, \text{ for } \xi \in \{\wedge, \vee\},$$
$$(*\wedge) \quad f_1 \wedge f_2 = \wedge_{X_1,X_2}(f_1, f_2), \text{ for } f_1 \colon U_1 \vdash X_1 \text{ and } f_2 \colon U_2 \vdash X_2,$$
$$(*\vee) \quad f_1 \vee f_2 = \vee_{X_1,X_2}(f_1, f_2), \text{ for } f_1 \colon X_1 \vdash U_1 \text{ and } f_2 \colon X_2 \vdash U_2,$$

provided f_1 and f_2 are Gentzen terms (the equations $(*\wedge)$ and $(*\vee)$ are trivial).

If for any form set X we have that $\mathbf{1}_X$ stands for a Gentzen term, then we have in **GDS**

$$d_{Y,X,Z} = cut_X(\mathbf{1}_{X \vee Z}, \mathbf{1}_{X \wedge Y}) = \wedge_{Y,X}(\mathbf{1}_Y, \mathbf{1}_{X \vee Z}) = \vee_{Z,X}(\mathbf{1}_Z, \mathbf{1}_{X \wedge Y}).$$

It remains only to note that, besides the equations $(*\wedge)$ and $(*\vee)$ above, we have in **GDS** the equation $g \circ f = cut_X(f, g)$ for the Gentzen terms $f \colon U \vdash X$ and $g \colon X \vdash W$. ⊣

A Gentzen term is *cut-free* when it has no subterm of the form $cut_X(f, g)$. A Gentzen term of the form $cut_X(f, g)$ such that f and g are cut-free is called a *topmost cut*.

We define inductively the *depth of a subterm* of a Gentzen term:

> f is a subterm of f of depth 0;
>
> if γ is cut_X or $\wedge_{X_1,X_2}$ or $\vee_{X_1,X_2}$, and $\gamma(f_1, f_2)$ is a subterm of f of depth n, then f_1 and f_2 are subterms of f of depth $n+1$.

For a topmost cut $cut_X(f, g)$ such that X is of colour $\wedge$ and is not a letter, we say that the $\wedge$-*rank* of $cut_X(f, g)$ is $n \geq 0$ when f has a subterm $\wedge_{X_1,X_2}(f_1, f_2)$ of depth n such that X is $X_1 \wedge X_2$. Because the objects of **GDS** are form sets, i.e., they are "diversified", there can be at most one subterm of f of that form. For a topmost cut $cut_X(f, g)$ such that X is of colour $\vee$ and is not a letter, we say that the $\vee$-*rank* of $cut_X(f, g)$ is $n \geq 0$ when g has a subterm $\vee_{X_1,X_2}(g_1, g_2)$ of depth n such that X is $X_1 \vee X_2$. For a topmost cut $cut_p(f, g)$, note that $\mathbf{1}_p$ must be a subterm of both f and g, which occurs in each of them exactly once, because of "diversification". We say that the p-*rank* of $cut_p(f, g)$ is $n \geq 0$ when n is the sum of the depth of $\mathbf{1}_p$ in f and of the depth of $\mathbf{1}_p$ in g.

The *rank* of a topmost cut $cut_X(f,g)$ is either its $\wedge$-rank, or $\vee$-rank, or p-rank, depending on X.

The *complexity* of a topmost cut $cut_X(f,g)$ is (m,n) where $m \geq 1$ is the number of letters in X and $n \geq 0$ is the rank of this cut. Complexities are ordered lexicographically; i.e., we have $(m_1,n_1) < (m_2,n_2)$ iff either $m_1 < m_2$, or $m_1 = m_2$ and $n_1 < n_2$.

We can prove the following theorem for **GDS**.

CUT-ELIMINATION THEOREM. *For every Gentzen term t there is a cut-free Gentzen term t' such that $t = t'$ in* **GDS**.

PROOF. By induction on the complexity of a topmost cut $cut_X(f,g)$, we prove that $cut_X(f,g)$ is equal in **GDS** to a cut-free Gentzen term. From this the theorem follows. In the remainder of this proof we assume that equality between arrow terms is equality in **GDS**.

For the basis we have that if the complexity of $cut_X(f,g)$ is $(1,0)$, then $cut_X(f,g)$ is of the form $cut_p(\mathbf{1}_p,\mathbf{1}_p)$, which is equal to $\mathbf{1}_p$.

Suppose now the complexity is $(m,0)$ for $m > 1$, and suppose X is of colour $\wedge$. Then $cut_X(f,g)$ is of the form $cut_{X_1\wedge X_2}(\wedge_{X_1,X_2}(f_1,f_2),g)$, and we have the following cases.

($\wedge$1.1) Consider the Gentzen term

$$\dfrac{\dfrac{f_1\colon U_1 \vdash X_1 \vee Z_1 \qquad f_2\colon U_2 \vdash X_2 \vee Z_2}{\wedge_{X_1,X_2}(f_1,f_2)\colon U_1 \wedge U_2 \vdash (X_1 \wedge X_2) \vee Z_1 \vee Z_2} \qquad g\colon X_1 \wedge X_2 \wedge Y \vdash W}{cut_{X_1\wedge X_2}(\wedge_{X_1,X_2}(f_1,f_2),g)\colon U_1 \wedge U_2 \wedge Y \vdash W \vee Z_1 \vee Z_2}$$

Then consider the Gentzen term

$$\dfrac{f_2\colon U_2 \vdash X_2 \vee Z_2 \qquad \dfrac{f_1\colon U_1 \vdash X_1 \vee Z_1 \qquad g\colon X_1 \wedge X_2 \wedge Y \vdash W}{cut_{X_1}(f_1,g)\colon U_1 \wedge X_2 \wedge Y \vdash W \vee Z_1}}{cut_{X_2}(f_2, cut_{X_1}(f_1,g))\colon U_1 \wedge U_2 \wedge Y \vdash W \vee Z_1 \vee Z_2}$$

We show that

$$(*) \quad cut_{X_1\wedge X_2}(\wedge_{X_1,X_2}(f_1,f_2),g) = cut_{X_2}(f_2, cut_{X_1}(f_1,g)),$$

and the Gentzen term on the right-hand side has a topmost cut $cut_{X_1}(f_1,g)$ of lower complexity (m',n') than the Gentzen term on the left-hand side;

here $m' < m$. Hence, by the induction hypothesis, it is equal to a cut-free Gentzen term h, and $cut_{X_2}(f_2, h)$ is a topmost cut of lower complexity, to which we can also apply the induction hypothesis.

To show $(*)$, we have to show

$$\begin{aligned}(g \vee \mathbf{1}_{Z_1 \vee Z_2}) \circ d_{Y,X_1 \wedge X_2, Z_1 \vee Z_2} \circ \\ \circ (((d_{X_2,X_1,Z_1} \vee \mathbf{1}_{Z_2}) \circ d_{X_1 \vee Z_1, X_2, Z_2} \circ (f_1 \wedge f_2)) \wedge \mathbf{1}_Y) = \\ (((g \vee \mathbf{1}_{Z_1}) \circ d_{Y \wedge X_2, X_1, Z_1} \circ (f_1 \wedge \mathbf{1}_{Y \wedge X_2})) \vee \mathbf{1}_{Z_2}) \circ d_{U_1 \wedge Y, X_2, Z_2} \circ (f_2 \wedge \mathbf{1}_{U_1 \wedge Y}),\end{aligned}$$

and to derive this equation for **GDS** we use essentially $(d^L \wedge)$ of §7.2.

$(\wedge 1.2)$ If we have $g \colon X_1 \wedge X_2 \vdash W$, while f_1 and f_2 are as in $(\wedge 1.1)$, then to show $(*)$ we have to show

$$\begin{aligned}(g \vee \mathbf{1}_{Z_1 \vee Z_2}) \circ (d_{X_2,X_1,Z_1} \vee \mathbf{1}_{Z_2}) \circ d_{X_1 \vee Z_1, X_2, Z_2} \circ (f_1 \wedge f_2) = \\ ((g \vee \mathbf{1}_{Z_1}) \circ d_{X_2,X_1,Z_1} \circ (f_1 \wedge \mathbf{1}_{X_2})) \vee \mathbf{1}_{Z_2}) \circ d_{U_1,X_2,Z_2} \circ (f_2 \wedge \mathbf{1}_{U_1}),\end{aligned}$$

which follows readily with the help of $(d^L\ nat)$.

$(\wedge 2.1)$ If we have $f_1 : U_1 \vdash X_1$, while f_2 and g are as in $(\wedge 1.1)$, then to show $(*)$ we have to show

$$\begin{aligned}(g \vee \mathbf{1}_{Z_2}) \circ d_{Y,X_1 \wedge X_2, Z_2} \circ ((d_{X_1,X_2,Z_2} \circ (f_1 \wedge f_2)) \wedge \mathbf{1}_Y) = \\ ((g \circ (f_1 \wedge \mathbf{1}_{Y \wedge X_2})) \vee \mathbf{1}_{Z_2}) \circ d_{U_1 \wedge Y, X_2, Z_2} \circ (f_2 \wedge \mathbf{1}_{U_1 \wedge Y}),\end{aligned}$$

which follows by using essentially $(d^L \wedge)$.

$(\wedge 2.2)$ If we have $f_1 \colon U_1 \vdash X_1$, $g \colon X_1 \wedge X_2 \vdash W$ and f_2 as in $(\wedge 1.1)$, then to show $(*)$ we have to show

$$\begin{aligned}(g \vee \mathbf{1}_{Z_2}) \circ d_{X_1,X_2,Z_2} \circ (f_1 \wedge f_2) = \\ ((g \circ (f_1 \wedge \mathbf{1}_{X_2})) \vee \mathbf{1}_{Z_2}) \circ d_{U_1,X_2,Z_2} \circ (f_2 \wedge \mathbf{1}_{U_1}),\end{aligned}$$

which follows readily with the help of $(d^L\ nat)$.

$(\wedge 3.1)$ If we have $f_1 : U_1 \vdash X_1$, $f_2 : U_2 \vdash X_2$ and g as in $(\wedge 1.1)$, then to show $(*)$ we have to show

$$g \circ (f_1 \wedge f_2 \wedge \mathbf{1}_Y) = g \circ (f_1 \wedge \mathbf{1}_{Y \wedge X_2}) \circ (f_2 \wedge \mathbf{1}_{U_1 \wedge Y}),$$

which follows from bifunctorial equations.

($\wedge$3.2) If we have $f_1 \colon U_1 \vdash X_1$, $f_2 \colon U_2 \vdash X_2$ and $g \colon X_1 \wedge X_2 \vdash W$, then to show ($*$) we have to show

$$g \circ (f_1 \wedge f_2) = g \circ (f_1 \wedge \mathbf{1}_{X_2}) \circ (f_2 \wedge \mathbf{1}_{U_1}),$$

which follows again from bifunctorial equations.

If the complexity of $cut_X(f,g)$ is $(m,0)$ for $m > 1$, and X is of colour $\vee$, then we proceed analogously.

Suppose now the complexity of $cut_X(f,g)$ is (m,n) with $m,n \geq 1$. Then we have the following cases.

($\wedge$4) The form set X is of colour $\wedge$ (it may be of the form $X_1 \wedge X_2$ or p) and f is $\wedge_{V_1,V_2}(f_1, f_2)$. So, since $n \geq 1$, we have

$$\dfrac{\dfrac{f_1 \colon U_1 \vdash V_1 \vee X \vee Z_1 \qquad f_2 \colon U_2 \vdash V_2 \vee Z_2}{\wedge_{V_1,V_2}(f_1, f_2) \colon U_1 \wedge U_2 \vdash (V_1 \wedge V_2) \vee X \vee Z_1 \vee Z_2} \qquad g \colon X \wedge Y \vdash W}{cut_X(\wedge_{V_1,V_2}(f_1, f_2), g) \colon U_1 \wedge U_2 \wedge Y \vdash (V_1 \wedge V_2) \vee Z_1 \vee Z_2 \vee W}$$

Then consider the Gentzen term

$$\dfrac{\dfrac{f_1 \colon U_1 \vdash V_1 \vee X \vee Z_1 \qquad g \colon X \wedge Y \vdash W}{cut_X(f_1, g) \colon U_1 \wedge Y \vdash V_1 \vee Z_1 \vee W} \qquad f_2 \colon U_2 \vdash V_2 \vee Z_2}{\wedge_{V_1,V_2}(cut_X(f_1, g), f_2) \colon U_1 \wedge U_2 \wedge Y \vdash (V_1 \wedge V_2) \vee Z_1 \vee Z_2 \vee W}$$

We show that

$$(**) \quad cut_X(\wedge_{V_1,V_2}(f_1, f_2), g) = \wedge_{V_1,V_2}(cut_X(f_1, g), f_2),$$

and the complexity $(m, n-1)$ of the topmost cut $cut_X(f_1, g)$ is lower than (m,n), so that we may apply the induction hypothesis.

To show the equation ($**$), we have to show

$$\begin{aligned}
&(g \vee \mathbf{1}_{(V_1 \wedge V_2) \vee Z_1 \vee Z_2}) \circ d_{Y,X,(V_1 \wedge V_2) \vee Z_1 \vee Z_2} \circ \\
&\qquad \circ (((d_{V_2,V_1,X \vee Z_1} \vee \mathbf{1}_{Z_2}) \circ d_{V_1 \vee X \vee Z_1, V_2, Z_2} \circ (f_1 \wedge f_2)) \wedge \mathbf{1}_Y) = \\
&(d_{V_2,V_1,Z_1 \vee W} \vee \mathbf{1}_{Z_2}) \circ d_{V_1 \vee Z_1 \vee W, V_2, Z_2} \circ \\
&\qquad \circ (((g \vee \mathbf{1}_{V_1 \vee Z_1}) \circ d_{Y,X,V_1 \vee Z_1} \circ (f_1 \wedge \mathbf{1}_Y)) \wedge f_2).
\end{aligned}$$

To derive this equation for **GDS**, we use essentially $(d\hat{b})$ and $(d\check{b})$, besides $(d^L\ nat)$, $(d^L\wedge)$ and $(d^L\vee)$ (see §7.2). We have also to consider cases where we have $f_1\colon U_1 \vdash V_1 \vee X$, or $f_2\colon U_2 \vdash V_2$, or $g\colon X \vdash W$ (analogously to what we had in $(\wedge 1.1)$-$(\wedge 3.2)$). In all of them, $(**)$ amounts to equations simpler than the equation above, which all hold in **GDS**.

$(\wedge 5)$ The form set X is of colour $\wedge$, and f is $\vee_{V_1,V_2}(f_1, f_2)$, so that we have $f_1\colon V_1 \wedge U_1 \vdash X \vee Z_1$, $f_2\colon V_2 \wedge U_2 \vdash Z_2$ and $g\colon X \wedge Y \vdash W$. Then we have to show the equation

$$(***) \quad cut_X(\vee_{V_1,V_2}(f_1, f_2), g) = \vee_{V_1,V_2}(cut_X(f_1, g), f_2)$$

with the complexity $(m, n-1)$ of $cut_X(f_1, g)$ lower than (m, n). To show this equation, we have to show

$$(g \vee \mathbf{1}_{Z_1 \vee Z_2}) \circ d_{Y,X,Z_1 \vee Z_2} \circ \\ \circ (((f_1 \vee f_2) \circ d_{U_2,V_2,V_1 \wedge U_1} \circ (d_{U_1,V_1,V_2} \wedge \mathbf{1}_{U_2})) \wedge \mathbf{1}_V) =$$

$$(((g \vee \mathbf{1}_{Z_1}) \circ d_{Y,X,Z_1} \circ (f_1 \wedge \mathbf{1}_Y)) \vee f_2) \circ d_{U_2,V_2,V_1 \wedge U_1 \wedge Y} \circ (d_{U_1 \wedge Y,V_1,V_2} \wedge \mathbf{1}_{U_2}),$$

and to derive that for **GDS** we use essentially $(d\hat{b})$. We have also to consider cases where we have $f_1 : V_1 \vdash X \vee Z_1$, or $f_1 : V_1 \wedge U_1 \vdash X$, or $f_1 : V_1 \vdash X$, or $f_2 : V_2 \vdash Z_2$, or $g : X \vdash W$. In all of them $(***)$, amounts to simpler equations, which all hold in **GDS**.

It remains to consider cases with complexity (m, n) where $m, n \geq 1$, and X is of colour $\vee$ (it may be of the form $X_1 \vee X_2$ or p). These additional cases are settled dually to cases $(\wedge 4)$ and $(\wedge 5)$. Note that in cases with complexity (m, n) where $m, n \geq 1$ and X is a letter (hence $m = 1$) we have that X is of both colours, and hence in these cases we can proceed either as in $(\wedge 4)$ and $(\wedge 5)$, or as in cases dual to $(\wedge 4)$ and $(\wedge 5)$ we have just mentioned. ⊣

§7.8. Invertibility in GDS

In this section, we prove the invertibility results announced at the end of §7.6. First, we cover some preliminary matters. We will need later the following equations of **GDS**:

$(\wedge\wedge 1)$ $\wedge_{W_1\wedge W_2,W_3}(\wedge_{W_1,W_2}(f_1,f_2),f_3)=\wedge_{W_1,W_2\wedge W_3}(f_1,\wedge_{W_2,W_3}(f_2,f_3))$
for f_i of type $H_i\vdash W_i\vee J_i$ or $H_i\vdash W_i$, where $i\in\{1,2,3\}$,

$(\wedge\wedge 2)$ $\wedge_{R_1,R_3}(\wedge_{W_1,W_2}(f_1,f_2),f_3)=\wedge_{W_1,W_2}(\wedge_{R_1,R_3}(f_1,f_3),f_2)$
for f_1 of type $H_1\vdash W_1\vee R_1\vee J_1$ or $H_1\vdash W_1\vee R_1$,
f_2 of type $H_2\vdash W_2\vee J_2$ or $H_2\vdash W_2$, and
f_3 of type $H_3\vdash R_3\vee J_3$ or $H_3\vdash R_3$,

$(\vee\vee 1)$ $\vee_{W_1\vee W_2,W_3}(\vee_{W_1,W_2}(f_1,f_2),f_3)=\vee_{W_1,W_2\vee W_3}(f_1,\vee_{W_2,W_3}(f_2,f_3))$
for f_i of type $W_i\wedge J_i\vdash H_i$ or $W_i\vdash H_i$, where $i\in\{1,2,3\}$,

$(\vee\vee 2)$ $\vee_{R_1,R_3}(\vee_{W_1,W_2}(f_1,f_2),f_3)=\vee_{W_1,W_2}(\vee_{R_1,R_3}(f_1,f_3),f_2)$,
for f_1 of type $W_1\wedge R_1\wedge J_1\vdash H_1$ or $W_1\wedge R_1\vdash H_1$,
f_2 of type $W_2\wedge J_2\vdash H_2$ or $W_2\vdash H_2$, and
f_3 of type $R_3\wedge J_3\vdash H_3$ or $R_3\vdash H_3$,

$(\wedge\vee)$ $\wedge_{W_2,W_3}(\vee_{V_1,V_2}(f_1,f_2),f_3)=\vee_{V_1,V_2}(f_1,\wedge_{W_2,W_3}(f_2,f_3))$,
for f_1 of type $V_1\wedge H_1\vdash J_1$ or $V_1\vdash J_1$,
f_2 of type $V_2\wedge H_2\vdash W_2\vee J_2$ or $V_2\wedge H_2\vdash W_2$ or
f_2 of type $V_2\vdash W_2\vee J_2$ or $V_2\vdash W_2$, and
f_3 of type $H_3\vdash W_3\vee J_3$ or $H_3\vdash W_3$.

The equations $(\wedge\wedge 1)$ and $(\wedge\wedge 2)$, or alternatively $(\vee\vee 1)$ and $(\vee\vee 2)$, are analogous to the two associativity equations for the cut operation one finds in multicategories (see [80] and [83], Section 3).

To derive these equations for **GDS** is a rather straightforward, though pretty lengthy, exercise. We always derive the most complex case, with all possible parameters present (for $(\wedge\wedge 1)$ this means that f_i is of type $H_i\vdash W_i\vee J_i$), and the remaining cases are obtained by simplifying this most complex case. For example, to derive the most complex case of $(\wedge\wedge 1)$ for **GDS** we use essentially $(d^L\wedge)$, $(d^L\ nat)$ and Net Coherence.

Let $let(X)$ be the set of letters occurring in the form set X. It is clear that we have the following.

BALANCE REMARK. *For every arrow* $f\colon X \vdash Y$ *of* **GDS**, *we have* $let(X) = let(Y)$.

A pair of form sets $(X_1 \wedge \ldots \wedge X_n, Y_1 \vee \ldots \vee Y_n)$, where $n \geq 2$, is *splittable* when $let(X_i) = let(Y_i)$ for every $i \in \{1, \ldots, n\}$. A sequence of form sets $X_1, \ldots, X_n, Y_1, \ldots, Y_n$ is a *total split* of the pair of form sets $(X_1 \wedge \ldots \wedge X_n, Y_1 \vee \ldots \vee Y_n)$ when $let(X_i) = let(Y_i)$ and none of the pairs (X_i, Y_i) is splittable. For every splittable pair of form sets there is a total split.

We say that an arrow $f\colon X \vdash Y$ of **GDS** is *splittable* when its type (X, Y) is splittable, and we say that a total split of (X, Y) is a *total split of* f.

SPLITTING REMARK. Take an arrow f of **GDS** of type

$$X \wedge X_1 \wedge \ldots \wedge X_n \vdash Z \vee Y_1 \vee \ldots \vee Y_n$$
$$\text{or} \quad X \wedge X_1 \wedge \ldots \wedge X_n \vdash Z \vee R' \vee Y_1 \vee \ldots \vee Y_n$$

and an arrow g of **GDS** of type

$$V \wedge V_1 \wedge \ldots \wedge V_m \vdash U \vee W_1 \vee \ldots \vee W_m$$
$$\text{or} \quad V \wedge V_1 \wedge \ldots \wedge V_m \vdash U \vee R'' \vee W_1 \vee \ldots \vee W_m$$

with $n+m \geq 1$ (if $n = 0$, then the subword $\wedge X_1 \wedge \ldots \wedge X_n$ is just omitted, and analogously in other cases).

Let $\wedge_{Z,U}(f, g)$ be splittable with the total split

$$X \wedge V, X_1, \ldots, X_n, V_1, \ldots, V_m, Y, Y_1, \ldots, Y_n, W_1, \ldots, W_m$$

where Y is $Z \wedge U$ or $(Z \wedge U) \vee R$, and R is R' or R'' or $R' \vee R''$. If $n \geq 1$, then f is splittable, and if $m \geq 1$, then g is splittable.

Here (X, Z) or $(X, Z \vee R')$ may be splittable, and hence the forms of the type of f above do not show the total split tied to this type. If the pair $(X, Z \vee R')$ is splittable and $S_1, \ldots, S_k, T_1, \ldots, T_k$ is its total split, then for every $j \in \{1, \ldots, k\}$ we have $let(Z) \cap let(T_j) \neq \emptyset$. (Otherwise, the total split of $\wedge_{Z,U}(f, g)$ mentioned above would not be a total split.) We have an analogous remark for (V, U), $(V, U \vee R'')$ and g.

An analogous remark holds for $\vee_{Z,U}(f, g)$.

It follows from the Splitting Remarks that, for $\xi \in \{\wedge, \vee\}$, if $\xi_{Z,U}(f, g)$ is splittable, then f or g is splittable. Since $\mathbf{1}_p$ is not splittable, we can easily conclude the following with the help of the Cut-Elimination Theorem of the preceding section.

SPLITTING COROLLARY. *No arrow of* **GDS** *is splittable.*

This corollary is related to the connectedness condition of proof nets (see [30]).

Next we prove the following lemma for **GDS**.

INVERTIBILITY LEMMA FOR $\wedge$. (i) *If* $f : U_1 \wedge U_2 \vdash (X_1 \wedge X_2) \vee Z_1 \vee Z_2$ *is a cut-free Gentzen term such that* $let(U_i) = let(X_i) \cup let(Z_i)$ *for* $i \in \{1, 2\}$, *then there are two cut-free Gentzen terms* $f_1 : U_1 \vdash X_1 \vee Z_1$ *and* $f_2 : U_2 \vdash X_2 \vee Z_2$ *such that* $f = \wedge_{X_1,X_2}(f_1, f_2)$.

(ii) *If* $f : U_1 \wedge U_2 \vdash (X_1 \wedge X_2) \vee Z_1$ *is a cut-free Gentzen term such that* $let(U_1) = let(X_1) \cup let(Z_1)$ *and* $let(U_2) = let(X_2)$, *then there are two cut-free Gentzen terms* $f_1 : U_1 \vdash X_1 \vee Z_1$ *and* $f_2 : U_2 \vdash X_2$ *such that* $f = \wedge_{X_1,X_2}(f_1, f_2)$.

(iii) *If* $f : U_1 \wedge U_2 \vdash X_1 \wedge X_2$ *is a cut-free Gentzen term such that* $let(U_i) = let(X_i)$ *for* $i \in \{1, 2\}$, *then there are two cut-free Gentzen terms* $f_1 : U_1 \vdash X_1$ *and* $f_2 : U_2 \vdash X_2$ *such that* $f = \wedge_{X_1,X_2}(f_1, f_2)$.

PROOF. We proceed by induction on the length of the cut-free Gentzen term f. If f is $\mathbf{1}_p$, the lemma holds trivially, since f cannot be of the required type.

Suppose next that f is $\wedge_{Y^a,Y^b}(f^a, f^b)$ for $f^a : W^a \vdash R^a$ and $f^b : W^b \vdash R^b$. Then, under the assumptions of (i), we have two cases:

$$\begin{array}{ll} (\wedge \text{ i I}) & X_1 \wedge X_2 \text{ is } Y^a \wedge Y^b, \\ (\wedge \text{ i II}) & X_1 \wedge X_2 \text{ is different from } Y^a \wedge Y^b. \end{array}$$

We deal first with ($\wedge$ i I).

The cases where $let(X_1) \cup let(X_2) \subseteq let(W^a)$ or $let(X_1) \cup let(X_2) \subseteq let(W^b)$ are impossible.

If $let(X_1) \subseteq let(W^a)$ and $let(X_2) \subseteq let(W^b)$, then we must have $let(Z_1) \subseteq let(W^a)$ and $let(Z_2) \subseteq let(W^b)$. All the other cases are excluded by the

Splitting Corollary. For example, if $let(Z_1) \cup let(Z_2) \subseteq let(W^a)$, then W^a must be $U_1 \wedge U_2^a$, R^a must be $X_1 \vee Z_1 \vee Z_2$, W^b must be U_2^b, and R^b must be X_2, where U_2 is $U_2^a \wedge U_2^b$. Then, since $let(U_2^a) = let(Z_2)$, the arrows f^a and f would be splittable, which contradicts the Splitting Corollary. In the only possible case mentioned above, we take f^a for f_1 and f^b for f_2.

The case where $let(X_1) \subseteq W^b$ and $let(X_2) \subseteq W^a$ is analogous to the case just settled.

Let $\rho(X, Y, Z)$ abbreviate the conjunction of the following conditions:

$$\begin{aligned} &let(X) \subseteq let(Y) \cup let(Z), \\ &let(X) \cap let(Y) \neq \emptyset, \\ &let(X) \cap let(Z) \neq \emptyset. \end{aligned}$$

If $let(X_1) \subseteq let(W^a)$ and $\rho(X_2, W^a, W^b)$, then we have as possible cases $let(Z_1) \subseteq let(W^a)$ together with

(1) $let(Z_2) \subseteq let(W^a)$, or
(2) $let(Z_2) \subseteq let(W^b)$, or
(3) $\rho(Z_2, W^a, W^b)$.

The remaining cases are excluded by the Splitting Corollary.

We deal first with (3). Then f^a is of the type $U_1 \wedge U_2^a \vdash (X_1 \wedge X_2^a) \vee Z_1 \vee Z_2^a$, while f^b is of the type $U_2^b \vdash X_2^b \vee Z_2^b$, where $U_2^a \wedge U_2^b$ is U_2, $X_2^a \wedge X_2^b$ is X_2 and $Z_2^a \vee Z_2^b$ is Z_2. By the induction hypothesis, $f^a = \wedge_{X_1, X_2^a}(f_1^a, f_2^a)$ for $f_1^a : U_1 \vdash X_1 \vee Z_1$ and $f_2^a : U_2^a \vdash X_2^a \vee Z_2^a$. Then by the equation $(\wedge \wedge 1)$ we have

$$\wedge_{X_1 \wedge X_2^a, X_2^b}(\wedge_{X_1, X_2^a}(f_1^a, f_2^a), f^b) = \wedge_{X_1, X_2}(f_1^a, \wedge_{X_2^a, X_2^b}(f_2^a, f^b)),$$

and we take that f_1 is f_1^a, while f_2 is $\wedge_{X_2^a, X_2^b}(f_2^a, f^b)$. In cases (1) and (2) we proceed analogously, using again $(\wedge \wedge 1)$ (less complex cases of this equation, with less parameters).

The three cases where we have $let(X_1) \subseteq let(W^b)$ and $\rho(X_2, W^a, W^b)$, or $let(X_2) \subseteq let(W^a)$ and $\rho(X_1, W^a, W^b)$, or $let(X_2) \subseteq let(W^b)$ and $\rho(X_1, W^a, W^b)$, are all settled analogously to the case we have just dealt with.

The remaining case of $(\wedge \text{ i I})$ is when $\rho(X_1, W^a, W^b)$ and $\rho(X_2, W^a, W^b)$. Then either $let(Z_i) \subseteq let(W^a)$, or $let(Z_i) \subseteq let(W^b)$, or $\rho(Z_i, W^a, W^b)$, and we always apply the induction hypothesis and equation $(\wedge\wedge 1)$ three times; namely, we use the equation

$$\begin{aligned}\wedge_{X_1^a\wedge X_2^a, X_1^b\wedge X_2^b}(\wedge_{X_1^a,X_2^a}(f_1^a,f_2^a),\wedge_{X_1^b,X_2^b}(f_1^b,f_2^b)) = \\ \wedge_{X_1^a\wedge X_1^b, X_2^a\wedge X_2^b}(\wedge_{X_1^a,X_1^b}(f_1^a,f_1^b),\wedge_{X_2^a,X_2^b}(f_2^a,f_2^b)).\end{aligned}$$

Under the assumption $(\wedge$ i II), we have the cases

$(\wedge$ i II.1) $\quad Z_1$ is $Z_1'\vee(Y^a\wedge Y^b)$,
$(\wedge$ i II.2) $\quad Z_1$ is $Y^a\wedge Y^b$,

and two more cases obtained by replacing the index $_1$ in Z_1 and Z_1' by $_2$. For $(\wedge$ i II.1), we have as possible cases $let(X_1\wedge X_2)\cup let(Z_2)\subseteq let(W^a)$ together with

(1) $\quad let(Z_1)\subseteq let(W^a)$, or
(2) $\quad let(Z_1')\subseteq let(W^b)$, or
(3) $\quad \rho(Z_1', W^a, W^b)$,

and three more cases with $let(X_1\wedge X_2)\cup let(Z_2)\subseteq let(W^b)$. All the remaining cases are excluded by the Splitting Corollary.

We deal first with (3). Then f^a is of the type $U_1^a\wedge U_2\vdash Y^a\vee(X_1\wedge X_2)\vee Z_1^a\vee Z_2$, while f^b is of the type $U_1^b\vdash Y^b\vee Z_1^b$, where $U_1^a\wedge U_1^b$ is U_1 and $Z_1^a\vee Z_1^b$ is Z_1'. By the induction hypothesis, $f^a=\wedge_{X_1,X_2}(f_1^a,f_2^a)$ for $f_1^a: U_1^a\vdash Y^a\vee X_1\vee Z_1^a$ and $f_2^a: U_2\vdash X_2\vee Z_2$. Then, by the equation $(\wedge\wedge 2)$, we have

$$\wedge_{Y^a,Y^b}(\wedge_{X_1,X_2}(f_1^a,f_2^a),f^b)=\wedge_{X_1,X_2}(\wedge_{Y^a,Y^b}(f_1^a,f^b),f_2^a)),$$

and we take that f_1 is $\wedge_{Y^a,Y^b}(f_1^a,f^b)$, while f_2 is f_2^b. In cases (1) and (2), and cases obtained by interchanging a and b, we proceed analogously.

For $(\wedge$ i II.2), we have as possible cases $let(X_1\wedge X_2)\cup let(Z_2)\subseteq let(W^a)$ and $let(X_1\wedge X_2)\cup let(Z_2)\subseteq let(W^b)$, for which we apply again the induction hypothesis and the equation $(\wedge\wedge 2)$. All the remaining cases are excluded by the Splitting Corollary.

We proceed analogously when we have $(\wedge$ i II) and Z_2 is $Z_2'\vee(Y^a\wedge Y^b)$ or $Y^a\wedge Y^b$. With that we have settled $(\wedge$ i II), and also (i).

Under the assumptions of (ii), we have again two cases:

(∧ ii I) $X_1 \wedge X_2$ is $Y^a \wedge Y^b$,
(∧ ii II) $X_1 \wedge X_2$ is different from $Y^a \wedge Y^b$.

We deal with these cases as above, with simplifications in cases already considered.

Under the assumptions of (iii), we must have that $X_1 \wedge X_2$ is $Y^a \wedge Y^b$, and we have cases simplifying again cases already considered. With that we have finished dealing with the assumption that $f = \wedge_{Y^a,Y^b}(f^a, f^b)$.

Suppose now f is $\vee_{Y^a,Y^b}(f^a, f^b)$ for $f^a : W^a \vdash R^a$ and $f^b : W^b \vdash R^b$. Then under the assumptions of (i) we have the cases:

(∨ i 1) U_1 is $U_1' \wedge (Y^a \vee Y^b)$,
(∨ i 2) U_1 is $Y^a \vee Y^b$,

and two more cases with the index $_1$ of U_1 and U_1' replaced by $_2$.

For (∨ i 1), we have as possible cases $let(X_1 \wedge X_2) \cup let(Z_2) \subseteq let(W^a)$ together with

(α) $let(Z_1) \subseteq let(W^b)$, or
(β) $\rho(Z_1, W^a, W^b)$,

and together with

(1) $let(U_1') \subseteq let(W^a)$, or
(2) $let(U_1') \subseteq let(W^b)$, or
(3) $\rho(U_1', W^a, W^b)$,

and six more analogous cases with $let(X_1 \wedge X_2) \cup let(Z_2) \subseteq let(W^b)$ together with

$let(Z_1) \subseteq let(W^a)$, or
$\rho(Z_1, W^a, W^b)$.

All the remaining cases are excluded by the Splitting Corollary, or because $let(Y^a \vee Y^b) \subseteq let(X_1 \wedge X_2) \cup let(Z_1)$.

We deal first with (β) together with (3). Then f^a is of the type $Y^a \wedge U_1^a \wedge U_2 \vdash (X_1 \wedge X_2) \vee Z_1^a \vee Z_2$, while f^b is of the type $Y^b \wedge U_1^b \vdash Z_1^b$, where $U_1^a \wedge U_1^b$ is U_1 and $Z_1^a \vee Z_1^b$ is Z_1. By the induction hypothesis, $f^a = \wedge_{X_1,X_2}(f_1^a, f_2^a)$

for $f_1^a : Y^a \wedge U_1^a \vdash X_1 \vee Z_1^a$ and $f_2^a : U_2 \vdash X_2 \vee Z_2$. Then, by the equation $(\wedge\vee)$, we have

$$\vee_{Y^a,Y^b}(f^b, \wedge_{X_1,X_2}(f_1^a, f_2^a)) = \wedge_{X_1,X_2}(\vee_{Y^b,Y^a}(f^b, f_1^a), f_2^a),$$

and we take that f_1 is $\vee_{Y^b,Y^a}(f^b, f_1^a)$, while f_2 is f_2^a. In all the remaining cases, we proceed analogously, as well as in ($\vee$ i 2). This settles (i).

Under the assumptions of (ii), we have cases analogous to those already treated with Z_2 omitted. So we apply again the equation $(\wedge\vee)$.

The assumptions of (iii) are excluded if $f = \vee_{Y^a,Y^b}(f^a, f^b)$. ⊣

We prove analogously the following lemma for **GDS**.

INVERTIBILITY LEMMA FOR $\vee$. (i) *If $f\colon (X_1 \vee X_2) \wedge Z_1 \wedge Z_2 \vdash U_1 \vee U_2$ is a cut-free Gentzen term such that $let(U_i) = let(X_i) \cup let(Z_i)$ for $i \in \{1, 2\}$, then there are two cut-free Gentzen terms $f_1\colon X_1 \wedge Z_1 \vdash U_1$ and $f_2\colon X_2 \wedge Z_2 \vdash U_2$ such that $f = \vee_{X_1,X_2}(f_1, f_2)$.*

(ii) *If $f : (X_1 \vee X_2) \wedge Z_1 \vdash U_1 \vee U_2$ is a cut-free Gentzen term such that $let(U_1) = let(X_1) \cup let(Z_1)$ and $let(U_2) = let(X_2)$, then there are two cut-free Gentzen terms $f_1 : X_1 \wedge Z_1 \vdash U_1$ and $f_2 : X_2 \vdash U_2$ such that $f = \vee_{X_1,X_2}(f_1, f_2)$.*

(iii) *If $f\colon X_1 \vee X_2 \vdash U_1 \vee U_2$ is a cut-free Gentzen term such that $let(U_i) = let(X_i)$ for $i \in \{1, 2\}$, then there are two cut-free Gentzen terms $f_1\colon X_1 \vdash U_1$ and $f_2\colon X_2 \vdash U_2$ such that $f = \vee_{X_1,X_2}(f_1, f_2)$.*

Let the *quantity of letters* in an arrow $f : X \vdash Y$ of **GDS** be the cardinality of $let(X)$ (which is equal to the cardinality of $let(Y)$). Then we can prove the following theorem for **GDS**.

CUT-FREE PREORDERING. *For every pair of cut-free Gentzen terms $f_1, f_2\colon X \vdash Y$ we have $f_1 = f_2$.*

PROOF. We proceed by induction on the quantity of letters in f_1 (which is equal to the quantity of letters in f_2). If $n = 1$, then $f_1 = f_2 = \mathbf{1}_p$.

Suppose $n > 1$. If f_1 is $\wedge_{Z_1,Z_2}(f_1', f_1'')$, then by the Invertibility Lemma for $\wedge$ we have that $f_2 = \wedge_{Z_1,Z_2}(f_2', f_2'')$ for f_2' and f_2'' of the same types as

f_1' and f_1'' respectively. By the induction hypothesis, $f_1' = f_2'$ and $f_1'' = f_2''$, and hence $f_1 = f_2$. We proceed analogously if f_1 is $\vee_{Z_1,Z_2}(f_1', f_1'')$. ⊣

As a corollary of the Cut-Elimination Theorem and of Cut-Free Preordering, we obtain that **GDS** is a preorder, which, under the assumption $Gf = Gg$, implies the assertion (div) of §7.6. This proves Symmetric Net Coherence.

Net Coherence of §7.3 could also have been proved via a Cut-Elimination Theorem and Cut-Free Preordering. Strictification, however, would be in the associative structure only, and not in the symmetric associative structure.

The category **DS** corresponds to the multiplicative conjunction-disjunction fragment of linear logic, for which proof nets were developed (see [58] and [30]). Proof nets, however, serve mainly to solve the theoremhood problem, while coherence in our sense is maybe implicitly presupposed with them. The theoremhood problem for **DS** can also be solved via our results for **GDS** in this and in the preceding section, based on cut elimination, and we do not find this solution in the style of Gentzen more complicated than the solution provided by proof nets.

§7.9. Linearly distributive categories

To obtain the natural logical category $\mathbf{DA}_{\top,\bot}$, we have that the logical system $\mathcal{C}(\mathbf{DA}_{\top,\bot})$ is in $\mathcal{L}_{\wedge,\vee,\top,\bot}$, with the transformations α included in $\mathbf{1}$, b, δ-σ and d. The specific equations of $\mathcal{E}(\mathbf{DA}_{\top,\bot})$ are obtained by taking the union of those of $\mathcal{E}(\mathbf{DA})$ and $\mathcal{E}(\mathbf{A}_{\top,\bot})$ plus

$$(\hat{\sigma}\, d^L) \quad d^L_{\top,B,C} = (\hat{\sigma}_{\overleftarrow{B}} \vee \mathbf{1}_C) \circ \hat{\sigma}_{\overrightarrow{B\vee C}},$$
$$(\check{\delta}\, d^L) \quad d^L_{A,B,\bot} = \check{\delta}_{\overleftarrow{A\wedge B}} \circ (\mathbf{1}_A \wedge \check{\delta}_{\overrightarrow{B}}),$$

$$(\hat{\delta}\, d^R) \quad d^R_{C,B,\top} = (\mathbf{1}_C \vee \hat{\delta}_{\overleftarrow{B}}) \circ \hat{\delta}_{\overrightarrow{C\vee B}},$$
$$(\check{\sigma}\, d^R) \quad d^R_{\bot,B,A} = \check{\sigma}_{\overleftarrow{B\wedge A}} \circ (\check{\sigma}_{\overrightarrow{B}} \wedge \mathbf{1}_A).$$

Natural $\mathbf{DA}_{\top,\bot}$-categories are called *linearly distributive* categories in [22] (the original name from [19] is *weakly distributive* categories). According to our nomenclature, they could be called *dissociative bimonoidal* categories. All of the specific equations above may be found in [19] (Section 2.1). (These equations should be compared with the equations $(d\hat{k})$ and $(d\check{k})$ of §11.1.)

We have still a functor G from $\mathbf{DA}_{\top,\bot}$ to *Rel*, but according to [8] (Section 4.2, pp. 275-278), for

$$\varepsilon_A =_{df} \check{\sigma}^{\rightarrow}_{A} \circ (\hat{\delta}^{\rightarrow}_{\bot} \vee \mathbf{1}_A) \circ d^L_{\bot,\top,A},$$
$$\eta_A =_{df} d^R_{\top,\bot,A} \circ (\check{\delta}^{\leftarrow}_{\top} \wedge \mathbf{1}_A) \circ \hat{\sigma}^{\leftarrow}_{A},$$

the equations

$$\eta_{\top\vee A} \circ (\mathbf{1}_\top \vee \varepsilon_A) = \mathbf{1}_{\top\vee(\bot\wedge(\top\vee A))},$$
$$(\mathbf{1}_\bot \wedge \eta_A) \circ \varepsilon_{\bot\wedge A} = \mathbf{1}_{\bot\wedge(\top\vee(\bot\wedge A))}$$

do not hold in $\mathbf{DA}_{\top,\bot}$, although, when f and g are respectively the left-hand side and right-hand side of one of these equations, we have $Gf = Gg$ in *Rel*. So G is not faithful, and coherence fails. The faithfulness of G in this case would yield preordering, and $\mathbf{DA}_{\top,\bot}$ is not a preorder.

Note that in $\mathbf{DA}_{\top,\bot}$ we have

$$(\mathbf{1}_\top \vee \varepsilon_A) \circ \eta_{\top\vee A} = \mathbf{1}_{\top\vee A},$$
$$\varepsilon_{\bot\wedge A} \circ (\mathbf{1}_\bot \wedge \eta_A) = \mathbf{1}_{\bot\wedge A},$$

which are the triangular equations of an adjunction (for the notion of adjunction see [94], Chapter 4; the functor $\bot\wedge$ is left-adjoint to the functor $\top\vee$). What fails is the isomorphism between $\top\vee(\bot\wedge(\top\vee A))$ and $\top\vee A$, and between $\bot\wedge(\top\vee(\bot\wedge A))$ and $\bot\wedge A$ ([113] deals with a related problem in symmetric monoidal closed categories). We do not know what other equations, if any, besides those that deliver these isomorphisms, should be added to the axioms of $\mathbf{DA}_{\top,\bot}$ in order to obtain coherence.

A sort of coherence for linearly distributive categories (symmetric and not symmetric, without the isomorphisms above) in the context of proof nets has been investigated in a number of papers (see [19], [8], [20] and [109]). Coherence in this sense is not quite foreign to what we mean by coherence, but it is not the same thing. The investigations of [19] appeal to a connection with the polycategories of [120].

CHAPTER 8

MIX CATEGORIES

In this chapter, we consider categories having what linear logicians call mix—namely, a natural transformation between the two bifunctors of the double monoidal structure. The double monoidal structure has or does not have associativity, symmetry and dissociativity. We prove coherence for such categories that lack unit objects. The mix principle is an important addition to Gentzen's plural sequent formulation of classical logic, and this is why we pay particular attention to it.

Our proofs are variations on the cut-elimination theme, and on the techniques of the preceding chapters. There are proofs based on composition-free languages for our categories, and a proof based on an extension of the cut-elimination procedure of the preceding chapter.

§8.1. Coherence of mix and mix-dissociative categories

To obtain the natural logical category **MI**, we have that the logical system $\mathcal{C}(\mathbf{MI})$ is in $\mathcal{L}_{\wedge,\vee}$, with the transformations α being $\mathbf{1}$ and m. The equations $\mathcal{E}(\mathbf{MI})$ are just those of $\mathcal{E}^{nat}_{\mathcal{C}(\mathrm{MI})}$ (see §4.1). We call natural **MI**-categories *mix* categories.

A logical principle called mix amounting to $m_{A,B} : A \wedge B \vdash A \vee B$ was considered in [58] (Section V.4), [30] (Section 3.3), [51], [7] and [20]. Gentzen called *Mischung* in German—which is usually translated as *mix*—a rule that generalizes the cut rule of sequent systems; an instance of *Mischung* is

$$\frac{\Gamma_1 \vdash \Delta_1, \Theta \qquad \Theta, \Gamma_2 \vdash \Delta_2}{\Gamma_1, \Gamma_2 \vdash \Delta_1, \Delta_2}$$

where Θ is a nonempty sequence of occurrences of the same formula (see [55], Section III.3.1). The mix principle of $m_{A,B}$ is related to the *Mischung* rule above where Θ is the empty sequence. (Gentzen did not envisage this mix principle because he could prove the conclusion from one of the premises with the help of the structural rules of thinning on the left and thinning on the right.)

In $\mathcal{C}(\mathbf{MI})$ we define as follows the binary total operation $\diamond$ on arrow terms:

$$\frac{f\colon A\vdash D \qquad g\colon B\vdash E}{f\diamond g =_{df} (f\vee g)\circ m_{A,B}\colon A\wedge B\vdash D\vee E}$$

for which in **MI** we have the equations

$$(\diamond)\quad (g_1\diamond g_2)\circ(f_1\wedge f_2)=(g_1\vee g_2)\circ(f_1\diamond f_2)=(g_1\circ f_1)\diamond(g_2\circ f_2).$$

From the equation (m *nat*) of §2.7, in **MI** we obtain immediately $f\diamond g = m_{D,E}\circ(f\wedge g)$, which gives an alternative definition of $\diamond$.

A syntactical system $\mathcal{C}(\diamond\mathbf{MI})$ synonymous with $\mathcal{C}(\mathbf{MI})$ is obtained by taking as objects the formulae of $\mathcal{L}_{\wedge,\vee}$, as primitive arrow terms identity arrow terms only, and as operations on arrow terms $\circ$, $\wedge$, $\vee$ and $\diamond$. The equational system $\mathcal{E}(\diamond\mathbf{MI})$ is obtained by assuming the categorial and bifunctorial equations for $\wedge$ and $\vee$, and the equations $(\diamond)$. The category $\diamond\mathbf{MI}$ is $\mathcal{C}(\diamond\mathbf{MI})/\mathcal{E}(\diamond\mathbf{MI})$.

With the definition

$$m_{A,B} =_{df} \mathbf{1}_A\diamond\mathbf{1}_B$$

in $\mathcal{C}(\mathbf{MI})$, we obtain two obvious functors from $\mathcal{C}(\mathbf{MI})$ to $\mathcal{C}(\diamond\mathbf{MI})$, and vice versa, which preserve the respective structures on the nose (see §2.8), and these functors induce functors that give the isomorphism of **MI** and $\diamond$**MI**. This means that $\mathcal{C}(\mathbf{MI})$ and $\mathcal{C}(\diamond\mathbf{MI})$ are synonymous (see the end of §2.4 for the notion of synonymity of syntactical systems). Note that officially $\mathcal{C}(\diamond\mathbf{MI})$ is not a logical system, because $\diamond$ is not of the ξ kind: it is not a bifunctor in **MI**.

We can prove the following proposition for $\diamond$**MI** simply by relying on the equations $(\diamond)$ and bifunctorial equations.

COMPOSITION ELIMINATION. *For every arrow term h there is a composition-free arrow term h' such that $h = h'$.*

This result is a simple kind of cut-elimination result, such as we had in §7.7.

The composition-free arrow term h' can be put into a unique normal form by applying the bifunctorial equations (ξ 1) so that every arrow term $\mathbf{1}_A$ in h' has a letter for A. From that, we obtain immediately that two different arrow terms in normal form must be of different types. So $\Diamond$**MI** is a preorder, from which we conclude the following.

MIX COHERENCE. *The category* **MI** *is a preorder.*

To obtain the natural logical category **MDI**, we have that the logical system $\mathcal{C}(\mathbf{MDI})$ is in $\mathcal{L}_{\wedge,\vee}$, with the transformations α included in $\mathbf{1}$, d and m. The equations $\mathcal{E}(\mathbf{MDI})$ are just those of $\mathcal{E}^{nat}_{\mathcal{C}(\mathbf{MDI})}$. We call natural **MDI**-categories *mix-dissociative* categories.

To prove that **MDI** is a preorder, we proceed as in §7.1 for **DI** by modifying the relations S^{ξ}_A. In $xS^{\xi}_A y$ we have as before that y is an occurrence of ξ in A, while x can be an occurrence of $\wedge$, $\vee$ or of a letter. With the help of this relation we proceed analogously to what we had in §7.1. So we have the following.

MIX-DISSOCIATIVE COHERENCE. *The category* **MDI** *is a preorder.*

§8.2. Coherence of mix-biassociative categories

To obtain the natural logical category **MA**, we have that the logical system $\mathcal{C}(\mathbf{MA})$ is in $\mathcal{L}_{\wedge,\vee}$, with the transformations α included in $\mathbf{1}$, b and m. The specific equations of $\mathcal{E}(\mathbf{MA})$ are those of $\mathcal{E}(\mathbf{A})$ plus

$$(bm) \quad (m_{A,B} \vee \mathbf{1}_C) \circ m_{A\wedge B,C} \circ \hat{b}{}^{\rightarrow}_{A,B,C} = \overset{\vee}{b}{}^{\rightarrow}_{A,B,C} \circ m_{A,B\vee C} \circ (\mathbf{1}_A \wedge m_{B,C}).$$

We call natural **MA**-categories *mix-biassociative* categories.

With $\Diamond$ primitive in a syntactical system synonymous with $\mathcal{C}(\mathbf{MA})$, the equation (bm) is replaced by

$$((f \Diamond g) \Diamond h) \circ \hat{b}{}^{\rightarrow}_{A,B,C} = \overset{\vee}{b}{}^{\rightarrow}_{D,E,F} \circ (f \Diamond (g \Diamond h)).$$

For $\mathcal{G}$ being **A** and $\mathcal{C}/\mathcal{E}$ being **MA**, we have that the conditions (IV$\mathcal{C}$) and (IV$\mathcal{G}$) of §3.1 are satisfied, and $\mathcal{G}$ is moreover a preorder. Thus we can apply the Direct-Strictification Theorem of §3.2 to obtain a category $\mathcal{C}_{\mathcal{G}}/\mathcal{E}_{\mathcal{G}}$, which we will call $\mathbf{MA}^{st}$, or $\Diamond\mathbf{MA}^{st}$ when $\Diamond$ is primitive. The categories $\mathbf{MA}^{st}$ and $\Diamond\mathbf{MA}^{st}$ are isomorphic, as $\mathbf{MI}$ and $\Diamond\mathbf{MI}$ are (see the preceding section). We call $\mathcal{C}_{\mathcal{G}}$ here $\mathcal{C}(\mathbf{MA}^{st})$, and $\mathcal{C}(\Diamond\mathbf{MA}^{st})$ is the synonymous syntactical system where $\Diamond$ is primitive.

We can easily prove the Composition Elimination proposition of the preceding section for $\mathbf{MA}^{st}$. Here is a sketch of how we proceed. By the Development Lemma of §2.7, there is for every arrow term of $\mathcal{C}(\mathbf{MA})$ a developed arrow term. For a $\hat{b}$-term f and a $\overset{\vee}{b}$-term or m-term g we have in $\mathbf{MA}$ that $f \circ g = g' \circ f'$ for a $\hat{b}$-term f' and a $\overset{\vee}{b}$-term or m-term g'. So we may say that $\hat{b}$-terms can be moved to the right. Analogously, $\overset{\vee}{b}$-terms can be moved to the left. Eventually, we obtain an arrow term of the form $f_1 \circ f_2 \circ f_3 \circ \mathbf{1}_A$, where in f_1 there are no $\hat{b}$-terms and m-terms, in f_2 there are no $\hat{b}$-terms and $\overset{\vee}{b}$-terms and in f_3 there are no $\overset{\vee}{b}$-terms and m-terms. Then it is enough to apply the Composition Elimination for $\Diamond\mathbf{MI}$ of the preceding section to the arrow term of $\mathcal{C}(\Diamond\mathbf{MA}^{st})$ corresponding to f_2 to obtain Composition Elimination for $\Diamond\mathbf{MA}^{st}$.

A composition-free arrow term of $\mathcal{C}(\Diamond\mathbf{MA}^{st})$ is *atomized* when for every occurrence of $\mathbf{1}_{[\![A]\!]}$ in it we have that A is a letter. We will write $\mathbf{1}_p$ instead of $\mathbf{1}_{[\![p]\!]}$.

For an atomized composition-free arrow term f of $\mathcal{C}(\Diamond\mathbf{MA}^{st})$ let $w(f)$ be the word obtained from f by deleting parentheses. We already defined $w(X)$ for a form sequence X in §7.3; it is, analogously, the word obtained from X by deleting parentheses.

To every pair of parentheses in a form sequence of letters X in natural notation (see §6.2 and §7.3), we can associate a pair of occurrences of letters (x, y) in X, where x is the first occurrence of a letter on the right-hand side of the left parenthesis and y is the first occurrence of a letter on the left-hand side of the right parenthesis. For example, in $p \wedge (p \vee (q \wedge r))$, to the outer pair of parentheses written down we associate (x, y) where x is the second p counting from the left and y is r. We suppose that atomized composition-free arrow terms of $\mathcal{C}(\Diamond\mathbf{MA}^{st})$ are written in natural notation (see §6.2), and we associate analogously pairs $(\mathbf{1}_x, \mathbf{1}_y)$ to pairs of parentheses in such arrow terms. (Such arrow terms correspond to a kind of form sequence of three colours: $\wedge$, $\vee$ and $\Diamond$.)

For every atomized composition-free arrow term $f\colon X \vdash Y$ of $\mathcal{C}(\Diamond\mathbf{MA}^{st})$ there are obvious bijections between occurrences of the same letters in X, Y and f, or in $w(X)$, $w(Y)$ and $w(f)$. We say that such occurrences *correspond obviously* to each other. For example, in $(\mathbf{1}_{p_1} \Diamond \mathbf{1}_{p_2}) \wedge \mathbf{1}_{p_3} : p_1 \wedge p_2 \wedge p_3 \vdash (p_1 \vee p_2) \wedge p_3$ the three occurrences of p_i, for $i \in \{1,2,3\}$, correspond obviously to each other.

For the proof of the following proposition we rely on the notion of immediate scope of §2.1.

PROPOSITION 1. *Let $\beta \in \{\wedge, \vee, \Diamond\}$, and let the atomized composition-free arrow term $f : X \vdash Y$ of $\mathcal{C}(\Diamond\mathbf{MA}^{st})$, where f, X and Y are written in natural notation, have a subterm $(f_1\beta\ldots\beta f_n)$, for $n \geq 2$, such that $(\mathbf{1}_{x_1}, \mathbf{1}_{x_2})$ is associated to the outermost parentheses of this subterm. Then there is a pair of parentheses in at least one of X and Y such that (y_1, y_2) is associated to this pair of parentheses, and, for $i \in \{1,2\}$, the occurrence of letter y_i corresponds obviously to the occurrence of the same letter x_i in $\mathbf{1}_{x_i}$.*

PROOF. If f is $(f_1\beta\ldots\beta f_n)$, then the assertion is trivial. (We usually omit, however, such outermost parentheses.) Suppose then that $(f_1\beta\ldots\beta f_n)$ is a proper subterm of f.

If β is $\wedge$ and is within the immediate scope of $\vee$, then we have both in X and in Y the required pair of parentheses. If β is $\wedge$ and is within the immediate scope of $\Diamond$, then we have in Y the required pair of parentheses.

If β is $\Diamond$ and is within the immediate scope of $\wedge$, then we have in Y the required pair of parentheses. If β is $\Diamond$ and is within the immediate scope of $\vee$, then we have in X the required pair of parentheses.

It remains to consider two cases where β is $\vee$, which are dual to the two cases above where it is $\wedge$. ⊣

PROPOSITION 2. *Let $f\colon X \vdash Y$ be an atomized composition-free arrow term of $\mathcal{C}(\Diamond\mathbf{MA}^{st})$, and let f, X and Y be written in natural notation. Then for every pair of parentheses in the form sequences X or Y to which (y_1, y_2) is associated there is a pair of parentheses in f to which $(\mathbf{1}_{x_1}, \mathbf{1}_{x_2})$ is associated such that, for $i \in \{1,2\}$, the occurrence of letter y_i corresponds obviously to the occurrence of the same letter x_i in $\mathbf{1}_{x_i}$.*

PROOF. If the pair of parentheses selected in X or Y is outermost (which we usually do not write), then the assertion is trivial. If the pair of parentheses selected is in X and belongs to $\wedge$ within the immediate scope of $\vee$, then in f we must have the required pair of parentheses, which belongs to $\wedge$ or $\diamond$ within the immediate scope of $\vee$. If the pair of parentheses selected is in Y and belongs to $\wedge$ within the immediate scope of $\vee$, then in f we must have the required pair of parentheses, which belongs to $\wedge$ within the immediate scope of $\vee$ or $\diamond$. The cases where $\vee$ is within the immediate scope of $\wedge$ are dual. ⊣

Building on Propositions 1 and 2, we can obtain a criterion for the existence of an arrow of $\diamond\mathbf{MA}^{st}$ of a given type $X \vdash Y$, which solves the theoremhood problem for $\diamond\mathbf{MA}^{st}$ (see §1.1). Let

(1) $w(X)$ and $w(Y)$ coincide save that in $w(Y)$ we can have an occurrence of $\vee$ at a place where in $w(X)$ we have an occurrence of $\wedge$.

Let u be obtained from $w(X)$ and $w(Y)$ by

(2.1) writing $\diamond$ at the places where $w(X)$ and $w(Y)$ differ,

(2.2) adding the parentheses of both X and Y in an obvious manner (here, two pairs of parentheses, one in X and the other in Y, associated to pairs of occurrences of letters that correspond obviously to each other yield a single pair of parentheses added to u),

(2.3) replacing every occurrence of a letter p by $\mathbf{1}_p$.

Then there is an arrow of $\diamond\mathbf{MA}^{st}$ of type $X \vdash Y$ iff (1) is fulfilled and u is an arrow term of $\mathcal{C}(\diamond\mathbf{MA}^{st})$ in natural notation. If u is such an arrow term, then it stands for the required arrow of $\diamond\mathbf{MA}^{st}$ of type $X \vdash Y$. This will yield a criterion for the existence of arrows of a given type in $\mathbf{MA}$.

We need the following proposition to prove coherence for $\mathbf{MA}$.

PROPOSITION 3. *If $f_1 : X_1 \vdash Y_1$ and $f_2 : X_2 \vdash Y_2$ are different atomized composition-free arrow terms of $\mathcal{C}(\diamond\mathbf{MA}^{st})$, then X_1 is different from X_2 or Y_1 is different from Y_2.*

PROOF. If $w(f_1)$ is different from $w(f_2)$, then it is clear that $w(X_1)$ is different from $w(X_2)$ or $w(Y_1)$ is different from $w(Y_2)$. If $w(f_1)$ coincides with $w(f_2)$, but f_1 and f_2 are different arrow terms, then f_1 and f_2 must differ with respect to parentheses. In that case, Propositions 1 and 2 yield the assertion. ⊣

From Composition Elimination and Proposition 3 we infer that $\Diamond\mathbf{MA}^{st}$ is a preorder. So we have the following.

MIX-BIASSOCIATIVE COHERENCE. *The category* **MA** *is a preorder.*

§8.3. Coherence of mix-net categories

To obtain the natural logical category **MDA**, we have that the logical system $\mathcal{C}(\mathbf{MDA})$ is in $\mathcal{L}_{\wedge,\vee}$, with the transformations α included in $\mathbf{1}$, b, d and m. The specific equations of $\mathcal{E}(\mathbf{MDA})$ are those of $\mathcal{E}(\mathbf{DA})$ plus

$$(\hat{b}\,mL) \quad m_{A\wedge B,C}\circ \hat{b}^{\rightarrow}_{A,B,C} = d^L_{A,B,C}\circ(\mathbf{1}_A\wedge m_{B,C}),$$

$$(\check{b}\,mL) \quad \check{b}^{\rightarrow}_{A,B,C}\circ m_{A,B\vee C} = (m_{A,B}\vee \mathbf{1}_C)\circ d^L_{A,B,C},$$

$$(\hat{b}\,mR) \quad m_{C,B\wedge A}\circ \hat{b}^{\leftarrow}_{C,B,A} = d^R_{C,B,A}\circ(m_{C,B}\wedge \mathbf{1}_A),$$

$$(\check{b}\,mR) \quad \check{b}^{\leftarrow}_{C,B,A}\circ m_{C\vee B,A} = (\mathbf{1}_C\vee m_{B,A})\circ d^R_{C,B,A}.$$

We call natural **MDA**-categories *mix-net* categories.

The specific equation (bm) of $\mathcal{E}(\mathbf{MA})$ is derived as follows for **MDA**:

$$\begin{aligned}(m_{A,B}\vee \mathbf{1}_C)\circ m_{A\wedge B,C}\circ \hat{b}^{\rightarrow}_{A,B,C} &\\ &= (m_{A,B}\vee \mathbf{1}_C)\circ d^L_{A,B,C}\circ(\mathbf{1}_A\wedge m_{B,C}),\ \text{by } (\hat{b}\,mL),\\ &= \check{b}^{\rightarrow}_{A,B,C}\circ m_{A,B\vee C}\circ(\mathbf{1}_A\wedge m_{B,C}),\ \text{by } (\check{b}\,mL).\end{aligned}$$

Alternatively, we could have used $(\hat{b}\,mR)$ and $(\check{b}\,mR)$.

For $\mathcal{G}$ being **A** and $\mathcal{C}/\mathcal{E}$ being **MDA**, we have that the conditions (IV$\mathcal{C}$) and (IV$\mathcal{G}$) of §3.1 are satisfied, and $\mathcal{G}$ is moreover a preorder. Thus we can apply the Direct-Strictification Theorem of §3.2 to obtain a category $\mathcal{C}_\mathcal{G}/\mathcal{E}_\mathcal{G}$, which we will call $\mathbf{MDA}^{st}$. We call $\mathcal{C}_\mathcal{G}$ here $\mathcal{C}(\mathbf{MDA}^{st})$.

For every object X of $\mathbf{MDA}^{st}$, i.e. for every form sequence of letters X in natural notation (see §6.2 and §7.3), we introduce a relation R'_X between the set of occurrences of $\wedge$ in X and the set of occurrences of $\wedge$, $\vee$ and letters in X. We define $xR'_X y$ as $xR_X y$ in §7.3 save that y need not be an occurrence of $\vee$, but may be an occurrence of $\wedge$, $\vee$ or of a letter (cf. the version of S^{ξ}_A in §8.1). More precisely, y is an occurrence of $\wedge$, $\vee$ or of a letter on the right-hand side of $l(x)$ and on the left-hand side of $r(x)$; here y can also be $l(x)$ or $r(x)$. We define the occurrences of letters $l(x)$ and $r(x)$ exactly as before (see §7.3).

Let $v(X)$ be the word obtained from a form sequence X by deleting every parenthesis and replacing every occurrence of $\wedge$ or $\vee$ in X by a single arbitrary new symbol γ. When for the form sequences X and Y we have that $v(X)$ and $v(Y)$ coincide, we say that X and Y are **MDA**-*comparable*.

It is clear that R'_X gives rise to a relation R'_{v_X} on occurrences of symbols in $v(X)$ such that we have $x'R'_{v_X}y'$ when x' is the occurrence of γ in $v(X)$ corresponding to an occurrence x of $\wedge$ in X, while y' is an occurrence of γ or of a letter p corresponding to an occurrence y of $\wedge$, $\vee$ or p in X, and we have $x'R'_X y'$.

Then it can be checked that for every arrow term $f\colon X \vdash Y$ of $\mathcal{C}(\mathbf{MDA}^{st})$ the form sequences X and Y are **MDA**-comparable and $R'_{v_Y} \subseteq R'_{v_X}$. Moreover, if d^L, d^R or m occurs in f, then R'_{v_Y} is a proper subset of R'_{v_X}; otherwise, $R'_{v_X} = R'_{v_Y}$. For example, with $m_{p,q} : p \wedge q \vdash p \vee q$ we have $v(p \wedge q) = v(p \vee q) = p\gamma q$, while $R'_{v_{p\wedge q}} = \{(\gamma, p), (\gamma, q)\}$ and $R'_{v_{p\vee q}} = \emptyset$.

A *place* in X is a subword of $v(X)$. We define when subwords of **MDA**-comparable form sequences are at the same place as in §7.3 and §2.1.

We have a Remark analogous to that of §7.3 with R replaced by R', and a lemma analogous to the Extraction Lemma of §7.3 with $\mathbf{DA}^{st}$ replaced by $\mathbf{MDA}^{st}$ and $(*)$ that reads:

$(*)$ *for every occurrence x of $\wedge$ in X_i and every occurrence y of $\wedge$, $\vee$ or of a letter in X_i, where $i \in \{1, 2\}$, if x' is an occurrence of $\wedge$ in X at the same place where $X_1\, z\, X_2$ has x, while y' is an occurrence of $\wedge$, $\vee$ or of a letter in X at the same place where $X_1\, z\, X_2$ has y, and $x'R'_X y'$, then $xR'_{X_i} y$.*

A more precise formulation of $(*)$ in the Extraction Lemma of §7.3 would be analogous to this version of $(*)$, but there we identified x and y with x' and y'. Note that here X and $X_1\, z\, X_2$ are comparable and not only **MDA**-

comparable. This means that y and y' are occurrences of the same symbol. The proofs of the Extraction Lemma of §7.3 and of its analogue for $\mathbf{MDA}^{st}$ do not involve the transformation m. They are based on considerations concerning $l(x)$ and $r(x)$, which are the same both for the relation R and for the relation R'.

With the help of this analogue of the Extraction Lemma, we can prove the following analogue of the Theoremhood Proposition of §7.3.

THEOREMHOOD PROPOSITION. *There is an arrow term* $f : X \vdash Y$ *of* $\mathcal{C}(\mathbf{MDA}^{st})$ *iff* X *and* Y *are* **MDA**-*comparable form sequences and* $R'_{vY} \subseteq R'_{vX}$.

PROOF. We enlarge the proof of the Theoremhood Proposition of §7.3. If $n > 1$ and Y is $Y_1\, z\, Y_2$ for z an occurrence of $\vee$, then there is no guarantee that X has an occurrence of $\vee$ at the same place. If it has it, then we proceed as before, applying the analogue of the Extraction Lemma. If, on the other hand, X has an occurrence u of $\wedge$ at that place, then we first take an arrow term $h\colon X \vdash X'$ of $\mathcal{C}(\mathbf{MDA}^{st})$ made of m, $\mathbf{1}$ and the operations $\wedge$ and $\vee$ on arrow terms such that X' differs from X just by having an occurrence of $\vee$ instead of u. It is clear that if we exclude from R'_{vX} all those pairs (u', y'), where u' corresponds to u in $v(X)$, then we obtain $R'_{vX'}$. We continue again as in the proof of the Theoremhood Proposition of §7.3, and f is $(f_1 \vee f_2) \circ g \circ h$. ⊣

For a form sequence X, let $d(X)$ be now the cardinality of R'_X. We can prove the following.

MIX-NET COHERENCE. *The category* **MDA** *is a preorder.*

PROOF. We proceed as in the proof of Net Coherence in §7.3 with the following additional cases.

(Lm) The head of f_1 is $d^L_{E,F,G}$ and the head of g_1 is $m_{H,I}$. Due to the presence of $(\check{b}\, mL)$, we may assume that I is not of the form $(X_1 \ldots X_n, \vee)$, and so I cannot be $F \vee G$. It remains to consider subcases analogous to $(LL1)$ and $(LL2)$, which are settled with the help of bifunctorial and naturality equations.

(Rm) The head of f_1 is $d^R_{G,F,E}$ and the head of g_1 is $m_{I,H}$. Here we invoke $(\check{b}\, mR)$, and deal as in (Lm).

(mm) The head of f_1 is $m_{E,F}$ and the head of g_1 is $m_{H,I}$. Due to the presence of $(\hat{b}\, mL)$ and $(\hat{b}\, mR)$, we may assume that E, F, H and I are not of the form $(X_1 \ldots X_n, \wedge)$. So, with the previous assumption based on $(\check{b}\, mL)$ and $(\check{b}\, mR)$, they must be occurrences of letters. We have the following subcase.

$(mm1)$ The occurrence of letter F coincides with H. Then by (bm), $(m\ nat)$, $(\check{b}\, mL)$ and $(\check{b}\, mR)$ we have

$$(\mathbf{1}_E \vee m_{F,I}) \circ d^R_{E,F,I} \circ (m_{E,F} \wedge \mathbf{1}_I) = (m_{E,F} \wedge \mathbf{1}_I) \circ d^L_{E,F,I} \circ (\mathbf{1}_E \vee m_{F,I}).$$

We continue reasoning by applying the Theoremhood Proposition of this section. The remaining subcases are settled with the help of bifunctorial and naturality equations. ⊣

§8.4. Coherence of mix-symmetric net categories

To obtain the natural logical category **MDS**, we have that the logical system $\mathcal{C}(\mathbf{MDS})$ is in $\mathcal{L}_{\wedge,\vee}$, with the transformations α included in $\mathbf{1}$, b, c, d and m. The specific equations of $\mathcal{E}(\mathbf{MDS})$ are obtained by taking the union of those of $\mathcal{E}(\mathbf{MDA})$ and $\mathcal{E}(\mathbf{DS})$ plus

$$(cm) \quad m_{B,A} \circ \hat{c}_{A,B} = \check{c}_{B,A} \circ m_{A,B}.$$

We call natural **MDS**-categories *mix-symmetric net* categories.

With $\diamond$ as in §8.1, the equation (cm) amounts to the equation

$$(g \diamond f) \circ \hat{c}_{A,B} = \check{c}_{E,D} \circ (f \diamond g).$$

In the arrow terms of $\mathcal{C}(\mathbf{MDS})$ we write d instead of d^L, as we did for $\mathcal{C}(\mathbf{DS})$, and we take d^R as defined by the equation $(d^R c)$ of §7.6. Among the specific equations $(\hat{b}\, mL)$, $(\check{b}\, mL)$, $(\hat{b}\, mR)$ and $(\check{b}\, mR)$ of $\mathcal{E}(\mathbf{MDA})$ (see §8.3), it is enough to keep $(\hat{b}\, mL)$ and $(\check{b}\, mL)$; the equations $(\hat{b}\, mR)$ and $(\check{b}\, mR)$ are derivable.

We build the syntactical category **GMDS** in the syntactical system $\mathcal{C}(\mathbf{GMDS})$ out of **MDS** as we built **GDS** out of **DS** in §7.6 and §7.7.

The only difference is that we replace everywhere **DS** by **MDS**. As before, it is enough to prove that **GMDS** is a preorder in order to infer the following.

MIX-SYMMETRIC NET COHERENCE. *The functor G from* **MDS** *to Rel is faithful.*

We define the *Gentzen terms* for arrows of **GMDS** as for **GDS**, in §7.7, with the following additional Gentzen operation:

$$\frac{f\colon U \vdash Z \qquad\qquad g\colon Y \vdash W}{mix(f,g) =_{dn} (f \vee g) \circ m_{U,Y}\colon U \wedge Y \vdash Z \vee W}$$

An alternative notation for $mix(f,g)$ is $f \diamondsuit g$, which we used in §8.1 and §8.2, but $mix(f,g)$ is handier in the present context. Note that due to (cm) we have $mix(f,g) = mix(g,f)$ in **GMDS**. So we will consider the terms on the two sides of this equation as the same Gentzen term.

Then by enlarging the proof of the Gentzenization Lemma of §7.7 we can prove that every arrow of **GMDS** is denoted by a Gentzen term. The only addition is that $m_{X,Y} = mix(\mathbf{1}_X, \mathbf{1}_Y)$.

In the definition of depth of §7.7, we now have that γ can be also mix. The notions of cut-free Gentzen term, topmost cut, rank and complexity of a topmost cut are exactly as in §7.7. We can then prove as follows the Cut-Elimination Theorem where **GDS** is replaced by **GMDS**.

PROOF OF THE CUT-ELIMINATION THEOREM FOR **GMDS**. We proceed as in the proof of §7.7 until we suppose that the complexity of the topmost cut $cut_X(f,g)$ is (m,n) with $m,n \geq 1$. Then we have the following additional case.

$(\wedge 6)$ The form set X is of colour $\wedge$ (it may be of the form $X_1 \wedge X_2$ or p) and f is $mix(f_1, f_2)$. So we have

$$\frac{\dfrac{f_1\colon U_1 \vdash X \vee V_1 \qquad f_2\colon U_2 \vdash V_2}{mix(f_1,f_2)\colon U_1 \wedge U_2 \vdash X \vee V_1 \vee V_2} \qquad g\colon X \wedge Y \vdash W}{cut_X(mix(f_1,f_2),g)\colon U_1 \wedge U_2 \wedge Y \vdash W \vee V_1 \vee V_2}$$

Then consider the Gentzen term

$$\dfrac{\dfrac{f_1 : U_1 \vdash X \vee V_1 \qquad g : X \wedge Y \vdash W}{cut_X(f_1, g) : U_1 \wedge Y \vdash W \vee V_1} \qquad f_2 : U_2 \vdash V_2}{mix(cut_X(f_1, g), f_2) : U_1 \wedge U_2 \wedge Y \vdash W \vee V_1 \vee V_2}$$

We show that

$$(***) \quad cut_X(mix(f_1, f_2), g) = mix(cut_X(f_1, g), f_2),$$

and the complexity $(m, n-1)$ of the topmost cut $cut_X(f_1, g)$ is lower than (m, n), so that we may apply the induction hypothesis.

To show the equation $(***)$, we have to show

$$(g \vee \mathbf{1}_{V_1 \vee V_2}) \circ d_{Y,X,V_1 \vee V_2} \circ ((m_{X \vee V_1, V_2} \circ (f_1 \wedge f_2)) \wedge \mathbf{1}_Y) =$$
$$m_{W \vee V_1, V_2} \circ (((g \vee \mathbf{1}_{V_1}) \circ d_{Y,X,V_1} \circ (f_1 \wedge \mathbf{1}_Y)) \wedge f_2).$$

To derive this equation for **GMDS**, we use essentially $(\hat{b}\, mL)$ of the preceding section and (cm) (for **GMDS**, the latter equation reads $m_{X,Y} = m_{Y,X}$).

We have to consider also subcases where we have $f_1 : U_1 \vdash X$ or $g : X \vdash W$. In all of them, $(***)$ amounts to equations simpler than the equation above, which all hold in **GMDS**.

If in $cut_X(f, g)$ with complexity (m, n) where $m, n \geq 1$ the form set X is of colour $\vee$ and g is $mix(g_1, g_2)$, then we have an additional case treated dually to $(\wedge 6)$. In that case, the equation $(\overset{\vee}{b}\, mL)$ of the preceding section, together with (cm), plays an essential role. ⊣

To prove the analogue of the Invertibility Lemmata of §7.8 for **GMDS** we need the following equations of **GMDS**:

$$(\wedge mix) \quad \wedge_{X,Y}(mix(f, h), g) = mix(\wedge_{X,Y}(f, g), h)$$

for f of type $U \vdash X \vee V$ or $U \vdash X$, and

$$(\vee mix) \quad \vee_{X,Y}(mix(f, h), g) = mix(\vee_{X,Y}(f, g), h)$$

for f of type $X \wedge V \vdash U$ or $X \vdash U$. To derive these equations for **GMDS** we use essentially $(\hat{b}\, mL)$ and $(\overset{\vee}{b}\, mL)$ of the preceding section.

Note that $(\wedge mix)$ and $(\vee mix)$ hold just with the types indicated for f. For other types, analogous equations need not hold; take, for example,

$$\wedge_{p\vee q,r}(mix(\mathbf{1}_p,\mathbf{1}_q),\mathbf{1}_r)\colon p\wedge q\wedge r\vdash(p\vee q)\wedge r.$$

We will need also the following equation:

$$(mix\,mix)\quad mix(mix(f,g),h)=mix(f,mix(g,h)),$$

which we derive for **GMDS** with the help of the equation (bm) of §8.2 (see the preceding section).

We define inductively the following abbreviation:

$$Mix(f)=_{df}f,$$
$$Mix(f_1,\ldots,f_{k-1},f_k)=_{df}mix(Mix(f_1,\ldots,f_{k-1}),f_k),\ \text{for } k\geq 2.$$

Due to the equation $(mix\,mix)$, for $k\geq 2$ we could have also

$$Mix(f_1,\ldots,f_{k-1},f_k)=_{df}mix(f_1,Mix(f_2,\ldots,f_k)).$$

With the help of $(\wedge mix)$, $(\vee mix)$ and $(mix\,mix)$ we can derive for **GMDS** the following equations:

$$(\wedge Mix)\quad \wedge_{X,Y}(Mix(f,f_1,\ldots,f_n),Mix(g,g_1,\ldots,g_m))=$$
$$Mix(\wedge_{X,Y}(f,g),f_1,\ldots,f_n,g_1,\ldots,g_m)$$

where $n,m\geq 0$ (if $n=0$, then $f_1,\ldots,f_n$ is just omitted, and analogously if $m=0$), while f is of type $U\vdash X\vee V$ or $U\vdash X$, and g is of type $W\vdash Y\vee S$ or $W\vdash Y$. We have also the equation $(\vee Mix)$ where $\wedge$ is replaced by $\vee$, while f is of type $X\wedge V\vdash U$ or $X\vdash U$, and g is of type $Y\wedge S\vdash W$ or $Y\vdash W$.

The Splitting Remark of §7.8 holds for **GMDS** as it holds for **GDS**. Note that the splittability of f need not entail the splittability of $\wedge_{Z,U}(f,g)$. A counterexample is provided with $mix(\mathbf{1}_p,\mathbf{1}_q)\colon p\wedge q\vdash p\vee q$, which is splittable, and $\wedge_{p\vee q,r}(mix(\mathbf{1}_p,\mathbf{1}_q),\mathbf{1}_r)$, mentioned above, which is not splittable.

A cut-free Gentzen term f for arrows of **GMDS** such that every subterm $\xi_{X,Y}(f_1,f_2)$ of f for $\xi\in\{\wedge,\vee\}$ is not splittable is called *split-normalized*. Let the quantity of letters for an arrow $f: X\vdash Y$ of **GMDS** be the cardinality of $let(X)$, as for **GDS** in §7.8. We can prove the following.

SPLIT-NORMALIZATION LEMMA. *For every Gentzen term h for arrows of* **GMDS** *there is a split-normalized Gentzen term h' such that $h = h'$ in* **GMDS**.

PROOF. We proceed by induction on the quantity of letters in h. In the basis of the induction, when this quantity is 1 and $h = \mathbf{1}_p : p \vdash p$, the lemma holds trivially.

Suppose that h is equal in **GMDS** to a cut-free Gentzen term $\wedge_{Z,U}(f, g)$ with f and g of the same types as in the Splitting Remark of §7.8, and with the total split mentioned there. Then we apply the induction hypothesis to f and g to obtain f' and g' split-normalized. If $\wedge_{Z,U}(f', g')$ is not splittable, we are done. If $\wedge_{Z,U}(f', g')$ is splittable, then we proceed as follows.

If $n \geq 1$, then the cut-free Gentzen term f' is splittable and can be written in the form $Mix(u, u_1, \ldots, u_n)$ with u of type $X \vdash Z$ or $X \vdash Z \vee R'$ and $u_i : X_i \vdash Y_i$, for $i \in \{1, \ldots, n\}$ and all of $u, u_1, \ldots, u_n$ split-normalized. To put f' in this form, we may need to use (*mix mix*), and we also use the fact that the subterms of a split-normalized Gentzen term are split-normalized. Here u_i is not splittable and is not of the form $mix(u', u'')$, but u may be splittable. If u is splittable, and is hence of the form $mix(u', u'')$ with $u' : X' \vdash Y'$ and $u'' : X'' \vdash Y''$, then we must have $let(Z) \cap let(Y') \neq \emptyset$ and $let(Z) \cap let(Y'') \neq \emptyset$. This follows from the Splitting Remark. If $n = 0$, then f' is not splittable, and can be written as $Mix(f)$.

We put g' analogously in the form $Mix(v, v_1, \ldots, v_m)$, and we apply ($\wedge Mix$) to obtain that $\wedge_{Z,U}(f, g)$ is equal to

$$Mix(\wedge_{Z,U}(u, v), u_1, \ldots, u_n, v_1, \ldots, v_m),$$

which is split-normalized, because $\wedge_{Z,U}(u, v)$ is not splittable. If $\wedge_{Z,U}(u, v)$ were splittable, then by the Splitting Remark, either u, which is split-normalized, would be of the form $mix(u', u'')$ with $u' : X' \vdash Y'$, $u'' : X'' \vdash Y''$ and $let(Z) \cap let(Y') = \emptyset$ or $let(Z) \cap let(Y'') = \emptyset$, or v would be of such a form, which is impossible, as we said above.

We proceed analogously if h is equal in **GMDS** to a cut-free Gentzen term $\vee_{Z,U}(f, g)$. If h is equal in **GMDS** to a cut-free Gentzen term $mix(f, g)$, then we just apply the induction hypothesis to f and g. ⊣

The Invertibility Lemma for $\wedge$ is formulated for **GMDS** as in §7.8, save that we assume for f not only that it is cut-free, but that it is split-normalized too. The proof of this lemma proceeds, as before, by induction on the length of f. In the induction step, when f is $\wedge_{Y^a,Y^b}(f^a, f^b)$ or $\vee_{Y^a,Y^b}(f^a, f^b)$, we work as in the proof in §7.8, save that when we eliminated some cases by appealing to the Split Corollary, now these cases are eliminated by appealing to the fact that f is split-normalized, and hence not splittable. It remains to consider the case where f is $mix(f^a, f^b)$. It is, however, easy to conclude that we may apply the induction hypothesis either to f^a or to f^b in order to obtain a cut-free Gentzen term $\wedge_{X_1,X_2}(g,h)$ equal to this term in **GMDS**. Then we apply ($\wedge mix$).

We proceed analogously to prove the Invertibility Lemma for $\vee$ for **GMDS**. We also have for **GMDS** the following new lemma of the same kind.

INVERTIBILITY LEMMA FOR *mix*. *If* $f : U_1 \wedge U_2 \vdash Z_1 \vee Z_2$ *is a split-normalized Gentzen term such that* $let(U_i) = let(Z_i)$ *for* $i \in \{1,2\}$, *then there are two split-normalized Gentzen terms* $f_1 : U_1 \vdash Z_1$ *and* $f_2 : U_2 \vdash Z_2$ *such that* $f = mix(f_1, f_2)$.

PROOF. We proceed by induction on the length of f. The basis of this induction, when f is $\mathbf{1}_p$, is trivial, as before.

For the induction step, because f is split-normalized, f must be of the form $mix(f^a, f^b)$ for $f^a : W^a \vdash R^a$ and $f^b : W^b \vdash R^b$. Then we have the following cases:

(1) W^a is U_1 and W^b is U_2,

(2) W^a is $U_1 \wedge U_2^a$ and W^b is U_2^b, for U_2 being $U_2^a \wedge U_2^b$,

(3) W^a is $U_1^a \wedge U_2^a$ and W^b is $U_1^b \wedge U_2^b$, for U_i being $U_i^a \wedge U_i^b$, where $i \in \{1,2\}$,

and cases analogous to these. In case (1), we take f^a for f_1 and f^b for f_2. In case (2), we apply the induction hypothesis to f^a to obtain a split-normalized Gentzen term $mix(f^{a\prime}, f^{a\prime\prime})$ equal to f^a in **GMDS** for $f^{a\prime}: U_1 \vdash Z_1$ and $f^{a\prime\prime}: U_2^a \vdash Z_2^a$. Then we take f_1 and f_2 to be $f^{a\prime}$ and $mix(f^{a\prime\prime}, f^b)$, respectively, and we apply ($mix\, mix$). In case (3), we proceed analogously, by applying ($mix\, mix$) three times (cf. the proof of the Invertibility Lemma for $\wedge$ in §7.8). ⊣

We can now prove Cut-Free Preordering of §7.8 for **GMDS**. The proof is analogous to the proof of §7.8, with an additional case when f_1 is $mix\,(f_1', f_1'')$, which is settled with the help of the Invertibility Lemma for mix. As a corollary of the Cut-Elimination Theorem for **GMDS** and of Cut-Free Preordering for **GMDS**, we obtain that the category **GMDS** is a preorder, which implies Mix-Symmetric Net Coherence.

§8.5. Coherence of mix-symmetric biassociative categories

To obtain the natural logical category **MS**, we have that the logical system $\mathcal{C}(\mathbf{MS})$ is in $\mathcal{L}_{\wedge,\vee}$, with the transformations α included in $\mathbf{1}$, b, c and m. The specific equations of $\mathcal{E}(\mathbf{MS})$ are obtained by taking the union of those of $\mathcal{E}(\mathbf{MA})$ and $\mathcal{E}(\mathbf{S})$ plus the equation (cm). So **MS** is analogous to **MDS**, but with d missing. We call natural **MS**-categories *mix-symmetric biassociative* categories.

We can prove the following.

MIX-SYMMETRIC BIASSOCIATIVE COHERENCE. *The functor G from* **MS** *to Rel is faithful.*

We prove this assertion as for **MDS** via a Cut-Elimination Theorem and Invertibility Lemmata. We keep in the proofs of the preceding section just the easy cases.

We do not consider here something that would be called mix-bimonoidal categories, symmetric or not symmetric, dissociative or not dissociative. (Some kinds of mix-dissociative bimonoidal categories are considered in [20], Sections 6-7.) We have left open in §7.9 the problem of what axioms should be added to those of linearly distributive categories in order to obtain coherence with respect to *Rel*. Mix brings in its own problems in the presence of $\top$ and $\bot$. These problems will disappear in the next two chapters.

CHAPTER 9

LATTICE CATEGORIES

This chapter is about coherence for categories with a double cartesian structure, i.e. with finite products and finite coproducts. We take this as a categorification of the notion of lattice. As before, we distinguish cases with and without special objects, which are here the empty product and the empty coproduct, i.e. the terminal and initial objects. The results presented are taken over from [42], [44] and [43].

We pay particular attention in this chapter to questions of maximality, i.e. to the impossibility of extending our axioms without collapse into preorder, and hence triviality. This maximality is a kind of syntactical completeness. (The sections on maximality, §9.3, §9.5 and §9.7, improve upon results reported in [42], [44] and [43].)

The techniques of this chapter are partly based on a composition elimination for conjunctive logic, related to normalization in natural deduction, and on a simple composition elimination for conjunctive-disjunctive logic, implicit in Gentzen's cut elimination.

§9.1. Coherence of semilattice categories

To obtain the natural logical category $\hat{\mathbf{L}}$, we have that the logical system $\mathcal{C}(\hat{\mathbf{L}})$ is in $\mathcal{L}_\wedge$, with the transformations α included in $\mathbf{1}$, $\hat{b}$, $\hat{c}$ and $\hat{w}$-$\hat{k}$. The specific equations of $\mathcal{E}(\hat{\mathbf{L}})$ are those of $\mathcal{E}(\hat{\mathbf{S}})$ plus

$$(\hat{b}\hat{w}) \qquad \hat{b}^{\rightarrow}_{A,A,A} \circ (\mathbf{1}_A \wedge \hat{w}_A) \circ \hat{w}_A = (\hat{w}_A \wedge \mathbf{1}_A) \circ \hat{w}_A,$$

$$(\hat{c}\hat{w}) \qquad \hat{c}_{A,A} \circ \hat{w}_A = \hat{w}_A,$$

for $\hat{c}^m_{A,B,C,D} =_{df} \hat{b}^{\rightarrow}_{A,C,B\wedge D} \circ (\mathbf{1}_A \wedge (\hat{b}^{\leftarrow}_{C,B,D} \circ (\hat{c}_{B,C} \wedge \mathbf{1}_D) \circ \hat{b}^{\rightarrow}_{B,C,D})) \circ \hat{b}^{\leftarrow}_{A,B,C\wedge D}$ of type $(A \wedge B) \wedge (C \wedge D) \vdash (A \wedge C) \wedge (B \wedge D)$,

$$(\hat{b}\hat{c}\hat{w}) \qquad \hat{w}_{A\wedge B} = \hat{c}^m_{A,A,B,B} \circ (\hat{w}_A \wedge \hat{w}_B),$$

$$(\hat{b}\hat{k}) \qquad (\hat{k}^1_{A,B} \wedge \mathbf{1}_C) \circ \hat{b}^{\rightarrow}_{A,B,C} = \mathbf{1}_A \wedge \hat{k}^2_{B,C},$$

$$(\hat{c}\hat{k}) \qquad \hat{k}^2_{A,B} = \hat{k}^1_{B,A} \circ \hat{c}_{A,B},$$

$$(\hat{w}\hat{k}) \qquad \hat{k}^i_{A,A} \circ \hat{w}_A = \mathbf{1}_A, \text{ for } i \in \{1,2\}.$$

We call natural $\hat{\mathbf{L}}$-categories *semilattice* categories. Usually, they are called categories with finite nonempty products. The objects of a semilattice category that is a partial order make a semilattice.

The equation $(\hat{b}\,\hat{c}\,\hat{w})$ is the *octagonal* equation of [39] (Section 2) and [40] (Section 1) (cf. [48], Proposition 3.29, p. 235, and, for $\hat{c}^m$, cf. [50], Section III.3, p. 517).

The equation $(\hat{b}\hat{k})$ is related to the equation $(\hat{b}\,\hat{\delta}\,\hat{\sigma})$ of §4.6. By using essentially this equation, we derive for $\hat{\mathbf{L}}$ the equations

$$(\hat{b}\hat{k}1) \quad \hat{k}^1_{A\wedge B,C} = (\mathbf{1}_A \wedge \hat{k}^1_{B,C}) \circ \hat{b}^{\leftarrow}_{A,B,C},$$
$$(\hat{b}\hat{k}2) \quad \hat{k}^2_{C,B\wedge A} = (\hat{k}^2_{C,B} \wedge \mathbf{1}_A) \circ \hat{b}^{\rightarrow}_{C,B,A},$$

which are related respectively to the equations $(\hat{b}\,\hat{\delta})$ and $(\hat{b}\hat{\sigma})$ of §4.6. To derive $(\hat{b}\hat{k}1)$, we derive first

$$\hat{k}^1_{A\wedge B,C} \wedge \mathbf{1}_{(A\wedge B)\wedge C} = ((\mathbf{1}_A \wedge \hat{k}^1_{B,C}) \circ \hat{b}^{\leftarrow}_{A,B,C}) \wedge \mathbf{1}_{(A\wedge B)\wedge C}$$

with the help of $(\hat{b}\ 5)$ of §4.2 and other equations; since for $f : E \vdash D$ we have $\hat{k}^1_{D,E} \circ (f \wedge \mathbf{1}_E) \circ \hat{w}_E = f$, we have $(\hat{b}\hat{k}1)$. We proceed analogously for $(\hat{b}\hat{k}2)$. Conversely, from $(\hat{b}\hat{k}1)$ and $(\hat{b}\hat{k}2)$ we can derive $(\hat{b}\hat{k})$ with the help of $(\hat{b}\,\hat{c})$ from §5.1 and $(\hat{c}\hat{k})$.

The equation $(\hat{c}\hat{k})$, which is related to the equation $(\hat{c}\,\hat{\delta}\,\hat{\sigma})$ of §5.3, says that $\hat{k}^1$ and $\hat{k}^2$ are interdefinable. In the presence of $(\hat{c}\hat{w})$ and $(\hat{c}\hat{k})$, we can derive $\hat{k}^2_{A,A} \circ \hat{w}_A = \mathbf{1}_A$ from $\hat{k}^1_{A,A} \circ \hat{w}_A = \mathbf{1}_A$, and vice versa, so that instead of $(\hat{w}\hat{k})$ we could have assumed just one of these two equations.

We can also derive for $\hat{\mathbf{L}}$ the equation

$$(\hat{w}\hat{k}\hat{k}) \quad (\hat{k}^1_{A,A} \wedge \hat{k}^2_{A,A}) \circ \hat{w}_{A\wedge A} = \mathbf{1}_{A\wedge A}.$$

For that we use $(\hat{b}\,\hat{c}\,\hat{w})$, $(\hat{b}\hat{k})$, $(\hat{b}\hat{k}1)$, $(\hat{b}\hat{k}2)$, $(\hat{c}\hat{k})$ and $(\hat{w}\hat{k})$.

For $f_i \colon C \vdash A_i$, where $i \in \{1,2\}$, we have in $\mathcal{C}(\hat{\mathbf{L}})$ the definition

$$(\langle\_,\_\rangle) \qquad \langle f_1, f_2\rangle =_{df} (f_1 \wedge f_2) \circ \hat{w}_C .$$

Then for $f \colon A \vdash D$ and $g \colon B \vdash E$ the following equations hold in $\hat{\mathbf{L}}$:

$$\begin{array}{ll}
(\wedge) & f \wedge g = \langle f \circ \hat{k}^1_{A,B}, g \circ \hat{k}^2_{A,B}\rangle, \\
(\hat{b}^{\rightarrow}) & \hat{b}^{\rightarrow}_{A,B,C} = \langle \mathbf{1}_A \wedge \hat{k}^1_{B,C}, \hat{k}^2_{B,C} \circ \hat{k}^2_{A,B\wedge C}\rangle, \\
(\hat{b}^{\leftarrow}) & \hat{b}^{\leftarrow}_{C,B,A} = \langle \hat{k}^1_{C,B} \circ \hat{k}^1_{C\wedge B,A}, \hat{k}^2_{C,B} \wedge \mathbf{1}_A\rangle, \\
(\hat{c}) & \hat{c}_{A,B} = \langle \hat{k}^2_{A,B}, \hat{k}^1_{A,B}\rangle, \\
(\hat{w}) & \hat{w}_A = \langle \mathbf{1}_A, \mathbf{1}_A\rangle.
\end{array}$$

This shows that with the operation $\langle\_,\_\rangle$ on arrow terms primitive, together with $\hat{k}^i$, where $i \in \{1,2\}$, we could take the arrow terms on the left-hand sides of these equations as defined. With these alternative primitives, all the equations of $\mathcal{E}(\hat{\mathbf{L}})$ can be derived from the categorial equations and the following equations of $\mathcal{E}(\hat{\mathbf{L}})$:

$$\begin{array}{ll}
(\wedge\beta) & \hat{k}^i_{A_1,A_2} \circ \langle f_1, f_2\rangle = f_i, \\
(\wedge\eta) & \langle \hat{k}^1_{A_1,A_2} \circ h, \hat{k}^2_{A_1,A_2} \circ h\rangle = h,
\end{array}$$

for $h \colon C \vdash A_1 \wedge A_2$ (for these equations see [85], Section I.3). In other words, we would obtain a syntactical system synonymous with $\mathcal{C}(\hat{\mathbf{L}})$ (see the end of §2.4 for this notion of synonymity).

Another alternative is to have $\hat{k}^i$ and $\hat{w}$ primitive, together with the operation $\wedge$ on arrow terms. Then we can assume, besides categorial, bifunctorial and naturality equations, the equations $(\hat{w}\hat{k})$ and $(\hat{w}\hat{k}\hat{k})$ in order to obtain a logical system synonymous with $\mathcal{C}(\hat{\mathbf{L}})$.

Synonymity with these alternative syntactical systems can be demonstrated directly, but this is a lengthy exercise. The coherence result for semilattice categories we are going to prove will easily yield this synonymity in an indirect way.

We introduce next still another syntactical system synonymous with $\mathcal{C}(\hat{\mathbf{L}})$, which will be formulated in the style of Gentzen, and will enable us to prove a composition-elimination result, i.e. a simple kind of cut-elimination result, such as we had in §8.1. Let $\mathcal{C}(\mathbf{G}\hat{\mathbf{L}})$ be the syntactical system with formulae of $\mathcal{L}_\wedge$ as objects, with the primitive arrow terms being only identity arrow terms, and with the following operations on arrow terms, besides the operation $\circ$:

$$\frac{f_1 : C \vdash A_1 \qquad f_2 : C \vdash A_2}{\langle f_1, f_2 \rangle : C \vdash A_1 \wedge A_2}$$

$$\frac{g_1 : A_1 \vdash C}{\hat{K}^1_{A_2} g_1 : A_1 \wedge A_2 \vdash C} \qquad\qquad \frac{g_2 : A_2 \vdash C}{\hat{K}^2_{A_1} g_2 : A_1 \wedge A_2 \vdash C}$$

To obtain the equations of $\mathcal{E}(\mathbf{G}\hat{\mathbf{L}})$, we assume the categorial equations and the following equations, for $i \in \{1, 2\}$:

$$\begin{array}{ll}
(\hat{K}1) & g \circ \hat{K}^i_A f = \hat{K}^i_A (g \circ f), \\
(\hat{K}2) & \hat{K}^i_A g \circ \langle f_1, f_2 \rangle = g \circ f_i, \\
(\hat{K}3) & \langle g_1, g_2 \rangle \circ f = \langle g_1 \circ f, g_2 \circ f \rangle, \\
(\hat{K}4) & \mathbf{1}_{A \wedge B} = \langle \hat{K}^1_B \mathbf{1}_A, \hat{K}^2_A \mathbf{1}_B \rangle,
\end{array}$$

with appropriate types assigned to f, g, f_i and g_i. The equation $(\hat{K}2)$ is related to $(\wedge\beta)$, while $(\hat{K}3)$ and $(\hat{K}4)$ are related to $(\wedge\eta)$. The syntactical category $\mathbf{G}\hat{\mathbf{L}}$ is $\mathcal{C}(\mathbf{G}\hat{\mathbf{L}})/\mathcal{E}(\mathbf{G}\hat{\mathbf{L}})$.

It is a straightforward, though somewhat lengthy, exercise to check that with the definitions corresponding to the equations $(\wedge)$, $(\hat{b}\rightarrow)$, $(\hat{b}\leftarrow)$, $(\hat{c})$, $(\hat{w})$, and the additional definitions

$$\hat{k}^1_{A_1,A_2} =_{df} \hat{K}^1_{A_2} \mathbf{1}_{A_1}, \qquad\qquad \hat{k}^2_{A_1,A_2} =_{df} \hat{K}^2_{A_1} \mathbf{1}_{A_2},$$

on the one hand, and the definitions $(\langle \_, \_ \rangle)$ and

$$\hat{K}^1_{A_2} g_1 =_{df} g_1 \circ \hat{k}^1_{A_1,A_2}, \qquad\qquad \hat{K}^2_{A_1} g_2 =_{df} g_2 \circ \hat{k}^2_{A_1,A_2},$$

on the other hand, we can prove that $\hat{\mathbf{L}}$ and $\mathbf{G}\hat{\mathbf{L}}$ are isomorphic categories, and that hence $\mathcal{C}(\hat{\mathbf{L}})$ and $\mathcal{C}(\mathbf{G}\hat{\mathbf{L}})$ are synonymous syntactical systems.

We can prove as follows Composition Elimination (see §8.1) for $\mathbf{G}\hat{\mathbf{L}}$.

PROOF OF COMPOSITION ELIMINATION FOR $\mathbf{G\hat{L}}$. Take a subterm $g \circ f$ of an arrow term of $\mathcal{C}(\mathbf{G\hat{L}})$ such that both f and g are composition-free. We call such a subterm a *topmost cut*. We show that $g \circ f$ is equal either to a composition-free arrow term or to an arrow term all of whose compositions occur in topmost cuts of strictly smaller length than the length of $g \circ f$. The possibility of eliminating composition in topmost cuts, and hence every composition, follows by induction on the length of topmost cuts.

The cases where f or g are $\mathbf{1}_A$ are taken care of by $(cat\ 1)$; the cases where f is $\hat{K}^i_A f'$ are taken care of by $(\hat{K}1)$; and the case where g is $\langle g_1, g_2\rangle$ is taken care of by $(\hat{K}3)$. The following cases remain.

If f is $\langle f_1, f_2\rangle$, then g is either of a form covered by cases above, or g is $\hat{K}^i_A g'$, and we apply $(\hat{K}2)$. ⊣

Note that we do not use the equations $(cat\ 2)$ and $(\hat{K}4)$ in this proof.

An arrow term of $\mathcal{C}(\mathbf{G\hat{L}})$ is said to be in *normal form* when it is composition-free and there are no subterms of it of the forms $\mathbf{1}_{A\wedge B}$ and $\hat{K}^i_A \langle f, g\rangle$. In $\mathbf{G\hat{L}}$ we have the equations $(\hat{K}4)$ and, for $i \in \{1, 2\}$,

$$(\hat{K}5) \quad \hat{K}^i_A \langle f, g\rangle = \langle \hat{K}^i_A f, \hat{K}^i_A g\rangle,$$

which is obtained with the help of $(cat\ 1)$, $(\hat{K}1)$ and $(\hat{K}3)$. With Composition Elimination and these equations, it can be shown that for every arrow term of $\mathcal{C}(\mathbf{G\hat{L}})$ there is an arrow term f' of $\mathcal{C}(\mathbf{G\hat{L}})$ in normal form such that $f = f'$ in $\mathbf{G\hat{L}}$. Namely, the following holds for $\mathbf{G\hat{L}}$.

NORMAL-FORM LEMMA. *Every arrow term is equal to an arrow term in normal form.*

(The proof of this lemma is incorporated in the proof of the Normal-Form Lemma for $\mathbf{G\hat{L}}_\top$ in the next section.)

The functor G from $\hat{\mathbf{L}}$ to *Rel* yields with the isomorphism from $\mathbf{G\hat{L}}$ to $\hat{\mathbf{L}}$ a functor, which we also call G, from $\mathbf{G\hat{L}}$ to *Rel*. We can then prove the following for arrow terms of $\mathcal{C}(\mathbf{G\hat{L}})$.

UNIQUENESS LEMMA. *For every arrow term f there is a unique arrow term f' in normal form such that $Gf = Gf'$.*

PROOF. Let f be of type $A \vdash B$. It follows from the Normal-Form Lemma and the functoriality of G that there is at least one arrow term f' in normal form such that $Gf = Gf'$. To show that f' is unique we proceed by induction on the number $n(B)$ of occurrences of $\wedge$ in B.

If $n(B) = 0$, then we make an auxiliary induction on $n(A)$. If $n(A) = 0$, then f' can be only of the form $\mathbf{1}_p$. If $n > 0$, then A must be of the form $A_1 \wedge A_2$, and f' can be only of the form $\hat{K}^1_{A_2}\, g$ or $\hat{K}^2_{A_1}\, h$. Since there are no g and h such that $G\hat{K}^1_{A_2}\, g = G\hat{K}^2_{A_1}\, h$, the arrow term f' is uniquely determined.

Suppose now $n(B) > 0$. Then B must be of the form $B_1 \wedge B_2$, and f' can be only of the form $\langle f_1, f_2 \rangle$ for $f_1 \colon A \vdash B_1$ and $f_2 \colon A \vdash B_2$ arrow terms in normal form. We have that

$$G\hat{K}^1_{B_2}\, f = G\hat{K}^1_{B_2}\, \langle f_1, f_2 \rangle = Gf_1,$$

and so, by the induction hypothesis, f_1 is unique. Analogously, $G\hat{K}^2_{B_1}\, f = Gf_2$, and f_2 is unique. So $\langle f_1, f_2 \rangle$ is unique. ⊣

(This Uniqueness Lemma is analogous, but not completely analogous, to the homonymous lemma of §5.2: in the former lemma we do not presuppose the Normal-Form Lemma, while in the formulation of the present one we do. Since, however, we have the Normal-Form Lemma in both cases, the difference is more in the style of exposition than in mathematical content.)

We can then prove the following (for references concerning this result, see the references mentioned in the next section before Cartesian Coherence).

SEMILATTICE COHERENCE. *The functor G from $\hat{\mathbf{L}}$ to Rel is faithful.*

PROOF. We prove that the functor G from $\mathbf{G}\hat{\mathbf{L}}$ to *Rel* is faithful. This yields Semilattice Coherence.

Suppose that for $f_1, f_2 \colon A \vdash B$ arrow terms of $\mathcal{C}(\mathbf{G}\hat{\mathbf{L}})$ we have $Gf_1 = Gf_2$. By the Normal-Form Lemma, for $i \in \{1, 2\}$, there is an arrow term f_i' in normal form such that $f_i = f_i'$. Then from $Gf_1' = Gf_1 = Gf_2 = Gf_2'$ and the Uniqueness Lemma we conclude that f_1' and f_2' are the same arrow term, and hence $f_1 = f_2$ in $\mathbf{G}\hat{\mathbf{L}}$. ⊣

§9.2. Coherence of cartesian categories

To obtain the natural logical category $\hat{\mathbf{L}}_\top$, we have that the logical system $\mathcal{C}(\hat{\mathbf{L}}_\top)$ is in $\mathcal{L}_{\wedge,\top}$, with the transformations α included in $\mathbf{1}$, $\hat{b}$, $\hat{c}$, $\hat{w}$-$\hat{k}$ and $\hat{\delta}$-$\hat{\sigma}$. The specific equations of $\mathcal{E}(\hat{\mathbf{L}}_\top)$ are obtained by taking the union of those of $\mathcal{E}(\hat{\mathbf{L}})$ and $\mathcal{E}(\hat{\mathbf{S}}_\top)$ plus

$$(\hat{k}\hat{\delta})\qquad \hat{k}^1_{A,\top} = \hat{\delta}^{\rightarrow}_A.$$

There are some redundancies in this axiomatization. The equation $(\hat{b}\hat{\delta}\hat{\sigma})$ is derivable with the help of $(\hat{b}\hat{k})$. In $\mathcal{E}(\hat{\mathbf{L}}_\top)$ we can derive the following

$$(\hat{w}\hat{\delta})\qquad \hat{w}_\top = \hat{\delta}^{\leftarrow}_\top.$$

Natural $\hat{\mathbf{L}}_\top$-categories are called *cartesian* categories. These are categories with all finite products, including the empty product. The objects of a cartesian category that is a partial order make a semilattice with unit.

In $\mathcal{C}(\hat{\mathbf{L}}_\top)$ we have the definition

$$\hat{\kappa}_A =_{df} \hat{k}^2_{A,\top} \circ \hat{\delta}^{\leftarrow}_A,$$

and for $\hat{\mathbf{L}}_\top$ we have the equations $(\hat{\kappa}\ nat)$ and

$$(\hat{\kappa}\mathbf{1})\qquad \hat{\kappa}_\top = \mathbf{1}_\top.$$

The equations $(\hat{\kappa}\ nat)$ and $(\hat{\kappa}\mathbf{1})$ amount to

$$(\hat{\kappa})\qquad \hat{\kappa}_A = f,\quad \text{for } f\colon A \vdash \top,$$

which says that $\top$ is a terminal object in $\hat{\mathbf{L}}_\top$ (see §2.2 for the notion of terminal object).

A logical system synonymous with $\mathcal{C}(\hat{\mathbf{L}}_\top)$ is obtained by taking as primitive $\hat{\kappa}$ instead of $\hat{\delta}$-$\hat{\sigma}$. This is based on the following equations of $\hat{\mathbf{L}}_\top$:

$$\hat{\delta}^{\leftarrow}_A = \langle \mathbf{1}_A, \hat{\kappa}_A\rangle,$$
$$\hat{\sigma}^{\leftarrow}_A = \langle \hat{\kappa}_A, \mathbf{1}_A\rangle.$$

Another alternative logical system synonymous with $\mathcal{C}(\hat{\mathbf{L}}_\top)$ is obtained by taking as primitives $\hat{\delta}$-$\hat{\sigma}$ and $\hat{\kappa}$ instead of $\hat{k}^1$ and $\hat{k}^2$. This is based on the following equations of $\hat{\mathbf{L}}_\top$:

$$\hat{k}^1_{A,B} = \hat{\delta}^{\rightarrow}_{A} \circ (\mathbf{1}_A \wedge \hat{\kappa}_B),$$
$$\hat{k}^2_{A,B} = \hat{\sigma}^{\rightarrow}_{B} \circ (\hat{\kappa}_A \wedge \mathbf{1}_B).$$

If the operation $\langle \_, \_ \rangle$ is primitive together with $\hat{k}^1$ and $\hat{k}^2$, then we can take $\hat{\kappa}$ as primitive, and assume besides the categorial equations, $(\wedge\beta)$ and $(\wedge\eta)$ just $(\hat{\kappa})$.

Let $\mathcal{C}(\mathbf{G}\hat{\mathbf{L}}_\top)$ be the syntactical system defined as $\mathcal{C}(\mathbf{G}\hat{\mathbf{L}})$ save that $\mathcal{L}_\wedge$ is replaced by $\mathcal{L}_{\wedge,\top}$, and besides identity arrow terms we have the arrow terms $\hat{\kappa}_A \colon A \vdash \top$ as primitive arrow terms. The equations of $\mathcal{E}(\mathbf{G}\hat{\mathbf{L}}_\top)$ are those of $\mathcal{E}(\mathbf{G}\hat{\mathbf{L}})$ plus $(\hat{\kappa})$. The syntactical category $\mathbf{G}\hat{\mathbf{L}}_\top$ is $\mathcal{C}(\mathbf{G}\hat{\mathbf{L}}_\top)/\mathcal{E}(\mathbf{G}\hat{\mathbf{L}}_\top)$. It is easy to ascertain that $\hat{\mathbf{L}}_\top$ and $\mathbf{G}\hat{\mathbf{L}}_\top$ are isomorphic, and that hence $\mathcal{C}(\hat{\mathbf{L}}_\top)$ and $\mathcal{C}(\mathbf{G}\hat{\mathbf{L}}_\top)$ are synonymous syntactical systems.

We can prove Composition Elimination for $\mathbf{G}\hat{\mathbf{L}}_\top$ by enlarging the proof of the preceding section. The additional cases are where g is $\hat{\kappa}_A$, which is taken care of by $(\hat{\kappa})$, and where f is $\hat{\kappa}_A$. In the latter case, g is either $\mathbf{1}_\top$, or $\hat{\kappa}_\top$, or $\langle g_1, g_2 \rangle$, which are cases already covered.

An arrow term of $\mathcal{C}(\mathbf{G}\hat{\mathbf{L}}_\top)$ is said to be in *normal form* when it is composition-free and has no subterms of the forms $\mathbf{1}_{A\wedge B}$, $\hat{K}^i_A \langle f, g \rangle$, $\mathbf{1}_\top$ and $\hat{K}^i_A \hat{\kappa}_B$. Then we can prove the Normal-Form Lemma of the preceding section for $\mathbf{G}\hat{\mathbf{L}}_\top$.

PROOF OF THE NORMAL-FORM LEMMA FOR $\mathbf{G}\hat{\mathbf{L}}_\top$. In $\mathbf{G}\hat{\mathbf{L}}_\top$ we have the equations $(\hat{K}4)$ and $(\hat{K}5)$ of the preceding section, and also

$$\mathbf{1}_\top = \hat{\kappa}_\top,$$
$$\hat{K}^1_B \hat{\kappa}_A = \hat{\kappa}_{A\wedge B},$$
$$\hat{K}^2_A \hat{\kappa}_B = \hat{\kappa}_{A\wedge B},$$

which are all instances of $(\hat{\kappa})$ (the first of these equations is $(\hat{\kappa}\mathbf{1})$).

For f a composition-free arrow term of $\mathcal{C}(\mathbf{G}\hat{\mathbf{L}}_\top)$, let n_1 be the number of occurrences of $\wedge$ and $\top$ in the indices of subterms of f that are identity arrow terms. Next, let n_2 be the number of subterms of f of the form $\langle f_1, f_2\rangle$ or $\hat{\kappa}_C$ such that there is a subterm $\hat{K}^i_D\, f'$ of f with $\langle f_1, f_2\rangle$ or $\hat{\kappa}_C$ a subterm of f'. Let the *grade* of f be (n_1, n_2), and let these grades be lexicographically ordered (see §7.7, before the Cut-Elimination Theorem). Then every replacement of subterms of f justified by one of the equations above reduces the grade of f, and so by induction we obtain that there is an arrow term f' of $\mathcal{C}(\mathbf{G}\hat{\mathbf{L}}_\top)$ in normal form such that $f = f'$ in $\mathbf{G}\hat{\mathbf{L}}_\top$. It remains only to appeal to Composition Elimination for $\mathbf{G}\hat{\mathbf{L}}_\top$ to obtain the Normal-Form Lemma. ⊣

As in the preceding section, we obtain the functor G from $\mathbf{G}\hat{\mathbf{L}}_\top$ to *Rel*, with which we can prove the Uniqueness Lemma of the preceding section for $\mathbf{G}\hat{\mathbf{L}}_\top$.

PROOF OF THE UNIQUENESS LEMMA FOR $\mathbf{G}\hat{\mathbf{L}}_\top$. We proceed by induction $n(B)$ with an auxiliary induction on $n(A)$ in the basis, as in the proof of the preceding section. If $n(B) = n(A) = 0$, then f' can be either of the form $\mathbf{1}_p$, or $\hat{\kappa}_p$, or $\hat{\kappa}_\top$, which exclude each other because of their types. If $n(B) = 0$ and $n(A) > 0$, then f' can be either of the form $\hat{K}^1_{A_2}\, g$, or $\hat{K}^2_{A_1}\, h$ or $\hat{\kappa}_A$, which exclude each other because of their types or for reasons mentioned in the proof of the preceding section. For that we use the fact, easily shown by induction on the length of A, that $\hat{\kappa}_A$ is the only arrow term in normal form of type $A \vdash \top$, and the fact that Gf is a function from GB to GA. For the rest of the proof we proceed as in the preceding section. ⊣

We can then infer as in the preceding section the following result, which stems from [72] (see p. 129, where the result is announced), [100] (Theorem 2.2), [122] (Theorem 8.2.3, p. 207), [102] (Section 7) and [42].

CARTESIAN COHERENCE. *The functor G from $\hat{\mathbf{L}}_\top$ to Rel is faithful.*

It is noteworthy that the functor G maps every arrow of $\hat{\mathbf{L}}_\top$ to a function from the target to the source. (We used that fact in the proof of the Uniqueness Lemma of this section.) The same holds, of course, for $\hat{\mathbf{L}}$.

§9.3. Maximality of semilattice and cartesian categories

A natural logical category $\mathcal{C}/\mathcal{E}$ is called *maximal* when every natural $\mathcal{C}/\mathcal{E}$-category that satisfies an equation between arrow terms of $\mathcal{C}$ that is not in $\mathcal{E}$ is a preorder. In other words, if $\mathcal{E}'$ is a proper extension of $\mathcal{E}$, then every natural $\mathcal{C}/\mathcal{E}'$-category is a preorder. Maximality is an interesting property when $\mathcal{C}/\mathcal{E}$ itself is not a preorder, and we will show in this section that $\hat{\mathbf{L}}$ and $\hat{\mathbf{L}}_{\top}$ are maximal in this sense—in the interesting way. (We take over these results from [42].)

The maximality property above is analogous to the property of usual formulations of the classical propositional calculus called *Post completeness.* That this calculus is Post complete means that if we add to it any new axiom-schema in the language of the calculus, then we can prove every formula. An analogue of Böhm's Theorem in the typed lambda calculus implies, similarly, that the typed lambda calculus cannot be extended without falling into triviality, i.e. without every equation (between terms of the same type) becoming derivable (see [111], [41] and references therein; see [4], Section 10.4, for Böhm's Theorem in the untyped lambda calculus).

Let us now consider several examples of common algebraic structures with analogous maximality properties. First, we have that semilattices are maximal in the following sense.

Let a and b be terms made exclusively of variables and of a binary operation $\cdot$, which we interpret as meet or join. That the equation $a = b$ holds in a semilattice S means that *every* instance of $a = b$ obtained by substituting names of elements of S for variables holds in S (cf. §2.3). Suppose $a = b$ does not hold in a free semilattice S_F (so it is not the case that $a = b$ holds in every semilattice). Hence there must be an instance of $a = b$ obtained by substituting names of elements of S_F for variables such that this instance does not hold in S_F. It is easy to conclude that in $a = b$ there must be at least two variables, and that S_F must have at least two free generators. Then every semilattice in which $a = b$ holds is trivial—namely, it has a single element.

Here is a short proof of that. If $a = b$ does not hold in S_F, then there must be a variable x in one of a and b that is not in the other. Then from

$a = b$, by substituting y for every variable in a and b different from x, and by applying the semilattice equations, we infer either $x = y$ or $x \cdot y = y$. If we have $x = y$, we are done, and, if we have $x \cdot y = y$, then we have also $y \cdot x = x$, and hence $x = y$.

Semilattices with unit, distributive lattices, distributive lattices with top and bottom, and Boolean algebras are maximal in the same sense. The equations $a = b$ in question are equations between terms made exclusively of variables and the operations of the kind of algebra we envisage: semilattices with unit, distributive lattices, etc. That such an equation holds in a particular structure means, as above, that every substitution instance of it holds. However, the number of variables in $a = b$ and the number of generators of the free structure mentioned need not always be at least two.

If we deal with semilattices with unit $\mathbf{1}$, then $a = b$ must have at least one variable, and the free semilattice with unit must have at least one free generator. We substitute $\mathbf{1}$ for every variable in a and b different from x in order to obtain $x = \mathbf{1}$, and hence triviality. So semilattices with unit are maximal in the same sense.

The same sort of maximality can be proven for distributive lattices, whose operations are $\wedge$ and $\vee$, which we call conjunction and disjunction, respectively. Then every term made of $\wedge$, $\vee$ and variables is equal to a term in disjunctive normal form (i.e. a multiple disjunction of multiple conjunctions of variables; see §10.2 for a precise definition), and to a term in conjunctive normal form (i.e. a multiple conjunction of multiple disjunctions of variables; see §10.2). These normal forms are not unique. If $a = b$, in which we must have at least two variables, does not hold in a free distributive lattice D_F with at least two free generators, then either $a \leq b$ or $b \leq a$ does not hold in D_F. Suppose $a \leq b$ does not hold in D_F. Let a' be a disjunctive normal form of a, and let b' be a conjunctive normal form of b. So $a' \leq b'$ does not hold in D_F. From that we infer that for a disjunct a'' of a' and for a conjunct b'' of b' we do not have $a'' \leq b''$ in D_F. This means that there is no variable in common in a'' and b''; otherwise, the conjunction of variables a'' would be lesser than or equal in D_F to the disjunction of variables b''. If in a distributive lattice $a = b$ holds, then $a'' \leq b''$ holds too, and hence, by substitution, we obtain $x \leq y$. So $x = y$.

For distributive lattices with top $\top$ and bottom $\bot$, we proceed analogously via disjunctive and conjunctive normal form. Here $a = b$ may be even without variables, and the free structure may have even an empty set of free generators. The additional cases to consider are when in $a'' \leq b''$ we have that a'' is $\top$ and b'' is $\bot$. In any case, we obtain $\top \leq \bot$, and hence our structure is trivial.

The same sort of maximality can be proven for Boolean algebras, i.e. complemented distributive lattices. Boolean algebras must have top and bottom. In a disjunctive normal form now the disjuncts are conjunctions of variables x or terms $\bar{x}$, where $\bar{\ }$ is complementation, or the disjunctive normal form is just $\top$ or $\bot$; analogously for conjunctive normal forms. Then we proceed as for distributive lattices with an equation $a = b$ that may be even without variables, until we reach that $a'' \leq b''$, which does not hold in a free Boolean algebra B_F, whose set of free generators may be even empty, holds in our Boolean algebra. If x is a conjunct of a'', then in b'' we cannot have a disjunct x; but we may have a disjunct $\bar{x}$. The same holds for the conjuncts $\bar{x}$ of a''. It is excluded that both x and $\bar{x}$ are conjuncts of a'', or disjuncts of b''; otherwise, $a'' \leq b''$ would hold in B_F. Then for every conjunct x in a'' and every disjunct $\bar{y}$ in b'' we substitute $\top$ for x and y, and for every other variable we substitute $\bot$. In any case, we obtain $\top \leq \bot$, and hence our Boolean algebra is trivial. This is essentially the proof of Post completeness for the classical propositional calculus, due to Bernays and Hilbert (see [129], Section 2.4, and [61], Section I.13), from which we can infer the ordinary completeness of this calculus with respect to valuations in the two-element Boolean algebra—namely, with respect to truth tables—and also completeness with respect to any nontrivial model.

As examples of common algebraic structures that are not maximal in the sense above, we have semigroups, commutative semigroups, lattices, and many others. What is maximal for semilattices and is not maximal for lattices is the equational theory of the structures in question. The equational theory of semilattices cannot be extended without falling into triviality, while the equational theory of lattices can be extended with the distributive law, for example.

The maximality of $\hat{\mathbf{L}}$ as defined at the beginning of the section differs from the maximality of semilattices, distributive lattices, etc., we have just considered, because in $\hat{\mathbf{L}}$ we have types, so that $f = g$ is excluded if f and g are of different types. Hence, the analogue of the trivial semilattice, which was a one-element structure, is for categories, like $\hat{\mathbf{L}}$, a preorder.

The maximality of $\hat{\mathbf{L}}$ is, of course, a quite separate result from the maximality of semilattices we have shown above. None of these results can be inferred from the other. After some strictification, any semilattice category yields a semilattice category that is a partial order, and whose objects will make a semilattice. The maximality of semilattices has to do with these objects, while the maximality of $\hat{\mathbf{L}}$ has to do with the arrows between these objects. We will now proceed with the proof of the latter maximality.

MAXIMALITY OF $\hat{\mathbf{L}}$. *The category* $\hat{\mathbf{L}}$ *is maximal.*

PROOF. Suppose A and B are formulae of $\mathcal{L}_\wedge$ in which only p occurs as a letter. Suppose $f_1, f_2 \colon A \vdash B$ are arrow terms of $\mathcal{C}(\hat{\mathbf{L}})$ such that $Gf_1 \neq Gf_2$. As we noted after Cartesian Coherence, at the end of the preceding section, Gf_1 and Gf_2 may be conceived as functions from GB to GA. So there must be an $n \in GB$ such that $Gf_1(n) \neq Gf_2(n)$. This means that we must have $GA \geq 2$ (i.e., there must be at least two occurrence of p in A), and we have, of course, $GB \geq 1$.

Then there is an arrow term $h^w : p \wedge p \vdash A$ of $\mathcal{C}(\hat{\mathbf{L}})$ made of possibly multiple occurrences of arrow terms in $\mathbf{1}$, $\hat{b}$, $\hat{c}$ and $\hat{w}$, together with the operations $\wedge$ and $\circ$ on arrow terms, such that $Gh^w(Gf_1(n)) = 0$ and $Gh^w(Gf_2(n)) = 1$. There is also an arrow term $h^k : B \vdash p$ of $\mathcal{C}(\hat{\mathbf{L}})$ that is either $\mathbf{1}_p$ or a possibly iterated composition of arrow terms in $\hat{k}^1$ and $\hat{k}^2$ such that $Gh^k(0) = n$. Then, for $i \in \{1, 2\}$, we have that $h^k \circ f_i \circ h^w$ is of type $p \wedge p \vdash p$ and $G(h^k \circ f_i \circ h^w) = G\,\hat{k}^i_{p,p}$. Therefore, by Composition Elimination for $\mathbf{G}\hat{\mathbf{L}}$ (see §9.1) and by the functoriality of G, we obtain that $h^k \circ f_i \circ h^w = \hat{k}^i_{p,p}$ in $\hat{\mathbf{L}}$. (This follows from Semilattice Coherence too.) So in $\mathcal{E}(\hat{\mathbf{L}})$ extended with $f_1 = f_2$ we can derive the equation

$$(\hat{k}\hat{k}) \qquad \hat{k}^1_{p,p} = \hat{k}^2_{p,p}\,.$$

If this equation holds in a semilattice category $\mathcal{A}$, then $\mathcal{A}$ is a preorder. This is shown as follows. For $f, g \colon a \vdash b$ in $\mathcal{A}$ we have

$$\hat{k}^1_{b,b} \circ \langle f, g\rangle = \hat{k}^2_{b,b} \circ \langle f, g\rangle,$$

and so $f = g$ in $\mathcal{A}$ by the equation $(\wedge\beta)$ of §9.1.

If for some arrow terms g_1 and g_2 of $\mathcal{C}(\hat{\mathbf{L}})$ of the same type we have that $g_1 = g_2$ is not in $\mathcal{E}(\hat{\mathbf{L}})$, then by Semilattice Coherence (see §9.1) we have $Gg_1 \neq Gg_2$. If we take the substitution instances g_1' of g_1 and g_2' of g_2 obtained by replacing every letter by a single letter p, then we obtain again $Gg_1' \neq Gg_2'$. If $g_1 = g_2$ holds in a semilattice category $\mathcal{A}$, then $g_1' = g_2'$ holds too, and $\mathcal{A}$ is a preorder, as we have shown above. ⊣

We have also the following.

MAXIMALITY OF $\hat{\mathbf{L}}_\top$. *The category* $\hat{\mathbf{L}}_\top$ *is maximal.*

To prove that we proceed as for $\hat{\mathbf{L}}$. The only modification is that in constructing h^w we envisage also arrow terms in $\hat{\delta}\!\leftarrow$ and $\hat{\sigma}\!\leftarrow$.

Note that the maximality of $\hat{\mathbf{L}}$ implies that in any semilattice category $\mathcal{A}$ that is not a preorder we can falsify any equation between arrow terms of $\mathcal{C}(\hat{\mathbf{L}})$ that does not hold in $\hat{\mathbf{L}}$. This does not mean, however, that there must be a faithful functor from $\hat{\mathbf{L}}$ to $\mathcal{A}$, which would falsify all such equations "simultaneously". The existence of such a functor is possible for particular semilattice categories $\mathcal{A}$, but it is another result, which does not follow from and does not imply maximality. In the case of $\hat{\mathbf{L}}$ and $\hat{\mathbf{L}}_\top$, the category *Set* of sets with functions, with $\wedge$ being cartesian product and $\top$ being a singleton, is an $\mathcal{A}$ such that there is a faithful functor from $\hat{\mathbf{L}}$ and $\hat{\mathbf{L}}_\top$ to $\mathcal{A}$ (see [27] and [111]).

Maximality holds also trivially for all logical categories $\mathcal{K}$ that are preorders, because we cannot extend $\mathcal{E}(\mathcal{K})$ properly in such cases. The logical categories that are not preorders that we have considered up to now are symmetric, i.e. they have $\hat{c}$ in $\mathcal{C}(\mathcal{K})$. Before $\hat{\mathbf{L}}$ and $\hat{\mathbf{L}}_\top$, however, the symmetric logical categories $\mathcal{K}$ from previous chapters that are coherent are not maximal, in spite of coherence, for the following reason.

All the types of arrow terms of $\mathcal{C}(\mathcal{K})$ are balanced (in the sense of §3.3). Let the *balance weight* of an equation $f = g$ where $f, g \colon A \vdash B$ are arrow terms of $\mathcal{C}(\mathcal{K})$ be the letter length of A or B. Then it can be shown that if $\mathcal{E}'$ is the extension of $\mathcal{E}(\mathcal{K})$ with an equation $f = g$ that is not in $\mathcal{E}(\mathcal{K})$, with a single letter occurring in A (and hence also in B), and the balance weight of $f = g$ is n, then all the equations $\mathcal{E}'$ that are not in $\mathcal{E}(\mathcal{K})$ must have a balance weight greater than or equal to n.

The notions of maximality envisaged in this section were extreme (or should we say "maximal"), in the sense that we envisaged collapsing only into preorder. (For semilattices, distributive lattices, etc., this is also preorder for a one-object category.) We may, however, envisage relativizing our notion of maximality by replacing preorder with a weaker property, such that structures possessing it are trivial, but not so trivial (cf. [34], Section 4.11). We will encounter maximality in such a relative sense in §9.7.

As an example of relative maximality in a common algebraic structure we can take symmetric groups. The axioms for the symmetric group $\mathcal{S}_n$, where $n \geq 2$, with the generators s_i, for $i \in \{1, \ldots, n{-}1\}$, were given in §5.1. If to $\mathcal{S}_n$ for $n \geq 5$ we add an equation $a = \mathbf{1}$ where a is built exclusively of the generators s_i of $\mathcal{S}_n$ with composition, and $a = \mathbf{1}$ does not hold in $\mathcal{S}_n$, then we can derive $s_i = s_j$. This does not mean that the resulting structure will be a one-element structure, i.e. the trivial one-element group. It will be such if a is an odd permutation, and if a is an even permutation, then we will obtain a two-element structure, which is $\mathcal{S}_2$. This can be inferred from facts about the normal subgroups of $\mathcal{S}_n$. Simple groups are maximal in the nonrelative sense, envisaged above for semilattices.

§9.4. Coherence of lattice categories

Let $\overset{\vee}{\mathbf{L}}$ be the natural logical category in $\mathcal{L}_\vee$ isomorphic to the category $\overset{\wedge}{\mathbf{L}}^{op}$ (which is $\overset{\wedge}{\mathbf{L}}$ with source and target functions interchanged; see §2.2). We just replace $\wedge$ by $\vee$, so that the primitive arrow terms of $\mathcal{C}(\overset{\vee}{\mathbf{L}})$ are included in $\mathbf{1}$, $\overset{\vee}{b}$, $\overset{\vee}{c}$ and $\overset{\vee}{w}$-$\overset{\vee}{k}$, while the equations of $\mathcal{E}(\overset{\vee}{\mathbf{L}})$ are duals of those of $\mathcal{E}(\overset{\wedge}{\mathbf{L}})$ (see the List of Equations and the List of Categories at the end of the book). Natural $\overset{\vee}{\mathbf{L}}$-categories would usually be called categories with finite nonempty coproducts.

Let $\mathcal{C}(\mathbf{G}\overset{\vee}{\mathbf{L}})$ be the syntactical system with formulae of $\mathcal{L}_\vee$ as objects, with the primitive arrow terms being only identity arrow terms, and with the following operations on arrow terms, dual to those of $\mathcal{C}(\mathbf{G}\overset{\vee}{\mathbf{L}})$, besides the operation $\circ$ (cf. §9.1):

$$\frac{g_1 : A_1 \vdash C \qquad g_2 : A_2 \vdash C}{[g_1, g_2] : A_1 \vee A_2 \vdash C}$$

$$\frac{f_1 : C \vdash A_1}{\check{K}^1_{A_2} f_1 : C \vdash A_1 \vee A_2} \qquad \frac{f_2 : C \vdash A_2}{\check{K}^2_{A_1} f_2 : C \vdash A_1 \vee A_2}$$

To obtain the equations of $\mathcal{E}(\mathbf{G}\check{\mathbf{L}})$, we assume the categorial equations and the following equations for $i \in \{1, 2\}$, obtained by dualizing the equations $(\hat{K}1)$-$(\hat{K}4)$ of §9.1:

$$\begin{array}{ll}
(\check{K}1) & \check{K}^i_A g \circ f = \check{K}^i_A (g \circ f), \\
(\check{K}2) & [g_1, g_2] \circ \check{K}^i_A f = g_i \circ f, \\
(\check{K}3) & g \circ [f_1, f_2] = [g \circ f_1, g \circ f_2], \\
(\check{K}4) & \mathbf{1}_{A \vee B} = [\check{K}^1_B \mathbf{1}_A, \check{K}^2_A \mathbf{1}_B],
\end{array}$$

with appropriate types assigned to f, g, f_i g_i. The syntactical category $\mathbf{G}\check{\mathbf{L}}$ is $\mathcal{C}(\mathbf{G}\check{\mathbf{L}})/\mathcal{E}(\mathbf{G}\check{\mathbf{L}})$. It is clear that $\mathbf{G}\check{\mathbf{L}}$ is isomorphic to $\check{\mathbf{L}}$, and also to $\hat{\mathbf{L}}^{op}$ and $\mathbf{G}\hat{\mathbf{L}}^{op}$. For later use, we note that in $\mathcal{C}(\check{\mathbf{L}})$ we have the definitions

$$[g_1, g_2] =_{df} \check{w}_C \circ (g_1 \vee g_2),$$

$$\check{K}^1_{A_2} f_1 =_{df} \check{k}^1_{A_1, A_2} \circ f_1, \qquad \check{K}^2_{A_1} f_2 =_{df} \check{k}^2_{A_1, A_2} \circ f_2.$$

(We introduce $\mathbf{G}(\check{\mathbf{L}})$ with so much detail for the sake of notation.)

To obtain the natural logical category $\mathbf{L}$, we have that the logical system $\mathcal{C}(\mathbf{L})$ is in $\mathcal{L}_{\wedge,\vee}$, with the transformations α included in $\mathbf{1}$, b, c and w-k. The specific equations of $\mathcal{E}(\mathbf{L})$ are obtained by taking the union of those of $\mathcal{E}(\hat{\mathbf{L}})$ and $\mathcal{E}(\check{\mathbf{L}})$.

We call natural $\mathbf{L}$-categories *lattice* categories. Usually, they would be called categories with finite nonempty products and finite nonempty coproducts. The objects of a lattice category that is a partial order make a lattice.

The syntactical system $\mathcal{C}(\mathbf{GL})$ has as objects the formulae of $\mathcal{L}_{\wedge,\vee}$, as primitive arrow terms the identity arrow terms, and as operations on arrow terms those of $\mathcal{C}(\mathbf{G}\hat{\mathbf{L}})$ and $\mathcal{C}(\mathbf{G}\check{\mathbf{L}})$. For the equations of $\mathcal{E}(\mathbf{GL})$ we assume the equations of $\mathcal{E}(\mathbf{G}\hat{\mathbf{L}})$ and $\mathcal{E}(\mathbf{G}\check{\mathbf{L}})$. The syntactical category $\mathbf{GL}$ is $\mathcal{C}(\mathbf{GL})/\mathcal{E}(\mathbf{GL})$, and it is isomorphic to $\mathbf{L}$. This isomorphism is based on the isomorphism of $\hat{\mathbf{L}}$ with $\mathbf{G}\hat{\mathbf{L}}$, and the isomorphism of $\check{\mathbf{L}}$ with $\mathbf{G}\check{\mathbf{L}}$.

We need as an auxiliary the following category. Let $\hat{\mathbf{L}}_\vee$ be the natural logical category in $\mathcal{L}_{\wedge,\vee}$ obtained as $\hat{\mathbf{L}}$. The only difference is that the arrow terms of $\mathcal{C}(\hat{\mathbf{L}}_\vee)$ are closed under the operation $\vee$ on arrow terms, besides being closed under the operations $\wedge$ and $\circ$, and for $\mathcal{E}(\hat{\mathbf{L}}_\vee)$ we have in addition to the equations assumed for $\mathcal{E}(\hat{\mathbf{L}})$ the bifunctorial equations for $\vee$. Let $\mathbf{G}\hat{\mathbf{L}}_\vee$ be the syntactical category whose objects are formulae of $\mathcal{L}_{\wedge,\vee}$, which is obtained as $\mathbf{G}\hat{\mathbf{L}}$ save that in addition to the operations on arrow terms of $\mathcal{C}(\mathbf{G}\hat{\mathbf{L}})$ we have also the operation $\vee$ on arrow terms, and for $\mathcal{E}(\mathbf{G}\hat{\mathbf{L}}_\vee)$ we assume the bifunctorial equations for $\vee$ in addition to what we had for $\mathcal{E}(\mathbf{G}\hat{\mathbf{L}})$. The categories $\hat{\mathbf{L}}_\vee$ and $\mathbf{G}\hat{\mathbf{L}}_\vee$ are isomorphic, and, hence, $\mathcal{C}(\hat{\mathbf{L}}_\vee)$ and $\mathcal{C}(\mathbf{G}\hat{\mathbf{L}}_\vee)$ are synonymous syntactical systems.

The categories $\check{\mathbf{L}}_\wedge$ and $\mathbf{G}\check{\mathbf{L}}_\wedge$ are isomorphic to $\hat{\mathbf{L}}_\vee^{op}$ and $\mathbf{G}\hat{\mathbf{L}}_\vee^{op}$, and to each other. In them, the $\wedge$ and $\vee$ of $\hat{\mathbf{L}}_\vee$ and $\mathbf{G}\hat{\mathbf{L}}_\vee$ are interchanged, and they are obtained by extending $\check{\mathbf{L}}$ and $\mathbf{G}\check{\mathbf{L}}$ with the bifunctor $\wedge$.

We can then prove Composition Elimination for $\mathbf{GL}$ and $\mathbf{G}\hat{\mathbf{L}}_\vee$ (and hence also for $\mathbf{G}\check{\mathbf{L}}_\wedge$) by enlarging the proof in §9.1.

Proofs of Composition Elimination for $\mathbf{GL}$ and $\mathbf{G}\hat{\mathbf{L}}_\vee$. For $\mathbf{GL}$, we have first the cases where f or g are $\mathbf{1}_A$, where f is $\hat{K}^i_A f'$ and where g is $\langle g_1, g_2\rangle$. For these cases we proceed as before. We have next cases dual to the last two, where g is $\check{K}^i_A g'$, which is taken care of by $(\check{K}1)$, and where f is $[f_1, f_2]$, which is taken care of by $(\check{K}3)$. In the remaining cases, if f is $\langle f_1, f_2\rangle$, then g is either of a form already covered by cases above, or g is $\hat{K}^i_A g'$, and we apply $(\hat{K}2)$. Finally, if f is $\check{K}^i_A f'$, then g is either of a form already covered by cases above, or g is $[g_1, g_2]$, and we apply $(\check{K}2)$.

For $\mathbf{G}\hat{\mathbf{L}}_\vee$, we do not have the cases where f is $[f_1, f_2]$ or $\check{K}^i_A f'$, but f can be $f_1 \vee f_2$. Then, if g is not of a form already covered by the proof in §9.1, it must be $g_1 \vee g_2$, and we apply the bifunctorial equation $(\vee 2)$. ⊣

Note that we do not use the equations (*cat* 2), ($\hat{K}4$) and ($\check{K}4$) in these proofs (which are taken over from [44], Section 3).

A composition-free arrow term of $\mathcal{C}(\mathbf{G}\hat{\mathbf{L}}_\vee)$ may be reduced to a unique normal form, which can then be used to demonstrate coherence for $\hat{\mathbf{L}}_\vee$, i.e. the fact that the functor G from $\hat{\mathbf{L}}_\vee$ to *Rel* is faithful (see [44], Section 4). We do not need, however, that coherence for the results below.

An arrow term of $\mathcal{C}(\mathbf{L})$ in which b and c do not occur is called *pure*. It is clear that with the help of the equations ($\hat{b}^{\rightarrow}$), ($\hat{b}^{\leftarrow}$) and ($\hat{c}$) of §9.1, and the dual equations with $\vee$, every arrow term of $\mathcal{C}(\mathbf{L})$ is equal to a pure arrow term.

An arrow term of $\mathcal{C}(\mathbf{L})$ is in *standard form* when it is of the form $g \circ f$ for f an arrow term $\mathcal{C}(\hat{\mathbf{L}}_\vee)$ and g an arrow term of $\mathcal{C}(\check{\mathbf{L}}_\wedge)$. We can then prove the following.

STANDARD-FORM LEMMA. *Every arrow term of* $\mathcal{C}(\mathbf{L})$ *is equal in* $\mathbf{L}$ *to an arrow term in standard form.*

PROOF. By categorial and bifunctorial equations, we may assume that we deal with a pure factorized arrow term f none of whose factors is a complex identity (see §2.7 and §2.6 for these notions). This will simplify our proof. Every factor of f is either an arrow term of $\mathcal{C}(\hat{\mathbf{L}}_\vee)$, and then we call it a $\wedge$-*factor*, or an arrow term of $\mathcal{C}(\check{\mathbf{L}}_\wedge)$, when we call it a $\vee$-*factor*.

Suppose $f\colon B \vdash C$ is a $\wedge$-factor and $g\colon A \vdash B$ is a $\vee$-factor. We show by induction on the length of $f \circ g$ that in $\mathbf{L}$

$$(*) \quad f \circ g = g' \circ f' \quad \text{or} \quad f \circ g = f' \quad \text{or} \quad f \circ g = g'$$

for f' a $\wedge$-factor and g' a $\vee$-factor.

We will consider various cases for f. In all such cases, if g is $\check{w}_B$, then we use ($\check{w}$ *nat*). If f is $\hat{w}_B$, then we use ($\hat{w}$ *nat*). If f is $\hat{k}^i_{D,E}$ and g is $g_1 \wedge g_2$, then we use ($\hat{k}^i$ *nat*). If f is $f_1 \wedge f_2$ and g is $g_1 \wedge g_2$, then we use bifunctorial and categorial equations and the induction hypothesis.

Finally, if f is $f_1 \vee f_2$, then we have the following cases. If g is $\check{k}^i_{B_1,B_2}$, then we use ($\check{k}^i$ *nat*). If g is $g_1 \vee g_2$, then we use bifunctorial and categorial equations and the induction hypothesis. This proves $(*)$, and it is clear that $(*)$ is sufficient to prove the lemma. $\dashv$

AUXILIARY LEMMA. *Let* $f : A_1 \wedge A_2 \vdash B$ *be an arrow term of* $\mathcal{C}(\hat{\mathbf{L}}_\vee)$. *If for every* $(x, y) \in Gf$ *we have that* $x \in GA_1$, *then* f *is equal in* $\hat{\mathbf{L}}_\vee$ *to an arrow term of the form* $f' \circ \hat{k}^1_{A_1,A_2}$, *and if for every* $(x, y) \in Gf$ *we have that* $x - GA_1 \in GA_2$, *then* f *is equal in* $\hat{\mathbf{L}}_\vee$ *to an arrow term of the form* $f' \circ \hat{k}^2_{A_1,A_2}$.

PROOF. We proceed by induction on the length of B. If B is a letter or $B_1 \vee B_2$, then by Composition Elimination for $\mathbf{G}\hat{\mathbf{L}}_\vee$ we have that f must be equal in $\hat{\mathbf{L}}_\vee$ to an arrow term of the form $f' \circ \hat{k}^i_{A_1,A_2}$. The condition on Gf dictates whether i here is 1 or 2.

If B is $B_1 \wedge B_2$, and for every $(x, y) \in Gf$ we have that $x \in GA_1$, then for $i \in \{1, 2\}$ and $\hat{k}^i_{B_1,B_2} \circ f : A_1 \wedge A_2 \vdash B_i$, for every $(x, z) \in G(\hat{k}^i_{B_1,B_2} \circ f)$ we have $x \in GA_1$. So, by the induction hypothesis,

$$\hat{k}^i_{B_1,B_2} \circ f = f_i \circ \hat{k}^1_{A_1,A_2}.$$

Hence, with the help of the equation $(\wedge\eta)$ of §9.1,

$$f = \langle \hat{k}^1_{B_1,B_2} \circ f, \hat{k}^2_{B_1,B_2} \circ f \rangle = \langle f_1, f_2 \rangle \circ \hat{k}^1_{A_1,A_2}.$$

We reason analogously if for every $(x, y) \in Gf$ we have that $x - GA_1 \in GA_2$. ⊣

Since $\check{\mathbf{L}}_\wedge$ is isomorphic to $\hat{\mathbf{L}}^{op}_\vee$, we have a dual Auxiliary Lemma for $\check{\mathbf{L}}_\wedge$. We can then prove the following result of [44] (Section 4).

LATTICE COHERENCE. *The functor* G *from* $\mathbf{L}$ *to* Rel *is faithful.*

PROOF. Suppose $f, g : A \vdash B$ are arrow terms of $\mathcal{C}(\mathbf{L})$ and $Gf = Gg$. We proceed by induction on the sum of the lengths of A and B to show that $f = g$ in $\mathbf{L}$. If A and B are both letters, then we conclude by Composition Elimination for $\mathbf{GL}$ that an arrow term of $\mathcal{C}(\mathbf{L})$ of the type $A \vdash B$ exists iff A and B are the same letter p, and we must have $f = g = \mathbf{1}_p$ in $\mathbf{L}$. Note that we do not need here the assumption $Gf = Gg$.

If B is $B_1 \wedge B_2$, then for $i \in \{1, 2\}$ we have that $\hat{k}^i_{B_1,B_2} \circ f$ and $\hat{k}^i_{B_1,B_2} \circ g$ are of type $A \vdash B_i$. We also have

$$G(\hat{k}^i_{B_1,B_2} \circ f) = G\hat{k}^i_{B_1,B_2} \circ Gf = G\hat{k}^i_{B_1,B_2} \circ Gg = G(\hat{k}^i_{B_1,B_2} \circ g),$$

whence, by the induction hypothesis, we have $\hat{k}^i_{B_1,B_2} \circ f = \hat{k}^i_{B_1,B_2} \circ g$ in $\mathbf{L}$. Then we infer

$$\langle \hat{k}^1_{B_1,B_2} \circ f, \hat{k}^2_{B_1,B_2} \circ f \rangle = \langle \hat{k}^1_{B_1,B_2} \circ g, \hat{k}^2_{B_1,B_2} \circ g \rangle,$$

from which $f = g$ follows with the help of the equation $(\wedge\eta)$ of §9.1. We proceed analogously if A is $A_1 \vee A_2$.

Suppose now that A is $A_1 \wedge A_2$ or a letter, and B is $B_1 \vee B_2$ or a letter, but A and B are not both letters. Then by Composition Elimination for **GL** we have that f is equal in $\mathbf{L}$ to an arrow term of $\mathcal{C}(\mathbf{L})$ that is either of the form $f' \circ \hat{k}^i_{A_1,A_2}$ or of the form $\check{k}^i_{B_1,B_2} \circ f'$. Suppose $f = f' \circ \hat{k}^1_{A_1,A_2}$. Then for every $(x, y) \in Gf$ we have $x \in GA_1$. (We reason analogously when $f = f' \circ \hat{k}^2_{A_1,A_2}$.)

By the Standard-Form Lemma, $g = g_2 \circ g_1$ in $\mathbf{L}$ for $g_1 : A_1 \wedge A_2 \vdash C$ an arrow term of $\mathcal{C}(\hat{\mathbf{L}}_\vee)$ and $g_2 : C \vdash B$ an arrow term of $\mathcal{C}(\check{\mathbf{L}}_\wedge)$. Since $\hat{k}^i$ does not occur in g_2, for every $z \in GC$ we have a $y \in GB$ such that $(z, y) \in Gg_2$. If for some $(x, z) \in Gg_1$ we had $x \notin GA_1$, then for some $(x, y) \in G(g_2 \circ g_1)$ we would have $x \notin GA_1$, but this is impossible, since $G(g_2 \circ g_1) = Gg = Gf$. So for every $(x, z) \in Gg_1$ we have $x \in GA_1$. Then, by the Auxiliary Lemma, $g = g_1' \circ \hat{k}^1_{A_1,A_2}$ in $\hat{\mathbf{L}}_\vee$, and hence in $\mathbf{L}$ too. Therefore, $g = g_2 \circ g_1' \circ \hat{k}^1_{A_1,A_2}$ in $\mathbf{L}$.

Because of the particular form of $G\hat{k}^1_{A_1,A_2}$, we can infer from $Gf = Gg$ that $Gf' = G(g_2 \circ g_1')$; but, since f' and $g_2 \circ g_1'$ are of type $A_1 \vdash B$, by the induction hypothesis we have $f' = g_2 \circ g_1'$ in $\mathbf{L}$, and hence $f = g$ in $\mathbf{L}$.

If $f = \check{k}^i_{B_1,B_2} \circ f'$, then we reason analogously, applying the dual Auxiliary Lemma for $\check{\mathbf{L}}_\wedge$. ⊣

With the help of Lattice Coherence we can easily verify that the following equation holds in $\mathbf{L}$:

$$(\textit{in-out}) \quad \langle [f, g], [h, j] \rangle = [\langle f, h \rangle, \langle g, j \rangle].$$

If in $\mathcal{C}(\mathbf{L})$ we define $c^k_{A,B,C,D} : (A \wedge B) \vee (C \wedge D) \vdash (A \vee C) \wedge (B \vee D)$ as follows:

$$c^k_{A,B,C,D} =_{df} \langle \hat{k}^1_{A,B} \vee \hat{k}^1_{C,D}, \hat{k}^2_{A,B} \vee \hat{k}^2_{C,D} \rangle,$$

then we can easily check that in **L** we have

$$c^k_{A,B,C,D} = [\check{k}^1_{A,C} \wedge \check{k}^1_{B,D}, \check{k}^2_{A,C} \wedge \check{k}^2_{B,D}],$$

which gives an alternative definition of $c^k_{A,B,C,D}$. One passes from one of these two definitions to the other with the help of the equations $(\wedge)$ of §9.1 and $(\vee)$ of the List of Equations, together with the equation (*in-out*):

$$\begin{aligned} c^k_{A,B,C,D} &= \langle [\check{k}^1_{A,C} \circ \hat{k}^1_{A,B}, \check{k}^2_{A,C} \circ \hat{k}^1_{C,D}], [\check{k}^1_{B,D} \circ \hat{k}^2_{A,B}, \check{k}^2_{B,D} \circ \hat{k}^2_{C,D}]\rangle \\ &= [\langle \check{k}^1_{A,C} \circ \hat{k}^1_{A,B}, \check{k}^1_{B,D} \circ \hat{k}^2_{A,B}\rangle, \langle \check{k}^2_{A,C} \circ \hat{k}^1_{C,D}, \check{k}^2_{B,D} \circ \hat{k}^2_{C,D}\rangle]. \end{aligned}$$

We can also show by Lattice Coherence that in **L** we have

$$\hat{w}_{A\vee B} = c^k_{A,A,B,B} \circ (\hat{w}_A \vee \hat{w}_B),$$
$$\check{w}_{A\wedge B} = (\check{w}_A \wedge \check{w}_B) \circ c^k_{A,B,A,B},$$

$$\hat{c}^m_{A,B,C,D} = \langle \hat{k}^1_{A,B} \wedge \hat{k}^1_{C,D}, \hat{k}^2_{A,B} \wedge \hat{k}^2_{C,D}\rangle,$$
$$\check{c}^m_{D,C,B,A} = [\check{k}^1_{D,C} \vee \check{k}^1_{B,A}, \check{k}^2_{D,C} \vee \check{k}^2_{B,A}]$$

(see §9.1 for $\hat{c}^m$, and the List of Equations for $\check{c}^m$). The last two equations should be compared with the definition of $c^k_{A,B,C,D}$ and its alternative definition. The arrows $c^k_{A,B,C,D}$ will be prominent in Chapter 11 (see also §13.2).

Arrows of the type of $c^k_{A,B,C,D}$ play in [3] an important role in the understanding of 2-fold loop spaces. In that paper, one finds a coherence result in our sense for bimonoidal categories where $\top = \bot$ to which c^k is added with appropriate specific equations. As a matter of fact, this coherence result, for which a long proof is presented, covers a hierarchy of c^k principles involving the binary connectives ξ_i and ξ_j where $1 \leq i < j \leq n$, which are needed for n-fold loop spaces. The role of arrows of the type of $c^k_{A,B,C,D}$ in understanding braiding is considered in [67] (Section 5). In that context, arrows of the type of $c^k_{A,B,C,D}$ may become arrows of the type of $\overset{\xi}{c}{}^m_{A,B,C,D}$ (cf. [3], Remarks 1.5-6).

§9.5. Maximality of lattice categories

In this section we prove that **L** is maximal in the sense of §9.3. (This result is taken over from [44], Section 5.)

Suppose A and B are formulae of $\mathcal{L}_{\wedge,\vee}$ in which only p occurs as a letter. If for some arrow terms $f_1, f_2 \colon A \vdash B$ of $\mathcal{C}(\mathbf{L})$ we have $Gf_1 \neq Gf_2$, then for some $x \in GA$ and some $y \in GB$ we have $(x,y) \in Gf_1$ and $(x,y) \notin Gf_2$, or vice versa. Suppose $(x,y) \in Gf_1$ and $(x,y) \notin Gf_2$.

For every subformula C of A and every formula D let A^C_D be the formula obtained from A by replacing the particular occurrence of the formula C in A by D. It can be shown that for every subformula $A_1 \vee A_2$ of A we have a $\check{k}^j$-term $h \colon A^{A_1 \vee A_2}_{A_j} \vdash A$ of $\mathcal{C}(\mathbf{L})$, whose head is $\check{k}^j_{A_1,A_2}$, such that there is an $x' \in GA^{A_1 \vee A_2}_{A_j}$ for which $(x',x) \in Gh$. Hence, for such an h, we have $(x',y) \in G(f_1 \circ h)$ and $(x',y) \notin G(f_2 \circ h)$.

We compose f_i repeatedly with such $\check{k}^j$-terms until we obtain the arrow terms $f'_i : p \wedge \ldots \wedge p \vdash B$ of $\mathcal{C}(\mathbf{L})$ such that parentheses are somehow associated in $p \wedge \ldots \wedge p$ and for some $z \in G(p \wedge \ldots \wedge p)$ we have $(z,y) \in Gf'_1$ and $(z,y) \notin Gf'_2$. The formula $p \wedge \ldots \wedge p$ may also be only p. We may further compose f'_i with $\hat{b}$-terms and $\hat{c}$-terms in order to obtain the arrow terms f''_i of type $p \wedge A' \vdash B$ or $p \vdash B$ such that A' is of the form $p \wedge \ldots \wedge p$ with parentheses somehow associated, and $(0,y) \in Gf''_1$ but $(0,y) \notin Gf''_2$.

By working dually on B with $\hat{k}^j$-terms, and by composing perhaps further with $\check{b}$-terms and $\check{c}$-terms, we obtain the arrow terms f'''_i of $\mathcal{C}(\mathbf{L})$ of type $p \wedge A' \vdash p \vee B'$, for A' of the form $p \wedge \ldots \wedge p$ and B' of the form $p \vee \ldots \vee p$, or of type $p \wedge A' \vdash p$, or of type $p \vdash p \vee B'$, such that $(0,0) \in Gf'''_1$ and $(0,0) \notin Gf'''_2$. (We cannot obtain that f'''_1 and f'''_2 are of type $p \vdash p$, since, otherwise, by Composition Elimination for **GL**, f'''_2 would not exist.)

There is an arrow term $h^\wedge : p \vdash p \wedge \ldots \wedge p$ of $\mathcal{C}(\mathbf{L})$ defined in terms of $\hat{w}$-terms such that for every $x \in G(p \wedge \ldots \wedge p)$ we have $(0,x) \in Gh^\wedge$. We define analogously with the help of $\check{w}$-terms an arrow term $h^\vee : p \vee \ldots \vee p \vdash p$ of $\mathcal{C}(\mathbf{L})$ such that for every $x \in G(p \vee \ldots \vee p)$ we have $(x,0) \in Gh^\vee$. The arrow terms $h^\wedge$ and $h^\vee$ may be $\mathbf{1}_p \colon p \vdash p$.

If f'''_i is of type $p \wedge A' \vdash p \vee B'$, let $f^\dagger_i : p \wedge p \vdash p \vee p$ be defined by

$$f^\dagger_i =_{df} (\mathbf{1}_p \vee h^\vee) \circ f'''_i \circ (\mathbf{1}_p \vee h^\wedge).$$

By Composition Elimination for **GL**, we have that $Gf^\dagger_i$ must be a singleton. If $(1,0)$ or $(1,1)$ belongs to $Gf^\dagger_2$, then for $f^*_i : p \wedge p \vdash p$ defined as $\check{w}_p \circ f^\dagger_i$ we have $(0,0) \in Gf^*_1$ and $(0,0) \notin Gf^*_2$. If $(0,1)$ or $(1,1)$ belongs to $Gf^\dagger_2$, then for $f_i^* : p \vdash p \vee p$ defined as $f_i^\dagger \circ \hat{w}_p$ we have $(0,0) \in Gf_1^*$ and $(0,0) \notin Gf_2^*$.

If f_i''' is of type $p \wedge A' \vdash p$, then for $f_i^* : p \wedge p \vdash p$ defined as $f_i''' \circ (\mathbf{1}_p \vee h^{\wedge})$ we have $(0,0) \in Gf_1^*$ and $(0,0) \notin Gf_2^*$.

If f_i''' is of type $p \vdash p \vee B'$, then for $f_i^* : p \vdash p \vee p$ defined as $(\mathbf{1}_p \vee h^{\vee}) \circ f_i'''$ we have $(0,0) \in Gf_1^*$ and $(0,0) \notin Gf_2^*$. In all that we have by Composition Elimination for **GL** that Gf_i^* must be a singleton.

In cases where f_i^* is of type $p \wedge p \vdash p$, by Composition Elimination for **GL**, by the conditions on Gf_1^* and Gf_2^*, and by the functoriality of G, we obtain in **L** the equation $f_i^* = \hat{k}^i_{p,p}$. (This follows from Lattice Coherence too.) So in $\mathcal{E}(\mathbf{L})$ extended with $f_1 = f_2$ we can derive $\hat{k}^1_{p,p} = \hat{k}^2_{p,p}$; namely, the equation $(\hat{k}\hat{k})$, mentioned in the proof of Maximality of $\hat{\mathbf{L}}$ in §9.3.

In cases where f_i^* is of type $p \vdash p \vee p$, we conclude analogously that we have in **L** the equation $f_i^* = \check{k}^i_{p,p}$, and so in $\mathcal{E}(\mathbf{L})$ extended with $f_1 = f_2$ we can derive

$$(\check{k}\check{k}) \qquad \check{k}^1_{p,p} = \check{k}^2_{p,p}\,.$$

If either of $(\hat{k}\hat{k})$ and $(\check{k}\check{k})$ holds in a lattice category $\mathcal{A}$, then $\mathcal{A}$ is a preorder. We use for that the equation $(\wedge\beta)$ of §9.1, or its dual with $\vee$ (see the proof of Maximality of $\hat{\mathbf{L}}$ in §9.3).

It remains to remark that if for some arrow terms g_1 and g_2 of $\mathcal{C}(\mathbf{L})$ of the same type we have that $g_1 = g_2$ is not in $\mathcal{E}(\mathbf{L})$, then by Lattice Coherence we have $Gg_1 \neq Gg_2$. If we take the substitution instances g_1' of g_1 and g_2' of g_2 obtained by replacing every letter by a single letter p, then we obtain again $Gg_1' \neq Gg_2'$. If $g_1 = g_2$ holds in a lattice category $\mathcal{A}$, then $g_1' = g_2'$ holds too, and $\mathcal{A}$ is a preorder, as we have shown above. This concludes the proof of maximality for **L**. (In the original presentation of this proof in [44], Section 5, there are some slight inaccuracies in the definition of f_i^*.)

§9.6. Coherence of dicartesian and sesquicartesian categories

Let $\check{\mathbf{L}}_\bot$ be the natural logical category in $\mathcal{L}_{\vee,\bot}$ isomorphic to the category $\hat{\mathbf{L}}_\top^{op}$. We just replace $\wedge$ and $\top$ by $\vee$ and $\bot$ respectively, so that the primitive arrow terms of $\mathcal{C}(\check{\mathbf{L}}_\bot)$ are included in $\mathbf{1}$, $\check{b}$, $\check{c}$, $\check{w}$-$\check{k}$ and $\check{\delta}$-$\check{\sigma}$, while the equations of $\mathcal{E}(\check{\mathbf{L}}_\bot)$ are duals of those of $\mathcal{E}(\hat{\mathbf{L}}_\top)$ (see the List of Equations and the List of Categories; cf. the beginning of §9.4). We have in $\mathcal{C}(\check{\mathbf{L}}_\bot)$ the definition

$$\check{\kappa}_A =_{df} \check{\delta}_A^{\rightarrow} \circ \hat{k}^2_{A,\bot},$$

and in $\check{\mathbf{L}}_\bot$ the equations $(\check{\kappa}\ nat)$ and

$$(\check{\kappa}\mathbf{1}) \quad \check{\kappa}_\bot = \mathbf{1}_\bot.$$

The equations $(\check{\kappa}\ nat)$ and $(\check{\kappa}\mathbf{1})$ amount to

$$(\check{\kappa}) \quad \check{\kappa}_A = f, \quad \text{for } f\colon \bot \vdash A,$$

which says that $\bot$ is an initial object in $\check{\mathbf{L}}_\bot$ (see §2.2 for the notion of initial object).

Natural $\check{\mathbf{L}}_\bot$-categories would usually be called categories with finite coproducts, including the empty coproduct. Another possible name would be *cocartesian* categories.

To obtain the natural logical category $\mathbf{L}_{\top,\bot}$, we have that the logical system $\mathcal{C}(\mathbf{L}_{\top,\bot})$ is in $\mathcal{L}_{\wedge,\vee,\top,\bot}$, with the transformations α included in $\mathbf{1}$, b, c, w-k and δ-σ. The specific equations of $\mathcal{E}(\mathbf{L}_{\top,\bot})$ are obtained by taking the union of those of $\mathcal{E}(\mathbf{L})$, $\mathcal{E}(\hat{\mathbf{L}}_\top)$ and $\mathcal{E}(\check{\mathbf{L}}_\bot)$ plus the equations $(\hat{c}\ \bot)$ and $(\check{c}\ \top)$ of §6.4.

We could replace the last two equations in this definition by their instances

$$(\hat{\bot}) \quad \hat{c}_{\bot,\bot} = \mathbf{1}_{\bot\wedge\bot},$$
$$(\check{\top}) \quad \check{c}_{\top,\top} = \mathbf{1}_{\top\vee\top}.$$

Another possibility is to have instead the following two equations:

$$(\hat{k}\ \bot) \quad \hat{k}^1_{\bot,\bot} = \hat{k}^2_{\bot,\bot},$$
$$(\check{k}\ \top) \quad \check{k}^1_{\top,\top} = \check{k}^2_{\top,\top}.$$

It is easy to see that from the last two equations we obtain that the pairs

$$\hat{k}^1_{\bot,\bot} = \hat{k}^2_{\bot,\bot}\colon \bot\wedge\bot \vdash \bot \quad \text{and} \quad \check{\kappa}_{\bot\wedge\bot} = \hat{w}_\bot\colon \bot \vdash \bot\wedge\bot,$$
$$\check{k}^1_{\top,\top} = \check{k}^2_{\top,\top}\colon \top \vdash \top\vee\top \quad \text{and} \quad \hat{\kappa}_{\top\vee\top} = \check{w}_\top\colon \top\vee\top \vdash \top$$

are inverses of each other. This shows that every letterless formula of $\mathcal{L}_{\wedge,\vee,\top,\bot}$ is isomorphic in $\mathbf{L}_{\top,\bot}$ either to $\top$ or to $\bot$. This is why above we could replace $(\hat{c}\ \bot)$ and $(\check{c}\ \top)$ by their instances $(\hat{\bot})$ and $(\check{\top})$.

We call natural $\mathbf{L}_{\top,\bot}$-categories *dicartesian* categories. The objects of a dicartesian category that is a partial order make a lattice with top and bottom. By omitting the equations $(\hat{c}\ \bot)$ and $(\check{c}\ \top)$ in the definition of $\mathbf{L}_{\top,\bot}$ we would obtain the natural logical category $\mathbf{L}'_{\top,\bot}$, and natural $\mathbf{L}'_{\top,\bot}$-categories are usually called *bicartesian* categories (cf. [85], Section I.8). Dicartesian categories were considered under the name *coherent bicartesian* categories in [44]. A study of equality of arrows in bicartesian categories may be found in [21].

Suppose that in the definition of $\mathbf{L}_{\top,\bot}$ we omit one of $\top$ and $\bot$ from $\mathcal{L}_{\wedge,\vee,\top,\bot}$, so that we have $\mathcal{L}_{\wedge,\vee,\bot}$ or $\mathcal{L}_{\wedge,\vee,\top}$. This means that in $\mathcal{C}(\mathbf{L}_{\top,\bot})$ and $\mathcal{E}(\mathbf{L}_{\top,\bot})$ we omit all the arrow terms and equations involving the omitted nullary connective. When we omit $\top$, we obtain the natural logical category $\mathbf{L}_{\bot}$, and when we omit $\bot$, we obtain the natural logical category $\mathbf{L}_{\top}$. It is clear that $\mathbf{L}_{\bot}$ is isomorphic to $\mathbf{L}_{\top}^{op}$. In [43] natural $\mathbf{L}_{\bot}$-categories were called *coherent sesquicartesian* categories. We call them here just *sequicartesian* categories.

The category *Set*, whose objects are sets and whose arrows are functions, with cartesian product $\times$ as $\wedge$, disjoint union $+$ as $\vee$, a singleton set $\{*\}$ as $\top$ and the empty set $\emptyset$ as $\bot$, is a bicartesian category, but not a dicartesian category. It is, however, a sesquicartesian category in the $\mathbf{L}_{\bot}$ sense, but not in the $\mathbf{L}_{\top}$ sense. This is because in *Set* we have that $\emptyset \times \emptyset$ is equal to $\emptyset$, but $\{*\} + \{*\}$ is not isomorphic to $\{*\}$.

The following results were proved in [43].

DICARTESIAN COHERENCE. *The functor G from $\mathbf{L}_{\top,\bot}$ to Rel is faithful.*

SESQUICARTESIAN COHERENCE. *The functor G from $\mathbf{L}_{\bot}$ to Rel is faithful.*

The proofs of these results are obtained by enlarging the proof of Lattice Coherence of §9.4, and we will give here just a summary of them. (Detailed proofs may be found in [43].)

The syntactical category $\mathbf{GL}_{\top,\bot}$ is obtained as $\mathbf{GL}$ save that we have in addition the primitive arrow terms $\hat{\kappa}_A\colon A \vdash \top$ and $\check{\kappa}_A\colon \bot \vdash A$, the equations $(\hat{\kappa})$ and $(\check{\kappa})$, and also the equations

$$\begin{array}{ll}(\hat{K}\bot) & \hat{K}^1_{\bot,\bot}\,\mathbf{1}_\bot = \hat{K}^2_{\bot,\bot}\,\mathbf{1}_\bot,\\ (\check{K}\top) & \check{K}^1_{\top,\top}\,\mathbf{1}_\top = \check{K}^2_{\top,\top}\,\mathbf{1}_\top.\end{array}$$

We can prove Composition Elimination for $\mathbf{GL}_{\top,\bot}$ by enlarging the proofs in §9.1 and §9.2. Note that we do not need the equations $(\hat{K}\bot)$ and $(\check{K}\top)$ for this proof, so that we have also Composition Elimination for $\mathbf{GL}'_{\top,\bot}$ based on $\mathbf{L}'_{\top,\bot}$.

We can then define an analogue of the standard form for arrow terms of $\mathcal{C}(\mathbf{L}_{\top,\bot})$ and prove an analogue of the Standard-Form Lemma of §9.4 for $\mathbf{L}_{\top,\bot}$. The analogue of $\hat{\mathbf{L}}_\vee$ involves also $\hat{\kappa}$ and the equations $(\hat{\kappa})$ and $(\hat{k}\bot)$, while the analogue of $\check{\mathbf{L}}_\wedge$ involves also $\check{\kappa}$ and the equations $(\check{\kappa})$ and $(\check{k}\top)$. Composition Elimination is provable for syntactical categories isomorphic to these analogues.

We can also prove Composition Elimination and an analogue of the Standard-Form Lemma for $\mathbf{L}_\bot$. Next we need the following lemmata for $\mathbf{L}_{\top,\bot}$ and $\mathbf{L}_\bot$.

LEMMA 1. *If in $f,g\colon A \vdash B$ either A or B is isomorphic to $\top$ or $\bot$, then $f = g$.*

PROOF. If A is isomorphic to $\bot$ or B is isomorphic to $\top$, then the matter is trivial. Suppose $i\colon B \vdash \bot$ is an isomorphism. Then from

$$\hat{k}^1_{\bot,\bot} \circ \langle i \circ f, i \circ g\rangle = \hat{k}^2_{\bot,\bot} \circ \langle i \circ f, i \circ g\rangle$$

we obtain $i \circ f = i \circ g$, which yields $f = g$. We proceed analogously if A is isomorphic to $\top$. ⊣

LEMMA 2. *If for $f,g\colon A \vdash B$ we have $Gf = Gg = \emptyset$, then $f = g$.*

PROOF. This proof depends on the analogue of the Standard-Form Lemma. We write down f in the standard form $f_2 \circ f_1$ for $f_1 : A \vdash C$ and g in the standard form $g_2 \circ g_1$ for $g_1 : A \vdash D$. Then we conclude from the assumption $Gf = Gg = \emptyset$ that C and D must be letterless formulae. If they are both isomorphic to $\top$ or $\bot$, then we have an isomorphism $i\colon C \vdash D$, and $f = f_2 \circ i^{-1} \circ i \circ f_1$. By Lemma 1, we have $i \circ f_1 = g_1$ and $f_2 \circ i^{-1} = g_2$, from which $f = g$ follows. If $i\colon C \vdash \bot$ and $j\colon \top \vdash D$ are isomorphisms, then we have

$$\begin{aligned} f_2 \circ f_1 &= g_2 \circ j \circ \hat{\kappa}_\bot \circ i \circ f_1, \quad \text{by Lemma 1,} \\ &= g_2 \circ g_1, \quad \text{by Lemma 1,} \end{aligned}$$

and so $f = g$. (Note that $\hat{\kappa}_\bot = \check{\kappa}_\top$.) ⊣

To prove now Dicartesian Coherence and Sesquicartesian Coherence we have Lemma 2 for the case when $Gf = Gg = \emptyset$, and when $Gf = Gg \neq \emptyset$, we imitate the proof of Lattice Coherence in §9.4. The Auxiliary Lemma of §9.4 is established for $\hat{\mathbf{L}}_\vee$ where $\top$ and $\bot$, or just $\bot$, are added, with the assumption that $Gf \neq \emptyset$. From Sesquicartesian Coherence we infer coherence for $\mathbf{L}_\top$, which is isomorphic to $\mathbf{L}_\bot^{op}$.

Note that if $\mathcal{K}$ is one of the categories $\hat{\mathbf{A}}$, $\hat{\mathbf{A}}_\top$, $\hat{\mathbf{S}}$ and $\hat{\mathbf{S}}_\top$, then $\mathcal{K}$ is isomorphic to $\mathcal{K}^{op}$, while if $\mathcal{K}$ is one of the categories $\hat{\mathbf{L}}$ and $\hat{\mathbf{L}}_\top$, then $\mathcal{K}$ is not isomorphic to $\mathcal{K}^{op}$. The categories $\hat{\mathbf{A}}$, $\hat{\mathbf{A}}_\top$, $\hat{\mathbf{S}}$ and $\hat{\mathbf{S}}_\top$ besides being isomorphic respectively to $\check{\mathbf{A}}^{op}$, $\check{\mathbf{A}}_\bot^{op}$, $\check{\mathbf{S}}^{op}$ and $\check{\mathbf{S}}_\bot^{op}$, are isomorphic respectively to $\check{\mathbf{A}}$, $\check{\mathbf{A}}_\bot$, $\check{\mathbf{S}}$ and $\check{\mathbf{S}}_\bot$ too, while $\hat{\mathbf{L}}$ and $\hat{\mathbf{L}}_\top$ are isomorphic respectively only to $\check{\mathbf{L}}^{op}$ and $\check{\mathbf{L}}_\bot^{op}$, and not to $\check{\mathbf{L}}$ and $\check{\mathbf{L}}_\bot$. So the symmetry between $\wedge$ and $\vee$ is deeper in $\mathbf{A}$, $\mathbf{A}_{\top,\bot}$, $\mathbf{S}$ and $\mathbf{S}_{\top,\bot}$ than in $\mathbf{L}$ and $\mathbf{L}_{\top,\bot}$.

§9.7. Relative maximality of dicartesian categories

The category $\mathbf{L}_{\top,\bot}$ is not maximal in the sense in which $\hat{\mathbf{L}}$, $\hat{\mathbf{L}}_\top$ and $\mathbf{L}$ are maximal (see §9.3 and §9.5). This is shown with the following counterexample.

Let Set_* be the category whose objects are sets with a distinguished element $*$, and whose arrows are $*$-preserving functions f between these sets; namely, $f(*) = *$. This category is isomorphic to the category of sets with partial functions. The following definitions serve to show that Set_* is a category of the $\mathcal{C}(\mathbf{L}_{\top,\bot})$ kind:

$$\mathrm{I} = \{*\}, \qquad a' = \{(x,*) \mid x \in a - \mathrm{I}\}, \qquad b'' = \{(*,y) \mid y \in b - \mathrm{I}\},$$

$$
\begin{aligned}
a \otimes b &= ((a - \mathrm{I}) \times (b - \mathrm{I})) \cup \mathrm{I}, \\
a \boxtimes b &= (a \otimes b) \cup a' \cup b'', \\
a \boxplus b &= a' \cup b'' \cup \mathrm{I}.
\end{aligned}
$$

Note that $a \boxtimes b$ is isomorphic in *Set* to the cartesian product $a \times b$; the element $*$ of $a \boxtimes b$ corresponds to the element $(*, *)$ of $a \times b$.

The functions $\hat{k}^i_{a_1,a_2}\colon a_1 \boxtimes a_2 \to a_i$, for $i \in \{1, 2\}$, are defined by

$$\hat{k}^i_{a_1,a_2}(x_1, x_2) = x_i, \qquad \hat{k}^i_{a_1,a_2}(*) = *;$$

for $f_i\colon c \to a_i$, the function $\langle f_1, f_2\rangle\colon c \to a_1 \boxtimes a_2$ is defined by

$$\langle f_1, f_2\rangle(z) = \begin{cases} (f_1(z), f_2(z)) & \text{if } f_1(z) \neq * \text{ or } f_2(z) \neq * \\ * & \text{if } f_1(z) = f_2(z) = *; \end{cases}$$

and the function $\hat{\kappa}_a\colon a \to \mathrm{I}$ is defined by $\hat{\kappa}_a(x) = *$. Having in mind the isomorphism between $a \boxtimes b$ and $a \times b$ mentioned above, the functions $\hat{k}^i_{a_1,a_2}\colon a_1 \boxtimes a_2 \to a_i$ correspond to the projection functions, while $\langle \_, \_\rangle$ corresponds to the usual pairing operation on functions.

The functions $\check{k}^i_{a_1,a_2}\colon a_i \to a_1 \boxplus a_2$ are defined by

$$
\begin{aligned}
&\check{k}^1_{a_1,a_2}(x) = (x, *), \quad \check{k}^2_{a_1,a_2}(x) = (*, x), \quad \text{for } x \neq *, \\
&\check{k}^i_{a_1,a_2}(*) = *;
\end{aligned}
$$

for $g_i\colon a_i \to c$, the function $[g_1, g_2]\colon a_1 \boxplus a_2 \to c$ is defined by

$$
\begin{aligned}
&[g_1, g_2](x_1, x_2) = g_i(x_i), \ \text{for } x_i \neq *, \\
&[g_1, g_2](*) = *;
\end{aligned}
$$

finally, the function $\check{\kappa}_a\colon \mathrm{I} \to a$ is defined by $\check{\kappa}_a(*) = *$.

If we take that $\wedge$ is $\boxtimes$ and $\vee$ is $\boxplus$, then it can be checked in a straightforward manner that Set_* and Set_* without I are lattice categories, and if in Set_* we take further that both $\top$ and $\bot$ are I, then Set_* is a dicartesian category.

Consider now the category $Set_*^{\emptyset}$, which is obtained by adding to Set_* the empty set $\emptyset$ as a new object, and the empty functions $\emptyset_a : \emptyset \to a$ as new arrows. The identity arrow $\mathbf{1}_\emptyset$ is $\emptyset_\emptyset$. For $Set_*^{\emptyset}$, we enlarge the definitions above by

$$\begin{aligned}
\emptyset \boxtimes a &= a \boxtimes \emptyset = \emptyset,\\
\emptyset \boxplus a &= a \boxplus \emptyset = a,\\
\hat{k}^i_{a_1,a_2} &= \emptyset_{a_i}, \text{ for } a_1 = \emptyset \text{ or } a_2 = \emptyset,\\
\langle \emptyset_{a_1}, \emptyset_{a_2} \rangle &= \emptyset_{a_1 \boxtimes a_2},\\
\hat{\kappa}_{\emptyset} &= \emptyset_{\mathrm{I}},\\
\check{k}^i_{a_1,a_2} &= \emptyset_{a_1 \boxplus a_2}, \text{ for } a_i = \emptyset,\\
[f_1, \emptyset_c] &= f_1, \qquad [\emptyset_c, f_2] = f_2,
\end{aligned}$$

and define now the function $\check{\kappa}_a \colon \emptyset \to a$ by $\check{\kappa}_a = \emptyset_a$. Then it can be checked that $Set^{\emptyset}_*$ where $\wedge$ is $\boxtimes$ and $\vee$ is $\boxplus$ as before, while $\top$ is I and $\bot$ is $\emptyset$, is a dicartesian category too.

In $\mathbf{L}_{\top,\bot}$ the equation $\hat{k}^1_{p,\bot} = \check{\kappa}_p \circ \hat{k}^2_{p,\bot}$ does not hold, because $G\hat{k}^1_{p,\bot} \neq \emptyset$ and $G(\check{\kappa}_p \circ \hat{k}^2_{p,\bot}) = \emptyset$, but in $Set^{\emptyset}_*$ this equation holds, because both sides are equal to $\emptyset_{\emptyset}$. Since $Set^{\emptyset}_*$ is not a preorder, we can conclude that $\mathbf{L}_{\top,\bot}$ is not maximal.

Although this maximality fails, the category $\mathbf{L}_{\top,\bot}$ may be shown maximal in a relative sense, which we mentioned at the end of §9.3. This relative maximality result, which we are going to demonstrate now, says that every dicartesian category that satisfies an equation between arrow terms of $\mathcal{C}(\mathbf{L}_{\top,\bot})$ that is not in $\mathcal{E}(\mathbf{L}_{\top,\bot})$ satisfies also some particular equations. These equations do not give preorder in general, but a kind of "contextual" preorder. Moreover, when $\mathcal{E}(\mathbf{L}_{\top,\bot})$ is extended with some of these equations we obtain a maximal natural logical category.

If for some arrow terms $f_1, f_2 \colon A \vdash B$ of $\mathcal{C}(\mathbf{L}_{\top,\bot})$ we have $Gf_1 \neq Gf_2$, then for some $x \in GA$ and some $y \in GB$ we have $(x,y) \in Gf_1$ and $(x,y) \notin Gf_2$, or vice versa. Suppose $(x,y) \in Gf_1$ and $(x,y) \notin Gf_2$. Suppose the $(x+1)$-th occurrence of letter in A, counting from the left, is an occurrence of p. So the $(y+1)$-th occurrence of letter in B must be an occurrence of p.

Let A' be the formula obtained from the formula A by replacing the $(x+1)$-th occurrence of letter in A by $p \wedge \bot$, and every other letter or $\top$ by $\bot$. Dually, let B' be the formula obtained from B by replacing the $(y+1)$-th occurrence of letter in B by $p \vee \top$, and every other letter or $\bot$ by $\top$. Then it can be shown that there is an arrow term $h^A : A' \vdash A$ of

$\mathcal{C}(\mathbf{L}_{\top,\bot})$ such that $Gh^A = \{(0,x)\}$, and an arrow term $h^B : B \vdash B'$ of $\mathcal{C}(\mathbf{L}_{\top,\bot})$ such that $Gh^B = \{(y,0)\}$. We build h^A with $\hat{k}^1_{p,\bot}\colon p \wedge \bot \vdash p$ and instances of $\check{\kappa}_C\colon \bot \vdash C$, with the help of the operations $\wedge$ and $\vee$ on arrow terms. Analogously, h^B is built with $\check{k}^1_{p,\top}\colon p \vdash p \vee \top$ and instances of $\hat{\kappa}_C\colon C \vdash \top$. It can also be shown that there are arrow terms $j^A\colon p \wedge \bot \vdash A'$ and $j^B : B' \vdash p \vee \top$ of $\mathcal{C}(\mathbf{L}_{\top,\bot})$ such that $Gj^A = Gj^B = \{(0,0)\}$. These arrow terms stand for isomorphisms of $\mathbf{L}_{\top,\bot}$.

Then it is clear that for f'_i being

$$j^B \circ h^B \circ f_i \circ h^A \circ j^A\colon p \wedge \bot \vdash p \vee \top,$$

with $i \in \{1,2\}$, we have $Gf'_1 = \{(0,0)\}$, while $Gf'_2 = \emptyset$. Hence, by Composition Elimination for $\mathbf{GL}_{\top,\bot}$ and by the functoriality of G, we obtain in $\mathbf{L}_{\top,\bot}$ the equations

$$\begin{aligned} f'_1 &= \check{k}^1_{p,\top} \circ \hat{k}^1_{p,\bot}, \\ f'_2 &= \check{\kappa}_{p\vee\top} \circ \hat{k}^2_{p,\bot} = \check{k}^2_{p,\top} \circ \hat{\kappa}_{p\wedge\bot}. \end{aligned}$$

(This follows from Dicartesian Coherence too.) If we write $\mathbf{0}_{\bot,\top}$ for $\hat{\kappa}_\bot$, which is equal to $\check{\kappa}_\top$ in $\mathbf{L}_{\top,\bot}$, then in $\mathbf{L}_{\top,\bot}$ we have

$$f'_2 = \check{k}^2_{p,\top} \circ \mathbf{0}_{\bot,\top} \circ \hat{k}^2_{p,\bot}.$$

So in $\mathcal{E}(\mathbf{L}_{\top,\bot})$ extended with $f_1 = f_2$ we can derive

$$(\hat{k}\check{k}) \quad \check{k}^1_{p,\top} \circ \hat{k}^1_{p,\bot} = \check{k}^2_{p,\top} \circ \mathbf{0}_{\bot,\top} \circ \hat{k}^2_{p,\bot}.$$

The equation

$$(\hat{k}\check{\kappa}) \quad \hat{k}^1_{p,\bot} = \check{\kappa}_p \circ \hat{k}^2_{p,\bot},$$

which holds in $Set^{\emptyset}_*$, and which we have used above for showing the nonmaximality of $\mathbf{L}_{\top,\bot}$, clearly yields $(\hat{k}\check{k})$, which hence holds in $Set^{\emptyset}_*$, and which hence we could have also used for showing this nonmaximality.

If we refine the procedure above by building A' and B' out of A and B more carefully, then in some cases we could derive $(\hat{k}\check{\kappa})$ or its dual

$$(\check{k}\hat{\kappa}) \quad \check{k}^1_{p,\top} = \check{k}^2_{p,\top} \circ \hat{\kappa}_p$$

instead of $(\hat{k}\check{k})$. We do not replace the $x+1$-th p by $p \wedge \bot$ in building A', and we can proceed more selectively with other occurrences of letters and $\top$ in A in order to obtain an A' isomorphic to p if possible. We can proceed analogously when we build B' out of B to obtain a B' isomorphic to p if possible.

Note that we have the following:

$$\begin{aligned} \check{\kappa}_{p\wedge\bot} \circ \hat{k}^2_{p,\bot} &= \langle \check{\kappa}_p, \mathbf{1}_\bot \rangle \circ \hat{k}^2_{p,\bot} \\ &= \langle \hat{k}^1_{p,\bot}, \hat{k}^2_{p,\bot} \rangle, \text{ with } (\hat{k}\check{\kappa}), \\ &= \mathbf{1}_{p\wedge\bot}. \end{aligned}$$

In the other direction, it is clear that the equation derived yields $(\hat{k}\check{\kappa})$. So with $(\hat{k}\check{\kappa})$ we have that $C \wedge \bot$ and $\bot$ are isomorphic, and, analogously, with $(\check{k}\hat{\kappa})$ we have that $C \vee \top$ and $\top$ are isomorphic. It can be shown that the natural logical category defined as $\mathbf{L}_{\top,\bot}$ save that we assume in addition both $(\hat{k}\check{\kappa})$ and $(\check{k}\hat{\kappa})$ is maximal. (This is achieved by eliminating letterless subformulae from C and D in $g_1, g_2 : C \vdash D$ such that $Gg_1 \neq Gg_2$, and falling upon the argument used for the maximality of $\mathbf{L}$ in §9.5.)

If $f : a \vdash b$ is any arrow of a dicartesian category $\mathcal{A}$ and $(\hat{k}\check{k})$ holds in $\mathcal{A}$, then we have in $\mathcal{A}$

$$\begin{aligned} \check{k}^1_{b,\top} \circ f \circ \hat{k}^1_{a,\bot} &= \check{k}^1_{b,\top} \circ \hat{k}^1_{b,\bot} \circ (f \wedge \mathbf{1}_\bot) \\ &= \check{k}^2_{b,\top} \circ \mathbf{0}_{\bot,\top} \circ \hat{k}^2_{a,\bot}, \end{aligned}$$

and hence for $f, g : a \vdash b$ we have in $\mathcal{A}$

$$(\hat{k}\check{k}\ fg) \qquad \check{k}^1_{b,\top} \circ f \circ \hat{k}^1_{a,\bot} = \check{k}^1_{b,\top} \circ g \circ \hat{k}^1_{a,\bot}.$$

So, although $\mathbf{L}_{\top,\bot}$ is not maximal, it is maximal in the relative sense that every dicartesian category that satisfies an equation between arrow terms of $\mathcal{C}(\mathbf{L}_{\top,\bot})$ that is not in $\mathcal{E}(\mathbf{L}_{\top,\bot})$ satisfies also $(\hat{k}\check{k})$ and $(\hat{k}\check{k}\ fg)$. Some of these dicartesian categories may satisfy more than just $(\hat{k}\check{k})$ and $(\hat{k}\check{k}\ fg)$. They may satisfy $(\hat{k}\check{\kappa})$ or $(\check{k}\hat{\kappa})$, which yields

$$f \circ \hat{k}^1_{a,\bot} = g \circ \hat{k}^1_{a,\bot} \qquad \text{or} \qquad \check{k}^1_{b,\top} \circ f = \check{k}^1_{b,\top} \circ g,$$

and some may be preorders.

CHAPTER 10

MIX-LATTICE CATEGORIES

In this chapter we consider categories with finite products and coproducts in which there is an operation of union on arrows with the same source and target, so that hom-sets are semilattices with this operation. This is what the mix principle of Chapter 8 amounts to in the present context. An example of such a category is the category of semilattices with semilattice homomorphisms.

We prove restricted coherence results for these categories, the restriction being on the sources and targets of arrows, which must be in disjunctive or conjunctive normal form. These coherence results are just an auxiliary for the proofs of coherence in the next chapter. The technique of proof here is again based on composition elimination.

§10.1. Mix-lattice categories and an example

To obtain the natural logical category **ML**, we have that the logical system $\mathcal{C}(\mathbf{ML})$ is in $\mathcal{L}_{\wedge,\vee}$, with the transformations α included in $\mathbf{1}$, b, c, w-k and m. The specific equations of $\mathcal{E}(\mathbf{ML})$ are obtained by taking the union of those of $\mathcal{E}(\mathbf{MS})$ and $\mathcal{E}(\mathbf{L})$ plus

$$(wm) \qquad \check{w}_A \circ m_{A,A} \circ \hat{w}_A = \mathbf{1}_A.$$

We call natural **ML**-categories *mix-lattice* categories.

Let $\mathcal{C}(\mathbf{GML})$ be the syntactical system with the formulae of $\mathcal{L}_{\wedge,\vee}$ as objects, with the primitive arrow terms being only identity arrow terms, and with the operations on arrow terms being those of $\mathcal{C}(\mathbf{GL})$ plus the following one:

$$\frac{f\colon A \vdash B \qquad g\colon A \vdash B}{f \cup g\colon A \vdash B}$$

To obtain the equations of $\mathcal{E}(\mathbf{GML})$, we assume the equations of $\mathcal{E}(\mathbf{GL})$ and the following equations:

$$(\cup\circ)\quad (f\cup g)\circ h=(f\circ h)\cup(g\circ h),\quad h\circ(f\cup g)=(h\circ f)\cup(h\circ g),$$

$$\begin{array}{ll} (\cup\ assoc) & f\cup(g\cup h)=(f\cup g)\cup h,\\ (\cup\ com) & f\cup g=g\cup f,\\ (\cup\ idemp) & f\cup f=f. \end{array}$$

The last equation, $(\cup\ idemp)$, can be replaced by $\mathbf{1}_A\cup\mathbf{1}_A=\mathbf{1}_A$. The syntactical category **GML** is $\mathcal{C}(\mathbf{GML})/\mathcal{E}(\mathbf{GML})$.

It is straightforward to show (by relying on Lattice Coherence of §9.4) that with the following definition in $\mathcal{C}(\mathbf{ML})$:

$$f\cup g=_{df}\check{w}_B\circ(f\diamond g)\circ\hat{w}_A$$

($f\diamond g$ is $(f\vee g)\circ m_{A,A}$, as in §8.1), and the following definition in $\mathcal{C}(\mathbf{GML})$:

$$m_{A,B}=_{df}\check{K}^1_B\hat{K}^1_B\mathbf{1}_A\cup\check{K}^2_A\hat{K}^2_A\mathbf{1}_B,$$

together with the definitions involved in showing the synonymity of $\mathcal{C}(\mathbf{L})$ and $\mathcal{C}(\mathbf{GL})$, we have that **ML** and **GML** are isomorphic categories, and that, hence, $\mathcal{C}(\mathbf{ML})$ and $\mathcal{C}(\mathbf{GML})$ are synonymous syntactical systems (see the end of §2.4 for this notion of synonymity).

It can be checked that for the functor G from **ML** to *Rel* we have

$$G(f\cup g)=Gf\cup Gg,$$

where $\cup$ on the left-hand side is defined in $\mathcal{C}(\mathbf{ML})$ as above, and $\cup$ on the right-hand side is union of relations with the same domain and codomain (remember that $Gm_{A,B}$ is an identity relation, i.e. identity function; see §2.9).

According to the equations $(\cup\ assoc)$, $(\cup\ com)$ and $(\cup\ idemp)$, the hom-sets in any mix-lattice category are semilattices with the operation $\cup$. In **ML** the following equations hold:

$$(\cup\,\xi)\quad (f_1\cup f_2)\,\xi\,(g_1\cup g_2)=(f_1\,\xi\,g_1)\cup(f_2\,\xi\,g_2),$$

for $\xi \in \{\wedge, \vee\}$. The derivation of these equations is based on the following equations of **ML**:

$$c^k_{A,C,B,D} \circ m_{A\wedge C,B\wedge D} \circ \hat{c}^m_{A,B,C,D} = m_{A,B} \wedge m_{C,D},$$
$$\check{c}^m_{A,B,C,D} \circ m_{A\vee C,B\vee D} \circ c^k_{A,B,C,D} = m_{A,B} \vee m_{C,D},$$

for whose checking we can use Semilattice Coherence of §9.1 (see §9.1 and the List of Equations for the definitions of $\hat{c}^m$ and $\check{c}^m$, and §9.4 for the definition of c^k).

As an example of a mix-lattice category, we have the category Set^{sl}_*, whose objects are semilattices with unit $\langle a, \cdot, *\rangle$ such that $x \cdot y = *$ iff $x = *$ and $y = *$, and whose arrows are homomorphisms f with trivial kernels; that is, $f(x) = *$ iff $x = *$. The unit $*$ may be conceived either as top or as bottom. This category is a subcategory of the category Set_* of §9.7.

We define $\langle a_1, \cdot, *\rangle \wedge \langle a_2, \cdot, *\rangle$ as the semilattice with unit $\langle a_1 \otimes a_2, \cdot, *\rangle$, where $\otimes$ is as in §9.7, and we have

$$(x_1, x_2) \cdot (y_1, y_2) = (x_1 \cdot y_1, x_2 \cdot y_2),$$
$$(x_1, x_2) \cdot * = * \cdot (x_1, x_2) = (x_1, x_2),$$
$$* \cdot * = *.$$

We define $\langle a_1, \cdot, *\rangle \vee \langle a_2, \cdot, *\rangle$ as the semilattice with unit $\langle a_1 \boxtimes a_2, \cdot, *\rangle$, where $\boxtimes$, which corresponds to cartesian product, is defined as in §9.7, and we have for $\cdot$ and $*$ the same clauses as above.

The functions $\hat{k}^i_{a_1,a_2} \colon a_1 \otimes a_2 \to a_i$, for $i \in \{1, 2\}$, are defined by

$$\hat{k}^i_{a_1,a_2}(x_1, x_2) = x_i, \qquad \hat{k}^i_{a_1,a_2}(*) = *;$$

for $f_i \colon c \to a_i$, the function $\langle f_1, f_2\rangle \colon c \to a_1 \otimes a_2$ is defined by

$$\langle f_1, f_2\rangle(z) = \begin{cases} (f_1(z), f_2(z)) & \text{if } z \neq * \\ * & \text{if } z = *. \end{cases}$$

The functions $\check{k}^i_{a_1,a_2} \colon a_i \to a_1 \boxtimes a_2$ are defined by

$$\check{k}^1_{a_1,a_2}(x) = (x, *), \quad \check{k}^2_{a_1,a_2}(x) = (*, x), \quad \text{for } x \neq *,$$
$$\check{k}^i_{a_1,a_2}(*) = *;$$

for $g_i \colon a_i \to c$, the function $[g_1, g_2] \colon a_1 \boxtimes a_2 \to c$ is defined by

$$[g_1, g_2](x_1, x_2) = g_1(x_1) \cdot g_2(x_2),$$
$$[g_1, g_2](*) = *.$$

(The clauses in the definitions of $\hat{k}{}^i_{a_1,a_2}$ and $\overset{\vee}{k}{}^i_{a_1,a_2}$ are taken over from §9.7, and we could have also taken over from there the clause for $\langle f_1, f_2\rangle$, but the operations in the domains and codomains are changed, and the functions defined are not the same; the clause for $[g_1, g_2]$ is new.)

We define the function $m_{a,b}\colon a \otimes b \to a \boxtimes b$ by

$$m_{a,b}(x_1, x_2) = (x_1, x_2), \qquad m_{a,b}(*) = *,$$

or for the functions $f, g\colon a \to b$ we define the function $f \cup g\colon a \to b$ by

$$(f \cup g)(x) = f(x) \cdot g(x).$$

It can be checked in a straightforward manner that with these definitions, and with $\wedge$ being $\otimes$ and $\vee$ being $\boxtimes$, the category Set_*^{sl} is a mix-lattice category (it is easier to rely on $\cup$ than on m in this context). A category isomorphic to Set_*^{sl} is the category *Semilat*, whose objects are semilattices and whose arrows are semilattice homomorphisms. We just reject $*$ from the domains of the objects of Set_*^{sl}, and the pairs $(*, *)$ from the sets of ordered pairs of the arrows of Set_*^{sl}. The mix-lattice structure of *Semilat* is then inherited from Set_*^{sl}. The domain of $\langle a_1, \cdot, *\rangle \wedge \langle a_2, \cdot, *\rangle$ is now $a_1 \times a_2$ instead of $a_1 \otimes a_2$, while the domain of $\langle a_1, \cdot, *\rangle \vee \langle a_2, \cdot, *\rangle$ is $(a_1 \times a_2) + a_1 + a_2$ instead of $a_1 \boxtimes a_2$, which corresponded to $a_1 \times a_2$ (here $+$ is disjoint union). It is, however, more practical to introduce the mix-lattice structure in Set_*^{sl}, with $*$ serving as an auxiliary, than directly in *Semilat*.

If we replace semilattices above by commutative semigroups, i.e., if we reject the idempotency law, then we will verify all the specific equations of **ML** except (wm) (which amounts to $(\cup\ idemp)$).

§10.2. Restricted coherence of mix-lattice categories

To prove a restricted coherence result for **ML**, we prove first Composition Elimination for **GML** by extending the proof for **GL** in §9.4. We use essentially here the equations $(\cup\, \circ)$.

Next, we define inductively formulae of $\mathcal{L}_{\wedge,\vee}$ in *disjunctive normal form* (*dnf*): every formula of $\mathcal{L}_{\wedge}$ is in *dnf*, and if A and B are both in *dnf*, then $A \vee B$ is in *dnf*. We define dually formulae of $\mathcal{L}_{\wedge,\vee}$ in *conjunctive normal form* (*cnf*): every formula of $\mathcal{L}_{\vee}$ is in *cnf*, and if A and B are both in *cnf*, then $A \wedge B$ is in *cnf*.

We define inductively composition-free arrow terms of $\mathcal{C}(\mathbf{GML})$ of type $A \vdash B$, for A in *dnf* and B in *cnf*, that are in *normal form*. We do that gradually, relying on two preliminary inductive definitions.

Arrow terms of the form $P_1 \ldots P_n Q_1 \ldots Q_m \mathbf{1}_p$, where $n, m \geq 0$, and P_i for $i \in \{1, \ldots, n\}$ is of the form $\overset{\vee}{K}{}^1_C$ or $\overset{\vee}{K}{}^2_C$, while Q_j for $j \in \{1, \ldots, m\}$ is of the form $\hat{K}{}^1_C$ or $\hat{K}{}^2_C$, are in *atomic bracket-free normal form*.

Every arrow term in atomic bracket-free normal form is in *bracket-free normal form*. If $f: D \vdash E$ and $g: D \vdash E$ are in bracket-free normal form, then $f \cup g: D \vdash E$ is in bracket-free normal form.

Every arrow term in bracket-free normal form is in *angle normal form*. If $f: D \vdash E$ and $g: D \vdash F$ are in angle normal form, then $\langle f, g \rangle: D \vdash E \wedge F$ is in angle normal form.

Every arrow term in angle normal form is in *normal form*. If $f: E \vdash D$ and $g: F \vdash D$ are in normal form, then $[f, g]: E \vee F \vdash D$ is in normal form.

We have also the following definitions. Let f be an arrow term of $\mathcal{C}(\mathbf{GML})$ in normal form, and let f' be a subterm of f such that f' is in atomic bracket-free normal form, and there is no subterm f'' of f in atomic bracket-free normal form with f' a proper subterm of f''. Then we say that f' is an *atomic component* of f.

An arrow term f of $\mathcal{C}(\mathbf{GML})$ in normal form is said to be in *settled* normal form when there are no subterms of f in bracket-free normal form in which an atomic component occurs more than once.

Let us illustrate all these definitions with an example. The following arrow terms of $\mathcal{C}(\mathbf{GML})$:

$$\begin{aligned}
\alpha_1 &=_{df} \hat{K}{}^1_q \mathbf{1}_p : p \wedge q \vdash p,\\
\alpha_2 &=_{df} \overset{\vee}{K}{}^1_s \hat{K}{}^2_p \mathbf{1}_q : p \wedge q \vdash q \vee s,\\
\alpha_3 &=_{df} \overset{\vee}{K}{}^1_{p\vee t} \hat{K}{}^1_q \mathbf{1}_p : p \wedge q \vdash p \vee (p \vee t),\\
\alpha_4 &=_{df} \overset{\vee}{K}{}^2_p \overset{\vee}{K}{}^1_t \hat{K}{}^1_q \mathbf{1}_p : p \wedge q \vdash p \vee (p \vee t),\\
\beta_1 &=_{df} \hat{K}{}^1_s \hat{K}{}^1_p \mathbf{1}_p : (p \wedge r) \wedge s \vdash p,\\
\beta_2 &=_{df} \overset{\vee}{K}{}^2_q \hat{K}{}^2_{p\wedge r} \mathbf{1}_s : (p \wedge r) \wedge s \vdash q \vee s,\\
\beta_3 &=_{df} \overset{\vee}{K}{}^1_{p\vee t} \hat{K}{}^1_s \hat{K}{}^1_r \mathbf{1}_p : (p \wedge r) \wedge s \vdash p \vee (p \vee t),\\
\beta_4 &=_{df} \overset{\vee}{K}{}^2_p \overset{\vee}{K}{}^1_t \hat{K}{}^1_s \hat{K}{}^1_r \mathbf{1}_p : (p \wedge r) \wedge s \vdash p \vee (p \vee t)
\end{aligned}$$

are all in atomic bracket-free normal form. The arrow terms $\alpha_2 \cup \alpha_2$, $\beta_3 \cup \beta_4$, $(\alpha_2 \cup \alpha_2) \cup (\beta_3 \cup \beta_4)$, etc., are in bracket-free normal form. Next,

$$\begin{aligned}
&\langle \alpha_1, \langle \alpha_2 \cup \alpha_2, \alpha_3 \rangle\rangle : p \wedge q \vdash p \wedge ((q \vee s) \wedge (p \vee (p \vee t))),\\
&\langle \beta_1, \langle \beta_2, \beta_3 \cup \beta_4 \rangle\rangle : (p \wedge r) \wedge s \vdash p \wedge ((q \vee s) \wedge (p \vee (p \vee t)))
\end{aligned}$$

are in angle normal form, and

$$\begin{aligned}
\gamma =_{df}\ & [[\langle \alpha_1, \langle \alpha_2, \alpha_3 \rangle\rangle, \langle \beta_1, \langle \beta_2, \beta_3 \cup \beta_4 \rangle\rangle], \langle \alpha_1, \langle \alpha_2, \alpha_4 \rangle\rangle] :\\
& ((p \wedge q) \vee ((p \wedge r) \wedge s)) \vee (p \wedge q) \vdash p \wedge ((q \vee s) \wedge (p \vee (p \vee t)))
\end{aligned}$$

is in settled normal form. This normal from would not be settled if, for example, α_2 in γ were replaced by $\alpha_2 \cup \alpha_2$. The set of occurrences of atomic components of γ is made of the two occurrences of α_1, the two occurrences of α_2, and of the occurrences of α_3, α_4, β_1, β_2, β_3 and β_4.

We can then prove the following.

NORMAL-FORM LEMMA. *Every arrow term $f : A \vdash B$ of $\mathcal{C}$(**GML**) for A in* dnf *and B in* cnf *is equal in* **GML** *to an arrow term in settled normal form.*

PROOF. We make an induction on the number of occurrences of $\vee$ in A and $\wedge$ in B. If there are no such occurrences of $\wedge$ and $\vee$, then we eliminate compositions, and by applying the following equations of **GML**:

$$\begin{aligned}
&\overset{\xi}{K}{}^i_A (f \cup g) = \overset{\xi}{K}{}^i_A f \cup \overset{\xi}{K}{}^i_A g,\\
&\hat{K}{}^i_A \overset{\vee}{K}{}^j_B f = \overset{\vee}{K}{}^j_B \hat{K}{}^i_A f,
\end{aligned}$$

we obtain an arrow term in bracket-free normal form equal to the original arrow term.

If there are no occurrences of $\vee$ in A, and B is $B_1 \wedge B_2$, then $f = \langle \hat{K}^1_{B_2} \mathbf{1}_{B_1} \circ f, \hat{K}^2_{B_1} \mathbf{1}_{B_2} \circ f \rangle$ in **GML**, and, by the induction hypothesis, we have that $\hat{K}^1_{B_2} \mathbf{1}_{B_1} \circ f$ and $\hat{K}^2_{B_1} \mathbf{1}_{B_2} \circ f$ must be equal respectively to f' and f'' in normal form, which must be in angle normal form, because $\vee$ does not occur in A. Hence $f = \langle f', f'' \rangle$, and $\langle f', f'' \rangle$ is in normal form.

If A is $A_1 \vee A_2$, then $f = [f \circ \check{K}^1_{A_2} \mathbf{1}_{A_1}, f \circ \check{K}^2_{A_1} \mathbf{1}_{A_2}]$ in **GML**, and, by the induction hypothesis, $f \circ \check{K}^1_{A_2} \mathbf{1}_{A_1}$ and $f \circ \check{K}^2_{A_1} \mathbf{1}_{A_2}$ must be equal respectively to f' and f'' in normal form. Hence $f = [f', f'']$, and $[f', f'']$ is in normal form.

We easily pass from the normal form to the settled normal form by applying ($\cup$ *assoc*), ($\cup$ *com*) and ($\cup$ *idemp*). ⊣

For an arrow term f of $\mathcal{C}$(**GML**) in settled normal form, there is a one-to-one correspondence between the set of occurrences of atomic components of f and the set of ordered pairs of Gf. For example, if f is the arrow term γ we had above, then we have the following correspondence:

left α_1	left α_2	α_3	β_1	β_2	β_3	β_4	right α_1	right α_2	α_4
$(0,0)$	$(1,1)$	$(0,3)$	$(2,0)$	$(4,2)$	$(2,3)$	$(2,4)$	$(5,0)$	$(6,1)$	$(5,4)$

which can be drawn as follows:

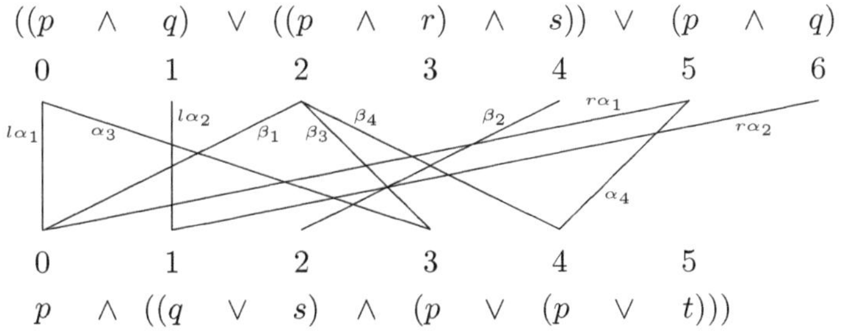

This one-to-one correspondence has a finer structure, which we are going to explain now. For A in *dnf* let a *minimal disjunct* of A be a subformula D of A that belongs to $\mathcal{L}_\wedge$ such that there is no subformula of A in $\mathcal{L}_\wedge$ of which D would be a proper subformula. We define analogously the *minimal conjuncts* of a formula B in *cnf*, by replacing $\wedge$ by $\vee$.

Let A be the source $((p \wedge q) \vee ((p \wedge r) \wedge s)) \vee (p \wedge q)$, and let B be the target $p \wedge ((q \vee s) \wedge (p \vee (p \vee t)))$, of the arrow term γ we had as an example above. Then the minimal disjuncts of A are A_1, which is the left occurrence of $p \wedge q$, next A_2, which is $(p \wedge r) \wedge s$, and A_3, which is the right occurrence of $p \wedge q$. The minimal conjuncts of B are B_1, which is the leftmost p, next B_2, which is $q \vee s$, and B_3, which is $p \vee (p \vee t)$.

For an arrow term $f \colon A \vdash B$ of $\mathcal{C}(\mathbf{GML})$ in normal form (not necessarily settled), consider subterms in bracket-free normal form that are not proper subterms of subterms of f in bracket-free normal form. We call such subterms the *molecular components* of f. There is a one-to-one correspondence between the set of occurrences of molecular components of f and the set of ordered pairs (A_i, B_j) for A_i a minimal disjunct of A and B_j a minimal conjunct of B. We call this correspondence the *molecular correspondence*. For example, the molecular component $\beta_3 \cup \beta_4$ of γ corresponds by the molecular correspondence to the ordered pair (A_2, B_3).

If $f \colon A \vdash B$ is an arrow term of $\mathcal{C}(\mathbf{GML})$ in settled normal form, then for every molecular component f' of f, the set of ordered pairs of Gf' is in one-to-one correspondence with the set of atomic components in f'. We call this correspondence the *atomic correspondence*. For example, if f' is the molecular component $\beta_3 \cup \beta_4$ of γ, then $Gf' = \{(0,0), (0,1)\}$, where $(0,0)$ corresponds by the atomic correspondence to β_3 and $(0,1)$ corresponds to β_4.

We can then prove the following.

RESTRICTED MIX-LATTICE COHERENCE. *Let $f, g \colon A \vdash B$ be arrow terms of $\mathcal{C}(\mathbf{ML})$ such that A is in* dnf *and B in* cnf. *If $Gf = Gg$, then $f = g$ in* **ML**.

PROOF. By the Normal-Form Lemma, we have that $f = f'$ and $g = g'$ in **GML** for f' and g' in settled normal form. Since $Gf = Gf'$ and $Gg = Gg'$, because G is a functor, we have that $Gf = Gg$ implies $Gf' = Gg'$. If $Gf' = Gg'$, then for the molecular components f'' of f' and g'' of g' such that f'' and g'' correspond by the molecular correspondence to the same ordered pair (A_i, B_j), for A_i a minimal disjunct of A and B_j a minimal conjunct of B, we must have $Gf'' = Gg''$. Hence, by the atomic correspondence, there is a one-to-one correspondence between the atomic components f''' in f'' and the atomic components g''' in g'' such that f''' and g''' correspond to the same ordered pair of Gf'', that is Gg''. Since $Gf''' = Gg'''$, we may conclude, by Lattice Coherence, that $f''' = g'''$ in **GL**, and hence also in **GML**. (As a matter of fact, f''' and g''' must be the same arrow term of $\mathcal{C}(\mathbf{GL})$.) Then, by using the equations ($\cup$ *assoc*) and ($\cup$ *com*), we must be able to show that $f'' = g''$ in **GML**. Since this holds for every pair f'' and g'' of corresponding molecular components, we obtain $f' = g'$, and so $f = g$ in **ML**. ⊣

We will not try to establish here an unrestricted coherence result for **ML**, or perhaps a category with $\mathcal{E}(\mathbf{ML})$ extended. The result we have above is sufficient for applications in the next chapter, which are our main concern.

§10.3. Restricted coherence of mix-dicartesian categories

To obtain the natural logical category $\mathbf{ML}_{\top,\bot}$, we have that the logical system $\mathcal{C}(\mathbf{ML}_{\top,\bot})$ is in $\mathcal{L}_{\wedge,\vee,\top,\bot}$, with the transformations α included in **1**, b, c, w-k, m and δ-σ. The specific equations of $\mathcal{E}(\mathbf{ML}_{\top,\bot})$ are obtained by taking the union of those of $\mathcal{E}(\mathbf{ML})$ and $\mathcal{E}(\mathbf{L}_{\top,\bot})$ plus

$$(m\top)\quad m_{A,\top} = \check{k}^1_{A,\top} \circ \hat{k}^1_{A,\top},$$
$$(m\bot)\quad m_{A,\bot} = \check{k}^1_{A,\bot} \circ \hat{k}^1_{A,\bot}.$$

It is easy to see that in $\mathcal{E}(\mathbf{ML}_{\top,\bot})$ we have the equations

$$m_{A,C} = \check{k}^1_{A,C} \circ \hat{k}^1_{A,C},$$
$$m_{C,A} = \check{k}^2_{C,A} \circ \hat{k}^2_{C,A}$$

for any letterless formula C of $\mathcal{L}_{\wedge,\vee,\top,\bot}$. It is clear that, by relying on the equation (cm) of §8.4, we could replace $(m\top)$ above by $m_{\top,A} = \check{k}^2_{\top,A} \circ \hat{k}^2_{\top,A}$, while $(m\bot)$ could be replaced by $m_{\bot,A} = \check{k}^2_{\bot,A} \circ \hat{k}^2_{\bot,A}$.

We call natural $\mathbf{ML}_{\top,\bot}$-categories *mix-dicartesian* categories.

The syntactical category $\mathbf{GML}_{\top,\bot}$ synonymous with $\mathbf{ML}_{\top,\bot}$ is obtained as $\mathbf{GL}_{\top,\bot}$ save that we have in addition the operation $\cup$ on arrow terms of the same type, and the equations ($\cup\,\circ$), ($\cup$ *assoc*), ($\cup$ *com*) and ($\cup$ *idemp*) of $\mathcal{E}(\mathbf{GML})$ (see §9.6 and §10.1), plus the equations

$$(\cup\mathbf{0}\top) \quad \mathbf{1}_{A\vee\top} \cup \check{K}^2_A \hat{\kappa}_{A\vee\top} = \mathbf{1}_{A\vee\top},$$
$$(\cup\mathbf{0}\bot) \quad \mathbf{1}_{A\wedge\bot} \cup \hat{K}^2_A \check{\kappa}_{A\wedge\bot} = \mathbf{1}_{A\wedge\bot}.$$

That $(\cup\mathbf{0}\top)$ holds in $\mathbf{ML}_{\top,\bot}$ is shown as follows:

$$\mathbf{1}_{A\vee\top} \cup (\check{k}^2_{A,\top} \circ \hat{\kappa}_{A\vee\top})$$
$$= \check{w}_{A\vee\top} \circ (\mathbf{1}_{A\vee\top} \vee \check{k}^2_{A,\top}) \circ m_{A\vee\top,\top} \circ (\mathbf{1}_{A\vee\top} \wedge \hat{\kappa}_{A\vee\top}) \circ \hat{w}_{A\vee\top}, \text{ by } (m\ nat),$$
$$= \mathbf{1}_{A\vee\top}, \text{ by } (m\top) \text{ and Dicartesian Coherence.}$$

We proceed analogously for $(\cup\mathbf{0}\bot)$ by using $(m\bot)$.

To show that $(m\top)$ holds in $\mathbf{GML}_{\top,\bot}$, we have

$$m_{A,\top} = \check{K}^1_\top \hat{K}^1_\top \mathbf{1}_A \cup \check{K}^2_A \hat{K}^2_A \mathbf{1}_\top$$
$$= (\mathbf{1}_{A\vee\top} \cup \check{K}^2_A \hat{\kappa}_{A\vee\top}) \circ \check{K}^1_\top \hat{K}^1_\top \mathbf{1}_A, \text{ by } (\cup\circ) \text{ and Dicartesian Coherence,}$$
$$= \check{K}^1_\top \hat{K}^1_\top \mathbf{1}_A, \text{ by } (\cup\mathbf{0}\top) \text{ and } (cat\ 1).$$

We proceed analogously for $(m\bot)$ by using $(\cup\mathbf{0}\bot)$.

The category Set^{sl}_* has a terminal object $\top$, which is the two-element semilattice $\{*, x\}$; this is the free semilattice with unit with a single free generator x. The initial object $\bot$ of Set^{sl}_* is the trivial semilattice with unit $\{*\}$; this is the free semilattice with unit with an empty set of generators. The function $\hat{\kappa}_a \colon a \to \top$ is defined by

$$\hat{\kappa}_a\ (y) = \begin{cases} x & \text{if } y \neq * \\ * & \text{if } y = *, \end{cases}$$

while for $\check{\kappa}_a \colon \bot \to a$ we have $\check{\kappa}_a\ (*) = *$. The category Set^{sl}_*, with the structure defined in §10.1 and here, is a sesquicartesian category in the $\mathcal{L}_\bot$ sense, but not a dicartesian category, because in Set^{sl}_* the object $\top \vee \top$ is not isomorphic to $\top$. Note that in Set^{sl}_* the equation $(m\bot)$ holds, but $(m\top)$ does not hold. In the category *Semilat*, which is isomorphic to Set^{sl}_*, the terminal object $\top$ is the trivial semilattice with a single element, while the initial object $\bot$ is the empty semilattice, i.e. the empty set.

To prove restricted coherence for $\mathbf{ML}_{\top,\bot}$, we need first some preliminary notions.

A *null term* is an arrow term $g\colon A \vdash B$ of $\mathcal{C}(\mathbf{L}_{\top,\bot})$ such that Gg is the empty relation. Let C be a formula of $\mathcal{L}_{\wedge,\top,\bot}$ and D a formula of $\mathcal{L}_{\vee,\top,\bot}$. Then $g\colon C \vdash D$ is a null term only if for some C' of $\mathcal{L}_{\wedge}$ and some D' of $\mathcal{L}_{\vee}$ we have that either C is isomorphic to $C' \wedge \bot$ or D is isomorphic to $D' \vee \top$ in $\mathbf{L}_{\top,\bot}$. This follows easily from Composition Elimination for $\mathbf{GL}_{\top,\bot}$ (see §9.6).

To show that for any such null term $g\colon C \vdash D$, where $i\colon C' \wedge \bot \vdash C$ is an isomorphism of $\mathbf{L}_{\top,\bot}$, and for any arrow term $f\colon C \vdash D$ of $\mathcal{C}(\mathbf{L}_{\top,\bot})$, we have in $\mathbf{ML}_{\top,\bot}$ the equation

$$(\cup\mathbf{0}g) \quad f \cup g = f,$$

we rely on the following instance of $(\cup\mathbf{0}\bot)$:

$$\mathbf{1}_{C'\wedge\bot} \cup (\check{\kappa}_{C'\wedge\bot} \circ \hat{k}^2_{C',\bot}) = \mathbf{1}_{C'\wedge\bot}.$$

From this equation we obtain

$$f \circ i \circ (\mathbf{1}_{C'\wedge\bot} \cup (\check{\kappa}_{C'\wedge\bot} \circ \hat{k}^2_{C',\bot})) \circ i^{-1} = f \circ i \circ i^{-1},$$

and $(\cup\mathbf{0}g)$ follows with the help of $(\cup\,\circ)$ and Dicartesian Coherence (we have $G(f \circ i \circ \check{\kappa}_{C'\wedge\bot} \circ \hat{k}^2_{C',\bot} \circ i^{-1}) = Gg$). We proceed analogously to derive $(\cup\mathbf{0}g)$ for a null term $g\colon C \vdash D$ where D is isomorphic to $D' \vee \top$ in $\mathbf{L}_{\top,\bot}$, and any arrow term $f\colon C \vdash D$ of $\mathcal{C}(\mathbf{L}_{\top,\bot})$.

We can then prove the following.

RESTRICTED MIX-DICARTESIAN COHERENCE. *Let $f, g\colon A \vdash B$ be arrow terms of $\mathcal{C}(\mathbf{ML}_{\top,\bot})$ such that A is in* dnf *and B in* cnf*. If $Gf = Gg$, then $f = g$ in $\mathbf{ML}_{\top,\bot}$.*

To prove this result, we proceed as follows. First, by extending the proof of Composition Elimination for $\mathbf{GL}_{\top,\bot}$ (see §9.6), we obtain Composition Elimination for $\mathbf{GML}_{\top,\bot}$.

We define inductively formulae of $\mathcal{L}_{\wedge,\vee,\top,\bot}$ in *dnf* and *cnf*. Every formula of $\mathcal{L}_{\wedge,\top,\bot}$ is in *dnf*; if A and B are both in *dnf*, then $A \vee B$ is in *dnf*. Every formula of $\mathcal{L}_{\vee,\top,\bot}$ is in *cnf*; if A and B are both in *cnf*, then $A \wedge B$ is in *cnf*.

We define inductively composition-free arrow terms of $\mathcal{C}(\mathbf{GML}_{\top,\bot})$ of type $A \vdash B$, for A in *dnf* and B in *cnf*, that are in *normal form*. The only difference with respect to the definition we had in the preceding section is that arrow terms in atomic bracket-free normal form can now have $\hat{\kappa}_p$, $\check{\kappa}_p$, $\hat{\kappa}_\top$, $\check{\kappa}_\bot$ or $\mathbf{0}_{\bot,\top}$ instead of $\mathbf{1}_p$; here $\mathbf{0}_{\bot,\top}$ stands for either $\hat{\kappa}_\bot$ or $\check{\kappa}_\top$, which are equal in $\mathbf{L}_{\top,\bot}$. Arrow terms in atomic bracket-free normal form in which we do not have $\mathbf{1}_p$, but $\hat{\kappa}_p$, $\check{\kappa}_p$ or $\mathbf{0}_{\bot,\top}$ are called *zero* atomic bracket-free terms, and those with $\mathbf{1}_p$ *nonzero* atomic bracket-free terms. We use the same terminology of "zero" and "nonzero" for atomic components. Zero atomic bracket-free terms are null terms in the sense specified above, and all such arrow terms of the same type are equal in $\mathbf{L}_{\top,\bot}$ by Dicartesian Coherence.

An arrow term f of $\mathcal{C}(\mathbf{GML}_{\top,\bot})$ in normal form is in *settled* normal form when, as before, there are no subterms of f in bracket-free normal form in which an atomic component occurs more than once, and, moreover, we do not have subterms of f of the form $g \cup h$ where one of g and h is a zero atomic component. There is a one-to-one correspondence between the set of occurrences of nonzero atomic components of an arrow term f of $\mathcal{C}(\mathbf{GML}_{\top,\bot})$ in settled normal form and the set of ordered pairs of Gf.

Then, by proceeding as in the preceding section, we can prove the Normal-Form Lemma where **GML** is replaced by $\mathbf{GML}_{\top,\bot}$. We use here also the equations $\hat{\kappa}_\top = \mathbf{1}_\top$ and $\check{\kappa}_\bot = \mathbf{1}_\bot$. To pass from the normal form to the settled normal form we apply the equations ($\cup$ *assoc*), ($\cup$ *com*), ($\cup$ *idemp*) and ($\cup \mathbf{0} g$). We can then prove Restricted Mix-Dicartesian Coherence as Restricted Mix-Lattice Coherence. As for **ML**, we will not try to establish here an unrestricted coherence result for $\mathbf{ML}_{\top,\bot}$, or perhaps a category with $\mathcal{E}(\mathbf{ML}_{\top,\bot})$ extended.

Since we do not have unrestricted coherence with respect to *Rel*, there is no point in discussing here the maximality of **ML** and $\mathbf{ML}_{\top,\bot}$. We conjecture, however, that **ML** is not maximal in the sense in which **L** was (see §9.5 and §9.3). For example, one could presumably add to $\mathcal{E}(\mathbf{ML})$ the equation $m_{p,p} = m_{p,p} \circ \hat{c}_{p,p}$, where $Gm_{p,p} \neq G(m_{p,p} \circ \hat{c}_{p,p})$, without falling into preorder. There are other such equations, but we will not go here into the problem of their classification.

CHAPTER 11

DISTRIBUTIVE LATTICE CATEGORIES

This is the central chapter of the book. We define in it the notion that we take as the categorification of the notion of distributive lattice. Distribution is here based on the dissociativity of Chapter 7, which delivers arrows corresponding to the common distributions of conjunction over disjunction and of disjunction over conjunction, but neither of these distributions happens to be an isomorphism (in bicartesian closed categories, the former distribution is an isomorphism, but the latter is not). For our categorification of distributive lattices, we prove coherence with respect to the category whose arrows are relations between finite ordinals, as before. We have this coherence both in the presence and in the absence of terminal and initial objects.

The essential ingredient of our proof is a cut-elimination theorem for a category corresponding to a plural sequent system for classical conjunctive-disjunctive logic. This category is obtained by strictifying the double cartesian structure so that arrows of the monoidal structure, i.e. associativity isomorphisms and isomorphisms tied to the terminal and initial objects, become identity arrows. This is very much in the spirit of Gentzen, who based his sequents on *sequences* of formulae, and Gentzen's intuition is here vindicated by the strictification results of Chapter 3. Our cut-elimination procedure differs, however, from Gentzen's in that it takes into account union of proofs. Gentzen's own procedure would lead to collapse, i.e. preorder. We also differ from Gentzen in how we deal with the structural rule of contraction. We eliminate cut directly, and do not introduce Gentzen's generalized cut rule, which may be understood as involving several cuts, or as blending contraction with cut. (Our approach here differs from previously published procedures of eliminating cut directly.)

We believe that one of the achievements of this chapter is notational. From the very beginning of categorial proof theory, equations imposed by cut elimination have been a guiding inspiration, but recording these equations precisely proved to be a rather difficult task.

§11.1. Distributive lattice categories and their Gentzenization

The categories we are going to investigate in this chapter, which we call distributive lattice categories, may be conceived as obtained by the categorification of the notion of distributive lattice. Freely generated categories of this kind may be conceived as codifying equality of derivations in the conjunction-disjunction fragment of logic (with or without the empty conjunction $\top$ and the empty disjunction $\bot$). This fragment of logic coincides in classical and intuitionistic logic, as far as provable sequents of the form $A \vdash B$ are concerned (cf. §1.3). Categories we have considered previously codify analogously equality of derivations in more restricted fragments of logic, which were sometimes fragments of nonclassical and nonintuitionistic logics. In particular, the free symmetric net category of §7.6 corresponds to a fragment of linear logic (in the jargon of that field, we have there the multiplicative conjunction-disjunction fragment of linear logic).

It is remarkable that equations between arrows in the free distributive lattice category cover a procedure of cut elimination in a plural, i.e. multiple-conclusion, sequent system. A sequent $\Gamma \vdash \Delta$ is a *singular*, or *single-conclusion*, sequent when Δ has a single member or is empty; without this restriction, it is a *plural*, or *multiple-conclusion*, sequent. The fact that we are within the realm of plural sequents for conjunctive-disjunctive logic allows us to assume that we are dealing with classical, rather than intuitionistic, logic (see the last part of §1.3).

Gentzen's cut-elimination theorem of [55] could be phrased as saying that for every term t coding a derivation of $\Gamma \vdash \Delta$ there is a term t' coding a cut-free derivation of $\Gamma \vdash \Delta$. As in the Cut-Elimination Theorem of §7.7, we prove something more in this chapter. We show also that $t = t'$ in a particular category $\mathcal{D}$. Gentzen did not care about equality between these terms—he did not even introduce terms to code his derivations. His intuition was, however, good in most cases, and we may copy his procedure to a great extent. But we cannot copy it completely, because if we did so, then our category $\mathcal{D}$ would be a preorder.

We want equality of arrows in $\mathcal{D}$ to correspond to equality of arrows in the freely generated distributive lattice category, and $\mathcal{D}$ should not be a preorder. Therefore our cut-elimination procedure will not be exactly Gentzen's procedure of [55] restricted to conjunctive-disjunctive logic, but a modification of it, and we will point out later where precisely we differ from Gentzen. The main difference is that we take into account the mix principle, which in this context yields the operation of union of derivations, corresponding to the operation $\cup$ on arrow terms of §10.1. The problematic situation in [55], mentioned in §1.6, where we introduce union was noted in [59] (Appendix B1, by Y. Lafont), where it was supposed that there is no alternative to Gentzen's way of dealing with it, and that preorder and triviality are inescapable in the proof theory of classical logic (see also [62], Section 1). We show that this is not the case, and obtain a coherence result for distributive lattice categories with respect to the category *Rel*.

Distributive lattice categories are not the only candidate for codifying equality of derivations in conjunctive-disjunctive logic. An alternative codification is in a fragment of bicartesian closed categories. The equations of these categories also cover a cut-elimination procedure in a single-conclusion sequent system (see [103]). With this alternative codification, we do not have, however, a coherence result with respect to *Rel* (see §1.2, [44], Section 1, and [103], Section 1; cf. [71], pp. 95-97). The distribution arrow of type $A \wedge (B \vee C) \vdash (A \wedge B) \vee (A \wedge C)$ is an isomorphism in bicartesian closed categories. In our distributive lattice categories we have an arrow of this type, and of the inverse type, but we need not have an isomorphism of this type. We pass now to the definition of distributive lattice category.

To obtain the natural logical category **DL**, we have that the logical system $\mathcal{C}(\mathbf{DL})$ is in $\mathcal{L}_{\wedge,\vee}$, with the transformations α included in $\mathbf{1}$, b, c, w-k, m and d. The specific equations of $\mathcal{E}(\mathbf{DL})$ are obtained by taking the union of those of $\mathcal{E}(\mathbf{DS})$ and $\mathcal{E}(\mathbf{ML})$ plus

$$(d\hat{k}) \qquad \hat{k}^2_{A,B\vee C} = (\hat{k}^2_{A,B} \vee \mathbf{1}_C) \circ d_{A,B,C},$$

$$(d\check{k}) \qquad \check{k}^1_{C\wedge B,A} = d_{C,B,A} \circ (\mathbf{1}_C \wedge \check{k}^1_{B,A}),$$

$$(dm) \qquad m_{A,C} = (\hat{k}^1_{A,B} \vee \mathbf{1}_C) \circ d_{A,B,C} \circ (\mathbf{1}_A \wedge \check{k}^2_{B,C}),$$

$$(m\,\hat{e}) \qquad c^k_{A,C,B,D} \circ \hat{e}'_{A,B,C,D} = m_{A,B} \wedge \mathbf{1}_{C \vee D},$$

$$(m\,\check{e}) \qquad \check{e}'_{D,C,B,A} \circ c^k_{D,C,B,A} = \mathbf{1}_{D \wedge C} \vee m_{B,A}$$

(see §7.6 for $\overset{\xi}{e}'$, and §9.4 for c^k). In the arrow terms of $\mathcal{C}(\mathbf{DL})$ we write d instead of d^L, as we did for $\mathcal{C}(\mathbf{DS})$ and $\mathcal{C}(\mathbf{MDS})$, and we take d^R as defined by the equation $(d^R c)$ of §7.6.

We call natural **DL**-categories *distributive lattice* categories. The objects of a distributive lattice category that is a partial order make a distributive lattice. In **DL**, the dissociativity arrows $d_{A,B,C}$ enable us to define arrows of the type of the common distribution principles of $\wedge$ over $\vee$ and of $\vee$ over $\wedge$ (see §11.3). These distribution arrows are, however, not isomorphisms. Note that our distributive lattice categories are not distributive categories in the sense of [89] (pp. 222-223 and Session 26) or [17], where distribution of $\wedge$ over $\vee$ must be an isomorphism.

The cartesian linearly distributive categories of [19] are symmetric net categories and are lattice categories, but they are not necessarily distributive lattice categories. The specific equations (wm), $(m\,\hat{e})$ and $(m\,\check{e})$ are not envisaged in that paper, nor in [20]. The equations $(d\,\hat{k})$ and $(d\,\check{k})$ hold in cartesian linearly distributive categories as a consequence of the presence of the equations $(\hat{\sigma}\; d^L)$ and $(\check{\delta}\; d^L)$ of §7.9 in these categories (see §11.5). The equations (dm), (bm) and (cm) hold in these categories (as can be gathered from the derivations for $\mathbf{DL}'$ below), though they are not explicitly mentioned in [19] and [20]. We know the equation (wm) does not hold in these categories (see §12.5), and we do not know how to derive $(m\,\hat{e})$ and $(m\,\check{e})$ from the remaining axioms of $\mathcal{E}(\mathbf{DL})$. We can, however, derive from the remaining axioms the following immediate consequences of $(m\,\hat{e})$ and $(m\,\check{e})$:

$$c^k_{A,C,B,D} \circ \hat{e}'_{A,B,C,D} \circ (\mathbf{1}_{A \wedge B} \wedge m_{C,D}) \;= m_{A,B} \wedge m_{C,D},$$

$$(m_{D,C} \vee \mathbf{1}_{B \vee A}) \circ \check{e}'_{D,C,B,A} \circ c^k_{D,C,B,A} = m_{D,C} \vee m_{B,A}$$

(see the derivation of $(m\,\hat{c}^m)$ in this section, and §13.2). Equations that have the same force as $(m\,\hat{e})$ and $(m\,\check{e})$ are

$$(\hat{k}^1_{A,C} \vee \hat{k}^1_{B,D}) \circ \hat{e}'_{A,B,C,D} = m_{A,B} \circ \hat{k}^1_{A\wedge B,C\vee D},$$
$$\check{e}'_{D,C,B,A} \circ (\check{k}^2_{D,B} \wedge \check{k}^2_{C,A}) = \check{k}^2_{D\wedge C,B\vee A} \circ m_{B,A}$$

(see §13.2 for further comments on $(m\,\hat{e})$ and $(m\,\check{e})$).

There are redundancies in our presentation of **DL**. A synonymous logical system $\mathcal{C}(\mathbf{DL}')$ is obtained by omitting m from $\mathcal{C}(\mathbf{DL})$. We introduce m in $\mathcal{C}(\mathbf{DL}')$ by (dm) understood as a definition. The specific equations of $\mathcal{E}(\mathbf{DL}')$ are obtained by taking the union of those of $\mathcal{E}(\mathbf{DS})$ and $\mathcal{E}(\mathbf{L})$ plus (wm), $(d\hat{k})$, $(d\check{k})$, $(m\,\hat{e})$ and $(m\,\check{e})$, where m is defined. The category $\mathbf{DL}'$ is $\mathcal{C}(\mathbf{DL}')/\mathcal{E}(\mathbf{DL}')$. To prove this synonymity, i.e. the isomorphism of **DL** with $\mathbf{DL}'$ (see the end of §2.4 for the notion of synonymity of syntactical systems), we have the following.

First, we derive for $\mathbf{DL}'$ the equation

$$(\hat{k}^1_{A,D} \vee \mathbf{1}_C) \circ d_{A,D,C} \circ (\mathbf{1}_A \wedge \check{k}^2_{D,C}) = (\hat{k}^1_{A,B} \vee \mathbf{1}_C) \circ d_{A,B,C} \circ (\mathbf{1}_A \wedge \check{k}^2_{B,C}),$$

which yields the equation (dm) for $\mathbf{DL}'$, because the left-hand side may be replaced by $m_{A,C}$. We have, by using naturality equations,

$$\begin{aligned}
&(\hat{k}^1_{A,D} \vee \mathbf{1}_C) \circ d_{A,D,C} \circ (\mathbf{1}_A \wedge \check{k}^2_{D,C})\\
&\quad = (\hat{k}^1_{A,D} \vee \mathbf{1}_C) \circ d_{A,D,C} \circ (\mathbf{1}_A \wedge (\hat{k}^1_{D,B} \vee \mathbf{1}_C)) \circ (\mathbf{1}_A \wedge \check{k}^2_{D\wedge B,C})\\
&\quad = (\hat{k}^1_{A,D\wedge B} \vee \mathbf{1}_C) \circ d_{A,D\wedge B,C} \circ (\mathbf{1}_A \wedge \check{k}^2_{D\wedge B,C})\\
&\quad = (\hat{k}^1_{A,B} \vee \mathbf{1}_C) \circ ((\mathbf{1}_A \wedge \hat{k}^2_{D\wedge B}) \vee \mathbf{1}_C) \circ d_{A,D\wedge B,C} \circ (\mathbf{1}_A \wedge \check{k}^2_{D\wedge B,C})\\
&\quad = (\hat{k}^1_{A,B} \vee \mathbf{1}_C) \circ d_{A,B,C} \circ (\mathbf{1}_A \wedge \check{k}^2_{B,C}).
\end{aligned}$$

Next, we derive the equation $(\hat{b}\;mL)$ (see §8.3) for $\mathbf{DL}'$:

$$\begin{aligned}
&d_{A,B,C} \circ (\mathbf{1}_A \wedge m_{B,C})\\
&\quad = d_{A,B,C} \circ (\mathbf{1}_A \wedge (\hat{k}^1_{B,D} \vee \mathbf{1}_C)) \circ (\mathbf{1}_A \wedge d_{B,D,C}) \circ\\
&\qquad\qquad \circ (\mathbf{1}_A \wedge (\mathbf{1}_B \wedge \check{k}^2_{D,C})), \text{ by } (dm),\\
&\quad = ((\mathbf{1}_A \wedge \hat{k}^1_{B,D}) \vee \mathbf{1}_C) \circ (\hat{b}^{\leftarrow}_{A,B,D} \vee \mathbf{1}_C) \circ d_{A\wedge B,D,C} \circ \hat{b}^{\rightarrow}_{A,B,D\vee C} \circ\\
&\qquad\qquad \circ (\mathbf{1}_A \wedge (\mathbf{1}_B \wedge \check{k}^2_{D,C})), \text{ by naturality and } (d^L\wedge) \text{ of §7.2,}
\end{aligned}$$

$$= (\hat{k}^1_{A\wedge B,D} \vee \mathbf{1}_C) \circ d_{A\wedge B,D,C} \circ (\mathbf{1}_{A\wedge B} \wedge \check{k}^2_{D,C}) \circ \hat{b}^{\rightarrow}_{A,B,C},$$

by $(\hat{b}\hat{k}1)$ of §9.1 and naturality,

$$= m_{A\wedge B,C} \circ \hat{b}^{\rightarrow}_{A,B,C}, \text{ by } (dm),$$

and we proceed analogously for $(\check{b}\ mL)$. Hence we have also $(\hat{b}\ mR)$ and $(\check{b}\, mR)$ (see §8.3). We can then derive the equation (bm) (see §8.2) for $\mathbf{DL}'$ as we derived it for **MDA** (see §8.3).

We derive as follows the equation (cm) (see §8.4) for $\mathbf{DL}'$. We have the equation

$$(\check{w}_A \vee \mathbf{1}_C) \circ ((\mathbf{1}_A \vee \hat{k}^1_{A,C}) \vee \mathbf{1}_C) \circ (d^R_{A,A,C} \vee \mathbf{1}_C) \circ d_{A\vee A,C,C} \circ (\check{k}^1_{A,A} \wedge \check{k}^2_{C,C}) =$$
$$(\mathbf{1}_A \vee \check{w}_C) \circ (\mathbf{1}_A \vee (\hat{k}^2_{A,C} \vee \mathbf{1}_C)) \circ (\mathbf{1}_A \vee d_{A,C,C}) \circ d^R_{A,A,C\vee C} \circ (\check{k}^1_{A,A} \wedge \check{k}^2_{C,C}),$$

by $(d\,\check{b})$ (see §7.2) and Lattice Coherence of §9.4. We obtain that the two sides of this equation are equal respectively to the two sides of the following equation:

$$(\hat{k}^1_{A,C} \vee \mathbf{1}_C) \circ d_{A,C,C} \circ (\mathbf{1}_A \wedge \check{k}^2_{C,C}) = (\mathbf{1}_A \vee \hat{k}^2_{A,C}) \circ d^R_{A,A,C} \circ (\check{k}^1_{A,A} \wedge \mathbf{1}_C),$$

by using $(d\,\hat{k})$, Lattice Coherence, naturality and bifunctorial equations, and from that equation we derive (cm).

It is easy to derive $(m\ nat)$ for defined m in $\mathcal{E}(\mathbf{DL}')$, so that we have in $\mathcal{E}(\mathbf{DL}')$ all the equations of $\mathcal{E}(\mathbf{MDS})$. We have also in $\mathcal{E}(\mathbf{DL}')$ all the equations of $\mathcal{E}(\mathbf{ML})$. Since all the equations of $\mathcal{E}(\mathbf{DL}')$ are clearly in $\mathcal{E}(\mathbf{DL})$, we obtain that **DL** and $\mathbf{DL}'$ are isomorphic.

Note that for **DL** we can derive

$$m_{A,C} = (\mathbf{1}_A \vee \hat{k}^2_{B,C}) \circ d^R_{A,B,C} \circ (\check{k}^1_{A,B} \wedge \mathbf{1}_C),$$

which is related to (cm). We can also derive for **DL** the following equations:

$$(m\,\hat{c}^m) \quad m_{A\wedge C,B\wedge D} \circ \hat{c}^m_{A,B,C,D} = \hat{e}'_{A,B,C,D} \circ (\mathbf{1}_{A\wedge B} \wedge m_{C,D}),$$
$$(m\,\check{c}^m) \quad \check{c}^m_{D,C,B,A} \circ m_{D\vee B,C\vee A} = (m_{D,C} \vee \mathbf{1}_{B\vee A}) \circ \check{e}'_{D,C,B,A},$$

which we will use in §11.2 (see also §13.2). Here is a derivation of $(m\,\hat{c}^m)$:

$$
\begin{aligned}
& m_{A\wedge C,B\wedge D} \circ \hat{c}^{m}_{A,B,C,D} \\
& \quad = m_{A\wedge C,B\wedge D} \circ \hat{b}^{\rightarrow}_{A,C,B\wedge D} \circ (\mathbf{1}_A \wedge \hat{b}^{\leftarrow}_{C,B,D}) \circ (\mathbf{1}_A \wedge (\hat{c}_{B,C} \wedge \mathbf{1}_D)) \circ \\
& \qquad\qquad \circ (\mathbf{1}_A \wedge \hat{b}^{\rightarrow}_{B,C,D}) \circ \hat{b}^{\leftarrow}_{A,B,C\wedge D} \\
& \quad = d_{A,C,B\wedge D} \circ (\mathbf{1}_A \wedge m_{C,B\wedge D}) \circ (\mathbf{1}_A \wedge \hat{c}_{B\wedge D,C}) \circ (\mathbf{1}_A \wedge \hat{b}^{\rightarrow}_{B,D,C}) \circ \\
& \qquad\qquad \circ \hat{b}^{\leftarrow}_{A,B,D\wedge C} \circ (\mathbf{1}_{A\wedge B} \wedge \hat{c}_{C,D}), \text{ by } (\hat{b}\, mL) \text{ and} \\
& \qquad\qquad \text{Symmetric Biassociative Coherence of §6.3,} \\
& \quad = d_{A,C,B\wedge D} \circ (\mathbf{1}_A \wedge \check{c}_{C,B\wedge D}) \circ (\mathbf{1}_A \wedge d_{B,D,C}) \circ (\mathbf{1}_A \wedge (\mathbf{1}_B \wedge m_{D,C})) \circ \\
& \qquad\qquad \circ \hat{b}^{\leftarrow}_{A,B,D\wedge C} \circ (\mathbf{1}_{A\wedge B} \wedge \hat{c}_{C,D}), \text{ by } (cm) \text{ and } (\hat{b}\, mL), \\
& \quad = \hat{e}'_{A,B,C,D} \circ (\mathbf{1}_{A\wedge B} \wedge m_{C,D}), \text{ by } (\hat{b}^{\leftarrow}\ nat) \text{ and } (cm),
\end{aligned}
$$

and we proceed analogously for $(m\,\check{c}^{m})$.

Let $\mathcal{C}$ and $\mathcal{C}'$ be respectively the logical systems $\mathcal{C}(\mathbf{DL})$ and $\mathcal{C}(\mathbf{A})$, while $\mathcal{E}$ is $\mathcal{E}(\mathbf{DL})$. Next, let $\mathcal{B}$ be $\mathcal{C}/\mathcal{E}$, that is $\mathbf{DL}$. Then it is easy to see that the conditions (IV$\mathcal{C}$) and (IV$\mathcal{B}$) of §3.1 are satisfied. Since the $\mathcal{C}'$-core of $\mathcal{C}/\mathcal{E}$ is the category $\mathbf{A}$, by Biassociative Coherence of §6.1, we have that the condition (IV$\mathcal{G}$) of §3.1 is also satisfied. So (IV) holds, and since $\mathbf{A}$ is a preorder by Biassociative Coherence, we can apply the Strictification Theorem of §3.1 to obtain that the category $\mathbf{DL}^{\mathbf{A}}$, that is $\mathbf{DL}^{\mathcal{G}}$, is equivalent to $\mathbf{DL}$ via two strong $\mathcal{C}(\mathbf{DL})$-functors. Remember that according to §6.2 the objects of $\mathbf{DL}^{\mathbf{A}}$ may be identified with form sequences of $\mathcal{L}_{\wedge,\vee}$ in the nonextended sense. (For understanding the category $\mathbf{DL}^{\mathbf{A}}$, see also §4.5.)

Let $\mathcal{D}$ be the category obtained as the disjoint union of $\mathbf{DL}^{\mathbf{A}}$ and of the trivial category with a single object $\emptyset$ (intuitively this is the empty form sequence of both colour) and a single arrow $\mathbf{1}_{\emptyset} : \emptyset \vdash \emptyset$. The adding of $\emptyset$ to $\mathbf{DL}^{\mathbf{A}}$ is made for practical reasons, to simplify the exposition of our cut-elimination proof by subsuming several cases under a single schema. We could also do without $\emptyset$ at the cost of considering more cases in the proof.

The operations $\xi'' \in \{\wedge'', \vee''\}$ on the objects of $\mathbf{DL}^{\mathbf{A}}$, i.e. on the form sequences of $\mathcal{L}_{\wedge,\vee}$, are extended in the following manner to operations that apply also to $\emptyset$. For X an object of $\mathbf{DL}^{\mathbf{A}}$ or $\emptyset$, we have

$$X\,\xi''\,\emptyset = \emptyset\,\xi''\,X = X.$$

So all the objects of $\mathcal{D}$ are closed under the operations ξ''. The operations

$\xi'' \in \{\wedge'', \vee''\}$ on arrows are extended to operations that apply also to $\mathbf{1}_\emptyset$ by stipulating that

$$f\ \xi''\ \mathbf{1}_\emptyset = \mathbf{1}_\emptyset\ \xi''\ f = f$$

(the variable f here ranges also over $\mathbf{1}_\emptyset$). So all the arrows of $\mathcal{D}$ are closed under the operations ξ''.

The category $\mathcal{D}$ will not have the structure of a **DL**-category. We lack in $\mathcal{D}$ the arrows $\overset{\xi}{k}{}^1_{\emptyset,X}$ and $\overset{\xi}{k}{}^2_{X,\emptyset}$ for X different from $\emptyset$. However, $\overset{\xi}{w}_\emptyset$ may be identified with $\mathbf{1}_\emptyset$, and the arrows $\overset{\xi}{b}{}^{\rightarrow}_{X,Y,Z}$, $\overset{\xi}{b}{}^{\leftarrow}_{X,Y,Z}$ and $\overset{\xi}{c}_{X,Y}$ where one of the subscripts stands for $\emptyset$ may also be identified with identity arrows. (So $\mathcal{D}$ would have the structure of something that could be called a *relevant* net category; see [39] and [102]; cf. §14.4).

A *basic sequence of colour* $\xi \in \{\wedge, \vee\}$ of $\mathcal{D}$ is either a form sequence of $\mathcal{L}_{\wedge,\vee}$ of the form $(A_1 \ldots A_n,\ \xi)$, for $n \geq 2$ and A_i, where $i \in \{1, \ldots, n\}$, a formula of $\mathcal{L}_{\wedge,\vee}$, or it is a formula of $\mathcal{L}_{\wedge,\vee}$, or it is $\emptyset$. So the object $\emptyset$, as well as the formulae of $\mathcal{L}_{\wedge,\vee}$, is both of colour $\wedge$ and of colour $\vee$. A *basic sequence* is a basic sequence of either colour. The *members* of a basic sequence $(A_1 \ldots A_n,\ \xi)$ are the occurrences of formulae $A_1, \ldots, A_n$; the only member of the basic sequence A is A; and the basic sequence $\emptyset$ has no members.

We use Γ and Φ, with or without indices, as variables for basic sequences of colour $\wedge$, and we use Δ and Ψ, with or without with indices, for basic sequences of colour $\vee$. For basic sequences in general, we use Θ and Ξ, with or without indices. We write $\mathbf{1}_\Theta$ for $(\mathbf{1}_{F\Theta}, \Theta, \Theta)$ when Θ is not $\emptyset$; otherwise, $\mathbf{1}_\Theta$ is $\mathbf{1}_\emptyset$.

A *sequent arrow* of $\mathcal{D}$ is an arrow whose type is $\Gamma \vdash \Delta$, for Γ and Δ basic sequences different from $\emptyset$. According to the convention above, Γ is of colour $\wedge$ and Δ is of colour $\vee$. The type of a sequent arrow is a *sequent* (this agrees with Gentzen's notion of sequent).

Let $\Theta_1{}^\xi\,\Theta_2 \ldots {}^\xi\,\Theta_n$ be an abbreviation for $(\ldots(\Theta_1\ \xi''\ \Theta_2)\ldots\ \xi''\ \Theta_n)$, and $f_1{}^\xi f_2 \ldots {}^\xi f_n$ an abbreviation for $(\ldots(f_1\ \xi''\ f_2)\ldots\ \xi''\ f_n)$, where $n \geq 2$. Next, let $\Theta\,{}^\xi f$ and $f\,{}^\xi\Theta$ be abbreviations for $\mathbf{1}_\Theta\,\xi''\,f$ and $f\,\xi''\,\mathbf{1}_\Theta$, respectively. Sometimes we will also write, ambiguously, $\mathbf{1}\,{}^\xi f$ and $f\,{}^\xi\mathbf{1}$ for $\mathbf{1}_\Theta\,{}^\xi f$ and $f\,{}^\xi\mathbf{1}_\Theta$, where Θ can be recovered from the context.

We use the following abbreviations:

$$\overset{\xi}{\boldsymbol{c}}_{\Theta,\Xi} =_{df} \begin{cases} \overset{\xi}{c}''_{\Theta,\Xi} & \text{if } \Theta \neq \emptyset \text{ and } \Xi \neq \emptyset \\ \mathbf{1}_{\Theta\xi\Xi} & \text{if } \Theta = \emptyset \text{ or } \Xi = \emptyset, \end{cases}$$

$$\overset{\xi}{\boldsymbol{w}}_{\Theta} =_{df} \begin{cases} \overset{\xi}{w}''_{\Theta} & \text{if } \Theta \neq \emptyset \\ \mathbf{1}_{\Theta} & \text{if } \Theta = \emptyset, \end{cases}$$

$$\hat{\boldsymbol{y}}_{\Gamma_1,\Gamma_2,\Gamma,\Theta} =_{df} (\Gamma_1{}^\wedge \hat{\boldsymbol{c}}_{\Gamma_2,\Theta}{}^\wedge\Theta{}^\wedge\Gamma) \circ (\Gamma_1{}^\wedge\Gamma_2{}^\wedge \hat{\boldsymbol{w}}_{\Theta}{}^\wedge\Gamma) : \\ \Gamma_1{}^\wedge\Gamma_2{}^\wedge\Theta{}^\wedge\Gamma \vdash \Gamma_1{}^\wedge\Theta{}^\wedge\Gamma_2{}^\wedge\Theta{}^\wedge\Gamma,$$

$$\check{\boldsymbol{y}}_{\Theta,\Delta,\Delta_2,\Delta_1} =_{df} (\Delta^\vee \check{\boldsymbol{w}}_{\Theta}{}^\vee\Delta_2{}^\vee\Delta_1) \circ (\Delta^\vee\Theta^\vee \check{\boldsymbol{c}}_{\Theta,\Delta_2}{}^\vee\Delta_1) : \\ \Delta^\vee\Theta^\vee\Delta_2{}^\vee\Theta^\vee\Delta_1 \vdash \Delta^\vee\Theta^\vee\Delta_2{}^\vee\Delta_1.$$

For $n \geq 3$, consider the abbreviations defined inductively as follows:

$$\hat{\boldsymbol{y}}_{\Gamma_1,\Gamma_2,\Gamma_3,\ldots,\Gamma_n,\Gamma,\Theta} =_{df} \hat{\boldsymbol{y}}_{\Gamma_1,\Gamma_2,\Gamma_3{}^\wedge\Theta{}^\wedge\ldots{}^\wedge\Gamma_n{}^\wedge\Theta{}^\wedge\Gamma,\Theta} \circ \hat{\boldsymbol{y}}_{\Gamma_1{}^\wedge\Gamma_2,\Gamma_3,\ldots,\Gamma_n,\Gamma,\Theta} : \\ \Gamma_1{}^\wedge\Gamma_2{}^\wedge\Gamma_3{}^\wedge\ldots{}^\wedge\Gamma_n{}^\wedge\Theta{}^\wedge\Gamma \vdash \Gamma_1{}^\wedge\Theta{}^\wedge\Gamma_2{}^\wedge\Theta{}^\wedge\Gamma_3{}^\wedge\Theta{}^\wedge\ldots{}^\wedge\Gamma_n{}^\wedge\Theta{}^\wedge\Gamma,$$

$$\check{\boldsymbol{y}}_{\Theta,\Delta,\Delta_n,\ldots,\Delta_3,\Delta_2,\Delta_1} =_{df} \check{\boldsymbol{y}}_{\Theta,\Delta,\Delta_n,\ldots,\Delta_3,\Delta_2{}^\vee\Delta_1} \circ \check{\boldsymbol{y}}_{\Theta,\Delta^\vee\Theta^\vee\Delta_n{}^\vee\ldots{}^\vee\Theta^\vee\Delta_3,\Delta_2,\Delta_1} : \\ \Delta^\vee\Theta^\vee\Delta_n{}^\vee\ldots{}^\vee\Theta^\vee\Delta_3{}^\vee\Theta^\vee\Delta_2{}^\vee\Theta^\vee\Delta_1 \vdash \Delta^\vee\Theta^\vee\Delta_n{}^\vee\ldots{}^\vee\Delta_3{}^\vee\Delta_2{}^\vee\Delta_1.$$

We also have the following abbreviations:

$$\boldsymbol{d}_{\Theta,A,\Xi} =_{df} \begin{cases} d''_{\Theta,A,\Xi} & \text{if } \Theta \neq \emptyset \text{ and } \Xi \neq \emptyset \\ \mathbf{1}_{\Theta^\wedge(A^\vee\Xi)} & \text{if } \Theta = \emptyset \text{ or } \Xi = \emptyset, \end{cases}$$

$$\hat{\boldsymbol{e}}'_{\Gamma_2,\Gamma_1,B,A} =_{df} \boldsymbol{d}_{\Gamma_2,B,\Gamma_1{}^\wedge A} \circ (\Gamma_2{}^\wedge \check{\boldsymbol{c}}_{B,\Gamma_1{}^\wedge A}) \circ (\Gamma_2{}^\wedge \boldsymbol{d}_{\Gamma_1,A,B}) \circ (\Gamma_2{}^\wedge\Gamma_1{}^\wedge \check{\boldsymbol{c}}_{A,B}) : \\ \Gamma_2{}^\wedge\Gamma_1{}^\wedge(B^\vee A) \vdash (\Gamma_2{}^\wedge B)^\vee(\Gamma_1{}^\wedge A),$$

$$\check{\boldsymbol{e}}'_{A,B,\Delta_1,\Delta_2} =_{df} (\hat{\boldsymbol{c}}_{B,A}{}^\vee\Delta_1{}^\vee\Delta_2) \circ (\boldsymbol{d}_{B,A,\Delta_1}{}^\vee\Delta_2) \circ (\hat{\boldsymbol{c}}_{A^\vee\Delta_1,B}{}^\vee\Delta_2) \circ \boldsymbol{d}_{A^\vee\Delta_1,B,\Delta_2} : \\ (A^\vee\Delta_1)^\wedge(B^\vee\Delta_2) \vdash (A^\wedge B)^\vee\Delta_1{}^\vee\Delta_2$$

(note that $\hat{\boldsymbol{e}}'_{\emptyset,\emptyset,B,A} = \mathbf{1}_{B^\vee A}$ and $\check{\boldsymbol{e}}'_{A,B,\emptyset,\emptyset} = \mathbf{1}_{A^\wedge B}$),

$$\boldsymbol{m}_{\Theta,\Xi} =_{df} (\hat{k}1''_{\Theta,B}{}^\vee\Xi) \circ \boldsymbol{d}_{\Theta,B,\Xi} \circ (\Theta^\wedge \check{k}2''_{B,\Xi})$$

(note that for $\boldsymbol{m}_{\Theta,\Xi}$ we must have $\Theta \neq \emptyset$ and $\Xi \neq \emptyset$),

$$\boldsymbol{c}^k_{\Theta_1,\Theta_2,\Theta_3,\Theta_4} =_{df} \check{\boldsymbol{w}}_{(\Theta_1{}^\vee\Theta_3)^\wedge(\Theta_2{}^\vee\Theta_4)} \circ ((\check{k}1''_{\Theta_1,\Theta_3}{}^\wedge \check{k}1''_{\Theta_2,\Theta_4})^\vee (\check{k}2''_{\Theta_1,\Theta_3}{}^\wedge \check{k}2''_{\Theta_2,\Theta_4}))$$

(note that none of Θ_1, Θ_2, Θ_3 and Θ_4 can be $\emptyset$ in the subscripts of $\boldsymbol{c}^k$).

Finally, let $\mathbf{1}^i_{A\,\xi\,B}$ stand for $(\mathbf{1}_{A\xi B}, A\,{}^\xi B, A\,\xi\,B)$, while $\mathbf{1}^e_{A\,\xi\,B}$ stands for $(\mathbf{1}_{A\xi B}, A\,\xi\,B, A\,{}^\xi B)$. We do not introduce the notation $\overset{\xi}{\boldsymbol{k}}{}^1_{\Theta,\Xi}$, because we could not interpret it when Θ is $\emptyset$ and Ξ is not $\emptyset$, and analogously with $\overset{\xi}{\boldsymbol{k}}{}^2_{\Xi,\Theta}$.

We will now define by induction a set of terms for sequent arrows of $\mathcal{D}$, which we call *Gentzen terms*. First, we stipulate that for every letter p the term $\mathbf{1}''_p : p \vdash p$, which denotes the arrow $(\mathbf{1}_p, p, p)$ of $\mathcal{D}$, is a Gentzen term. The remaining Gentzen terms are obtained by closing under the following operations on Gentzen terms, which we call *Gentzen operations*. As in §7.7, we present these operations by inductive clauses in fractional notation, which are interpreted as saying that if the terms above the horizontal line are Gentzen terms, then the term below the horizontal line is a Gentzen term (cf. §2.2). The schema on the left-hand side of the $=_{dn}$ sign stands for the Gentzen term, while the schema on the right-hand side of $=_{dn}$ is the arrow denoted by this term. First, we have the operations that correspond to Gentzen's *structural* rules:

$$\frac{f : \Gamma_1{}^\wedge\Gamma_2{}^\wedge\Gamma_3{}^\wedge\Gamma_4 \vdash \Delta}{c^L_{\Gamma_1,\Gamma_2,\Gamma_3,\Gamma_4} f =_{dn} f \circ (\Gamma_1{}^\wedge\, \hat{c}_{\Gamma_3,\Gamma_2}{}^\wedge\Gamma_4) : \Gamma_1{}^\wedge\Gamma_3{}^\wedge\Gamma_2{}^\wedge\Gamma_4 \vdash \Delta}$$

provided Γ_2 and Γ_3 are not $\emptyset$ (one of the indices Γ_1 and Γ_4 is superfluous as an index of c^L; it is recoverable from the source of f),

$$\frac{f : \Gamma \vdash \Delta_4{}^\vee\Delta_3{}^\vee\Delta_2{}^\vee\Delta_1}{c^R_{\Delta_4,\Delta_3,\Delta_2,\Delta_1} f =_{dn} (\Delta_4{}^\vee\, \check{c}_{\Delta_2,\Delta_3}{}^\vee\Delta_1) \circ f : \Gamma \vdash \Delta_4{}^\vee\Delta_2{}^\vee\Delta_3{}^\vee\Delta_1}$$

provided Δ_2 and Δ_3 are not $\emptyset$ (one of Δ_1 and Δ_4 is superfluous as an index of c^R),

$$\frac{f : \Gamma \vdash \Delta}{k^L_A f =_{dn} f \circ \hat{k}1''_{\Gamma,A} : \Gamma{}^\wedge A \vdash \Delta}$$

$$\frac{f\colon \Gamma \vdash \Delta}{k^R_A f =_{dn} \check{k}2''_{A,\Delta} \circ f\colon \Gamma \vdash A \vee \Delta}$$

$$\frac{f\colon \Gamma_1 \wedge A \wedge \ldots \wedge \Gamma_n \wedge A \wedge \Gamma \vdash \Delta}{w^L_{\Gamma_1,\ldots,\Gamma_n,\Gamma} f =_{dn} f \circ \hat{\boldsymbol{y}}_{\Gamma_1,\ldots,\Gamma_n,\Gamma,A}\colon \Gamma_1 \wedge \ldots \wedge \Gamma_n \wedge A \wedge \Gamma \vdash \Delta},\ n \geq 2,$$

$$\frac{f\colon \Gamma \vdash \Delta \vee A \vee \Delta_n \vee \ldots \vee A \vee \Delta_1}{w^R_{\Delta,\Delta_n,\ldots,\Delta_1} f =_{dn} \check{\boldsymbol{y}}_{A,\Delta,\Delta_n,\ldots,\Delta_1} \circ f\colon \Gamma \vdash \Delta \vee A \vee \Delta_n \vee \ldots \vee \Delta_1},\ n \geq 2,$$

$$\frac{f\colon \Gamma_3 \vdash \Delta_2 \vee A \vee \Delta_1 \qquad g\colon \Gamma_1 \wedge A \wedge \Gamma_2 \vdash \Delta_3}{cut_{\Gamma_2,\Delta_2}(f,g)\colon \Gamma_1 \wedge \Gamma_2 \wedge \Gamma_3 \vdash \Delta_3 \vee \Delta_2 \vee \Delta_1}$$

where $cut_{\Gamma_2,\Delta_2}(f,g)$ denotes

$$(g \vee \Delta_2 \vee \Delta_1) \circ ((\Gamma_1 \wedge \hat{\boldsymbol{c}}_{\Gamma_2,A}) \vee \Delta_2 \vee \Delta_1) \circ \boldsymbol{d}_{\Gamma_1 \wedge \Gamma_2,A,\Delta_2 \vee \Delta_1} \circ \\ \circ (\Gamma_1 \wedge \Gamma_2 \wedge (\check{\boldsymbol{c}}_{A,\Delta_2} \vee \Delta_1)) \circ (\Gamma_1 \wedge \Gamma_2 \wedge f),$$

and A is called the *cut formula* of $cut_{\Gamma_2,\Delta_2}(f,g)$,

$$\frac{f\colon \Gamma_1 \vdash \Delta_1 \qquad g\colon \Gamma_2 \vdash \Delta_2}{mix(f,g) =_{dn} (g \vee \Delta_1) \circ \boldsymbol{m}_{\Gamma_2,\Delta_1} \circ (\Gamma_2 \wedge f)\colon \Gamma_2 \wedge \Gamma_1 \vdash \Delta_2 \vee \Delta_1}$$

Note that by $(m\ nat)$ in $\mathcal{D}$ we have

$$mix(f,g) = \boldsymbol{m}_{\Delta_2,\Delta_1} \circ (g \wedge f) = (g \vee f) \circ \boldsymbol{m}_{\Gamma_2,\Gamma_1}.$$

If we write $\boldsymbol{d}_{\Gamma_2,\emptyset,\Delta_1}$ instead of $\boldsymbol{m}_{\Gamma_2,\Delta_1}$, then $mix(f,g)$ can be conceived as $cut_{\emptyset,\emptyset}(f,g)$ where the cut formula A is replaced by $\emptyset$.

Had we favoured d^R, rather than d^L, for $f : \Gamma_1 \vdash \Delta_3 \vee A \vee \Delta_2$ and $g : \Gamma_2 \wedge A \wedge \Gamma_3 \vdash \Delta_1$ we could take that $cut_{\Gamma_1,\Delta_1}(f,g)\colon \Gamma_1 \wedge \Gamma_2 \wedge \Gamma_3 \vdash \Delta_3 \vee \Delta_2 \vee \Delta_1$ denotes

$$(\Delta_3 \vee \Delta_2 \vee g) \circ (\Delta_3 \vee \Delta_2 \vee (\hat{\boldsymbol{c}}_{A,\Gamma_2} \wedge \Gamma_3)) \circ \boldsymbol{d}^R_{\Delta_3 \vee \Delta_2,A,\Gamma_2 \wedge \Gamma_3} \circ \\ \circ ((\Delta_3 \vee \check{\boldsymbol{c}}_{\Delta_2,A}) \wedge \Gamma_2 \wedge \Gamma_3) \circ (f \wedge \Gamma_2 \wedge \Gamma_3)$$

where $\boldsymbol{d}^R$ stands for $d^{R''}$ if $\Delta_3{}^\vee\Delta_2$ and $\Gamma_2{}^\wedge\Gamma_3$ are not $\emptyset$, and otherwise for $\mathbf{1}$. This would prevent the Γ's and Δ's of f and g to switch from right to left, as in our official definition of the Gentzen operation *cut*. But since we favour d^L, we have to tolerate this switch, which does not cause serious trouble, anyway. We have made the same switch in our Gentzen operation *mix*, to make it parallel to our *cut*.

Here are the remaining Gentzen operations, which correspond to rules for ∧ and ∨:

$$\frac{f : \Gamma{}^\wedge A{}^\wedge B \vdash \Delta}{\wedge^L f =_{dn} f \circ (\Gamma{}^\wedge \mathbf{1}^e_{A^\wedge B}) : \Gamma{}^\wedge(A \wedge B) \vdash \Delta}$$

$$\frac{f : \Gamma_1 \vdash A{}^\vee\Delta_1 \qquad\qquad g : \Gamma_2 \vdash B{}^\vee\Delta_2}{\wedge^R(f,g) =_{dn} (\mathbf{1}^i_{A^\wedge B}{}^\vee\Delta_1{}^\vee\Delta_2) \circ \overset{\vee}{\boldsymbol{e}}{}'_{A,B,\Delta_1,\Delta_2} \circ (f{}^\wedge g) : \Gamma_1{}^\wedge\Gamma_2 \vdash (A \wedge B){}^\vee\Delta_1{}^\vee\Delta_2}$$

$$\frac{g : \Gamma_2{}^\wedge B \vdash \Delta_2 \qquad\qquad f : \Gamma_1{}^\wedge A \vdash \Delta_1}{\vee^L(g,f) =_{dn} (g{}^\vee f) \circ \hat{\boldsymbol{e}}{}'_{\Gamma_2,\Gamma_1,B,A} \circ (\Gamma_2{}^\wedge\Gamma_1{}^\wedge\mathbf{1}^e_{B\vee A}) : \Gamma_2{}^\wedge\Gamma_1{}^\wedge(B \vee A) \vdash \Delta_2{}^\vee\Delta_1}$$

$$\frac{f : \Gamma \vdash B{}^\vee A{}^\vee\Delta}{\vee^R f =_{dn} (\mathbf{1}^i_{B\vee A}{}^\vee\Delta) \circ f : \Gamma \vdash (B \vee A){}^\vee\Delta}$$

This concludes the list of Gentzen operations.

For $n \geq 2$, we introduce the following abbreviations by induction:

$$\begin{aligned}
w^{L\emptyset}_{\Gamma_1,\dots,\Gamma_n,\Gamma} f &=_{df} f, \quad \text{for } f : \Gamma_1{}^\wedge \dots {}^\wedge\Gamma_n{}^\wedge\Gamma \vdash \Delta,\\
w^{LA^\wedge\Phi}_{\Gamma_1,\dots,\Gamma_n,\Gamma} f &=_{df} w^L_{\Gamma_1,\dots,\Gamma_n,\Phi\Gamma} w^{L\Phi}_{\Gamma_1{}^\wedge A,\dots,\Gamma_n{}^\wedge A,\Gamma} f,\\
&\qquad \text{for } f : \Gamma_1{}^\wedge A{}^\wedge\Phi{}^\wedge \dots {}^\wedge\Gamma_n{}^\wedge A{}^\wedge\Phi{}^\wedge\Gamma \vdash \Delta,\\
w^{R\emptyset}_{\Delta,\Delta_n,\dots,\Delta_1} f &=_{df} f, \quad \text{for } f : \Gamma \vdash \Delta{}^\vee\Delta_n{}^\vee \dots {}^\vee\Delta_1,\\
w^{R\Psi^\vee A}_{\Delta,\Delta_n,\dots,\Delta_1} f &=_{df} w^R_{\Delta\Psi,\Delta_n,\dots,\Delta_1} w^{R\Psi}_{\Delta,A^\vee\Delta_n,\dots,A^\vee\Delta_1} f,\\
&\qquad \text{for } f : \Gamma \vdash \Delta{}^\vee\Psi{}^\vee A{}^\vee\Delta_n{}^\vee \dots {}^\vee\Psi{}^\vee A{}^\vee\Delta_1.
\end{aligned}$$

By Semilattice Coherence of §9.1 (in fact, we use here the relevant coherence result of [102], Section 5), we have in $\mathcal{D}$ the equations

$$(w\,\boldsymbol{y})\quad w^{L\Phi}_{\Gamma_1,\ldots,\Gamma_n,\Gamma}f = f\circ \hat{\boldsymbol{y}}_{\Gamma_1,\ldots,\Gamma_n,\Gamma,\Phi},\quad w^{R\Psi}_{\Delta,\Delta_n,\ldots,\Delta_1}f = \check{\boldsymbol{y}}_{\Psi,\Delta,\Delta_n,\ldots,\Delta_1}\circ f.$$

To lighten the burden of notation, in proofs we will sometimes omit subscripts in Gentzen terms or other terms for arrows of $\mathcal{D}$. A reader checking the proofs should be able to restore these subscripts. We will also sometimes take for granted the subscripts of Gentzen operations, and omit them. We do this in cases where no confusion is likely, and the subscripts serve no particular purpose. We use γ, γ_1, $\gamma_2, \ldots$ as variables for Gentzen operations (with subscripts omitted or not).

Note that Gentzen terms codify derivations in a plural sequent system for conjunctive-disjunctive classical propositional logic. (We have mentioned at the beginning of the section that we believe that we are within classical, rather than intuitionistic, logic; cf. also §1.3.) We have in this sequent system rules for connectives of the multiplicative kind, to use the terminology of linear logic. In this terminology, Gentzen's rules for conjunction and disjunction of [55] would be called additive. This is not, however, an essential difference. We could have worked with additive rules as well. We took multiplicative rules for practical reasons, which have to do with our way of dealing with the structural rule of contraction. This difference does not bar comparing our cut-elimination procedure with Gentzen's, and it will turn out that, though the two procedures have much in common, they are not the same.

The main difference is that we take into account the mix principle, which yields union of derivations (see §8.1 and §10.1). Gentzen did not take this principle into account, because, for his more limited purposes, he did not need to do so. This mix principle should not be confused with Gentzen's generalized cut, *Mischung*, also called *mix* in English (see §8.1), which is is derivable in our system with the help of contractions, that is w^L and w^R.

We also differ from Gentzen in the way how we deal with contraction, embodied in the operations w^L and w^R. We eliminate cut directly, and do not introduce as Gentzen his generalized cut *Mischung*, which he used to deal with problems caused by contraction (see [11], Sections 1 and 2).

Eliminating cut directly is handier for notational reasons, because Gentzen's *Mischung* is more difficult to code in our categorial setting. (Our procedure of direct cut elimination differs from similar procedures in [9], [11], [124] and [10]; except for [11], where categories are mentioned occasionally, these papers are not concerned with categorial proof theory and the difficulties of notation for arrow terms.)

Another difference with Gentzen is that we distinguish "conjunctive commas", our $^\wedge$ (which abbreviates $\wedge''$), from "disjunctive commas", our $^\vee$ (which abbreviates $\vee''$), whereas Gentzen has just one kind of comma. In other words, we have two-coloured form sequences, whereas Gentzen has just ordinary sequences. Indeed, if we stay at the level of sequent arrows of $\mathcal{D}$, then the fact that a sequence is on the left-hand side or on the right-hand side of $\vdash$ dictates whether it is of colour $\wedge$ or $\vee$, and then we could do as Gentzen. But we do not pay attention only to sequents, as Gentzen does. For example, in building a sequent arrow denoted by $cut_{\Gamma_2,\Delta_2}(f,g)$ we refer to arrows of $\mathcal{D}$ like $\hat{\boldsymbol{c}}_{\Gamma_2,A}$, or $\boldsymbol{d}_{\Gamma_1{}^\wedge\Gamma_2,A,\Delta_2{}^\vee\Delta_1}$, etc., which are not sequent arrows. With $\boldsymbol{d}$, we even have that conjunctive commas are nested within disjunctive commas and vice versa. Gentzen did not have these problems because he was not considering explicitly arrows and equality between them, but only types of arrows and, moreover, just sequent types. Gentzen stays somewhere near the lowest level of $\mathcal{D}$, while we take somewhat more of $\mathcal{D}$ into account when we compute equality of sequent arrows.

In principle, we could have worked with directly strictified **DL** in the sense of §3.2, but then we would be less close to Gentzen. In that case, we would not have operations corresponding to $\wedge^L$ and $\vee^R$, but at the price of complications in the computation of rank. We do this computation below very much in the style of Gentzen. (Cut elimination in something corresponding to our directly strictified **DL**, but without taking into account equality of derivations, may be found in [13].)

Instead of directly strictifying, we have produced $\mathcal{D}$ according to the recipe of §3.1, §4.5 and §6.2. We find it is interesting to locate Gentzen's sequents within this strictified biassociative structure constructed in the style of Joyal, Street and Mac Lane (the last author was close to Gentzen in his youth). This tells us that Gentzen had a sound premonition that nothing is lost by strictifying with respect to associativity.

Every arrow of $\mathcal{D}$ denoted by a Gentzen term is a sequent arrow. We show in the following lemma that these are all the sequent arrows of $\mathcal{D}$.

GENTZENIZATION LEMMA. *Every sequent arrow of $\mathcal{D}$ is denoted by a Gentzen term.*

PROOF. We prove first that every sequent arrow (f, A, B) of $\mathcal{D}$ is denoted by a Gentzen term. After that we will pass to the sequent arrows (f, Γ, Δ) for Γ and Δ with more than one member.

We show by induction on the lenght of A that $(\mathbf{1}_A, A, A)$ is denoted by a Gentzen term. If A is p, then $(\mathbf{1}_p, p, p)$ is denoted by $\mathbf{1}''_p$. If A is $A_1 \wedge A_2$, and $(\mathbf{1}_{A_i}, A_i, A_i)$ is denoted by the Gentzen term $\mathbf{1}''_{A_i}$, for $i \in \{1, 2\}$, then $(\mathbf{1}_A, A, A)$ is denoted by $\wedge^L \wedge^R(\mathbf{1}''_{A_1}, \mathbf{1}''_{A_2})$. If A is $A_1 \vee A_2$, then $(\mathbf{1}_A, A, A)$ is denoted by $\vee^R \vee^L(\mathbf{1}''_{A_1}, \mathbf{1}''_{A_2})$. We write, in general, $\mathbf{1}''_A$ for the Gentzen term denoting $(\mathbf{1}_A, A, A)$.

We have that $(\hat{b}{}^{\rightarrow}_{A,B,C}, A \wedge (B \wedge C), (A \wedge B) \wedge C)$ is denoted by

$$\wedge^L \wedge^L \wedge^R(\wedge^R(\mathbf{1}''_A, \mathbf{1}''_B), \mathbf{1}''_C),$$

according to Associative Coherence of §4.3.

The inverse arrow $(\hat{b}{}^{\leftarrow}_{A,B,C}, (A \wedge B) \wedge C, A \wedge (B \wedge C))$ is denoted by

$$\wedge^L c^L_{\emptyset,C,A\wedge B,\emptyset} \wedge^L c^L_{\emptyset,A\wedge B,C,\emptyset} \wedge^R(\mathbf{1}''_A, \wedge^R(\mathbf{1}''_B, \mathbf{1}''_C)),$$

according to Symmetric Associative Coherence of §5.1.

We have that $(\hat{c}_{A,B}, A \wedge B, B \wedge A)$ is denoted by $\wedge^L c^L_{\emptyset,B,A,\emptyset} \wedge^R(\mathbf{1}''_B, \mathbf{1}''_A)$, and $(\hat{w}_A, A, A \wedge A)$ is denoted by $w^L \wedge^R(\mathbf{1}''_A, \mathbf{1}''_A)$. Next, $(\hat{k}{}^1_{A,B}, A \wedge B, A)$ is denoted by $\wedge^L k^L_B \mathbf{1}''_A$, while $(\hat{k}{}^2_{A,B}, A \wedge B, B)$ is denoted by $\wedge^L c^L_{\emptyset,B,A,\emptyset} k^L_A \mathbf{1}''_B$. We proceed analogously for $\overset{\vee}{b}{}^{\rightarrow}$, $\overset{\vee}{b}{}^{\leftarrow}$, $\check{c}$, $\check{w}$ and $\overset{\vee}{k}{}^i$ by using $\vee^L$ and $\vee^R$.

We have that $(d_{A,B,C}, A \wedge (B \vee C), (A \wedge B) \vee C)$ is denoted by

$$\vee^R \wedge^L \vee^L(\wedge^R(\mathbf{1}''_A, \mathbf{1}''_B), \mathbf{1}''_C) \quad \text{or} \quad \wedge^L \vee^R \wedge^R(\mathbf{1}''_A, \vee^L(\mathbf{1}''_B, \mathbf{1}''_C)).$$

If the Gentzen terms f'' and g'' denote the sequent arrows (f, A, B) and (g, C, D) respectively, then $\wedge^L \wedge^R(f'', g'')$ denotes $(f \wedge g, A \wedge C, B \wedge D)$, while $\vee^R\vee^L(f'', g'')$ denotes $(f \vee g, A \vee C, B \vee D)$. If the Gentzen terms f'' and g'' denote the sequent arrows (f, A, B) and (g, B, C) respectively, then $cut_{\emptyset,\emptyset}(f'', g'')$ denotes $(g \circ f, A, C)$.

Take now a sequent arrow $(f, A_1{}^{\wedge} \ldots {}^{\wedge} A_n, B_m{}^{\vee} \ldots {}^{\vee} B_1)$ of $\mathcal{D}$ where $n, m \geq 2$. We have proved above that, for F defined as in §4.5, the sequent arrow $(f, F(A_1{}^{\wedge} \ldots {}^{\wedge} A_n), F(B_m{}^{\vee} \ldots {}^{\vee} B_1))$ is denoted by a Gentzen term f''. Then for g and h being respectively $\wedge^R(\ldots \wedge^R(\mathbf{1}''_{A_1}, \mathbf{1}''_{A_2}) \ldots, \mathbf{1}''_{A_n})$ and $\vee^L(\mathbf{1}''_{B_m}, \ldots \vee^L(\mathbf{1}''_{B_2}, \mathbf{1}''_{B_1}) \ldots)$, the Gentzen term $cut_{\emptyset,\emptyset}(cut_{\emptyset,\emptyset}(g, f''), h)$ denotes the sequent arrow $(f, A_1{}^{\wedge} \ldots {}^{\wedge} A_n, B_m{}^{\vee} \ldots {}^{\vee} B_1)$. ⊣

§11.2. Cut elimination in $\mathcal{D}$

In this section we will prove a cut-elimination theorem for the Gentzen terms of $\mathcal{D}$. Before stating and proving this result, we introduce some technical notions and prove some auxiliary results.

A *cut* is a Gentzen term of the form $cut_{\Gamma,\Delta}(f, g)$. A *cut-free* Gentzen term is a Gentzen term none of whose subterms is a cut. A cut $cut_{\Gamma,\Delta}(f, g)$ is called *topmost* when f and g are cut-free.

We say that a Gentzen term is *k-atomized* when for every subterm of it of the form k^L_A or k^R_A we have that A is an atomic formula, which here means that it is a letter. Then we have the following lemma.

ATOMIC-k LEMMA. *For every Gentzen term g there is a k-atomized Gentzen term g' such that $g = g'$ in $\mathcal{D}$. Moreover, if g is cut-free, then g' is cut-free.*

PROOF. By Semilattice Coherence of §9.1, in $\mathcal{D}$ we have

$$k^L_{A\wedge B} f = \wedge^L k^L_B k^L_A f.$$

We show next that for $f : \Gamma \vdash \Delta$ we have in $\mathcal{D}$

$$k^L_{A\vee B} f = w^{R\Delta}_{\emptyset,\emptyset,\emptyset}\, w^{L\Gamma}_{\emptyset,\emptyset,A\vee B} \vee^L(k^L_A f, k^L_B f).$$

The right-hand side (RHS) of this equation is equal to

$$\check{w}_\Delta \circ ((f \circ \hat{k}1''_{\Gamma,A})^{\vee}(f \circ \hat{k}1''_{\Gamma,B})) \circ \hat{e}'_{\Gamma,\Gamma,A,B} \circ (\Gamma^{\wedge}\Gamma^{\wedge}\mathbf{1}^e_{A^{\vee}B}) \circ (\hat{w}_\Gamma{}^{\wedge}(A \vee B)),$$

by the equations (w $\boldsymbol{y}$) of the preceding section. Next we have

$$RHS = f \circ \check{w}_{\Gamma} \circ \hat{k}1''_{\Gamma, A^{\vee}B} \circ \boldsymbol{c}^{k}_{\Gamma,A,\Gamma,B} \circ \hat{\boldsymbol{e}}'_{\Gamma,\Gamma,A,B} \circ (\hat{w}_{\Gamma} \,{}^{\wedge}(A \,{}^{\vee} B)) \circ (\Gamma \,{}^{\wedge} \mathbf{1}^{e}_{A^{\vee}B}),$$
by Lattice Coherence of §9.4,

$$= f \circ \hat{k}1''_{\Gamma, A^{\vee}B} \circ (\check{w}_{\Gamma} \,{}^{\wedge}(A \,{}^{\vee} B)) \circ (\boldsymbol{m}_{\Gamma,\Gamma} {}^{\wedge}(A \,{}^{\vee} B)) \circ (\hat{w}_{\Gamma} \,{}^{\wedge}(A \,{}^{\vee} B)) \circ$$
$$\circ (\Gamma \,{}^{\wedge} \mathbf{1}^{e}_{A^{\vee}B}),$$ by $(\hat{k}1\ nat)$ and $(m\,\hat{e})$,

$$= f \circ \hat{k}1''_{\Gamma, A \vee B},$$ by (mw) and $(\hat{k}1\ nat)$.

For $k^{R}_{A \vee B}$ and $k^{R}_{A \wedge B}$ we proceed analogously. ⊣

We call *leaf formulae* of a Gentzen term h the following occurrences of formulae in the type of h:

when h is $\mathbf{1}''_{p}$, the two occurrences of p in the type $p \vdash p$ of h,

when h is $k^{L}_{A} f$, or $\wedge^{L} f$, or $\vee^{L}(f, g)$, the rightmost member of the source of h,

when h is $k^{R}_{A} f$, or $\wedge^{R}(f, g)$, or $\vee^{R} f$, the leftmost member of the target of h.

For example, the rightmost occurrence of A in the source $\Gamma \,{}^{\wedge} A$ of $k^{L}_{A} f :$ $\Gamma \,{}^{\wedge} A \vdash \Delta$ is a leaf formula of $k^{L}_{A} f$.

The occurrence of A in the type of $w^{L}_{\Gamma_1,\ldots,\Gamma_n,\Gamma} f$ (i.e. in the source of $w^{L}_{\Gamma_1,\ldots,\Gamma_n,\Gamma} f$), recognized according to the index Γ, is called the *lower contraction formula* of $w^{L}_{\Gamma_1,\ldots,\Gamma_n,\Gamma} f$. For every lower contraction formula A of $w^{L} f$ there are two or more occurrences of A in the type of f (i.e. in the source of f), recognized according to the indices $\Gamma_1, \ldots, \Gamma_n, \Gamma$, which we call the *upper contraction formulae* of $w^{L}_{\Gamma_1,\ldots,\Gamma_n,\Gamma} f$. We determine analogously the lower and upper contraction formulae of $w^{R}_{\Delta,\Delta_n,\ldots,\Delta_1} f$ (the difference is that they are now in the targets of $w^{R}_{\Delta,\Delta_n,\ldots,\Delta_1} f$ and f).

For every unary Gentzen operation γ_1 and every Gentzen term $\gamma_1 f$, according to the indices of γ_1 we can recognize in the type of $\gamma_1 f$ what basic sequences are the basic sequences Γ, $\Gamma_1, \ldots, \Gamma_n$, Δ, $\Delta_1, \ldots, \Delta_n$ mentioned in the inductive clause for γ_1. We call these basic sequences the *lower parametric basic sequences* of $\gamma_1 f$. Our inductive clauses for unary Gentzen operations are such that for every lower parametric basic sequence Θ of $\gamma_1 f$ there is a unique basic sequence Θ in the type of f, recognized according to the indices of γ_1 and the inductive clause for γ_1, which we call an *upper parametric basic sequence* of $\gamma_1 f$. We determine analogously the lower and upper parametric basic sequences of $\gamma_2(f,g)$ for a binary Gentzen operation γ_2. Note that our inductive clauses for binary Gentzen operations are such that every lower parametric basic sequence Θ of $\gamma_2(f,g)$ leads unambiguously to a unique upper parametric basic sequence Θ of $\gamma_2(f,g)$ in the type of f or in the type of g. (In terms of linear logic, these clauses correspond to rules for connectives of the multiplicative kind.)

For any Gentzen term $h : \Gamma \vdash \Delta$, and x a member of Γ or Δ, we have that x is either a leaf formula of h, or a lower contraction formula of h, or a member of a lower parametric basic sequence of h. We define the notion of cluster of x in the following manner (this notion, called *Bund* in German, stems from Gentzen; see [56], Section 3.41, [97], Section 2.621, [32], Section 5, and [40], Section 2).

The *cluster* of x in h is a finite tree whose nodes are occurrences of the same formula in the types of subterms of h. We assign to every node a *label*, which is a subterm h' of h such that the node occurs in the type of h'. The root of the cluster of x in h is x, and the label of the root is h. If a node y is a leaf formula of its label, then y is a leaf; i.e., it has no successors. If a node y is the lower contraction formula of its label $w^L_{\Gamma_1,\ldots,\Gamma_n,\Gamma} f$ or $w^R_{\Delta,\Delta_n,\ldots,\Delta_1} f$, then y has as successors the upper contraction formulae of $w^L_{\Gamma_1,\ldots,\Gamma_n,\Gamma} f$ or $w^R_{\Delta,\Delta_n,\ldots,\Delta_1} f$. These successors, of which there are at least two, all have f as labels. If a node y is a member of a lower parametric basic sequence Θ of its label h', then y has a single successor, which is the occurrence of the same formula as y, at the same place, as a member of the upper parametric basic sequence Θ of h'. If h' is here $\gamma_1 f$, then the label of the successor is f, and if h' is $\gamma_2(f,g)$, then the label of the successor is f or g, depending on whether the upper parametric basic sequence Θ occurs in the type of f or in the type of g. With that, we have defined the cluster of x in h.

For a cut $cut_{\Gamma_2,\Delta_2}(f,g)$, the two occurrences of the cut formula A in the target of f and in the source of g, recognized according to the indices Γ_2 and Δ_2, are called respectively the *left cut formula* and the *right cut formula* of $cut_{\Gamma_2,\Delta_2}(f,g)$. (Note that the left cut formula is on the right-hand side of $\vdash$, while the right cut formula is on the left-hand side of $\vdash$.)

For any Gentzen term $h\colon \Gamma \vdash \Delta$, and x a member of Γ or Δ, let $\rho_h(x)$ be the number of nodes in the cluster of x in h. The *left rank* of a cut $cut_{\Gamma_2,\Delta_2}(f,g)$ is $\rho_f(x)$ where x is the left cut formula of $cut_{\Gamma_2,\Delta_2}(f,g)$, and the *right rank* of $cut_{\Gamma_2,\Delta_2}(f,g)$ is $\rho_g(y)$ where y is the right cut formula of $cut_{\Gamma_2,\Delta_2}(f,g)$. The *rank* of a cut is the sum of its left and right ranks. The least rank of a cut is 2, and in that case the left rank and the right rank are both 1. This definition of rank is analogous to Gentzen's, except that Gentzen counts the number of nodes in the longest path, while we count the total number of nodes—either measure is good. (A very formal definition of rank may be found in [11], Section 3.) As a matter of fact, we are interested only in ranks of topmost cuts, but our definition applies to any cut.

We announced in the preceding section (after the equations (w $\boldsymbol{y}$)) that we will sometimes omit the subscripts of Gentzen operations. In the definition below, and sometimes later on, we take for granted the subscripts of $w^L_{\Gamma_1,\dots,\Gamma_n,\Gamma}$, and write just w^L. We do the same with w^R, and other Gentzen operations, when their subscripts are cumbersome, but not important.

We say that a Gentzen term of the form $w^L f$ is a w^L *term*. Subterms that are w^L terms are called w^L *subterms*. We have an analogous terminology with w^R. The *rank* of a w^L term $w^L f$ is $\rho_{w^L f}(x)$ for x being the lower contraction formula of $w^L f$, and analogously for w^R terms. We are interested below only in ranks of cut-free w^L terms, but our definition applies to any w^L term.

Let x be the left cut formula and y the right cut formula of the cut $cut_{\Gamma,\Delta}(f,g)$. Then we say that a w^L subterm h of g is *tied* to $cut_{\Gamma,\Delta}(f,g)$ when h is the label of a node of n-ary branching for $n \geq 2$ in the cluster of y in g, and we say analogously that a w^R subterm h of f is *tied* to $cut_{\Gamma,\Delta}(f,g)$ when h is the label of a node of n-ary branching for $n \geq 2$ in the cluster of x in f.

We say that a w^L term is *blocked* when it is of one of the following forms:

($w\,\gamma\,1$) $w^L_{\Gamma''_1,\Gamma''_2{}^\wedge\Gamma'_1,\Gamma'_2}\, mix(f,g)\colon \Gamma''_1{}^\wedge\Gamma''_2{}^\wedge\Gamma'_1{}^\wedge C^\wedge\Gamma'_2 \vdash \Delta''{}^\vee\Delta'$
for $f\colon \Gamma'_1{}^\wedge C^\wedge\Gamma'_2 \vdash \Delta'$ and $g\colon \Gamma''_1{}^\wedge C^\wedge\Gamma''_2 \vdash \Delta''$,

($w\,\gamma\,2$) $w^L_{\Gamma'_1,\Gamma'_2{}^\wedge\Gamma''_1,\Gamma''_2} \wedge^R(f,g)\colon \Gamma'_1{}^\wedge\Gamma'_2{}^\wedge\Gamma''_1{}^\wedge C^\wedge\Gamma''_2 \vdash (A \wedge B)^\vee\Delta'{}^\vee\Delta''$
for $f\colon \Gamma'_1{}^\wedge C^\wedge\Gamma'_2 \vdash A^\vee\Delta'$ and $g\colon \Gamma''_1{}^\wedge C^\wedge\Gamma''_2 \vdash B^\vee\Delta''$,

$(w\,\gamma\,3)$ $w^L_{\Gamma_1',\Gamma_2'{}^\wedge\Gamma_1'',\Gamma_2''{}^\wedge(A\vee B)}\vee^L(f,g)\colon \Gamma_1'{}^\wedge\Gamma_2'{}^\wedge\Gamma_1''{}^\wedge C^\wedge\Gamma_2''{}^\wedge(A\vee B)\vdash \Delta'{}^\vee\Delta''$
for $f\colon \Gamma_1'{}^\wedge C^\wedge\Gamma_2'{}^\wedge A\vdash\Delta'$ and $g\colon \Gamma_1''{}^\wedge C^\wedge\Gamma_2''{}^\wedge B\vdash\Delta''$,

$(w\wedge 1)$ $w^L_{\Gamma_1,\Gamma_2,\emptyset}\wedge^L f\colon \Gamma_1{}^\wedge\Gamma_2{}^\wedge(A\wedge B)\vdash\Delta$
for $f\colon \Gamma_1{}^\wedge(A\wedge B)^\wedge\Gamma_2{}^\wedge A^\wedge B\vdash\Delta$,

$(w\vee 1)$ $w^L_{\Gamma_1',\Gamma_2'{}^\wedge\Gamma'',\emptyset}\vee^L(f,g)\colon \Gamma_1'{}^\wedge\Gamma_2'{}^\wedge\Gamma''{}^\wedge(A\vee B)\vdash\Delta'{}^\vee\Delta''$
for $f\colon \Gamma_1'{}^\wedge(A\vee B)^\wedge\Gamma_2'{}^\wedge A\vdash\Delta'$ and $g\colon \Gamma''{}^\wedge B\vdash\Delta''$,

$(w\vee 2)$ $w^L_{\Gamma'{}^\wedge\Gamma_1'',\Gamma_2'',\emptyset}\vee^L(f,g)\colon \Gamma'{}^\wedge\Gamma_1''{}^\wedge\Gamma_2''{}^\wedge(A\vee B)\vdash\Delta'{}^\vee\Delta''$
for $f\colon \Gamma'{}^\wedge A\vdash\Delta$ and $g\colon \Gamma_1''{}^\wedge(A\vee B)^\wedge\Gamma_2''{}^\wedge B\vdash\Delta''$,

$(w\vee 3)$ $w^L_{\Gamma_1',\Gamma_2'{}^\wedge\Gamma_1'',\Gamma_2'',\emptyset}\vee^L(f,g)\colon \Gamma_1'{}^\wedge\Gamma_2'{}^\wedge\Gamma_1''{}^\wedge\Gamma_2''{}^\wedge(A\vee B)\vdash\Delta'{}^\vee\Delta''$
for $f\colon \Gamma_1'{}^\wedge(A\vee B)^\wedge\Gamma_2'{}^\wedge A\vdash\Delta'$ and $g\colon \Gamma_1''{}^\wedge(A\vee B)^\wedge\Gamma_2''{}^\wedge B\vdash\Delta''$.

A w^L subterm $w^L h$ of f_2 tied to a topmost cut $cut_{\Gamma,\Delta}(f_1,f_2)$ is *reducible* when it is not blocked and every w^L subterm $w^L t$ of f_2 tied to $cut_{\Gamma,\Delta}(f_1,f_2)$ such that $w^L h$ is a subterm of t is blocked. We can prove the following lemma.

REDUCIBILITY LEMMA. *For every reducible $w^L h$ there is a cut-free Gentzen term h' such that $w^L h = h'$ in $\mathcal{D}$ and after replacing $w^L h$ in $cut_{\Gamma,\Delta}(f_1,f_2)$ by h' all the reducible w^L subterms of h' are of rank lesser than the rank of $w^L h$.*

PROOF. We proceed by cases depending on the form of h.

$(W1)$ First we have cases where $w^L h$ is $w^L_{\Gamma_1',\ldots,\Gamma_n',\Gamma'}h$ while h is $\gamma f\colon \Gamma_1'{}^\wedge A^\wedge\ldots{}^\wedge\Gamma_n'{}^\wedge A^\wedge\Gamma'\vdash\Delta'$ for $f\colon \Gamma_1{}^\wedge A^\wedge\ldots{}^\wedge\Gamma_n{}^\wedge A^\wedge\Gamma\vdash\Delta$ and γ is either one of c^L, c^R, k^R, w^R and $\vee^R$, or k^L, w^L and $\wedge^L$ with the occurrences of A displayed in $\Gamma_1'{}^\wedge A^\wedge\ldots{}^\wedge\Gamma_n'{}^\wedge A^\wedge\Gamma'\vdash\Delta'$ members of the lower parametric basic sequences of γf. Then, by Semilattice Coherence of §9.1, we have either $w^L\gamma f=\gamma w^L f$ or $w^L c^L f = w^L f$.

$(W2)$ For $f\colon \Gamma_1{}^\wedge A^\wedge\ldots{}^\wedge\Gamma_n{}^\wedge A^\wedge\Gamma\vdash\Delta$, by Semilattice Coherence, we have that

$$w^L_{\Gamma_1,\ldots,\Gamma_n,\Gamma,\emptyset}k^L_A f = c^L_{\Gamma_1{}^\wedge\ldots{}^\wedge\Gamma_n,\Gamma,A,\emptyset}w^L_{\Gamma_1,\ldots,\Gamma_n,\Gamma}f.$$

$(W3)$ Suppose we have $f\colon \Gamma_1{}^\wedge A^\wedge \ldots {}^\wedge\Gamma_n{}^\wedge A^\wedge\Gamma \vdash \Delta$ and

$$w^L_{\Gamma_1,\ldots,\Gamma_n,\Gamma} f\colon \Gamma'_1{}^\wedge A^\wedge \ldots {}^\wedge\Gamma'_m{}^\wedge A^\wedge\Gamma' \vdash \Delta,$$

where $\Gamma'_1{}^\wedge A^\wedge \ldots {}^\wedge\Gamma'_m{}^\wedge A^\wedge\Gamma'$ and $\Gamma_1{}^\wedge \ldots {}^\wedge\Gamma_n{}^\wedge A^\wedge\Gamma$ are designations of the same basic sequence, and one of the occurrences of A displayed in the first designation is the occurrence of A displayed in the second. Then we have that

$$w^L_{\Gamma'_1,\ldots,\Gamma'_m,\Gamma'} w^L_{\Gamma_1,\ldots,\Gamma_n,\Gamma} f = w^L f$$

with appropriate subscripts for w^L in $w^L f$. If the rank of $w^L_{\Gamma_1,\ldots,\Gamma_n,\Gamma} f$ is $k+1$, and the rank of $w^L_{\Gamma'_1,\ldots,\Gamma'_m,\Gamma'} w^L_{\Gamma_1,\ldots,\Gamma_n,\Gamma} f$ is $k+1+l+m$, then the rank of $w^L f$ on the right-hand side is $k+l+1$. (Here we have $m \geq 2$.)

$(W4)$ For $f\colon \Gamma_1{}^\wedge(A \wedge B)^\wedge \ldots {}^\wedge\Gamma_n{}^\wedge(A \wedge B)^\wedge\Gamma{}^\wedge A^\wedge B \vdash \Delta$ we have that

$$w^L_{\Gamma_1,\ldots,\Gamma_n,\Gamma,\emptyset} \wedge^L f = w^L_{\Gamma_1{}^\wedge\ldots{}^\wedge\Gamma_n,\Gamma,\emptyset} \wedge^L w^L_{\Gamma_1,\ldots,\Gamma_n,\Gamma{}^\wedge A^\wedge B} f$$

by Semilattice Coherence. Here, the right-hand side is blocked according to $(w \wedge 1)$. The w^L term $w^L_{\Gamma_1,\ldots,\Gamma_n,\Gamma{}^\wedge A^\wedge B} f$ need not be blocked and may be reducible, but it is of lower rank than the left-hand side (the difference is $n+1$).

$(W5)$ For $f\colon \Phi \vdash \Psi$ and $g\colon \Gamma_1{}^\wedge A^\wedge \ldots {}^\wedge\Gamma_n{}^\wedge A^\wedge\Gamma \vdash \Delta$ we have that

$$w^L_{\Gamma_1,\ldots,\Gamma_n,\Gamma{}^\wedge\Phi}\, mix\,(f,g) = mix\,(f, w^L_{\Gamma_1,\ldots,\Gamma_n,\Gamma} g)$$

by Semilattice Coherence and $(m\ nat)$. We proceed analogously when $mix\,(g,f)$ replaces $mix\,(f,g)$.

For $f\colon \Phi_1{}^\wedge A^\wedge \ldots {}^\wedge\Phi_m{}^\wedge A^\wedge\Phi \vdash \Psi$ and g as above we have that

$$\begin{aligned} &w^L_{\Gamma_1,\ldots,\Gamma_n,\Gamma{}^\wedge\Phi_1,\Phi_2,\ldots,\Phi_m,\Phi}\, mix\,(f,g) = \\ &\qquad w^L_{\Gamma_1{}^\wedge\ldots{}^\wedge\Gamma_n,\Gamma{}^\wedge\Phi_1{}^\wedge\ldots{}^\wedge\Phi_m,\Phi}\, mix\,(w^L_{\Phi_1,\ldots,\Phi_m,\Phi} f, w^L_{\Gamma_1,\ldots,\Gamma_n,\Gamma} g) \end{aligned}$$

by Semilattice Coherence and $(m\ nat)$. Here, the right-hand side is blocked according to $(w\,\gamma\,1)$. The w^L terms $w^L_{\Phi_1,\ldots,\Phi_m,\Phi} f$ and $w^L_{\Gamma_1,\ldots,\Gamma_n,\Gamma} g$ need not be blocked and may be reducible, but they are both of lower rank than the left-hand side.

We proceed as in case $(W5)$ when $w^L h$ is $w^L\wedge^R(f,g)$ (one of these cases involves a blocked w^L term according to $(w\,\gamma\,2)$). We have cases analogous to $(W5)$ also when $w^L h$ is $w^L\vee^L(f,g)$ (here we apply Semilattice Coherence and $(d^L\ nat)$, and one of these cases involves a blocked w^L term according to $(w\,\gamma\,3)$). We have three additional cases when $w^L h$ is $w^L\vee^L(f,g)$, which all yield blocked w^L terms according to $(w\vee 1)$, $(w\vee 2)$ and $(w\vee 3)$. One of these cases is the following.

$(W6)$ For the Gentzen terms $f\colon \Gamma'_1{}^\wedge(A\vee B)^\wedge\ldots{}^\wedge\Gamma'_n{}^\wedge(A\vee B)^\wedge\Gamma'^\wedge A\vdash\Delta'$ and $g\colon \Gamma''_1{}^\wedge(A\vee B)^\wedge\ldots{}^\wedge\Gamma''_m{}^\wedge(A\vee B)^\wedge\Gamma''^\wedge B\vdash\Delta''$ we have that

$$w^L_{\Gamma'_1,\ldots,\Gamma'_n,\Gamma'^\wedge\Gamma''_1,\Gamma''_2,\ldots,\Gamma''_m,\Gamma'',\emptyset}\vee^L(f,g) =$$
$$w^L_{\Gamma'_1{}^\wedge\ldots{}^\wedge\Gamma'_n,\Gamma'^\wedge\Gamma''_1{}^\wedge\ldots{}^\wedge\Gamma''_m,\Gamma'',\emptyset}\vee^L(w^L_{\Gamma'_1,\ldots,\Gamma'_n,\Gamma'^\wedge A}f,w^L_{\Gamma''_1,\ldots,\Gamma''_m,\Gamma''^\wedge B}g)$$

by Semilattice Coherence and $(d^L\ nat)$. Here, the right-hand side is blocked according to $(w\vee 3)$. The w^L terms $w^L_{\Gamma'_1,\ldots,\Gamma'_n,\Gamma'^\wedge A}f$ and $w^L_{\Gamma''_1,\ldots,\Gamma''_m,\Gamma''^\wedge B}g$ need not be blocked and may be reducible, but they are both of lower rank than the left-hand side.

To conclude the proof of the lemma we have only to check that the condition on ranks is satisfied in all cases, even in those where we have not noted the fact. ⊣

We have an analogous definition of *blocked* w^R terms and of *reducible* w^R subterms. With that, we prove for w^R terms a lemma exactly analogous to the Reducibility Lemma. As a corollary of these two Reducibility Lemmata, we have the following lemma.

BLOCKED-w LEMMA. *Every topmost* $cut_{\Gamma,\Delta}(f,g)$ *is equal in* $\mathcal{D}$ *to a topmost* $cut_{\Gamma,\Delta}(f',g')$ *in which all* w^L *and* w^R *subterms tied to* $cut_{\Gamma,\Delta}(f',g')$ *are blocked.*

The proof of this corollary is based on a multiset-ordering induction, which stems from Gentzen (see [56] and [31]).

The *degree* of a cut is the number of occurrences of connectives (in this case, the number of occurrences of $\wedge$ and $\vee$) in the cut formula.

The *complexity* of a topmost cut is a pair (d,r) where d is the degree of this cut and r is its rank. These complexities are ordered lexicographically (i.e., we have $(d_1, r_1) < (d_2, r_2)$ iff either $d_1 < d_2$, or $d_1 = d_2$ and $r_1 < r_2$; cf. §7.7).

According to the Atomic-k Lemma and the Blocked-w Lemma, every topmost cut $cut_{\Gamma,\Delta}(f, g)$ is equal to a topmost cut $cut_{\Gamma,\Delta}(f', g')$ with the same cut formula such that f' and g' are k-atomized and every w^L or w^R subterm of f' and g' tied to $cut_{\Gamma,\Delta}(f', g')$ is blocked. We call topmost cut such as $cut_{\Gamma,\Delta}(f', g')$ *clean cuts.*

We can then prove the following theorem.

CUT-ELIMINATION THEOREM. *For every Gentzen term t there is a cut-free Gentzen term t' such that $t = t'$ in $\mathcal{D}$.*

PROOF. We show by induction on the complexity of clean cuts that they are equal in $\mathcal{D}$ to cut-free Gentzen terms. This will suffice to prove the theorem.

For the basis of this induction, take a clean cut of complexity $(0, 2)$. This means that this clean cut is of one of the forms displayed on the left-hand side of the following equations of $\mathcal{D}$:

$$cut_{\emptyset,\emptyset}(\mathbf{1}''_p, \mathbf{1}''_p) = \mathbf{1}''_p,$$
$$cut_{\emptyset,\emptyset}(\mathbf{1}''_p, k^L_p g) = k^L_p g,$$
$$cut_{\emptyset,\emptyset}(k^R_p f, \mathbf{1}''_p) = k^R_p f,$$
$$cut_{\emptyset,\emptyset}(k^R_p f, k^L_p g) = mix(f, g).$$

For the first three equations we use (cut 1), while the fourth holds by definition. With that, we have proved the basis of the induction.

Note that with the first three equations we proceed as Gentzen, but not with the fourth. Instead of reducing the left-hand side of this equation to the right-hand side, Gentzen would reduce it to a cut-free term obtained either from f with a number of k^L, c^L and k^R operations, or from g with a number of k^R, c^R and k^L operations (cf. [55], Section III.3.113.1-2). Such reductions are, however, not supported by equations of $\mathcal{D}$.

We pass now to the induction step. Suppose first that the complexity of our clean cut is $(d, 2)$ for $d > 0$. When the cut formula is of the form $A \wedge B$, our clean cut must be of the form

$$cut_{\emptyset,\emptyset}(\wedge^R(f, g), \wedge^L h),$$

for $f\colon \Gamma_1 \vdash A^\vee\Delta_1$, $g\colon \Gamma_2 \vdash B^\vee\Delta_2$ and $h\colon \Gamma^\wedge A^\wedge B \vdash \Delta$. Then we have in $\mathcal{D}$ the equation

$$cut_{\emptyset,\emptyset}(\wedge^R(f,g),\wedge^L h) = c^R_{\Delta,\Delta_2,\Delta_1,\emptyset} c^L_{\Gamma,\Gamma_2,\Gamma_1,\emptyset} cut_{\Gamma_2,\emptyset}(f, cut_{\emptyset,\emptyset}(g,h)).$$

To show that this equation holds in $\mathcal{D}$, we have that, with subscripts omitted, the left-hand side is equal to

$$(h^\vee\mathbf{1})\circ((\mathbf{1}^\wedge\mathbf{1}^e)^\vee\mathbf{1})\circ\boldsymbol{d}\circ(\mathbf{1}^\wedge(\mathbf{1}^{i\vee}\mathbf{1}))\circ(\mathbf{1}^\wedge\check{\boldsymbol{e}}')\circ(\mathbf{1}^\wedge f^\wedge g) =$$
$$(h^\vee\mathbf{1})\circ\boldsymbol{d}\circ(\mathbf{1}^\wedge\check{\boldsymbol{e}}')\circ(\mathbf{1}^\wedge f^\wedge g),$$

by $(d^L\ nat)$ and the fact that $\mathbf{1}^i$ and $\mathbf{1}^e$ are isomorphisms, while, again with subscripts omitted, for the right-hand side we have

$$(\mathbf{1}^\vee\check{\boldsymbol{c}})\circ(h^\vee\mathbf{1})\circ(\boldsymbol{d}^\vee\mathbf{1})\circ((\mathbf{1}_\Gamma{}^\wedge\mathbf{1}^\wedge g)^\vee\mathbf{1})\circ((\mathbf{1}_\Gamma{}^\wedge\hat{\boldsymbol{c}})^\vee\mathbf{1})\circ\boldsymbol{d}\circ(\mathbf{1}^\wedge f)\circ(\mathbf{1}^\wedge\hat{\boldsymbol{c}}) =$$
$$(h^\vee\mathbf{1})\circ(\mathbf{1}^\vee\check{\boldsymbol{c}})\circ(\boldsymbol{d}^\vee\mathbf{1})\circ((\mathbf{1}_\Gamma{}^\wedge\hat{\boldsymbol{c}})^\vee\mathbf{1})\circ\boldsymbol{d}\circ(\mathbf{1}^\wedge\hat{\boldsymbol{c}})\circ(\mathbf{1}^\wedge f^\wedge g),$$

by the bifunctorial equation $(\vee 2)$ of §2.7, $(\hat{c}\ nat)$ and $(d^L\ nat)$. It suffices to note now that

$$\boldsymbol{d}\circ(\mathbf{1}^\wedge\check{\boldsymbol{e}}') = (\mathbf{1}^\wedge\check{\boldsymbol{c}})\circ(\boldsymbol{d}^\vee\mathbf{1})\circ((\mathbf{1}^\wedge\hat{\boldsymbol{c}})^\vee\mathbf{1})\circ\boldsymbol{d}\circ(\mathbf{1}^\wedge\hat{\boldsymbol{c}})$$

holds by Symmetric Net Coherence of §7.6. When Γ is $\emptyset$, we have essentially a case of the equation $(\check{e})$ of §7.6.

We replaced a clean cut $cut_{\emptyset,\emptyset}(\wedge^R(f,g),\wedge^L h)$ of complexity $(d,2)$ by

$$c^R_{\Delta,\Delta_2,\Delta_1,\emptyset} c^L_{\Gamma,\Gamma_2,\Gamma_1,\emptyset} cut_{\Gamma_2,\emptyset}(f, cut_{\emptyset,\emptyset}(g,h)).$$

According to the Atomic-k Lemma and the Blocked-w Lemma, the topmost cut $cut_{\emptyset,\emptyset}(g,h)$ is equal to a clean cut $cut_{\emptyset,\emptyset}(g',h')$ of complexity (d',r) with $d' < d$, because the cut formula is now B instead of $A\wedge B$. By the induction hypothesis, $cut_{\emptyset,\emptyset}(g',h') = s$ for a cut-free Gentzen term s. The topmost cut $cut_{\Gamma_2,\emptyset}(f,s)$ is equal to a clean cut $cut_{\Gamma_2,\emptyset}(f',s')$ of complexity (d'',r) with $d'' < d$, because the cut formula is now A instead of $A\wedge B$. So we can apply again the induction hypothesis.

We proceed analogously when the cut formula is of the form $A\vee B$. With that, we are over with the cases where the complexity of our clean cut is $(d,2)$ for $d>0$. We dealt with them in the spirit of Gentzen.

Suppose now that the complexity of our clean cut is (d,r) for $r > 2$, and suppose the right rank of this clean cut is greater than 1. We proceed analogously if the left rank is greater than 1, and we need not consider this case separately.

Suppose first that in our clean cut $cut_{\Gamma_2,\Delta_2}(f,\gamma g)$ the cut formula occurs in a lower parametric basic sequence of γg. Depending on various cases for the unary Gentzen operation γ, we want to show that one of the following two equations holds in $\mathcal{D}$:

$$(\dagger) \quad cut_{\Gamma_2,\Delta_2}(f,\gamma g) = \gamma cut_{\Gamma_2',\Delta_2}(f,g),$$
$$(\dagger\dagger) \quad cut_{\Gamma_2,\Delta_2}(f,\gamma g) = c^L \gamma cut_{\Gamma_2',\Delta_2}(f,g)$$

for a clean cut $cut_{\Gamma_2',\Delta_2}(f,g)$ of complexity (d,r') with $r' = r-1$. The subscripts omitted in γ need not be the same on the two sides of $(\dagger)$ or $(\dagger\dagger)$, and often they are not such. We have the following cases.

(1) If γ is c^L, then $(\dagger)$ holds by Symmetric Net Coherence.

(2) If γ is c^R, or k^R, or w^R, or $\vee^R$, then $(\dagger)$ holds by the bifunctorial equation $(\vee 2)$.

(3) If γ is k^L, then $(\dagger\dagger)$ holds. To show that, we distinguish two cases. In both cases, we have $g\colon \Gamma_1{}^\wedge A{}^\wedge \Gamma_2' \vdash \Delta_3$ and $k^L_B g\colon \Gamma_1{}^\wedge A{}^\wedge\Gamma_2'{}^\wedge B \vdash \Delta_3$ where $\Gamma_2'{}^\wedge B$ is Γ_2. In the first case, the basic sequence $\Gamma_1{}^\wedge\Gamma_2'$ is not $\emptyset$, and in the second case it is $\emptyset$.

In the first case, for the left-hand side of $(\dagger\dagger)$ we have

$$(g^\vee \mathbf{1}) \circ (\hat{k}1''_{\Gamma_1{}^\wedge A{}^\wedge\Gamma_2',B}{}^\vee \mathbf{1}) \circ ((\mathbf{1}{}^\wedge \hat{\boldsymbol{c}}_{\Gamma_2'{}^\wedge B,A})^\vee \mathbf{1}) \circ \boldsymbol{d} \circ (\mathbf{1}{}^\wedge(\check{\boldsymbol{c}}{}^\vee \mathbf{1})) \circ (\mathbf{1}{}^\wedge f),$$

while for the right-hand side of $(\dagger\dagger)$ we have

$$\begin{aligned}
&(g^\vee \mathbf{1}) \circ ((\mathbf{1}{}^\wedge \hat{\boldsymbol{c}})^\vee \mathbf{1}) \circ \boldsymbol{d} \circ (\mathbf{1}{}^\wedge(\check{\boldsymbol{c}}{}^\vee \mathbf{1})) \circ (\mathbf{1}{}^\wedge f) \circ \hat{k}1''_{\Gamma_1{}^\wedge\Gamma_2'{}^\wedge\Gamma_3,B} \circ (\mathbf{1}{}^\wedge \hat{\boldsymbol{c}}_{B,\Gamma_3})\\
&\quad = (g^\vee \mathbf{1}) \circ ((\mathbf{1}{}^\wedge \hat{\boldsymbol{c}})^\vee \mathbf{1}) \circ \boldsymbol{d} \circ (\mathbf{1}{}^\wedge(\check{\boldsymbol{c}}{}^\vee \mathbf{1})) \circ \hat{k}1''_{\Gamma_1{}^\wedge\Gamma_2'{}^\wedge(\Delta_2{}^\vee A{}^\vee\Delta_1),B} \circ\\
&\qquad\qquad \circ (\mathbf{1}{}^\wedge \hat{\boldsymbol{c}}_{B,\Delta_2{}^\vee A{}^\vee\Delta_1}) \circ (\mathbf{1}{}^\wedge f)\\
&\quad = (g^\vee \mathbf{1}) \circ ((\mathbf{1}{}^\wedge \hat{\boldsymbol{c}})^\vee \mathbf{1}) \circ \boldsymbol{d} \circ (\mathbf{1}{}^\wedge(\check{\boldsymbol{c}}{}^\vee \mathbf{1})) \circ (\hat{k}1''_{\Gamma_1{}^\wedge\Gamma_2',B}{}^\wedge \mathbf{1}_{\Delta_2{}^\vee A{}^\vee\Delta_1}) \circ (\mathbf{1}{}^\wedge f),
\end{aligned}$$

by Semilattice Coherence (this step cannot be made if $\Gamma_1{}^\wedge\Gamma_2'$ is $\emptyset$, since $\hat{k}1''_{\Gamma_1{}^\wedge\Gamma_2',B}$ would not be defined),

$$\quad = (g^\vee \mathbf{1}) \circ ((\mathbf{1}{}^\wedge \hat{\boldsymbol{c}}_{\Gamma_2',A})^\vee \mathbf{1}) \circ ((\hat{k}1''_{\Gamma_1{}^\wedge\Gamma_2',B}{}^\wedge \mathbf{1})^\vee \mathbf{1}) \circ \boldsymbol{d} \circ (\mathbf{1}{}^\wedge(\check{\boldsymbol{c}}{}^\vee \mathbf{1})) \circ (\mathbf{1}{}^\wedge f),$$

which is equal to the left-hand side by Semilattice Coherence.

In the second case, when $\Gamma_1{}^\wedge\Gamma_2'$ is $\emptyset$, for the left-hand side of (††) we have

$$(g^\vee \mathbf{1})\circ(\hat{k}^1_{A,B}{}^\vee \mathbf{1})\circ(\hat{\boldsymbol{c}}_{B,A}{}^\vee \mathbf{1})\circ \boldsymbol{d}\circ(\mathbf{1}^\wedge(\check{\boldsymbol{c}}{}^\vee \mathbf{1}))\circ(\mathbf{1}^\wedge f),$$

while for the right-hand side we have

$$\begin{aligned}(g^\vee \mathbf{1})&\circ(\check{\boldsymbol{c}}{}^\vee \mathbf{1})\circ f\circ \hat{k}1''_{\Gamma_3,B}\circ\ \hat{\boldsymbol{c}}_{B,\Gamma_3}\\ &=(g^\vee \mathbf{1})\circ \hat{k}2''_{B,A^\vee\Delta_2{}^\vee\Delta_1}\circ(\mathbf{1}^\wedge(\check{\boldsymbol{c}}{}^\vee \mathbf{1}))\circ(\mathbf{1}^\wedge f)\\ &=(g^\vee \mathbf{1})\circ(\hat{k}2''_{B,A}{}^\vee \mathbf{1}_{\Delta_2{}^\vee\Delta_1})\circ \boldsymbol{d}_{B,A,\Delta_2{}^\vee\Delta_1}\circ(\mathbf{1}^\wedge(\check{\boldsymbol{c}}{}^\vee \mathbf{1}))\circ(\mathbf{1}^\wedge f),\ \text{by } (d\hat{k}),\end{aligned}$$

which is equal to the left-hand side by $(\hat{c}\hat{k})$ (see §9.1).

(4) If γ is w^L or $\wedge^L$, then (†) holds by various bifunctorial and naturality equations.

Suppose next that in our clean cut $cut_{\Gamma_2,\Delta_2}(f,\gamma(g,h))$, the cut formula A occurs in a lower parametric basic sequence of $\gamma(g,h)$ (here, γ is a binary Gentzen operation).

(5) If γ is mix, then for $f\colon \Gamma_3\vdash \Delta_2{}^\vee A^\vee\Delta_1$, $g\colon \Gamma_2''\vdash\Delta_3''$ and $h\colon \Gamma_1{}^\wedge A^\wedge\Gamma_2'\vdash \Delta_3'$ we have in $\mathcal{D}$ the equation

$$cut_{\Gamma_2'{}^\wedge\Gamma_2'',\Delta_2}(f, mix(g,h)) =$$
$$c^R_{\Delta_3',\Delta_2{}^\vee\Delta_1,\Delta_3'',\emptyset}c^L_{\Gamma_1{}^\wedge\Gamma_2',\Gamma_3,\Gamma_2'',\emptyset}mix(g, cut_{\Gamma_2',\Delta_2}(f,h)),$$

where the complexity of the clean cut $cut_{\Gamma_2',\Delta_2}(f,h)$ is (d,r') with $r'=r-1$.

By bifunctorial and naturality equations, the left-hand side of this equation is equal to $(h^\vee\mathbf{1})\circ LHS^*\circ(\mathbf{1}^\wedge g^\wedge f)$ where

$$LHS^* = (\boldsymbol{m}_{\Gamma_1{}^\wedge A^\wedge\Gamma_2',\Delta_3''}{}^\vee \mathbf{1})\circ((\mathbf{1}^\wedge\hat{\boldsymbol{c}})^\vee\mathbf{1})\circ\boldsymbol{d}\circ(\mathbf{1}^\wedge(\check{\boldsymbol{c}}{}^\vee\mathbf{1})),$$

while the right-hand side is equal to $(h^\vee\mathbf{1})\circ RHS^*\circ(\mathbf{1}^\wedge g^\wedge f)$ where

$$\begin{aligned}RHS^* = (\mathbf{1}^\vee\check{\boldsymbol{c}})\circ((\mathbf{1}^\wedge\hat{\boldsymbol{c}})^\vee\mathbf{1})\circ(\boldsymbol{d}^\vee\mathbf{1})\circ((\mathbf{1}^\wedge(\check{\boldsymbol{c}}{}^\vee\mathbf{1}))^\vee\mathbf{1})\circ\\ \circ\,\boldsymbol{m}_{\Gamma_1{}^\wedge\Gamma_2'{}^\wedge(\Delta_2{}^\vee A^\vee\Delta_1),\Delta_3''}\circ(\mathbf{1}^\wedge\hat{\boldsymbol{c}}).\end{aligned}$$

We have $LHS^* = RHS^*$ by Mix-Symmetric Net Coherence of §8.4. We proceed analogously if g and h have types interchanged.

(6) If γ is $\wedge^R$, then we have in $\mathcal{D}$ one of the following two equations:

$$cut(f, \wedge^R(g,h)) = c^R c^L \wedge^R(cut(f,g), h),$$
$$cut(f, \wedge^R(g,h)) = \wedge^R(f, cut(g,h)),$$

where the complexity of the clean cuts $cut(f,g)$ and $cut(g,h)$ is (d,r') with $r' = r-1$, and appropriate subscripts are assigned to *cut*. Both of these equations are justified by Symmetric Net Coherence. We proceed analogously if γ is $\vee^L$.

With that, we are over with the cases of a clean cut $cut_{\Gamma_2,\Delta_2}(f, \gamma g)$ or $cut_{\Gamma_2,\Delta_2}(f, \gamma(g,h))$ of complexity (d,r) with $r > 2$ where the cut formula occurs in a lower parametric basic sequence of γg or $\gamma(g,h)$. All these cases are dealt with in the spirit of Gentzen, except for the case with *mix*, which Gentzen did not envisage.

Now we proceed with the cases of complexity (d,r) with $r > 2$ where the cut formula does not occur in this manner in a lower parametric basic sequence. Then our clean cut must be of the form $cut(f, w^L g)$ with blocked $w^L g$ tied to our clean cut, and we have to go through the cases $(w\,\gamma\,1)$, $(w\,\gamma\,2), \ldots, (w \vee 3)$ for blocked w^L terms.

$(w\,\gamma\,1)$ For $f\colon \Gamma_1' {}^\wedge C {}^\wedge \Gamma_2' \vdash \Delta'$, $g\colon \Gamma_1'' {}^\wedge C {}^\wedge \Gamma_2'' \vdash \Delta''$ and $h\colon \Gamma \vdash \Delta_1 {}^\vee C {}^\vee \Delta_2$, where Γ' is $\Gamma_1' {}^\wedge \Gamma_2'$, while Γ'' is $\Gamma_1'' {}^\wedge \Gamma_2''$, and Δ is $\Delta_1 {}^\vee \Delta_2$, we have in $\mathcal{D}$ the equation

$$\begin{aligned} &cut_{\Gamma_2'',\Delta_1}(h, w^L_{\Gamma_1',\Gamma_2'{}^\wedge\Gamma_1'',\Gamma_2''}(mix(g,f))) = \\ &\qquad c^R_{\Delta',\Delta,\Delta'',\emptyset} w^{R\Delta}_{\Delta',\Delta'',\emptyset} w^{L\Gamma}_{\Gamma',\Gamma'',\emptyset} mix(cut_{\Gamma_2'',\Delta_1}(h,g), cut_{\Gamma_2',\Delta_1}(h,f)), \end{aligned}$$

where the complexity of the clean cut $cut_{\Gamma_2',\Delta_1}(h,f)$ is (d,r') with $r' < r$, and analogously for the clean cut $cut_{\Gamma_2'',\Delta_1}(h,g)$.

If $f'\colon \Gamma'{}^\wedge C \vdash \Delta'$ is $f \circ (\mathbf{1}^\wedge\, \hat{\boldsymbol{c}}_{\Gamma_2',C})$, while $g'\colon \Gamma''{}^\wedge C \vdash \Delta''$ is $g \circ (\mathbf{1}^\wedge\, \hat{\boldsymbol{c}}_{\Gamma_2'',C})$, and h' is $(\check{\boldsymbol{c}}_{C,\Delta_1}{}^\vee \mathbf{1}) \circ h$, then, by Semilattice Coherence and bifunctorial and naturality equations, the left-hand side of our equation is equal to

$$(\boldsymbol{m}_{\Delta',\Delta''}{}^\vee \mathbf{1}) \circ ((f'\,{}^\wedge g')^\vee \mathbf{1}) \circ LHS^* \circ (\mathbf{1}^\wedge h')$$

where LHS^* is $((\mathbf{1}_{\Gamma'}{}^\wedge\, \hat{\boldsymbol{c}}_{\Gamma'',C}\, {}^\wedge \mathbf{1}_C)^\vee \mathbf{1}_\Delta) \circ ((\mathbf{1}^\wedge\, \hat{\boldsymbol{w}}_C)^\vee \mathbf{1}) \circ \boldsymbol{d}$, while the right-hand side is equal to

$$(\mathbf{1}^{\vee}\,\check{\boldsymbol{w}}_{\Delta})\circ(\mathbf{1}_{\Delta'}{}^{\vee}\,\check{\boldsymbol{c}}_{\Delta'',\Delta}{}^{\vee}\mathbf{1}_{\Delta})\circ\boldsymbol{m}_{\Delta'^{\vee}\Delta,\Delta''^{\vee}\Delta}\circ((f'\,^{\vee}\mathbf{1})^{\wedge}(g'\,^{\vee}\mathbf{1}))\circ(\boldsymbol{d}^{\wedge}\boldsymbol{d})\circ$$
$$\circ(\mathbf{1}^{\wedge}\,\hat{\boldsymbol{c}}\,^{\wedge}\mathbf{1})\circ(\mathbf{1}^{\wedge}\,\hat{\boldsymbol{w}}_{C^{\vee}\Delta})\circ(\mathbf{1}^{\wedge}h').$$

By the equation $(m\,\check{c}^{m})$ of the preceding section, we have

$$(\mathbf{1}_{\Delta'}{}^{\vee}\,\check{\boldsymbol{c}}_{\Delta'',\Delta}{}^{\vee}\mathbf{1}_{\Delta})\circ\boldsymbol{m}_{\Delta'^{\vee}\Delta,\Delta''^{\vee}\Delta}=(\boldsymbol{m}_{\Delta',\Delta''}{}^{\vee}\mathbf{1}_{\Delta^{\vee}\Delta})\circ\check{\boldsymbol{e}}'_{\Delta',\Delta'',\Delta,\Delta}.$$

So the right-hand side is equal to

$$(\boldsymbol{m}_{\Delta',\Delta''}{}^{\vee}\mathbf{1})\circ((f'\,^{\wedge}g')^{\vee}\mathbf{1})\circ RHS^{*}\circ(\mathbf{1}^{\wedge}h')$$

where

$$\begin{aligned}
RHS^{*} &= (\mathbf{1}^{\vee}\,\check{\boldsymbol{w}}_{\Delta})\circ\check{\boldsymbol{e}}'_{\Gamma'^{\wedge}C,\Gamma''^{\wedge}C,\Delta,\Delta}\circ(\boldsymbol{d}^{\wedge}\boldsymbol{d})\circ(\mathbf{1}^{\wedge}\,\hat{\boldsymbol{c}}\,^{\wedge}\mathbf{1})\circ(\mathbf{1}^{\wedge}\,\hat{\boldsymbol{w}}_{C^{\vee}\Delta})\\
&= (\mathbf{1}^{\vee}\,\check{\boldsymbol{w}}_{\Delta})\circ\check{\boldsymbol{e}}'\circ(\boldsymbol{d}^{\wedge}\boldsymbol{d})\circ(\mathbf{1}^{\wedge}\,\hat{\boldsymbol{c}}\,^{\wedge}\mathbf{1})\circ(\mathbf{1}^{\wedge}\boldsymbol{c}^{k})\circ(\mathbf{1}^{\wedge}(\mathbf{1}^{\vee}\,\hat{\boldsymbol{w}}_{\Delta}))\circ\\
&\qquad\circ(\mathbf{1}^{\wedge}(\hat{\boldsymbol{w}}_{C}\,^{\vee}\mathbf{1})),\ \text{by Lattice Coherence, provided } \Delta \text{ is not } \emptyset,\\
&= (\mathbf{1}^{\vee}\,\check{\boldsymbol{w}}_{\Delta})\circ((\mathbf{1}^{\wedge}\,\hat{\boldsymbol{c}}\,^{\wedge}\mathbf{1})^{\vee}\mathbf{1})\circ\boldsymbol{d}\circ(\mathbf{1}^{\wedge}\check{\boldsymbol{e}}')\circ(\mathbf{1}^{\wedge}\boldsymbol{c}^{k})\circ(\mathbf{1}^{\wedge}(\mathbf{1}^{\vee}\,\hat{\boldsymbol{w}}_{\Delta}))\circ\\
&\qquad\circ(\mathbf{1}^{\wedge}(\hat{\boldsymbol{w}}_{C}\,^{\vee}\mathbf{1})),\ \text{by Symmetric Net Coherence},\\
&= (\mathbf{1}^{\vee}\,\check{\boldsymbol{w}}_{\Delta})\circ((\mathbf{1}^{\wedge}\,\hat{\boldsymbol{c}}\,^{\wedge}\mathbf{1})^{\vee}\mathbf{1})\circ\boldsymbol{d}\circ(\mathbf{1}^{\wedge}(\mathbf{1}^{\vee}\boldsymbol{m}_{\Delta,\Delta}))\circ(\mathbf{1}^{\wedge}(\mathbf{1}^{\vee}\,\hat{\boldsymbol{w}}_{\Delta}))\circ\\
&\qquad\circ(\mathbf{1}^{\wedge}(\hat{\boldsymbol{w}}_{C}\,^{\vee}\mathbf{1})),\ \text{by } (m\,\check{e}),\\
&= ((\mathbf{1}^{\wedge}\,\hat{\boldsymbol{c}}\,^{\wedge}\mathbf{1})^{\vee}\mathbf{1})\circ\boldsymbol{d}\circ(\mathbf{1}^{\wedge}(\hat{\boldsymbol{w}}_{C}\,^{\vee}\mathbf{1})),\ \text{by bifunctorial and naturality}\\
&\qquad\text{equations, and } (wm),\\
&= LHS^{*},\ \text{by } (d^{L}\ nat).
\end{aligned}$$

If Δ is $\emptyset$, then $LHS^{*}=RHS^{*}$ by Semilattice Coherence.

$(w\,\gamma\,2)$ For $f:\Gamma'_1{}^{\wedge}C^{\wedge}\Gamma'_2\vdash A^{\vee}\Delta'$, $g:\Gamma''_1{}^{\wedge}C^{\wedge}\Gamma''_2\vdash B^{\vee}\Delta''$ and $h:\Gamma\vdash\Delta_1{}^{\vee}C^{\vee}\Delta_2$, with Γ', Γ'' and Δ as in $(w\,\gamma\,1)$, we have in $\mathcal{D}$ the equation

$$cut_{\Gamma''_2,\Delta_1}(h,w^{L}_{\Gamma'_1,\Gamma'_2{}^{\wedge}\Gamma''_1,\Gamma''_2}{}^{\wedge R}(f,g))=$$
$$c^{R}_{A\wedge B\,^{\vee}\Delta',\Delta,\Delta'',\emptyset}w^{R\Delta}_{A\wedge B\,^{\vee}\Delta',\Delta'',\emptyset}w^{L\Gamma}_{\Gamma',\Gamma'',\emptyset}{}^{\wedge R}(cut_{\Gamma'_2,\Delta_1}(h,f),cut_{\Gamma''_2,\Delta_1}(h,g)),$$

where the complexity of the clean cut $cut_{\Gamma'_2,\Delta_1}(h,f)$ is (d,r') with $r'<r$, and analogously for the clean cut $cut_{\Gamma''_2,\Delta_1}(h,g)$.

For $f':\Gamma'^{\wedge}C\vdash A^{\vee}\Delta'$, $g':\Gamma''^{\wedge}C\vdash B^{\vee}\Delta''$ and $h':\Gamma\vdash C^{\vee}\Delta$ defined as in $(w\,\gamma\,1)$, we have by Semilattice Coherence and bifunctorial and naturality equations that the left-hand side of our equation is equal to

$$(\mathbf{1}^{i}_{A\wedge B}{}^{\vee}\mathbf{1})\circ(\check{\boldsymbol{e}}'_{A,B,\Delta',\Delta''}{}^{\vee}\mathbf{1})\circ((f'{}^{\wedge}g')^{\vee}\mathbf{1})\circ LHS^{*}\circ(\mathbf{1}^{\wedge}h')$$

where LHS^{*} is $((\mathbf{1}^{\wedge}\hat{\boldsymbol{c}}{}^{\wedge}\mathbf{1})^{\vee}\mathbf{1})\circ((\mathbf{1}^{\wedge}\hat{\boldsymbol{w}}_{C})^{\vee}\mathbf{1})\circ\boldsymbol{d}$, while the right-hand side is equal to

$$(\mathbf{1}^{\vee}\check{\boldsymbol{w}}_{\Delta})\circ(\mathbf{1}^{\vee}\check{\boldsymbol{c}}{}^{\vee}\mathbf{1})\circ(\mathbf{1}^{i\vee}\mathbf{1})\circ\check{\boldsymbol{e}}'_{A,B,\Delta'\vee\Delta,\Delta''\vee\Delta}\circ((f'{}^{\vee}\mathbf{1})^{\wedge}(g'{}^{\vee}\mathbf{1}))\circ(\boldsymbol{d}^{\wedge}\boldsymbol{d})\circ$$
$$\circ(\mathbf{1}^{\wedge}h'{}^{\wedge}\mathbf{1}^{\wedge}h')\circ(\mathbf{1}^{\wedge}\hat{\boldsymbol{c}}{}^{\wedge}\mathbf{1})\circ(\mathbf{1}^{\wedge}\hat{\boldsymbol{w}}_{\Gamma}).$$

We have by Symmetric Net Coherence that

$$\check{\boldsymbol{e}}'_{A,B,\Delta'\vee\Delta,\Delta''\vee\Delta}=(\mathbf{1}_{(A\wedge B)\vee\Delta'}{}^{\vee}\check{\boldsymbol{c}}_{\Delta,\Delta''}{}^{\vee}\mathbf{1}_{\Delta})\circ(\check{\boldsymbol{e}}'_{A,B,\Delta',\Delta''}{}^{\vee}\mathbf{1}_{\Delta\vee\Delta})\circ$$
$$\circ\check{\boldsymbol{e}}'_{A\vee\Delta',B\vee\Delta'',\Delta,\Delta}.$$

Then, by bifunctorial and naturality equations, and $(\check{c}\check{c})$ (see the List of Equations), the right-hand side is equal to

$$(\mathbf{1}^{i}_{A\wedge B}{}^{\vee}\mathbf{1})\circ(\check{\boldsymbol{e}}'_{A,B,\Delta',\Delta''}{}^{\vee}\mathbf{1})\circ((f'{}^{\wedge}g')^{\vee}\mathbf{1})\circ RHS^{*}\circ(\mathbf{1}^{\wedge}h')$$

where

$$\begin{aligned}RHS^{*}&=(\mathbf{1}^{\vee}\check{\boldsymbol{w}}_{\Delta})\circ\check{\boldsymbol{e}}'_{\Gamma'\wedge C,\Gamma''\wedge C,\Delta,\Delta}\circ(\boldsymbol{d}^{\wedge}\boldsymbol{d})\circ(\mathbf{1}^{\wedge}\hat{\boldsymbol{c}}{}^{\wedge}\mathbf{1})\circ(\mathbf{1}^{\wedge}\hat{\boldsymbol{w}}_{C\vee\Delta})\\&=LHS^{*},\text{ as in }(w\,\gamma\,1).\end{aligned}$$

$(w\,\gamma\,3)$ For $f:\Gamma_1'{}^{\wedge}C^{\wedge}\Gamma_2'{}^{\wedge}A\vdash\Delta'$, $g:\Gamma_1''{}^{\wedge}C^{\wedge}\Gamma_2''{}^{\wedge}B\vdash\Delta''$ and $h:\Gamma\vdash\Delta_1{}^{\vee}C^{\vee}\Delta_2$, with Γ', Γ'' and Δ as in $(w\,\gamma\,1)$, we have in $\mathcal{D}$ the equation

$$\begin{aligned}&cut_{\Gamma_2''\wedge A\vee B,\Delta_1}(h,w^{L}_{\Gamma_1',\Gamma_2'\wedge\Gamma_1'',\Gamma_2''\wedge A\vee B}\vee^{L}(f,g))=\\&\quad c^{L}_{\Gamma'\wedge\Gamma'',\Gamma,A\vee B,\emptyset}c^{R}_{\Delta',\Delta,\Delta'',\emptyset}w^{R\Delta}_{\Delta',\Delta'',\emptyset}w^{L\Gamma}_{\Gamma',\Gamma'',A\vee B}\\&\qquad\vee^{L}(c^{L}_{\Gamma',A,\Gamma,\emptyset}cut_{\Gamma_2'\wedge A,\Delta_1}(h,f),c^{L}_{\Gamma'',B,\Gamma,\emptyset}cut_{\Gamma_2''\wedge B,\Delta_1}(h,g)),\end{aligned}$$

where the complexity of the clean cut $cut_{\Gamma_2'\wedge A,\Delta_1}(h,f)$ is (d,r') with $r<r'$, and analogously for the clean cut $cut_{\Gamma_2''\wedge B,\Delta_1}(h,g)$.

For $h':\Gamma\vdash C^{\vee}\Delta$ defined as in $(w\,\gamma\,1)$, we have by Semilattice Coherence and bifunctorial and naturality equations that the left-hand side of our equation is equal to

$$(f\,{}^{\vee}g^{\vee}\mathbf{1})\circ LHS^{*}\circ(\mathbf{1}^{\wedge}h')\circ(\mathbf{1}^{\wedge}\mathbf{1}^{e}_{A^{\vee}B}{}^{\wedge}\mathbf{1})$$

where LHS^* is $(\hat{\boldsymbol{e}}'^{\vee}\mathbf{1})\circ((\mathbf{1}^{\wedge}\,\hat{\boldsymbol{c}}\,{}^{\wedge}\mathbf{1})^{\vee}\mathbf{1})\circ((\mathbf{1}^{\wedge}\,\hat{\boldsymbol{c}})^{\vee}\mathbf{1})\circ\boldsymbol{d}\circ(\mathbf{1}^{\wedge}(\hat{\boldsymbol{w}}_C\,{}^{\vee}\mathbf{1}))$, while the right-hand side is equal to

$$(f\,{}^{\vee}g^{\vee}\mathbf{1})\circ RHS^{*}\circ(\mathbf{1}^{\wedge}h')\circ(\mathbf{1}^{\wedge}\mathbf{1}^{e}_{A^{\vee}B}{}^{\wedge}\mathbf{1})$$

where

$$\begin{aligned}
RHS^{*} &= (\mathbf{1}^{\vee}\,\check{\boldsymbol{w}}_{\Delta})\circ(\mathbf{1}^{\vee}\,\check{\boldsymbol{c}}\,{}^{\vee}\mathbf{1})\circ((\mathbf{1}^{\wedge}\,\hat{\boldsymbol{c}})^{\vee}\mathbf{1}^{\vee}(\mathbf{1}^{\wedge}\,\hat{\boldsymbol{c}})^{\vee}\mathbf{1})\circ(\boldsymbol{d}\,{}^{\vee}\boldsymbol{d})\,\circ\\
&\qquad\circ((\mathbf{1}^{\wedge}\,\hat{\boldsymbol{c}})^{\vee}(\mathbf{1}^{\wedge}\,\hat{\boldsymbol{c}}))\circ\hat{\boldsymbol{e}}'\circ(\mathbf{1}^{\wedge}\,\hat{\boldsymbol{c}})\circ(\mathbf{1}^{\wedge}\,\hat{\boldsymbol{c}}\,{}^{\wedge}\mathbf{1})\circ(\mathbf{1}^{\wedge}\,\hat{\boldsymbol{w}}_{C^{\vee}\Delta})\\
&= (\mathbf{1}^{\vee}\,\check{\boldsymbol{w}}_{\Delta})\circ(\hat{\boldsymbol{e}}'^{\vee}\mathbf{1})\circ((\mathbf{1}^{\wedge}\,\hat{\boldsymbol{c}}\,{}^{\wedge}\mathbf{1})^{\vee}\mathbf{1})\circ((\mathbf{1}^{\wedge}\,\hat{\boldsymbol{c}})^{\vee}\mathbf{1})\circ\boldsymbol{d}\circ(\mathbf{1}^{\wedge}\check{\boldsymbol{e}}')\,\circ\\
&\qquad\circ(\mathbf{1}^{\wedge}\,\hat{\boldsymbol{w}}_{C^{\vee}\Delta}),\ \text{by Symmetric Net Coherence},\\
&= (\mathbf{1}^{\vee}\,\check{\boldsymbol{w}}_{\Delta})\circ(\hat{\boldsymbol{e}}'^{\vee}\mathbf{1})\circ((\mathbf{1}^{\wedge}\,\hat{\boldsymbol{c}}\,{}^{\wedge}\mathbf{1})^{\vee}\mathbf{1})\circ((\mathbf{1}^{\wedge}\,\hat{\boldsymbol{c}})^{\vee}\mathbf{1})\circ\boldsymbol{d}\circ(\mathbf{1}^{\wedge}\check{\boldsymbol{e}}')\,\circ\\
&\qquad\circ(\mathbf{1}^{\wedge}\boldsymbol{c}^{k})\circ(\mathbf{1}^{\wedge}(\mathbf{1}^{\vee}\,\hat{\boldsymbol{w}}_{\Delta}))\circ(\mathbf{1}^{\wedge}(\hat{\boldsymbol{w}}_C\,{}^{\vee}\mathbf{1})),\\
&\qquad\text{by Lattice Coherence, provided } \Delta \text{ is not } \emptyset,\\
&= LHS^{*},\ \text{by } (m\,\check{e}),\ (wm) \text{ and bifunctorial and naturality equations}\\
&\qquad(\text{cf. the case } (w\,\gamma\,1)).
\end{aligned}$$

If Δ is $\emptyset$, then we obtain that $LHS^* = RHS^*$ by Semilattice Coherence, in a simplified version of the derivation above.

$(w \wedge 1)$ For $f\colon \Gamma_1'{}^{\wedge}A \wedge B^{\wedge}\Gamma_2'{}^{\wedge}A^{\wedge}B \vdash \Delta'$ and $h\colon \Gamma \vdash \Delta_1{}^{\vee}A \wedge B^{\vee}\Delta_2$, with Γ' and Δ as in $(w\,\gamma\,1)$, we have in $\mathcal{D}$ the equation

$$\begin{aligned}
&cut_{\emptyset,\Delta_1}(h, w^{L}_{\Gamma_1',\Gamma_2',\emptyset}\wedge^{L} f) =\\
&\qquad w^{R\Delta}_{\Delta',\emptyset,\emptyset}w^{L\Gamma}_{\Gamma',\emptyset,\emptyset}cut_{\emptyset,\Delta_1}(h, \wedge^{L} c^{L}_{\Gamma',A^{\wedge}B,\Gamma,\emptyset}cut_{\Gamma_2{}^{\wedge}A^{\wedge}B,\Delta_1}(h,f)),
\end{aligned}$$

where the complexity of the clean cut $cut_{\Gamma_2{}^{\wedge}A^{\wedge}B,\Delta_1}(h,f)$ is (d,r') with $r' < r$.

Then, by the induction hypothesis, $cut_{\Gamma_2{}^{\wedge}A^{\wedge}B,\Delta_1}(h,f) = f'$, for a cut-free Gentzen term f', which by the Atomic-k Lemma we may assume to be k-atomized. (As a matter of fact, our procedure of cut elimination is such that it produces out of a clean cut a k-atomized cut-free Gentzen term.)

The topmost cut $cut_{\emptyset,\Delta_1}(h, \wedge^L c^L_{\Gamma',A\wedge B,\Gamma,\emptyset} f')$, whose right rank is 1, is clean, and its complexity is (d, r') with $r' < r$.

To justify the equation displayed above, we proceed as follows. For $h'\colon \Gamma \vdash A \wedge B \vee \Delta$ defined as in $(w\,\gamma\,1)$, we have by Semilattice Coherence and bifunctorial and naturality equations that the left-hand side of our equation is equal to

$$(f \vee \mathbf{1}) \circ ((\mathbf{1} \wedge \mathbf{1}^e_{A\wedge B}) \vee \mathbf{1}) \circ LHS^* \circ (\mathbf{1} \wedge h'),$$

where LHS^* is $((\mathbf{1} \wedge \hat{\boldsymbol{c}} \wedge \mathbf{1}) \vee \mathbf{1}) \circ \boldsymbol{d} \circ (\mathbf{1} \wedge (\hat{\boldsymbol{w}}_{A\wedge B} \vee \mathbf{1}))$, while the right-hand side is equal to

$$(f \vee \mathbf{1}) \circ ((\mathbf{1} \wedge \mathbf{1}^e_{A\wedge B}) \vee \mathbf{1}) \circ RHS^* \circ (\mathbf{1} \wedge h')$$

where

$$\begin{aligned} RHS^* &= (\mathbf{1} \vee \check{\boldsymbol{w}}_\Delta) \circ ((\mathbf{1} \wedge \hat{\boldsymbol{c}}) \vee \mathbf{1}) \circ (\boldsymbol{d} \vee \mathbf{1}) \circ ((\mathbf{1} \wedge \hat{\boldsymbol{c}}) \vee \mathbf{1}) \circ \boldsymbol{d} \circ (\mathbf{1} \wedge \hat{\boldsymbol{w}}_{A\wedge B \vee \Delta}) \\ &= (\mathbf{1} \vee \check{\boldsymbol{w}}_\Delta) \circ ((\mathbf{1} \wedge \hat{\boldsymbol{c}} \wedge \mathbf{1}) \vee \mathbf{1}) \circ \boldsymbol{d} \circ (\mathbf{1} \wedge \check{\boldsymbol{e}}') \circ (\mathbf{1} \wedge \hat{\boldsymbol{w}}_{A\wedge B \vee \Delta}), \\ &\hspace{18em} \text{by Symmetric Net Coherence,} \\ &= (\mathbf{1} \vee \check{\boldsymbol{w}}_\Delta) \circ ((\mathbf{1} \wedge \hat{\boldsymbol{c}} \wedge \mathbf{1}) \vee \mathbf{1}) \circ \boldsymbol{d} \circ (\mathbf{1} \wedge \check{\boldsymbol{e}}') \circ (\mathbf{1} \wedge \boldsymbol{c}^k) \circ (\mathbf{1} \wedge (\mathbf{1} \vee \hat{\boldsymbol{w}}_\Delta)) \circ \\ &\qquad \circ (\mathbf{1} \wedge (\hat{\boldsymbol{w}}_{A\wedge B} \vee \mathbf{1})), \text{ by Lattice Coherence, provided } \Delta \text{ is not } \emptyset, \\ &= LHS^*, \text{ by } (m\,\check{e}),\ (wm) \text{ and bifunctorial and naturality equations} \\ &\hspace{20em} (\text{cf. the case } (w\,\gamma\,1)). \end{aligned}$$

If Δ is $\emptyset$, then we obtain that $LHS^* = RHS^*$ by Semilattice Coherence, in a simplified version of the derivation above.

$(w \vee 1)$ For $f\colon \Gamma'_1 \wedge A \vee B \wedge \Gamma_2 \wedge A \vdash \Delta'$, $g\colon \Gamma'' \wedge B \vdash \Delta''$ and $h\colon \Gamma \vdash \Delta_1 \vee A \vee B \vee \Delta_2$, with Γ' and Δ as in $(w\,\gamma\,1)$, we have in $\mathcal{D}$ the equation

$$\begin{aligned} &cut_{\emptyset,\Delta_1}(h, w^L_{\Gamma'_1,\Gamma'_2\wedge\Gamma'',\emptyset} \vee^L(f,g)) = \\ &\qquad c^R_{\Delta',\Delta,\Delta'',\emptyset} w^{R\Delta}_{\Delta',\Delta'',\emptyset} w^{L\Gamma}_{\Gamma',\Gamma'',\emptyset} cut_{\emptyset,\Delta_1}(h, \vee^L(c^L_{\Gamma',A,\Gamma,\emptyset} cut_{\Gamma'_2\wedge A,\Delta_1}(h,f), g)), \end{aligned}$$

where the complexity of the clean cut $cut_{\Gamma'_2\wedge A,\Delta_1}(h,f)$ is (d, r') with $r' < r$.

Then, by the induction hypothesis, $cut_{\Gamma'_2\wedge A,\Delta_1}(h,f) = f'$ for a cut-free Gentzen term f', which by the Atomic-k Lemma we may assume to be k-atomized. The topmost cut $cut_{\emptyset,\Delta_1}(h, \vee^L(c^L_{\Gamma',A,\Gamma,\emptyset} f', g))$, whose rank is 1, is clean, and its complexity is (d, r') with $r' < r$.

To justify the equation displayed above, we proceed as follows. For $h' : \Gamma \vdash A \vee B\,{}^{\vee}\Delta$ defined as in $(w\,\gamma\,1)$, we have by Semilattice Coherence and bifunctorial and naturality equations that the left-hand side of our equation is equal to $(f\,{}^{\vee}g\,{}^{\vee}\mathbf{1}) \circ LHS^* \circ (\mathbf{1}^{\wedge}h')$ where LHS^* is

$$(\hat{\boldsymbol{e}}'\,{}^{\vee}\mathbf{1})((\mathbf{1}^{\wedge}\,\hat{\boldsymbol{c}}\,{}^{\wedge}\mathbf{1})^{\vee}\mathbf{1}) \circ \boldsymbol{d} \circ (\mathbf{1}^{\wedge}((\mathbf{1}^{\wedge}\mathbf{1}^{e}_{A\vee B})^{\vee}\mathbf{1})) \circ (\mathbf{1}^{\wedge}(\hat{\boldsymbol{w}}_{A\vee B}\,{}^{\vee}\mathbf{1})),$$

while the right-hand side is equal to $(f\,{}^{\vee}g\,{}^{\vee}\mathbf{1}) \circ RHS^* \circ (\mathbf{1}^{\wedge}h')$ where

$$\begin{aligned}
RHS^* &= (\mathbf{1}^{\vee}\,\check{\boldsymbol{w}}_{\Delta}) \circ (\mathbf{1}^{\vee}\,\check{\boldsymbol{c}}\,{}^{\vee}\mathbf{1}) \circ ((\mathbf{1}^{\wedge}\,\hat{\boldsymbol{c}})^{\vee}\mathbf{1}) \circ (\boldsymbol{d}\,{}^{\vee}\mathbf{1}) \circ ((\mathbf{1}^{\wedge}\,\hat{\boldsymbol{c}})^{\vee}\mathbf{1}) \circ (\hat{\boldsymbol{e}}'\,{}^{\vee}\mathbf{1}) \circ \boldsymbol{d} \circ \\
&\qquad \circ (\mathbf{1}^{\wedge}\,\hat{\boldsymbol{c}}\,{}^{\wedge}\mathbf{1}) \circ (\mathbf{1}^{\wedge}(\mathbf{1}^{e}_{A\vee B}\,{}^{\vee}\mathbf{1})) \circ (\mathbf{1}^{\wedge}\,\hat{\boldsymbol{w}}_{A\vee B\,{}^{\vee}\Delta}) \\
&= (\mathbf{1}^{\vee}\,\check{\boldsymbol{w}}_{\Delta}) \circ (\hat{\boldsymbol{e}}'\,{}^{\vee}\mathbf{1}) \circ ((\mathbf{1}^{\wedge}\,\hat{\boldsymbol{c}}\,{}^{\wedge}\mathbf{1})^{\vee}\mathbf{1}) \circ \boldsymbol{d} \circ (\mathbf{1}^{\wedge}\check{\boldsymbol{e}}') \circ (\mathbf{1}^{\wedge}(\mathbf{1}^{e}_{A\vee B}\,{}^{\vee}\mathbf{1})) \circ \\
&\qquad \circ (\mathbf{1}^{\wedge}\,\hat{\boldsymbol{w}}_{A\vee B\,{}^{\vee}\Delta}), \text{ by Symmetric Net Coherence,} \\
&= (\mathbf{1}^{\vee}\,\check{\boldsymbol{w}}_{\Delta}) \circ (\hat{\boldsymbol{e}}'\,{}^{\vee}\mathbf{1}) \circ ((\mathbf{1}^{\wedge}\,\hat{\boldsymbol{c}}\,{}^{\wedge}\mathbf{1})^{\vee}\mathbf{1}) \circ \boldsymbol{d} \circ (\mathbf{1}^{\wedge}\check{\boldsymbol{e}}') \circ (\mathbf{1}^{\wedge}(\mathbf{1}^{e}_{A\vee B}\,{}^{\vee}\mathbf{1})) \circ \\
&\qquad \circ (\mathbf{1}^{\wedge}\boldsymbol{c}^{k}) \circ (\mathbf{1}^{\wedge}(\mathbf{1}^{\vee}\,\hat{\boldsymbol{w}}_{\Delta})) \circ (\mathbf{1}^{\wedge}(\hat{\boldsymbol{w}}_{A\vee B}\,{}^{\vee}\mathbf{1})), \\
&\qquad \text{by Lattice Coherence, provided } \Delta \text{ is not } \emptyset, \\
&= LHS^*, \text{ by } (m\,\check{e}),\ (wm) \text{ and bifunctorial and naturality equations} \\
&\qquad \text{(cf. the case } (w\,\gamma\,1)).
\end{aligned}$$

If Δ is $\emptyset$, then we obtain that $LHS^* = RHS^*$ by Semilattice Coherence, in a simplified version of the derivation above.

We proceed analogously in the case $(w \vee 2)$.

$(w \vee 3)$ For $f : \Gamma_1'{}^{\wedge}A \vee B\,{}^{\wedge}\Gamma_2'{}^{\wedge}A \vdash \Delta'$, $g : \Gamma_1''{}^{\wedge}A \vee B\,{}^{\wedge}\Gamma_2''{}^{\wedge}B \vdash \Delta''$ and $h : \Gamma \vdash \Delta_1{}^{\vee}A \vee B\,{}^{\vee}\Delta_2$, with Γ', Γ'' and Δ as in $(w\,\gamma\,1)$, we have in $\mathcal{D}$ the equation

$$\begin{aligned}
cut_{\emptyset,\Delta_1}(h, w^{L}_{\Gamma_1',\Gamma_2'{}^{\wedge}\Gamma_1'',\Gamma_2'',\emptyset}\vee^{L}(f,g)) &= c^{R}_{\Delta',\Delta,\Delta'',\emptyset}\, w^{R\Delta}_{\Delta',\Delta'',\emptyset,\emptyset}\, w^{L\Gamma}_{\Gamma',\Gamma'',\emptyset,\emptyset} \\
&\quad cut_{\emptyset,\Delta_1}(h, \vee^{L}(c^{L}_{\Gamma',A,\Gamma,\emptyset}\, cut_{\Gamma_2'{}^{\wedge}A,\Delta_1}(h,f), c^{L}_{\Gamma'',B,\Gamma,\emptyset}\, cut_{\Gamma_2''{}^{\wedge}B,\Delta_1}(h,g)),
\end{aligned}$$

where the complexity of the clean cut $cut_{\Gamma_2'{}^{\wedge}A,\Delta_1}(h,f)$ is (d, r') with $r' < r$, and analogously for the clean cut $cut_{\Gamma_2''{}^{\wedge}B,\Delta_1}(h,g)$.

Then, by the induction hypothesis, these two cuts are equal to the cut-free Gentzen terms f' and g', respectively, which by the Atomic-k Lemma we may assume to be k-atomized. The topmost cut

$$cut_{\emptyset,\Delta_1}(h, \vee^{L}(c^{L}_{\Gamma',A,\Gamma,\emptyset}f', c^{L}_{\Gamma'',B,\Gamma,\emptyset}g')),$$

whose right rank is 1, is clean, and its complexity is (d, r') with $r' < r$.

To justify the equation displayed above we proceed as follows. For $h' : \Gamma \vdash A \vee B\,{}^{\vee}\Delta$ defined as in $(w\,\gamma\,1)$, we have by Semilattice Coherence and bifunctorial and naturality equations that the left-hand side of our equation is equal to $(f\,{}^{\vee}g\,{}^{\vee}\mathbf{1})\circ LHS^{*}\circ(\mathbf{1}^{\wedge}h')$ where LHS^{*} is

$$(\hat{\boldsymbol{e}}'{}^{\vee}\mathbf{1})\circ((\mathbf{1}^{\wedge}\hat{\boldsymbol{c}}\,^{\wedge}\mathbf{1})^{\vee}\mathbf{1})\circ((\mathbf{1}^{\wedge}\hat{\boldsymbol{c}}\,^{\wedge}\mathbf{1})^{\vee}\mathbf{1})\circ\boldsymbol{d}\circ(\mathbf{1}^{\wedge}((\mathbf{1}^{\wedge}\mathbf{1}^{e}_{A\vee B})^{\vee}\mathbf{1}))\circ$$
$$\circ(\mathbf{1}^{\wedge}((\mathbf{1}^{\wedge}\hat{\boldsymbol{w}}_{A\vee B})^{\vee}\mathbf{1}))\circ(\mathbf{1}^{\wedge}(\hat{\boldsymbol{w}}_{A\vee B}{}^{\vee}\mathbf{1})),$$

while the right-hand side is equal to $(f\,{}^{\vee}g\,{}^{\vee}\mathbf{1})\circ RHS^{*}\circ(\mathbf{1}^{\wedge}h')$ where

$$\begin{aligned}
RHS^{*} &= (\mathbf{1}^{\vee}\check{\boldsymbol{w}}_{\Delta})\circ(\mathbf{1}^{\vee}\check{\boldsymbol{w}}_{\Delta})\circ(\mathbf{1}^{\vee}\check{\boldsymbol{c}}\,^{\vee}\mathbf{1})\circ((\mathbf{1}^{\wedge}\hat{\boldsymbol{c}})^{\vee}\mathbf{1}^{\vee}(\mathbf{1}^{\wedge}\hat{\boldsymbol{c}})^{\vee}\mathbf{1})\circ(\boldsymbol{d}^{\vee}\boldsymbol{d}^{\vee}\mathbf{1})\circ\\
&\qquad\circ((\mathbf{1}^{\wedge}\hat{\boldsymbol{c}})^{\vee}(\mathbf{1}^{\wedge}\hat{\boldsymbol{c}})^{\vee}\mathbf{1})\circ(\hat{\boldsymbol{e}}'{}^{\vee}\mathbf{1})\circ\boldsymbol{d}\circ(\mathbf{1}^{\wedge}\hat{\boldsymbol{c}}\,^{\wedge}\mathbf{1})\circ(\mathbf{1}^{\wedge}(\mathbf{1}^{e}_{A\vee B}{}^{\vee}\mathbf{1}))\circ\\
&\qquad\circ(\mathbf{1}^{\wedge}\,\hat{\boldsymbol{w}}_{A\vee B\,{}^{\vee}\Delta})\circ(\mathbf{1}^{\wedge}\,\hat{\boldsymbol{w}}_{A\vee B\,{}^{\vee}\Delta})\\
&= (\mathbf{1}^{\vee}\check{\boldsymbol{w}}_{\Delta})\circ(\mathbf{1}^{\vee}\check{\boldsymbol{w}}_{\Delta})\circ(\hat{\boldsymbol{e}}'{}^{\vee}\mathbf{1})\circ((\mathbf{1}^{\wedge}\hat{\boldsymbol{c}}\,^{\wedge}\mathbf{1})^{\vee}\mathbf{1})\circ((\mathbf{1}^{\wedge}\hat{\boldsymbol{c}}\,^{\wedge}\mathbf{1})^{\vee}\mathbf{1})\circ\boldsymbol{d}\circ\\
&\qquad\circ(\mathbf{1}^{\wedge}\check{\boldsymbol{e}}')\circ(\mathbf{1}^{\wedge}\check{\boldsymbol{e}}')\circ(\mathbf{1}^{\wedge}(\mathbf{1}^{e}_{A\vee B}{}^{\vee}\mathbf{1}))\circ(\mathbf{1}^{\wedge}\,\hat{\boldsymbol{w}}_{A\vee B\,{}^{\vee}\Delta})\circ(\mathbf{1}^{\wedge}\,\hat{\boldsymbol{w}}_{A\vee B\,{}^{\vee}\Delta}),\\
&\qquad\text{by Symmetric Net Coherence,}\\
&= LHS^{*},\ \text{by Lattice Coherence, } (m\,\check{e}),\ (wm) \text{ and bifunctorial and}\\
&\qquad\text{naturality equations, provided } \Delta \text{ is not } \emptyset \text{ (cf. the case } (w\,\gamma\,1)).
\end{aligned}$$

If Δ is $\emptyset$, then we obtain that $LHS^{*} = RHS^{*}$ by Semilattice Coherence, in a simplified version of the derivation above.

Note that the cases $(w\,\gamma\,2)$, $(w\,\gamma\,3)$, $(w \wedge 1)$, $(w \vee 1)$, $(w \vee 2)$ and $(w \vee 3)$ are dealt with in the spirit of Gentzen. The case $(w\,\gamma\,1)$, which involves mix, was not envisaged by him. This concludes the proof. ⊣

§11.3. Coherence of distributive lattice categories

The essential ingredient in our proof of coherence for the category **DL** is the Cut-Elimination Theorem of the preceding section. Another ingredient is Restricted Mix-Lattice Coherence of §10.2. Before proving coherence for **DL**, we consider some matters that serve to connect **DL** with the category **ML**, but are also of an independent interest.

With the abbreviations

$$\hat{s}_{A,C,D} =_{df} \hat{e}'_{A,A,C,D}\circ(\hat{\boldsymbol{w}}_{A}\wedge\mathbf{1}_{C\vee D})\colon A\wedge(C\vee D)\vdash(A\wedge C)\vee(A\wedge D),$$
$$\check{s}_{D,C,A} =_{df} (\mathbf{1}_{D\wedge C}\vee\check{\boldsymbol{w}}_{A})\circ\check{e}'_{D,C,A,A}\colon (D\vee A)\wedge(C\vee A)\vdash(D\wedge C)\vee A,$$

$$\hat{t}_{A,C,D} =_{df} (\check{w}_A \wedge \mathbf{1}_{A\vee D}) \circ c^k_{A,C,A,D}\colon\; (A\wedge C)\vee(A\wedge D) \vdash A\wedge(C\vee D),$$
$$\check{t}_{D,C,A} =_{df} c^k_{D,C,A,A} \circ (\mathbf{1}_{D\wedge C} \vee \hat{w}_A)\colon\; (D\wedge C)\vee A \vdash (D\vee A)\wedge(C\vee A),$$

we obtain the following equations in **DL** as an immediate consequence of $(m\,\hat{e})$, $(m\,\check{e})$ and (wm):

$$\hat{t}_{A,C,D} \circ \hat{s}_{A,C,D} = \mathbf{1}_{A\wedge(C\vee D)},$$
$$\check{s}_{D,C,A} \circ \check{t}_{D,C,A} = \mathbf{1}_{(D\wedge C)\vee A}.$$

This means that $\hat{s}_{A,C,D}$ is a right inverse (i.e. section) of $\hat{t}_{A,C,D}$, while $\check{s}_{D,C,A}$ is a left inverse (i.e. retraction) of $\check{t}_{D,C,A}$ (see [94], Section I.5). It is easy to see that $\hat{t}_{A,C,D}$ and $\hat{s}_{A,C,D}$ are not inverse to each other in **DL**, since $G(\hat{s}_{A,C,D} \circ \hat{t}_{A,C,D})$ is different from $G(\mathbf{1}_{(A\wedge C)\vee(A\wedge D)})$; analogously, $\check{t}_{D,C,A}$ and $\check{s}_{D,C,A}$ are not inverse to each other. The types of the arrow terms in the families $\hat{s}$ and $\check{s}$ give what is usually called distribution of $\wedge$ over $\vee$ and distribution of $\vee$ over $\wedge$. However, these arrow terms do not stand for isomorphisms in **DL**.

For every formula A of $\mathcal{L}_{\wedge,\vee}$, let A^{dnf} be any formula of $\mathcal{L}_{\wedge,\vee}$ in disjunctive normal form (*dnf*; see §10.2) such that there is an arrow term $\hat{t}^A\colon A^{dnf} \vdash A$ of $\mathcal{C}(\mathbf{L})$ and an arrow term $\hat{s}^A\colon A \vdash A^{dnf}$ of $\mathcal{C}(\mathbf{DL})$ for which in **DL** we have $\hat{t}^A \circ \hat{s}^A = \mathbf{1}_A$. (We do not require the uniqueness of $\hat{t}^A$ and $\hat{s}^A$, as we did not require the uniqueness of A^{dnf}.) That for every formula A of $\mathcal{L}_{\wedge,\vee}$ there is a formula A^{dnf} is shown by an easy induction on the number of occurrences of $\vee$ in the scope of an occurrence of $\wedge$.

Dually, for every formula A of $\mathcal{L}_{\wedge,\vee}$, let A^{cnf} be any formula in conjunctive normal form (*cnf*; see §10.2) such that there is an arrow term $\check{t}^A\colon A \vdash A^{cnf}$ of $\mathcal{C}(\mathbf{L})$ and an arrow term $\check{s}^A\colon A^{cnf} \vdash A$ of $\mathcal{C}(\mathbf{DL})$ for which in **DL** we have $\check{s}^A \circ \check{t}^A = \mathbf{1}_A$. That for every formula A of $\mathcal{L}_{\wedge,\vee}$ there is a formula A^{cnf} is shown by an easy induction, as above.

For $\xi \in \{\wedge, \vee\}$, the arrow terms $\overset{\xi}{t}{}^A$ are built out of arrow terms of the form $\overset{\xi}{t}_{B,C,D}$ and arrow terms of $\mathcal{C}(\mathbf{S})$ with the help of the operations $\wedge$, $\vee$ and $\circ$ on arrow terms, while the arrow terms $\overset{\xi}{s}{}^A$ are built out of arrow terms of the form $\overset{\xi}{s}_{B,C,D}$ and arrow terms of $\mathcal{C}(\mathbf{S})$ with the help of the operations $\wedge$, $\vee$, and $\circ$ on arrow terms. For example, if A is $p\wedge((q\vee(r\wedge s))\vee q)$, and

A^{dnf} is $((p \wedge q) \vee ((p \wedge r) \wedge s)) \vee (p \wedge q)$ (this is the source of the arrow term γ we had as an example in §10.2), then we can take that $\hat{t}^{A}$ is

$$\hat{t}_{p,q\vee(r\wedge s),q} \circ (\hat{t}_{p,q,r\wedge s} \vee \mathbf{1}_{p\wedge q}) \circ ((\mathbf{1}_{p\wedge q} \vee \hat{b}^{\leftarrow}_{p,r,s}) \vee \mathbf{1}_{p\wedge q}),$$

while $\hat{s}^{A}$ is

$$((\mathbf{1}_{p\wedge q} \vee \hat{b}^{\rightarrow}_{p,r,s}) \vee \mathbf{1}_{p\wedge q}) \circ (\hat{s}_{p,q,r\wedge s} \vee \mathbf{1}_{p\wedge q}) \circ \hat{s}_{p,q\vee(r\wedge s),q}.$$

It is easy to verify, by referring to definitions, that a cut-free Gentzen term of $\mathcal{D}$ of the type $A \vdash B$ for A in $\mathcal{L}_{\wedge}$ and B in $\mathcal{L}_{\vee}$ denotes an arrow (f, A, B) of $\mathcal{D}$ such that f is an arrow term of $\mathcal{C}(\mathbf{ML})$. As a consequence of that and of the Cut-Elimination Theorem of the preceding section, we obtain the following.

PROPOSITION. *For A in $\mathcal{L}_{\wedge}$ and B in $\mathcal{L}_{\vee}$, every arrow term of $\mathcal{C}(\mathbf{DL})$ of type $A \vdash B$ is equal in* **DL** *to an arrow term of $\mathcal{C}(\mathbf{ML})$.*

Then we can prove the following lemma, which appeals to the notion of settled normal form of §10.2.

NORMAL-FORM LEMMA. *Every arrow term $f : A^{dnf} \vdash B^{cnf}$ of $\mathcal{C}(\mathbf{DL})$ is equal in* **DL** *to an arrow term of $\mathcal{C}(\mathbf{GML})$ in settled normal form.*

PROOF. As in the proof of the Normal-Form Lemma of §10.2, we make an induction on the number of occurrences of $\vee$ in A and $\wedge$ in B. If there are no such occurrences of $\wedge$ and $\vee$, then we apply the Proposition above and the Normal-Form Lemma of §10.2. For the remainder of the proof we proceed as in the proof of that lemma in §10.2. ⊣

Then we can prove the following.

DISTRIBUTIVE LATTICE COHERENCE. *The functor G from* **DL** *to Rel is faithful.*

PROOF. Suppose $f, g \colon A \vdash B$ are arrow terms of $\mathcal{C}(\mathbf{DL})$. If $Gf = Gg$, then $G(\check{t}^{B} \circ f \circ \hat{t}^{A}) = G(\check{t}^{B} \circ g \circ \hat{t}^{A})$. By the Normal-Form Lemma above, we have in **DL** that $\check{t}^{B} \circ f \circ \hat{t}^{A} = f'$ and $\check{t}^{B} \circ g \circ \hat{t}^{A} = g'$ for f' and g' arrow terms of $\mathcal{C}(\mathbf{ML})$. By Restricted Mix-Lattice Coherence of §10.2, we have that $f' = g'$ in **ML**, and hence also in **DL**. So in **DL** we have

$$\check{s}^B \circ \check{t}^B \circ f \circ \hat{t}^A \circ \hat{s}^A = \check{s}^B \circ \check{t}^B \circ g \circ \hat{t}^A \circ \hat{s}^A,$$

and hence $f = g$. ⊣

A logical system synonymous with $\mathcal{C}(\mathbf{DL})$ may be obtained by taking $\hat{e}$ or $\hat{e}'$ as primitive transformations instead of d, since in **DL** we have the equation

$$d_{A,B,C} = (\mathbf{1}_{A\wedge B} \vee \hat{k}^2_{A,C}) \circ \hat{e}'_{A,A,B,C} \circ (\hat{w}_A \wedge \mathbf{1}_{B\vee C}),$$

which can easily be checked by Distributive Lattice Coherence. Analogously, we could take $\check{e}$ or $\check{e}'$ as primitive, since in **DL** we have

$$d_{C,B,A} = (\mathbf{1}_{C\wedge B} \vee \check{w}_A) \circ \check{e}'_{C,B,A,A} \circ (\check{k}^1_{C,A} \wedge \mathbf{1}_{B\vee A}).$$

Alternative primitive transformations are $\hat{s}$ and $\check{s}$, whose members occur in the two equations above. With such alternative primitives, however, we have not managed to find an axiomatization simpler than what we have for $\mathcal{E}(\mathbf{DL})$ and $\mathcal{E}(\mathbf{DL}')$.

A primitive of the same type as $\hat{s}$ was considered in [86] and [87] as an addition to $\mathbf{S}_{\top,\bot}$ extended with the isomorphism of $A \wedge \bot$ with $\bot$. In the presence of this isomorphism, we cannot expect coherence with respect to *Rel* with a functor such as our functors G. The coherence result of [86] is a restricted coherence result in the sense of preorder, while the coherence result of [87] is a result about a faithful functor into *Rel*, which differs from our functor G with respect to $\wedge$. The equations of those papers without $\top$ and $\bot$ hold, however, in **DL**.

If to $\mathcal{E}(\mathbf{DL})$ we add the equation $\hat{s}_{A,C,D} \circ \hat{t}_{A,C,D} = \mathbf{1}_{(A\wedge C)\vee(A\wedge D)}$ or the equation $\check{t}_{D,C,A} \circ \check{s}_{D,C,A} = \mathbf{1}_{(D\vee A)\wedge(C\vee A)}$, then we can derive that all arrow terms of the same type are equal. Here is a proof of that fact for the first equation. (We proceed analogously with the second equation.)

We have that

$$\alpha = \hat{s}_{A,B,B} \circ \hat{t}_{A,B,B} = \mathbf{1}_{(A\wedge B)\vee(A\wedge B)},$$

$$\beta = (\hat{c}_{B,A} \vee \hat{c}_{B,A}) \circ \hat{s}_{B,A,A} \circ \hat{t}_{B,A,A} \circ (\hat{c}_{A,B} \vee \hat{c}_{A,B}) = \mathbf{1}_{(A\wedge B)\vee(A\wedge B)}.$$

So $\alpha \cup \beta = \mathbf{1}_{(A\wedge B)\vee(A\wedge B)}$, by ($\cup$ *idemp*). By Distributive Lattice Coherence, we infer that

$$[\check{k}^1_{A\wedge B,A\wedge B} \cup \check{k}^2_{A\wedge B,A\wedge B}, \check{k}^1_{A\wedge B,A\wedge B} \cup \check{k}^2_{A\wedge B,A\wedge B}] = \mathbf{1}_{(A\wedge B)\vee(A\wedge B)};$$

therefore, with ($\vee\beta$) (see the List of Equations), we have

$$\check{k}^1_{A\wedge B,A\wedge B} = \check{k}^1_{A\wedge B,A\wedge B} \cup \check{k}^2_{A\wedge B,A\wedge B} = \check{k}^2_{A\wedge B,A\wedge B}.$$

For $f, g \colon A \vdash B$, we have

$$\begin{aligned}[f \wedge \mathbf{1}_A, g \wedge \mathbf{1}_A] \circ \check{k}^1_{A\wedge A,A\wedge A} &= [f \wedge \mathbf{1}_A, g \wedge \mathbf{1}_A] \circ \check{k}^2_{A\wedge A,A\wedge A} \\ f \wedge \mathbf{1}_A &= g \wedge \mathbf{1}_A, \text{ by } (\vee\beta), \\ f \circ \hat{k}^1_{A,A} &= g \circ \hat{k}^1_{A,A}, \text{ by } (\hat{k}^1\ nat),\end{aligned}$$

from which we infer $f = g$ with $(\hat{w}\hat{k})$ (see §9.1). So $\overset{\xi}{s}$ and $\overset{\xi}{t}$ cannot be inverses of each other in the context of **DL** without preorder, i.e. triviality. (For a result of the same kind, see [19], Proposition 3.1.)

The category *Set* of sets with functions is a lattice category with $\wedge$ being cartesian product $\times$ and $\vee$ being disjoint union $+$ (cf. §9.6). Products and coproducts are unique up to isomorphism (see [94], Sections IV.1-2), and so there is no alternative lattice-category structure in *Set*. Since *Set* is, of course, not a preorder, we can conclude, according to what we said above, that it is not a distributive lattice category with $\hat{s}_{a,b,c} \colon a \times (b+c) \vdash (a \times b)+(a \times c)$ and $\hat{t}_{a,b,c} \colon (a \times b)+(a \times c) \vdash a \times (b+c)$ being identity arrows. With $d_{a,b,c}$ defined as $(\mathbf{1}_{a\times b} \vee \hat{k}^2_{a,c}) \circ \hat{s}_{a,b,c}$, the equation $(d\,\check{b})$ of §7.2 does not hold in *Set*, as noted in [19] (Section 3; the remaining specific equations of $\mathcal{E}(\mathbf{DA})$ hold in *Set*). At the same place, an argument is presented that *Set* with $\wedge$ and $\vee$ being $\times$ and $+$ cannot satisfy $(d\,\check{b})$ for any definition of d^L and d^R. Here is another argument to the same effect.

The category *Set* is not a distributive lattice category. If it were that, then we would have in it for every set a a function $m_{a,a} \colon a \times a \to a + a$ that satisfies the following instance of the equation (cm) of §8.4:

$$m_{a,a} \circ \hat{c}_{a,a} = \check{c}_{a,a} \circ m_{a,a}.$$

If $a = \{x\}$, then, since $a \times a = \{(x,x)\}$ is a terminal object, we obtain

$$m_{a,a} = \check{c}_{a,a} \circ m_{a,a}.$$

There are only two functions from $a \times a$ to $a + a = \{(x,*),(*,x)\}$, and none of them satisfies the last equation, because $\check{c}_{a,a}\,(x,*) = (*,x)$ and $\check{c}_{a,a}\,(*,x) = (x,*)$. This argument shows also that *Set* is not a mix-lattice category.

§11.4. Legitimate relations

At the end of §9.2, we made a brief comment on the image under the functor G of the categories $\hat{\mathbf{L}}_\top$ and $\hat{\mathbf{L}}$. Once we have proved Distributive Lattice Coherence, it is of some interest to consider the image under G of the category **DL**. We will devote the present section to this matter.

For A and B formulae of $\mathcal{L}_{\wedge,\vee}$, we will say that a relation $R \subseteq GA \times GB$ is *legitimate* when there is an arrow term $f : A \vdash B$ of $\mathcal{C}(\mathbf{DL})$ such that $Gf = R$. We will prove two propositions that will enable us to decide whether a relation is legitimate.

For k and l finite ordinals, let k^{+l} be the set $\{n{+}l \mid n \in k\}$. Other notions mentioned in the statements and proofs of our two propositions are defined in the preceding section and in §10.2.

PROPOSITION 1. *In $GA^{dnf} = k_1 + \ldots + k_n$, for $n \geq 1$, and $GB^{cnf} = l_1 + \ldots + l_m$, for $m \geq 1$, let k_i, for $i \in \{1,\ldots,n\}$, be GA_i for a minimal disjunct A_i of A^{dnf}, and let l_j, for $j \in \{1,\ldots,m\}$, be GB_j for a minimal conjunct B_j of B^{cnf}. Then a relation $R \subseteq GA^{dnf} \times GB^{cnf}$ is legitimate iff for every $i \in \{1,\ldots,n\}$ and every $j \in \{1,\ldots,m\}$ the relation $R \cap (k_i^{+k_1+\ldots+k_{i-1}} \times l_j^{+l_1+\ldots+l_{j-1}})$ is not empty.*

PROOF. Suppose R is legitimate; i.e., there is an arrow term $f\colon A^{dnf} \vdash B^{cnf}$ of $\mathcal{C}(\mathbf{DL})$ such that $Gf = R$. By the Normal-Form Lemma of the preceding section, f is equal in **DL** to an arrow term of $\mathcal{C}(\mathbf{GML})$ in normal form. Then the molecular correspondence (see §10.2) is enough to prove the proposition from left to right, because for every molecular component f' we have that Gf' is not empty. For the other direction, we build out of the relations $R \cap (k_i^{+k_1+\ldots+k_{i-1}} \times l_j^{+l_1+\ldots+l_{j-1}})$ arrow terms of $\mathcal{C}(\mathbf{GML})$ in bracket-free normal form, which we then combine, as molecular components, to build an f in normal form such that $Gf = R$. ⊣

We can check that the relation $G\gamma$, which we have drawn in §10.2, satisfies the condition equivalent to legitimacy stated in Proposition 1. For example, for the couple (A_2, B_3) we have $\{(2,3),(2,4)\} \subseteq G\gamma \cap (3^{+2} \times 3^{+1+2})$. The molecular component of γ corresponding to (A_2, B_3) is $\beta_3 \cup \beta_4$.

For the second of our two propositions, remember that, according to the definition of §2.9, we write composition of relations from right to left, as composition of functions.

PROPOSITION 2. *The relation* $R \subseteq GA \times GB$ *is legitimate iff the relation* $G\overset{\vee}{t}{}^{B} \circ R \circ G\hat{t}{}^{A} \subseteq GA^{dnf} \times GB^{cnf}$ *is legitimate.*

PROOF. The left-to-right direction of the proposition is trivial. For the other direction, suppose $G\overset{\vee}{t}{}^{B} \circ R \circ G\hat{t}{}^{A}$ is legitimate. Then

$$G\overset{\vee}{s}{}^{B} \circ G\overset{\vee}{t}{}^{B} \circ R \circ G\hat{t}{}^{A} \circ G\hat{s}{}^{A},$$

which is equal to R, is legitimate. ⊣

By combining Propositions 1 and 2 we can decide whether any relation $R \subseteq GA \times GB$ is legitimate.

For the *maximal relation* $R_{\max} \subseteq GA \times GB$ we have $(i, j) \in R_{\max}$ when the i+1-th occurrence of letter in A (counting from the left) and the j+1-th occurrence of letter in B are occurrences of the same letter. For $\gamma\colon A \vdash B$ being our arrow term of §10.2, we have that $G\gamma$ is not $R_{\max} \subseteq GA \times GB$. If γ' is obtained from γ by replacing α_3 and α_4 by $\alpha_3 \cup \alpha_4$, then $G\gamma'$ coincides with $R_{\max} \subseteq GA \times GB$.

We can use maximal relations to solve the theoremhood problem for the category **DL**. If any relation $R \subseteq GA \times GB$ is legitimate, then $R_{\max} \subseteq GA \times GB$ is legitimate. So, to check whether there is an arrow of type $A \vdash B$ in **DL**, it is enough to check whether $R_{\max} \subseteq GA \times GB$ is legitimate. The theoremhood problem for **DL** is, however, solved in a much more familiar way by noting that there is an arrow of type $A \vdash B$ in **DL** iff the implication $A \rightarrow B$ is a tautology.

§11.5. Coherence of distributive dicartesian categories

To obtain the natural logical category $\mathbf{DL}_{\top,\bot}$, we have that the logical system $\mathcal{C}(\mathbf{DL}_{\top,\bot})$ is in $\mathcal{L}_{\wedge,\vee,\top,\bot}$, with the transformations α included in $\mathbf{1}$, b, c, w-k, m, d and δ-σ. The specific equations of $\mathcal{E}(\mathbf{DL}_{\top,\bot})$ are obtained by taking the union of those of $\mathcal{E}(\mathbf{DL})$ and $\mathcal{E}(\mathbf{L}_{\top,\bot})$. We call natural $\mathbf{DL}_{\top,\bot}$-categories *distributive dicartesian* categories. The objects of a distributive dicartesian category that is a partial order make a distributive lattice with top and bottom.

Note that the equations $(\hat{\sigma}\, d^L)$, $(\check{\delta}\, d^L)$, $(\hat{\delta}\, d^R)$ and $(\check{\sigma}\, d^R)$ of §7.9 hold in $\mathbf{DL}_{\top,\bot}$. It suffices to derive the first two of these equations:

$$(\hat{\sigma}\, d^L) \quad d_{\top,B,C} = (\hat{\sigma}^{\leftarrow}_{B} \vee \mathbf{1}_C) \circ \hat{\sigma}^{\rightarrow}_{B\vee C},$$
$$(\check{\delta}\, d^L) \quad d_{A,B,\bot} = \check{\delta}^{\leftarrow}_{A\wedge B} \circ (\mathbf{1}_A \wedge \check{\delta}^{\rightarrow}_{B});$$

the remaining two equations then follow easily. For $(\hat{\sigma}\, d^L)$, since $\hat{\sigma}^{\rightarrow}_{B\vee C} = \hat{k}^2_{\top,B\vee C}$, we have that the right-hand side is equal to

$$(\hat{\sigma}^{\leftarrow}_{B} \vee \mathbf{1}_C) \circ (\hat{k}^2_{\top,B} \vee \mathbf{1}_C) \circ d_{\top,B,C}$$

by $(d\hat{k})$, and this is equal to $d_{\top,B,C}$. Conversely, as we noted in §11.1, one can derive $(d\hat{k})$ from $(\hat{\sigma}\, d^L)$ by precomposing with $\hat{\kappa}_A \wedge \mathbf{1}_{B\vee C}$. We proceed analogously for $(\check{\delta}\, d^L)$.

With the help of $(d\,\hat{k})$, $(d\,\check{k})$ and Dicartesian Coherence we obtain the following equations of $\mathbf{DL}_{\top,\bot}$:

$$(d\top\top) \quad d_{A,\top,\top} = \check{k}^1_{A\wedge\top,\top} \circ (\mathbf{1}_A \wedge \hat{\kappa}_{\top\vee\top}),$$
$$(d\bot\bot) \quad d_{\bot,\bot,C} = (\check{\kappa}_{\bot\wedge\bot} \vee \mathbf{1}_C) \circ \hat{k}^2_{\bot,\bot\vee C}.$$

With the help of $(\hat{\sigma}\, d^L)$ or $(d\top\top)$, together with Dicartesian Coherence, we obtain that the equation $(m\top)$ of $\mathbf{ML}_{\top,\bot}$ (see §10.3) holds in $\mathbf{DL}_{\top,\bot}$. For $(m\bot)$ (see §10.3), we rely on $(\check{\delta}\, d^L)$ or $(d\bot\bot)$, and Dicartesian Coherence. So in $\mathcal{E}(\mathbf{DL}_{\top,\bot})$ we have all the equations of $\mathcal{E}(\mathbf{ML}_{\top,\bot})$.

Let $\mathcal{C}$ and $\mathcal{C}'$ be respectively the logical systems $\mathcal{C}(\mathbf{DL}_{\top,\bot})$ and $\mathcal{C}(\mathbf{A}_{\top,\bot})$, while $\mathcal{E}$ is $\mathcal{E}(\mathbf{DL}_{\top,\bot})$. Next, let $\mathcal{B}$ be $\mathcal{C}/\mathcal{E}$, that is $\mathbf{DL}_{\top,\bot}$. Then it is easy to see that the conditions (IV$\mathcal{C}$) and (IV$\mathcal{B}$) of §3.1 are satisfied. Since the $\mathcal{C}'$-core of $\mathcal{C}/\mathcal{E}$ is the category $\mathbf{A}_{\top,\bot}$, by Bimonoidal Coherence of §6.1, we have that the condition (IV$\mathcal{G}$) of §3.1 is also satisfied. So (IV) holds, and since $\mathbf{A}_{\top,\bot}$ is a preorder, by Bimonoidal Coherence, we can apply the Strictification Theorem of §3.1 to obtain that the category $\mathbf{DL}_{\top,\bot}^{\mathbf{A}_{\top,\bot}}$, that is $\mathbf{DL}_{\top,\bot}^{\mathcal{G}}$, is equivalent to $\mathbf{DL}_{\top,\bot}$ via two strong $\mathcal{C}(\mathbf{DL}_{\top,\bot})$-functors. According to §6.2, the objects of $\mathbf{DL}_{\top,\bot}^{\mathbf{A}_{\top,\bot}}$ may be identified with form sequences of $\mathcal{L}_{\wedge,\vee,\top,\bot}$ in the extended sense.

Let $\mathcal{D}$ be now the category $\mathbf{DL}_{\top,\bot}^{\mathbf{A}_{\top,\bot}}$. We use the terminology of §11.1 with the following changes.

A *basic sequence* of colour $\xi \in \{\wedge, \vee\}$ of $\mathcal{D}$ is a form sequence of $\mathcal{L}_{\wedge,\vee,\top,\bot}$ in the extended sense that is either of the form $(A_1 \ldots A_n,\ \xi)$, for $n \geq 2$ and A_i, where $i \in \{1, \ldots, n\}$, a formula of $\mathcal{L}_{\wedge,\vee,\top,\bot}$, or it is a formula of $\mathcal{L}_{\wedge,\vee,\top,\bot}$, or it is $(\emptyset, \xi)$. The basic sequence $(\emptyset, \xi)$ has no members.

We define Gentzen terms by induction as in §11.1 with the following additions. Besides $\mathbf{1}''_p \colon p \vdash p$, which, as before, denotes the arrow $(\mathbf{1}_p, p, p)$ of $\mathcal{D}$, we put among atomic Gentzen terms $\mathbf{1}^i_\top : (\emptyset, \wedge) \vdash \top$, which denotes $(\mathbf{1}_\top, (\emptyset, \wedge), \top)$, and $\mathbf{1}^e_\bot : \bot \vdash (\emptyset, \vee)$, which denotes $(\mathbf{1}_\bot, \bot, (\emptyset, \vee))$. The Gentzen operations c^L, c^R, k^L, k^R, w^L, w^R, *cut*, *mix*, $\wedge^L$, $\wedge^R$, $\vee^L$ and $\vee^R$ are defined as before, save that $\boldsymbol{c}$, $\boldsymbol{w}$ and $\boldsymbol{d}$ are now replaced by c'', w'' and d''. In $\mathcal{D}$ we have now the equations

$$\overset{\xi}{c}{}''_{\Theta,\Xi} = \mathbf{1}_{\Theta\xi\Xi},\ \text{if } \Theta = (\emptyset, \xi) \text{ or } \Xi = (\emptyset, \xi),\ \text{by } (\overset{\xi}{c}\overset{\xi}{\delta}\overset{\xi}{\sigma}) \text{ (see §5.3)},$$

$$\overset{\xi}{w}{}''_{(\emptyset,\xi)} = \mathbf{1}_{(\emptyset,\xi)},\ \text{by } (\overset{\xi}{w}\overset{\xi}{\delta}) \text{ (see §9.2)},$$

$$d''_{\Theta,A,\Xi} = \mathbf{1}_{\Theta\wedge(A\vee\Xi)},\ \text{if } \Theta = (\emptyset, \wedge) \text{ or } \Xi = (\emptyset, \vee),\ \text{by } (\hat{\sigma}\, d^L) \text{ or } (\check{\delta}\, d^L).$$

Note that now $\overset{\xi}{k}{}^{1''}_{(\emptyset,\xi),\Xi}$ and $\overset{\xi}{k}{}^{2''}_{\Xi,(\emptyset,\xi)}$ are defined in $\mathcal{D}$, and they are equal to $\overset{\xi}{\kappa}{}''_{\Xi}$. With this in mind, we may continue using the other abbreviations we had before.

If $\mathbf{1}^e_\top$ denotes $(\mathbf{1}_\top, \top, (\emptyset, \wedge))$, and $\mathbf{1}^i_\bot$ denotes $(\mathbf{1}_\bot, (\emptyset, \vee), \bot)$, then for $f \colon \Gamma \vdash \Delta$ we have in $\mathcal{D}$ the equations

$$(\top)\quad k^L_\top f = f \circ \hat{k}^{1''}_{\Gamma,\top} = f \circ (\Gamma \wedge \hat{\kappa}{}''_\top) = f \circ (\Gamma \wedge \mathbf{1}^e_\top),$$

$$(\bot)\quad k^R_\bot f = \check{k}^{2''}_{\bot,\Delta} \circ f = (\check{\kappa}{}''_\bot \vee \Delta) \circ f = (\mathbf{1}^i_\bot \vee \Delta) \circ f.$$

For $(\top)$, we rely on the fact that $\hat{k}^2_{A,B} = \hat{\sigma}_{\vec{B}} \circ (\hat{\kappa}_A \wedge \mathbf{1}_B)$ in $\hat{\mathbf{L}}_\top$ and$\hat{\sigma}\text{-}\!\!\to''_X = \mathbf{1}_X$ in $\mathcal{D}$, and analogously for $(\bot)$.

We prove the Gentzenization Lemma as in §11.1, with the following additions. We have that $(\mathbf{1}_\top, \top, \top)$ is denoted by $k^L_\top \mathbf{1}^i_\top$, while $(\mathbf{1}_\bot, \bot, \bot)$ is denoted by $k^R_\bot \mathbf{1}^e_\bot$. To show that, we rely on the equations $(\top)$ and $(\bot)$. We also have that $(\hat{\kappa}_A, A, \top)$ is denoted by $k^L_A \mathbf{1}^i_\top$, while $(\check{\kappa}_A, \bot, A)$ is denoted by $k^R_A \mathbf{1}^e_\bot$. (We can define the arrow terms in the family δ-σ in terms of those in the family κ; see §9.2.) If we have a sequent arrow (f, Γ, Δ) and Γ is $(\emptyset, \wedge)$ or Δ is $(\emptyset, \vee)$, then we proceed as before by using $\mathbf{1}^i_\top$ or $\mathbf{1}^e_\bot$.

As before, a Gentzen term is k-atomized when for every subterm of it of the form k^L_A or k^R_A we have that A is a an atomic formula, which now means that it is a letter or $\top$ or $\bot$. We prove the Atomic-k Lemma of §11.2 exactly as before.

Note that we keep the Gentzen operations $k^L_\bot$ and $k^R_\top$, which need not be in the spirit of Gentzen. For $f \colon \Gamma \vdash \Delta$, Gentzen would perhaps equate $k^L_\bot f$ with an arrow obtained from $\mathbf{1}^e_\bot : \bot \vdash (\emptyset, \vee)$ by thinning with Γ on the left and Δ on the right, and he would equate $k^R_\top f$ with an arrow obtained from $\mathbf{1}^i_\top : (\emptyset, \wedge) \vdash \top$ by thinning with Γ on the left and Δ on the right. We do not do that.

To define clusters and rank, we now count among the leaf formulae of h also the occurrences of $\top$ and $\bot$ in the types of $\mathbf{1}^i_\top : (\emptyset, \wedge) \vdash \top$ and $\mathbf{1}^e_\bot : \bot \vdash (\emptyset, \vee)$ when h is one of these Gentzen terms.

With the definition of blocked w^L and w^R subterms copied from what we had in §11.2, we can prove the Blocked-w Lemma as before. The degree of a cut is as before the number of occurrences of connectives in the cut formula; here we count $\top$ and $\bot$ among these connectives. We can then prove the Cut-Elimination Theorem for the new category $\mathcal{D}$.

PROOF OF THE CUT-ELIMINATION THEOREM. We enlarge the proof in §11.2 with the following cases.

If the complexity of our clean cut is $(d, 2)$ for $d > 0$ and the cut formula is $\top$, then our clean cut can be of the form $cut_{\emptyset,\emptyset}(\mathbf{1}^i_\top, k^L_\top f)$, which for $f \colon \Gamma \vdash \Delta$ is equal to f in $\mathcal{D}$ by relying on $(\top)$ and the equation $d''_{\Gamma,\top,(\emptyset,\vee)} = \mathbf{1}_{\Gamma\wedge\top}$ of $\mathcal{D}$, which we obtain by using the equation $(\check{\delta}\, d^L)$. The remaining possible form of our clean cut can be $cut_{\emptyset,\emptyset}(k^R_\top g, k^L_\top f)$, which is equal to $mix(f,g)$ in $\mathcal{D}$. (This last step is not in the spirit of Gentzen, who did not envisage our mix.)

If the complexity of our clean cut is $(d,2)$ for $d>0$ and the cut formula is $\bot$, then we proceed analogously. We rely now on the equations $(\bot)$ and $(\hat{\sigma}\, d^L)$.

If the complexity of our clean cut is (d,r) for $r>2$ and the right rank of this clean cut is greater than 1, then we proceed as in the proof of the Cut-Elimination Theorem in §11.2. Note that $\hat{k}1''_{(\emptyset,\wedge),B}$ is defined in $\mathcal{D}$ and is equal to $\hat{k}''_B$. So case (3) can now be handled without distinguishing cases as in §11.2.

We proceed analogously if the left rank is greater than 1, and the remainder of the proof follows the proof of the Cut-Elimination Theorem of §11.2. ⊣

We can then prove the following.

DISTRIBUTIVE DICARTESIAN COHERENCE. *The functor G from* $\mathbf{DL}_{\top,\bot}$ *to* *Rel is faithful.*

We proceed as for the proof of Distributive Lattice Coherence in §11.3. In the Proposition and in the Normal-Form Lemma of §11.3, and at other appropriate places, we replace $\mathcal{L}_\wedge$, $\mathcal{L}_\vee$, $\mathbf{DL}$ and $\mathbf{ML}$ by $\mathcal{L}_{\wedge,\top,\bot}$, $\mathcal{L}_{\vee,\top,\bot}$, $\mathbf{DL}_{\top,\bot}$ and $\mathbf{ML}_{\top,\bot}$, respectively. We now use Restricted Mix-Dicartesian Coherence of §10.3 instead of Restricted Mix-Lattice Coherence.

When we look for conditions of legitimacy of relations, we take over for $\mathbf{DL}_{\top,\bot}$ Proposition 2 of the preceding section as it stands. Proposition 1 of that section can now have $R\cap(k_i^{+k_1+\ldots+k_{i-1}} \times l_j^{+l_1+\ldots+l_{j-1}})$ empty, provided either $\bot$ is in A_i or $\top$ is in B_j.

As far as the maximality of $\mathbf{DL}$ and $\mathbf{DL}_{\top,\bot}$ is concerned, we conjecture that $\mathbf{DL}$ is not maximal. We conjectured at the end of §10.3 that we could extend $\mathcal{E}(\mathbf{ML})$ with $m_{p,p} = m_{p,p}\circ \hat{c}_{p,p}$ without falling into preorder. We conjecture the same thing for $\mathcal{E}(\mathbf{DL})$. There are other such equations, which we will not try to classify here. For $\mathbf{DL}_{\top,\bot}$ we can show that it is relatively maximal in the same sense in which $\mathbf{L}_{\top,\bot}$ is maximal (see §9.7).

Namely, every distributive dicartesian category that satisfies an equation between arrow terms of $\mathcal{C}(\mathbf{DL}_{\top,\bot})$ that is not in $\mathcal{E}(\mathbf{DL}_{\top,\bot})$ satisfies also the equations $(\hat{k}\check{k})$ and $(\hat{k}\check{k}\ fg)$. The argument in §9.7 can be transferred to the present context to demonstrate this fact. Some of these distributive dicartesian categories may, of course, satisfy more, as indicated at the end of §9.7.

It can be shown that the arrows of **DL** that are isomorphisms are denoted by arrow terms of $\mathcal{C}(\mathbf{S})$. So **S** catches the isomorphisms fragment of **DL** (cf. [40] for an analogous result showing that $\hat{\mathbf{S}}$ catches the isomorphisms fragment of $\hat{\mathbf{L}}$, and $\hat{\mathbf{S}}_\top$ the isomorphisms fragment of $\hat{\mathbf{L}}_\top$). That can be established by an argument based on coherence and on distinguished disjunctive and conjunctive normal forms.

CHAPTER 12

ZERO-LATTICE CATEGORIES

A kind of dual of the operation of union of proofs is the notion of zero proof. With zero proofs, which are mapped into empty relations in establishing coherence, we disregard provability in logic. With a zero proof we can pass from any premise to any conclusion.

We first prove coherence for categories with finite products and coproducts to which we add zero arrows, i.e. arrows that correspond to zero proofs. We call such categories zero-lattice categories. Zero arrows amount in this context to the inverse of the mix principle of Chapter 8. Our technique for the proof of coherence is based on composition elimination. Maximality, i.e. the impossibility to extend axioms without collapse into preorder, is easy to establish for zero-lattice categories.

As an example of a zero-lattice category in which the operations corresponding to conjunction and disjunction are not isomorphic, we have the category Set_* of sets with a distinguished object $*$ and $*$-preserving functions. By inverting the operations corresponding to conjunction and disjunction, we have as a subcategory in every zero-lattice category, and in Set_* in particular, a symmetric double monoidal category with dissociativity, and without unit objects, such as those for which we proved coherence in Chapter 7.

We also consider adding only zero arrows that correspond to proofs in conjunctive-disjunctive logic, in the sense that we have also non-zero proofs with the same premises and conclusions. We call such zero arrows zero-identity arrows. We prove coherence when zero-identity arrows are added to the categories of Chapters 9 and 11, and restricted coherence when they are added to the categories of Chapter 10. These categories are interesting because Gentzen's procedure can be modified to incorporate zero-identity arrows. The modified procedure can yield coherence not only with respect to the category whose arrows are relations between finite ordinals, but also with respect to the category whose arrows are matrices, where composition is matrix multiplication.

§12.1. Zero-lattice and zero-dicartesian categories

To obtain the natural logical category **ZL**, we have that the logical system $\mathcal{C}(\mathbf{ZL})$ is in $\mathcal{L}_{\wedge,\vee}$, with the transformations α included in $\mathbf{1}$, b, c, w-k and m^{-1}. The specific equations of $\mathcal{E}(\mathbf{ZL})$ are those of $\mathcal{E}(\mathbf{L})$ plus

$$
\begin{aligned}
&(m^{-1}0) \quad \hat{k}^2_{A,B} \circ m^{-1}_{A,B} \circ \check{k}^1_{A,B} = \hat{k}^1_{B,A} \circ m^{-1}_{B,A} \circ \check{k}^2_{B,A},\\
&(m^{-1}1) \quad \hat{k}^1_{A,B} \circ m^{-1}_{A,B} \circ \check{k}^1_{A,B} = \hat{k}^2_{B,A} \circ m^{-1}_{B,A} \circ \check{k}^2_{B,A} = \mathbf{1}_A.
\end{aligned}
$$

We call natural **ZL**-categories *zero-lattice* categories. The reason for this name will become clear below.

Note first that in **ZL** we have the equation

$$(cm^{-1}) \quad m^{-1}_{B,A} \circ \check{c}_{B,A} = \hat{c}_{A,B} \circ m^{-1}_{A,B},$$

which is dual to the equation (cm) of §8.4. We derive (cm^{-1}) as follows, with subscripts omitted:

$$
\begin{aligned}
m^{-1} \circ \check{c} &= [m^{-1} \circ \check{k}^2, m^{-1} \circ \check{k}^1], \text{ by } (\check{c}) \text{ and } (\check{K}\ 3),\\
&= \langle [\hat{k}^1 \circ m^{-1} \circ \check{k}^2, \hat{k}^1 \circ m^{-1} \circ \check{k}^1], [\hat{k}^2 \circ m^{-1} \circ \check{k}^2, \hat{k}^2 \circ m^{-1} \circ \check{k}^1]\rangle,\\
&\hspace{8cm} \text{by } (\wedge\eta) \text{ and } (\check{K}\ 3),\\
&= \langle [\hat{k}^2 \circ m^{-1} \circ \check{k}^1, \hat{k}^2 \circ m^{-1} \circ \check{k}^2], [\hat{k}^1 \circ m^{-1} \circ \check{k}^1, \hat{k}^1 \circ m^{-1} \circ \check{k}^2]\rangle,\\
&\hspace{6cm} \text{by } (m^{-1}0) \text{ and } (m^{-1}1),\\
&= [\langle \hat{k}^2 \circ m^{-1} \circ \check{k}^1, \hat{k}^1 \circ m^{-1} \circ \check{k}^1\rangle, \langle \hat{k}^2 \circ m^{-1} \circ \check{k}^2, \hat{k}^1 \circ m^{-1} \circ \check{k}^2\rangle],\\
&\hspace{8cm} \text{by } (\mathit{in\text{-}out}),\\
&= \hat{c} \circ m^{-1}, \text{ by } (\hat{K}\ 3), (\vee\eta) \text{ and } (\hat{c})
\end{aligned}
$$

(see the List of Equations at the end of the book for all the equations mentioned in this derivation).

We can derive analogously the following dual of the equation (bm) of §8.2:

$$(bm^{-1})\quad (m^{-1}_{A,B}\wedge \mathbf{1}_C)\circ m^{-1}_{A\vee B,C}\circ \check{b}^{\rightarrow}_{A,B,C}=\hat{b}^{\rightarrow}_{A,B,C}\circ m^{-1}_{A,B\wedge C}\circ(\mathbf{1}_A\vee m^{-1}_{B,C}).$$

The equations (bm) and (cm) are specific equations for the category **MS** of §8.5. The remaining specific equations of $\mathcal{E}(\mathbf{MS})$ are delivered by the equations of $\mathcal{E}(\mathbf{L})$. We can then infer that if $f=g$ in $\mathbf{MS}^{op}$, then $f=g$ in **ZL**. Since we have a functor G from **ZL** to *Rel*, and we have coherence for **MS**, i.e. the faithfulness of G from **MS** to *Rel*, we can conclude that **ZL** has a subcategory isomorphic to $\mathbf{MS}^{op}$.

If we assume the equation (cm^{-1}) as primitive for $\mathcal{E}(\mathbf{ZL})$, then the equation $(m^{-1}0)$ becomes superfluous, and from the equations $(m^{-1}1)$ it is enough to keep either $\hat{k}^1_{A,B}\circ m^{-1}_{A,B}\circ \check{k}^1_{A,B}=\mathbf{1}_A$ or $\hat{k}^2_{B,A}\circ m^{-1}_{B,A}\circ \check{k}^2_{B,A}=\mathbf{1}_A$.

A logical system $\mathcal{C}(\mathbf{0ZL})$ synonymous with $\mathcal{C}(\mathbf{ZL})$ (see the end of §2.4 for this notion of synonymity) is obtained by having as primitive instead of m^{-1} the transformation $\mathbf{0}$ whose members $\mathbf{0}_{A,B}\colon A\vdash B$ are called *zero arrow terms*. Zero arrow terms, which denote *zero arrows*, are defined in terms of m^{-1} by

$$\begin{aligned}\mathbf{0}_{A,B} &=_{df} \hat{k}^2_{A,B}\circ m^{-1}_{A,B}\circ \check{k}^1_{A,B}\\ &= \hat{k}^1_{B,A}\circ m^{-1}_{B,A}\circ \check{k}^2_{B,A},\ \text{by } (m^{-1}0),\end{aligned}$$

and m^{-1} is defined in terms of $\mathbf{0}$ by

$$\begin{aligned}m^{-1}_{A,B} &=_{df} \langle[\mathbf{1}_A,\mathbf{0}_{B,A}],[\mathbf{0}_{A,B},\mathbf{1}_B]\rangle\\ &= [\langle\mathbf{1}_A,\mathbf{0}_{A,B}\rangle,\langle\mathbf{0}_{B,A},\mathbf{1}_B\rangle],\ \text{by } (\textit{in-out}).\end{aligned}$$

It is clear that $G\mathbf{0}_{A,B}$ is the empty relation $\emptyset\subseteq GA\times GB$.

The specific equations of $\mathcal{E}(\mathbf{0ZL})$ are those of $\mathcal{E}(\mathbf{L})$ plus

$$f\circ\mathbf{0}_{A,A}=\mathbf{0}_{B,B}\circ f=\mathbf{0}_{A,B},$$

for $f\colon A\vdash B$. These equations deliver immediately the more general equations

$$(\mathbf{0})\quad f\circ\mathbf{0}_{C,A}=\mathbf{0}_{C,B},\qquad \mathbf{0}_{B,C}\circ f=\mathbf{0}_{A,C},$$

and they also deliver that the arrows $\mathbf{0}_{A,A}$ make a natural transformation from the identity functor to the identity functor; namely, one of these equations is

$$(\mathbf{0}\ nat) \quad f \circ \mathbf{0}_{A,A} = \mathbf{0}_{B,B} \circ f,$$

which becomes $(\mathbf{1}\ nat)$ when $\mathbf{0}_{A,A}$ and $\mathbf{0}_{B,B}$ are replaced by $\mathbf{1}_A$ and $\mathbf{1}_B$ respectively. It is straightforward to check that **ZL** and **0ZL** are isomorphic categories. (The notion of a zero arrow satisfying $(\mathbf{0})$ is considered in [94], Section VIII.2, and [89], p. 279.)

Note that in **0ZL**, and hence also in **ZL**, we have the equations

$$(\mathbf{0}\,\xi) \quad \mathbf{0}_{A,C}\ \xi\ \mathbf{0}_{B,D} = \mathbf{0}_{A\xi B, C\xi D},$$

for $\xi \in \{\wedge, \vee\}$. To derive $(\mathbf{0}\wedge)$ we have

$$\begin{aligned} \mathbf{0}_{A,C} \wedge \mathbf{0}_{B,D} &= \langle \mathbf{0}_{A,C} \circ \hat{k}^1_{A,B}, \mathbf{0}_{B,D} \circ \hat{k}^2_{A,B} \rangle, \text{ by } (\wedge), \\ &= \langle \hat{k}^1_{C,D}, \hat{k}^2_{C,D} \rangle \circ \mathbf{0}_{A\wedge B, C\wedge D}, \text{ by } (\mathbf{0}) \text{ and } (\hat{K}\ 3), \\ &= \mathbf{0}_{A\wedge B, C\wedge D}, \text{ by } (\wedge\eta) \text{ and } (cat\ 1), \text{ or by } (\mathbf{0}), \end{aligned}$$

and we proceed analogously for $(\mathbf{0}\vee)$. The equations $(\mathbf{0}\,\xi)$, which are analogous to the bifunctorial equations $(\,\xi\ 1)$ of §2.7, are null cases of the equations $(\cup\,\xi)$ of §10.1.

Another logical system $\mathcal{C}(d^{-1}\mathbf{ZL})$ synonymous with $\mathcal{C}(\mathbf{ZL})$, and hence also with $\mathcal{C}(\mathbf{0ZL})$, is obtained by having as primitive instead of m^{-1}, or $\mathbf{0}$, the transformation d^{-1} whose members are

$$d^{-1}_{A,B,C} \colon A \vee (B \wedge C) \vdash (A \vee B) \wedge C.$$

The type of $d^{-1}_{A,B,C}$ is converse to the type of $d^{R}_{A,B,C}$. The specific equations of $\mathcal{E}(d^{-1}\mathbf{ZL})$ are those of $\mathcal{E}(\mathbf{L})$ plus

$$\begin{aligned} &(d^{-1}\,1) \quad \hat{k}^1_{A\vee B,C} \circ d^{-1}_{A,B,C} \circ \check{k}^1_{A,B\wedge C} = \check{k}^1_{A,B}, \\ &(d^{-1}\,2) \quad \hat{k}^2_{A\vee B,C} \circ d^{-1}_{A,B,C} \circ \check{k}^2_{A,B\wedge C} = \hat{k}^2_{B,C}, \\ &(d^{-1}\,3) \quad \hat{k}^1_{A\vee B,C} \circ d^{-1}_{A,B,C} \circ \check{k}^2_{A,B\wedge C} = \check{k}^2_{A,B} \circ \hat{k}^1_{B,C}\,. \end{aligned}$$

That **0ZL** and d^{-1}**ZL** are isomorphic is demonstrated with the following definitions:

$$\mathbf{0}_{A,C} =_{df} \hat{k}^2_{A\vee B,C} \circ d^{-1}_{A,B,C} \circ \check{k}^1_{A,B\wedge C},$$

$$d^{-1}_{A,B,C} =_{df} \langle \mathbf{1}_A \vee \hat{k}^1_{B,C}, [\mathbf{0}_{A,C}, \hat{k}^2_{B,C}]\rangle$$
$$= [\langle \check{k}^1_{A,B}, \mathbf{0}_{A,C}\rangle, \check{k}^2_{A,B} \wedge \mathbf{1}_C].$$

This demonstration is quite straightforward; we will here just check that in $d^{-1}\mathbf{ZL}$ we have

$$d^{-1}_{A,B,C} = \langle \mathbf{1}_A \vee \hat{k}^1_{B,C}, [\hat{k}^2_{A\vee B,C} \circ d^{-1}_{A,B,C} \circ \check{k}^1_{A,B\wedge C}, \check{k}^2_{B,C}]\rangle.$$

We show first with the help of $(d^{-1}1)$, $(d^{-1}2)$ and $(d^{-1}3)$ that the right-hand side *RHS* of this equation is equal to the following arrow term with subscripts omitted:

$$\langle [\hat{k}^1 \circ d^{-1} \circ \check{k}^1, \hat{k}^1 \circ d^{-1} \circ \check{k}^2], [\hat{k}^2 \circ d^{-1} \circ \check{k}^1, \hat{k}^2 \circ d^{-1} \circ \check{k}^2]\rangle.$$

Then it is enough to establish that

$$\hat{k}^i_{A\vee B,C} \circ RHS = \hat{k}^i_{A\vee B,C} \circ d^{-1}_{A,B,C},$$

for $i \in \{1,2\}$, and use $(\wedge\eta)$.

In $\mathcal{C}(\mathbf{ZL})$ we define $d^{-1}_{A,B,C}$ by combining the definition of d^{-1} in terms of $\mathbf{0}$ and the definition of $\mathbf{0}$ in terms of m^{-1}, or by the following arrow term:

$$(\mathbf{1}_{A\vee B} \wedge (\hat{k}^2_{A,C} \circ m^{-1}_{A,C})) \circ c^k_{A,A,B,C} \circ (\hat{w}_A \vee \mathbf{1}_{B\wedge C}).$$

That we can do so will be clear after we have established coherence for **ZL** with respect to *Rel.* Note that $Gd^{-1}_{A,B,C}$ is an identity relation, i.e. identity function, in *Rel.*

The dissociativity principle of the type $A\vee(B\wedge C) \vdash (A\vee B)\wedge C$ of $d^{-1}_{A,B,C}$ is contained in the trivial part of the modularity law, which is satisfied in any lattice:

$$\text{if } a \leq c, \text{ then } a \vee (b \wedge c) \leq (a \vee b) \wedge c$$

(see [6], Sections I.5 and I.7). In the presence of an arrow $f: A \vdash C$, we have in the logical category **L** an arrow $g: A \vee (B \wedge C) \vdash (A \vee B) \wedge C$, which is defined like $d^{-1}_{A,B,C}$ in terms of $\mathbf{0}$ save that $\mathbf{0}_{A,C}$ in this definition is replaced by f. The relation Gg, however, is not an identity relation in that case, while $Gd^{-1}_{A,B,C}$ is an identity relation.

With **ZL** we have abandoned the realm of conjunctive-disjunctive logic as far as provability is concerned. The type $A \vee B \vdash A \wedge B$ of m^{-1} does not correspond in general to a logical consequence; namely, the implication $A \vee B \rightarrow A \wedge B$ is not a tautology. In **ZL** we have $\mathbf{0}_{A,B} : A \vdash B$ for any formulae A and B. This does not mean, however, that we have abandoned the realm of logic as far as equality of proofs is concerned. Our coherence results will show that adding zero arrows will not enable us to demonstrate new equations between arrow terms in which $\mathbf{0}$ does not occur; namely, the extension with zero arrows is conservative. And this extension can be useful to facilitate calculations (cf. §13.1).

Bits of zero arrows already existed in all our categories whenever we had $\top$ and $\bot$, and this not only in the δ-σ and κ families, but in other families α as well. For example, $\mathbf{1}_\top : \top \vdash \top$, $\hat{k}^2_{A,\top}\colon A \wedge \top \vdash \top$ and $\hat{w}_\top\colon \top \vdash \top \wedge \top$ all have an empty image in *Rel* by G, and behave like zero arrows.

We have seen in §8.1 that m is like Gentzen's mix (*Mischung*) where Θ is the empty sequence. In a similar vein, m^{-1} is related to the following version of Gentzen's mix:

$$\frac{\Gamma_1 \vdash \Delta_1, \Theta \qquad \Theta, \Gamma_2 \vdash \Delta_2}{\Gamma_1, \Gamma_2 \vdash \Delta_1, \Delta_2}$$

where either Δ_1 or Γ_2 is empty, while Θ is any nonempty sequence of formulae, and not necessarily a sequence of occurrences of the same formula, as Gentzen requires. Such a principle is not logically valid as far as provability is concerned, as we have seen above. It is, however, safe to introduce it if we are interested not in provability, but in equality of proofs.

Note that the following arrow term of $\mathcal{C}(\mathbf{L})$:

$$\langle \check{w}_A, \check{w}_A \rangle \colon A \vee A \vdash A \wedge A,$$

which is equal in **L** to $[\hat{w}_A, \hat{w}_A]$ and to $(\check{w}_A \wedge \check{w}_A) \circ c^k_{A,A,A,A} \circ (\hat{w}_A \vee \hat{w}_A)$, stands behind Gentzen's mix, with Θ a sequence of occurrences of the same formula. However, $G\langle \check{w}_A, \check{w}_A \rangle$ is different from $Gm^{-1}_{A,A}$.

It is pointless to add m^{-1} to the categories **I**, **A** and **S**, with appropriate equations that guarantee coherence, since the resulting categories would be isomorphic to **MI**, **MA** and **MS**. We just interchange $\wedge$ and $\vee$. On the other hand, the categories **ZL** and **ML** are not isomorphic. (Compare this with the remark on the symmetries of **A**, **S** and **L**, made at the end of §9.6.)

To obtain the natural logical category $\mathbf{ZL}_{\top,\bot}$, we have that the logical system $\mathcal{C}(\mathbf{ZL}_{\top,\bot})$ is in $\mathcal{L}_{\wedge,\vee,\top,\bot}$, with the transformations α included in $\mathbf{1}$, b, c, w-k, δ-σ and m^{-1}. The specific equations of $\mathcal{E}(\mathbf{ZL}_{\top,\bot})$ are obtained by taking the union of those of $\mathcal{E}(\mathbf{ZL})$ and $\mathcal{E}(\mathbf{L}_{\top,\bot})$. We call natural $\mathbf{ZL}_{\top,\bot}$-categories *zero-dicartesian* categories.

Note that in $\mathbf{ZL}_{\top,\bot}$ the following equations hold:

$$\hat{\kappa}_A = f = \mathbf{0}_{A,\top}, \qquad \text{for } f\colon A \vdash \top,$$
$$\check{\kappa}_A = f = \mathbf{0}_{\bot,A}, \qquad \text{for } f\colon \bot \vdash A,$$

$$\hat{\kappa}_\bot = \check{\kappa}_\top = \mathbf{0}_{\bot,\top},$$

$$\hat{\kappa}_\top = \mathbf{1}_\top = \mathbf{0}_{\top,\top}, \qquad \check{\kappa}_\bot = \mathbf{1}_\bot = \mathbf{0}_{\bot,\bot},$$
$$\hat{k}^1_{\bot,\bot} = \hat{k}^2_{\bot,\bot} = \mathbf{0}_{\bot\wedge\bot,\bot}, \qquad \check{k}^1_{\top,\top} = \check{k}^2_{\top,\top} = \mathbf{0}_{\top,\top\vee\top}.$$

The arrow $\mathbf{0}_{\top,\bot} : \top \vdash \bot$ is the inverse of $\mathbf{0}_{\bot,\top} : \bot \vdash \top$, and so $\top$ and $\bot$ are isomorphic in $\mathbf{ZL}_{\top,\bot}$. Hence $\top$ and $\bot$ are both terminal and initial objects in $\mathbf{ZL}_{\top,\bot}$, which means that they are *null* objects in the sense of [94] (Section I.5).

We also have in $\mathbf{ZL}_{\top,\bot}$ the equation

$$(\mathbf{0}\top\bot) \quad \mathbf{0}_{A,B} = \check{\kappa}_B \circ \mathbf{0}_{\top,\bot} \circ \hat{\kappa}_A,$$

according to which $\mathbf{0}_{\top,\bot}$ could be taken as an alternative primitive. The equations $(\mathbf{0})$ are derivable from $(\mathbf{0}\top\bot)$. So $\mathbf{ZL}_{\top,\bot}$ can be conceived as obtained from $\mathbf{L}_{\top,\bot}$ just by adding the arrow $\mathbf{0}_{\top,\bot}$, without any new equation, the equation $(\mathbf{0}\top\bot)$ being taken as a definition. In this context, the equations $(\hat{k}\ \bot)$ and $(\check{k}\ \top)$ of §9.6 become derivable from the remaining equations.

We can also conceive of $\mathbf{ZL}_{\top,\bot}$ as being obtained from **ZL** whose objects are formulae of $\mathcal{L}_{\wedge,\vee,\top,\bot}$ with the additional equations $\mathbf{1}_\top = \mathbf{0}_{\top,\top}$ and $\mathbf{1}_\bot = \mathbf{0}_{\bot,\bot}$, and the definitions $\hat{\kappa}_A =_{df} \mathbf{0}_{A,\top}$ and $\check{\kappa}_A =_{df} \mathbf{0}_{\bot,A}$.

§12.2. Coherence of zero-lattice and zero-dicartesian categories

Our purpose now is to prove the following.

ZERO-LATTICE COHERENCE. *The functor G from* **ZL** *to Rel is faithful.*

We proceed by enlarging the proof of Lattice Coherence in §9.4. Throughout this section, we assume that **ZL** stands for **0ZL**, where **0** is primitive.

The syntactical system $\mathcal{C}(\mathbf{GZL})$ is defined as $\mathcal{C}(\mathbf{GL})$ of §9.4, save that it has as primitive arrow terms also the arrow terms $\mathbf{0}_{A,B}\colon A \vdash B$. The equations of $\mathcal{E}(\mathbf{GZL})$ are obtained by adding the equations $(\mathbf{0})$ to the equations of $\mathcal{E}(\mathbf{GL})$. The syntactical category **GZL**, which is $\mathcal{C}(\mathbf{GZL})/\mathcal{E}(\mathbf{GZL})$, is isomorphic to **ZL**. The syntactical category $\mathbf{GZL}_{\top,\bot}$, isomorphic to $\mathbf{ZL}_{\top,\bot}$, is defined as **GZL**: we just replace **L** everywhere by $\mathbf{L}_{\top,\bot}$.

The categories $\mathbf{Z}\hat{\mathbf{L}}_{\vee}$, $\mathbf{GZ}\hat{\mathbf{L}}_{\vee}$, $\mathbf{Z}\check{\mathbf{L}}_{\wedge}$ and $\mathbf{GZ}\check{\mathbf{L}}_{\wedge}$ are introduced as $\hat{\mathbf{L}}_{\vee}$, $\mathbf{G}\hat{\mathbf{L}}_{\vee}$, $\check{\mathbf{L}}_{\wedge}$ and $\mathbf{G}\check{\mathbf{L}}_{\wedge}$ in §9.4, save that we have in addition $\mathbf{0}_{A,B}$ and the equations $(\mathbf{0})$. We have also variants of these systems with $\top$ and $\bot$, involving either $\hat{\kappa}$ with the equations $(\hat{\kappa})$ and $(\hat{k}\ \bot)$, or $\check{\kappa}$ with the equations $(\check{\kappa})$ and $(\check{k}\ \top)$, respectively (see §9.6).

We can prove Composition Elimination for **GZL**, $\mathbf{GZL}_{\top,\bot}$, $\mathbf{GZ}\hat{\mathbf{L}}_{\vee}$, $\mathbf{GZ}\check{\mathbf{L}}_{\wedge}$, and for the variants of the last two categories with $\top$ and $\bot$, by enlarging the proofs of Composition Elimination in §9.1, §9.2 and §9.4. The equations $(\mathbf{0})$ take care of all the additional cases.

Let a *zero term* of $\mathcal{C}(\mathbf{ZL})$ be defined inductively by: $\mathbf{1}_A$ and $\mathbf{0}_{A,B}$ are zero terms for every A and B; if f and g are zero terms, then $f\ \xi\ g$ for $\xi \in \{\wedge, \vee\}$ is a zero term. A *proper* zero term is a zero term in which **0** occurs at least once.

It is easy to show by induction on the sum of the lengths of $h_1\colon A \vdash B$ and $h_2\colon B \vdash C$ that if h_1 and h_2 are zero terms, then $h_2 \circ h_1$ is equal in **ZL** to a zero term. (If at least one of h_1 and h_2 is a proper zero term, then $h_2 \circ h_1$ is equal to a proper zero term.)

An arrow term $g \circ h \circ f$ of $\mathcal{C}(\mathbf{ZL})$ is in *standard form* when f is an arrow term of $\mathcal{C}(\hat{\mathbf{L}}_{\vee})$ and g is an arrow term of $\mathcal{C}(\check{\mathbf{L}}_{\wedge})$, while h is a zero term. Note that zero terms are arrow terms of both $\mathcal{C}(\mathbf{Z}\hat{\mathbf{L}}_{\vee})$ and $\mathcal{C}(\mathbf{Z}\check{\mathbf{L}}_{\wedge})$ (cf. §9.4). Then we can prove the following.

STANDARD-FORM LEMMA. *Every arrow term of* $\mathcal{C}(\mathbf{ZL})$ *is equal in* **ZL** *to an arrow term in standard form.*

PROOF. We proceed as in the proof of the Standard-Form Lemma of §9.4, save for the following additional cases involving $\mathbf{0}$. If in the proof of $(*)$ we have for $f \circ g$ that $f : B \vdash C$ is $\mathbf{0}_{B,C}$ or $g : A \vdash B$ is $\mathbf{0}_{A,B}$, then we apply the equations $(\mathbf{0})$. Here we treat zero terms first as $\wedge$-factors, and next as $\vee$-factors, or vice versa. We also appeal to the fact noted above that the composition of two zero terms is equal in **ZL** to a zero term. ⊣

We can also prove the analogue of the Auxiliary Lemma of §9.4 where $\hat{\mathbf{L}}_{\vee}$ is replaced by $\mathbf{Z}\hat{\mathbf{L}}_{\vee}$, and it is assumed that $Gf \neq \emptyset$. We have, of course, a dual lemma for $\mathbf{Z}\check{\mathbf{L}}_{\wedge}$. We can also prove the following lemma.

ZERO-TERM LEMMA. *For* $h : A \vdash B$ *a zero term of* $\mathcal{C}(\mathbf{ZL})$, *if* $Gh = \emptyset$, *then* $h = \mathbf{0}_{A,B}$ *in* **ZL**.

PROOF. We proceed by induction on the length of h. If h is $\mathbf{0}_{A,B}$, then we are done. Here h cannot be $\mathbf{1}_A$, for if it were $\mathbf{1}_A$, then Gh would not be empty. If h is $h_1 \xi h_2$ for $\xi \in \{\wedge, \vee\}$, then from $Gh = \emptyset$ we infer $Gh_1 = Gh_2 = \emptyset$, and by the equation $(\mathbf{0}\, \xi)$ of the preceding section we obtain $h = \mathbf{0}_{A,B}$. ⊣

We have also the following strengthening of this lemma.

EMPTY-RELATION LEMMA. *For* $h : A \vdash B$ *an arrow term of* $\mathcal{C}(\mathbf{ZL})$, *if* $Gh = \emptyset$, *then* $h = \mathbf{0}_{A,B}$ *in* **ZL**.

PROOF. By the Standard-Form Lemma above, we have $h = h_3 \circ h_2 \circ h_1$ in **ZL**, where h_1 is an arrow term of $\mathcal{C}(\hat{\mathbf{L}}_{\vee})$ and h_3 is an arrow term of $\mathcal{C}(\check{\mathbf{L}}_{\wedge})$, while $h_2 : C \vdash D$ is a zero term. From $Gh = \emptyset$ we conclude that Gh_2 must also be empty. So $h_2 = \mathbf{0}_{C,D}$ by the Zero-Term Lemma, and hence, by $(\mathbf{0})$, we obtain that $h = \mathbf{0}_{A,B}$. ⊣

This last lemma entails Lemma 2 of §9.6 for **ZL**; namely, the assertion that if for $f, g : A \vdash B$ we have $Gf = Gg = \emptyset$, then $f = g$ in **ZL**. If $Gf = Gg \neq \emptyset$, then we imitate the proof of Lattice Coherence in §9.4 to obtain $f = g$ in **ZL**. So we have Zero-Lattice Coherence.

We can also prove the following in an analogous manner.

ZERO-DICARTESIAN COHERENCE. *The functor G from $\mathbf{ZL}_{\top,\bot}$ to Rel is faithful.*

Zero terms are defined as before, and, as before, the composition of two zero terms is equal in $\mathbf{ZL}_{\top,\bot}$ to a zero term. We have a definition of standard form for arrow terms of $\mathcal{C}(\mathbf{ZL}_{\top,\bot})$ analogous to that for $\mathcal{C}(\mathbf{ZL})$, and the analogue of the Standard-Form Lemma for $\mathbf{ZL}_{\top,\bot}$.

We can also prove the Zero-Term Lemma with $\mathbf{ZL}$ replaced by $\mathbf{ZL}_{\top,\bot}$. In the basis of the induction, we have to consider the case where h is $\mathbf{1}_C$ for a letterless formula C. This C is isomorphic in $\mathbf{ZL}_{\top,\bot}$ both to $\top$ and to $\bot$. For $i\colon C \vdash \top$ and $i^{-1}\colon \top \vdash C$ being inverse to each other, in $\mathbf{ZL}_{\top,\bot}$ we have

$$\begin{aligned}\mathbf{1}_C &= i^{-1} \circ i\\ &= i^{-1} \circ \mathbf{0}_{\top,\top} \circ i, \quad \text{since } \mathbf{1}_\top = \mathbf{0}_{\top,\top},\\ &= \mathbf{0}_{C,C}, \;\text{ by } (\mathbf{0}).\end{aligned}$$

From the Standard-Form Lemma and the Zero-Term Lemma for $\mathbf{ZL}_{\top,\bot}$ we infer as above the Empty-Relation Lemma with $\mathbf{ZL}$ replaced by $\mathbf{ZL}_{\top,\bot}$. For the remainder of the proof of Zero-Dicartesian Coherence, we imitate the proof of Zero-Lattice Coherence.

§12.3. Maximality of zero-lattice and zero-dicartesian categories

In this section, we show that $\mathbf{ZL}$ and $\mathbf{ZL}_{\top,\bot}$ are maximal in the sense of §9.3. We deal first with $\mathbf{ZL}$.

Suppose that for some arrow terms $f_1, f_2 : A \vdash B$ of $\mathcal{C}(\mathbf{ZL})$ we have $Gf_1 \neq Gf_2$. Suppose that for some $x \in GA$ and some $y \in GB$ we have $(x,y) \in Gf_1$ and $(x,y) \notin Gf_2$. Then with the help of the arrow terms $\overset{\vee}{k}{}^i_{A_1,A_2}\colon A_i \vdash A_1 \vee A_2$, for $i \in \{1,2\}$, together with

$$\begin{aligned}\langle \mathbf{1}_{A_1}, \mathbf{0}_{A_1,A_2}\rangle &\colon A_1 \vdash A_1 \wedge A_2,\\ \langle \mathbf{0}_{A_2,A_1}, \mathbf{1}_{A_2}\rangle &\colon A_2 \vdash A_1 \wedge A_2\end{aligned}$$

and the operation of composition, we can build an arrow term $h_1\colon p \vdash A$ of $\mathcal{C}(\mathbf{ZL})$ such that $Gh_1 = \{(0,x)\}$. The $(x+1)$-th occurrence of letter in A

counting from the left is an occurrence of p. Analogously, with the help of $\hat{k}^i_{B_1,B_2}\colon B_1 \wedge B_2 \vdash B_i$, together with

$$[\mathbf{1}_{B_1}, \mathbf{0}_{B_2,B_1}]\colon B_1 \vee B_2 \vdash B_1,$$
$$[\mathbf{0}_{B_1,B_2}, \mathbf{1}_{B_2}]\colon B_1 \vee B_2 \vdash B_2$$

and composition, we build an arrow term $h_2 : B \vdash p$ of $\mathcal{C}(\mathbf{ZL})$ such that $Gh_2 = \{(y,0)\}$. The $(y+1)$-th occurrence of letter in B counting from the left is an occurrence of the same p we had for $h_1 : p \vdash A$. This must be the case because $(x,y) \in Gf_1$. Then for $h_2 \circ f_i \circ h_1 : p \vdash p$, where $i \in \{1,2\}$, we have that $G(h_2 \circ f_1 \circ h_1) = \{(0,0)\}$, while $G(h_2 \circ f_2 \circ h_1) = \emptyset$. It follows from Composition Elimination for **GZL** and from the functoriality of G that in **ZL** we have

$$h_2 \circ f_1 \circ h_1 = \mathbf{1}_p,$$
$$h_2 \circ f_2 \circ h_1 = \mathbf{0}_{p,p}.$$

(This follows from Zero-Lattice Coherence too.) So, if we extend $\mathcal{E}(\mathbf{ZL})$ with $f_1 = f_2$, then we can derive $\mathbf{1}_p = \mathbf{0}_{p,p}$.

If an equation $f_1 = f_2$ that is not in $\mathcal{E}(\mathbf{ZL})$ holds in a zero-lattice category $\mathcal{A}$, then by Zero-Lattice Coherence we have $Gf_1 \neq Gf_2$. So $\mathbf{1}_p = \mathbf{0}_{p,p}$ holds in $\mathcal{A}$. Then for $f, g\colon a \vdash b$ in $\mathcal{A}$, with $\mathbf{1}_a = \mathbf{0}_{a,a}$ and the equations $(\mathbf{0})$ we obtain

$$f = f \circ \mathbf{0}_{a,a} = \mathbf{0}_{a,b} = g \circ \mathbf{0}_{a,a} = g,$$

and hence $\mathcal{A}$ is a preorder. This proves the maximality of **ZL**.

Exactly the same argument serves to prove the maximality of $\mathbf{ZL}_{\top,\bot}$. We just replace **ZL** by $\mathbf{ZL}_{\top,\bot}$, and appeal to Zero-Dicartesian Coherence.

§12.4. Zero-lattice and symmetric net categories

The category **ZL** has a subcategory isomorphic to the category **DS** of §7.6. We define a functor F from **DS** to **ZL** in the following manner:

$$Fp = p,$$

$$F(A \wedge B) = FA \vee FB, \qquad F(A \vee B) = FA \wedge FB,$$

$$F\mathbf{1}_A = \mathbf{1}_{FA},$$

$$F\hat{b}^{\rightarrow}_{A,B,C} = \overset{\vee}{b}^{\rightarrow}_{FA,FB,FC}, \qquad F\overset{\vee}{b}^{\rightarrow}_{A,B,C} = \hat{b}^{\rightarrow}_{FA,FB,FC}$$
$$F\hat{b}^{\leftarrow}_{A,B,C} = \overset{\vee}{b}^{\leftarrow}_{FA,FB,FC}, \qquad F\overset{\vee}{b}^{\leftarrow}_{A,B,C} = \hat{b}^{\leftarrow}_{FA,FB,FC},$$
$$F\hat{c}_{A,B} = \overset{\vee}{c}_{FB,FA}, \qquad F\overset{\vee}{c}_{A,B} = \hat{c}_{FB,FA},$$
$$Fd_{A,B,C} = d^{-1}_{FA,FB,FC},$$
$$F(f \wedge g) = Ff \vee Fg, \qquad F(f \vee g) = Ff \wedge Fg,$$
$$F(g \circ f) = Fg \circ Ff.$$

To show that F is indeed a functor, we have to check that if $f = g$ holds in **DS**, then $Ff = Fg$ holds in **ZL** (cf. the penultimate paragraph of §2.4). So suppose that $f = g$ holds in **DS**; then $Gf = Gg$ in *Rel*, and we have $GFf = Gf = Gg = GFg$ in *Rel*. By Zero-Lattice Coherence, we obtain that $Ff = Fg$ holds in **ZL**.

It is clear that F establishes a one-to-one correspondence on objects. To show that F is faithful, which here implies that F is one-one on arrows, suppose that for $f, g\colon A \vdash B$ arrow terms of $\mathcal{C}(\mathbf{DS})$ we have $Ff = Fg$ in **ZL**. Hence in *Rel* we have $Gf = GFf = GFg = Gg$, and, by Symmetric Net Coherence, we obtain that $f = g$ in **DS**. So the subcategory of **ZL** that is the image under F of **DS** is isomorphic to **DS**.

It was shown in §9.7 that Set_* is a lattice category with $\wedge$ being $\boxtimes$ and $\vee$ being $\boxplus$. If we define the function $\mathbf{0}_{a,b}\colon a \to b$ by $\mathbf{0}_{a,b}(x) = *$, then it is easy to check that Set_* is a zero-lattice category. It is also a zero-dicartesian category with both $\top$ and $\bot$ being $\{*\}$.

By what we have shown above, a subcategory of Set_* is a symmetric net category with $\wedge$ being $\boxplus$ and $\vee$ being $\boxtimes$. The claim made in [19] (p. 22) that Set_* with $\wedge$ being $\boxtimes$ and $\vee$ being $\boxplus$ is a cartesian linearly (*alias* weakly) distributive category, which would imply that it is a symmetric net category, is not correct. The functions $d_{a,b,c} : a \boxtimes (b \boxplus c) \to (a \boxtimes b) \boxplus c$ defined at that place, which are there called δ^L_L, and for which one has $d_{a,b,c}(x,(*,z)) = *$ and $d_{a,b,c}(x,*) = ((x,*),*)$ for every $x \in a$ and every $z \in c$, do not make a natural transformation d. Take $a = \{x,*\}$, $b = \{*\}$ and $c = \{z,*\}$, and let the function $h\colon c \to c$ be defined by $h(z) = h(*) = *$. Then we have

$$d_{a,b,c}((\mathbf{1}_a \boxtimes (\mathbf{1}_b \boxplus h))(x,(*,z))) = ((x,*),*),$$
$$((\mathbf{1}_a \boxtimes \mathbf{1}_b) \boxplus h)(d_{a,b,c}(x,(*,z))) = *.$$

The category Set_* with $\wedge$ being $\boxtimes$ and $\vee$ being $\boxplus$ is trivially a symmetric net category when we take that $d_{a,b,c}$ is defined as $\mathbf{0}_{a\boxtimes(b\boxplus c),a\boxtimes(b\boxplus c)}$. With that definition, however, it is neither a distributive lattice category, in our sense, nor a linearly distributive category, in the sense of §7.9, because the equations $(d\hat{k})$ and $(d\overset{\vee}{k})$, or $(\hat{\sigma}\, d^L)$ and $(\check{\delta}\, d^L)$, would not hold. For the same reason, it is also not a cartesian linearly distributive category in the sense of [19]. That no other definition of $d_{a,b,c}\colon a \boxtimes (b \boxplus c)) \to (a \boxtimes b) \boxplus c$ can make of Set_* a distributive lattice category or a cartesian linearly distributive category is shown in §13.2.

§12.5. Zero-identity arrows

Let the natural logical category **ZIL** in $\mathcal{L}_{\wedge,\vee}$ be defined as **0ZL** save that the transformation $\mathbf{0}$ has as members $\mathbf{0}_A\colon A \vdash A$ (we write here $\mathbf{0}_A$ instead of $\mathbf{0}_{A,A}$), which are *zero-identity arrow terms* that stand for *zero-identity arrows*, and instead of the equations $(\mathbf{0})$ we have only the following consequence of $(\mathbf{0})$:

$$(\mathbf{0}\mathrm{I}) \quad f \circ \mathbf{0}_A = \mathbf{0}_B \circ g$$

for $f,g\colon A \vdash B$. By putting g for f, this equation delivers that $\mathbf{0}$ is a natural transformation from the identity functor to the identity functor; namely, we obtain the equation $(\mathbf{0}\ nat)$ of §12.1 with $\mathbf{0}_{A,A}$ and $\mathbf{0}_{B,B}$ replaced by $\mathbf{0}_A$ and $\mathbf{0}_B$ respectively. As other consequences of $(\mathbf{0}\mathrm{I})$, we have

$$f \circ \mathbf{0}_A = g \circ \mathbf{0}_A,$$
$$\mathbf{0}_B \circ f = \mathbf{0}_B \circ g,$$

and the following equation:

$$(\mathbf{00}) \quad \mathbf{0}_A \circ \mathbf{0}_A = \mathbf{0}_A.$$

In **ZIL** we define $\mathbf{0}_{A,B}$ by

$$\mathbf{0}_{A,B} =_{df} f \circ \mathbf{0}_A$$

for some arrow term $f\colon A \vdash B$ of $\mathcal{C}(\mathbf{ZIL})$. This definition is correct because we have $f \circ \mathbf{0}_A = g \circ \mathbf{0}_A$, as remarked above. We do not have, however, $\mathbf{0}_{A,B}$ in **ZIL** for every A and B of $\mathcal{L}_{\wedge,\vee}$, as we had it in **ZL**, but only for those pairs (A, B) where there is an arrow of **L** of type $A \vdash B$. It is easy to show, as in §12.1, that the following instance of $(\mathbf{0}\,\xi)$:

$$(\mathbf{0}\mathrm{I}\,\xi) \quad \mathbf{0}_A \,\xi\, \mathbf{0}_B = \mathbf{0}_{A\xi B}$$

holds in **ZIL**, and if $\mathbf{0}_{A,C}$ and $\mathbf{0}_{B,D}$ are defined, then $(\mathbf{0}\,\xi)$ holds too.

We cannot define in **ZIL** every $m^{-1}_{A,B}\colon A \vee B \vdash A \wedge B$ of **ZL**, but only those where there are arrows of the types $A \vdash B$ and $B \vdash A$ in **L**; and we cannot define in **ZIL** every $d^{-1}_{A,B,C}\colon A \vee (B \wedge C) \vdash (A \vee B) \wedge C$ of **ZL**, but only those where there is an arrow of type $A \vdash C$ in **L**.

The natural logical category $\mathbf{ZIL}_{\top,\bot}$ in $\mathcal{L}_{\wedge,\vee,\top,\bot}$ is defined as **ZIL** save that it is based on $\mathbf{L}_{\top,\bot}$ instead of **L**. With **ZIL** and $\mathbf{ZIL}_{\top,\bot}$, contrary to what we had with **ZL** and $\mathbf{ZL}_{\top,\bot}$, we stay within the realm of conjunctive-disjunctive logic as far as provability is concerned (see §12.1). We also stay within the realm of logic as far as equality of proofs is concerned, as it is shown by the coherence results below.

By assuming, as before, that $G\mathbf{0}_A$ is an empty relation, and by going over our proof of Zero-Lattice Coherence and Zero-Dicartesian Coherence in §12.2, we can easily establish the following.

ZERO-IDENTITY LATTICE COHERENCE. *The functor G from* **ZIL** *to Rel is faithful.*

ZERO-IDENTITY DICARTESIAN COHERENCE. *The functor G from* $\mathbf{ZIL}_{\top,\bot}$ *to Rel is faithful.*

We establish the maximality of **ZIL** as we established the maximality of **ZL** in §12.3. The only difference in the proof is that for $f_1, f_2\colon A \vdash B$ we assume that only one letter p occurs in A and B (cf. the proofs of maximality for $\hat{\mathbf{L}}$ and **L** in §9.3 and §9.5) . On the other hand, $\mathbf{ZIL}_{\top,\bot}$ is maximal only in the relative sense in which $\mathbf{L}_{\top,\bot}$ is maximal (see §9.7).

Let the natural logical categories **ZIML** and $\mathbf{ZIML}_{\top,\bot}$ be obtained from **ML** and $\mathbf{ML}_{\top,\bot}$ respectively by adding the zero-identity arrow terms and the equations $(\mathbf{0}\mathrm{I})$ and

$$(\cup\mathbf{0}) \quad f \cup \mathbf{0}_{A,B} = f$$

for every arrow term $f: A \vdash B$ of $\mathcal{C}(\mathbf{ZIML})$ or $\mathcal{C}(\mathbf{ZIML}_{\top,\bot})$. The arrow terms $\mathbf{0}_{A,B}$ are defined in terms of $\mathbf{0}_A$ as above. Instead of $(\cup\mathbf{0})$, we could alternatively assume its instance

$$\mathbf{1}_A \cup \mathbf{0}_A = \mathbf{1}_A,$$

which yields $(\cup\mathbf{0})$. In **ZIML** and $\mathbf{ZIML}_{\top,\bot}$, a hom-set whose arrows are of type $A \vdash B$ is a semilattice with the unit $\mathbf{0}_{A,B}$ (which can be conceived as either top or bottom).

Restricted Zero-Identity Mix-Lattice Coherence is formulated as Restricted Mix-Lattice Coherence in §10.2, save that **ZIML** replaces **ML**. To prove this coherence result for **ZIML** we proceed as for **ML**, with the following modifications.

The syntactical category **GZIML** differs from the category **GML** by having in $\mathcal{C}(\mathbf{GZIML})$ the primitive arrow terms $\mathbf{0}_A : A \vdash A$ besides $\mathbf{1}_A : A \vdash A$; moreover, we assume for it in addition to the equations of $\mathcal{E}(\mathbf{GML})$ the equations $(\mathbf{0}\mathrm{I})$ and $(\cup\mathbf{0})$.

For A in *dnf* and B in *cnf*, arrow terms of $\mathcal{C}(\mathbf{GZIML})$ of type $A \vdash B$ that are in normal form are defined as in §10.2 save that we allow $\mathbf{0}_p$ to replace $\mathbf{1}_p$ in arrow terms in atomic bracket-free normal form. Arrow terms in atomic bracket-free normal form where instead of $\mathbf{1}_p$ we have $\mathbf{0}_p$ are called *zero* atomic bracket-free terms, and those with $\mathbf{1}_p$ *nonzero* atomic bracket-free terms. We use the same terminology of "zero" and "nonzero" for atomic components. An analogous terminology was used in §10.3. The settled normal form is defined as for $\mathbf{GML}_{\top,\bot}$ in §10.3.

We prove Composition Elimination for **GZIML** by extending the proof for **GML**. In that proof we apply the equation $(\mathbf{00})$ and the following equations of **GZIML**:

$$\hat{K}^i_{A_j} f \circ \mathbf{0}_{A_1 \wedge A_2} = \hat{K}^i_{A_j}(\mathbf{0}_B \circ f),$$
$$\mathbf{0}_{A_1 \vee A_2} \circ \check{K}^i_{A_j} g = \check{K}^i_{A_j}(g \circ \mathbf{0}_B),$$

for $i, j \in \{1,2\}$ such that $i \neq j$, which are easily derived with the help of $(\mathbf{0}\mathrm{I})$ and $(\overset{\xi}{K}\ 1)$, as well as

$$\mathbf{0}_{A_1 \wedge A_2} \circ \langle g_1, g_2 \rangle = \langle g_1 \circ \mathbf{0}_C, g_2 \circ \mathbf{0}_C \rangle,$$
$$[f_1, f_2] \circ \mathbf{0}_{A_1 \vee A_2} = [\mathbf{0}_C \circ f_1, \mathbf{0}_C \circ f_2],$$

which are easily derived with the help of $(\mathbf{0}\mathrm{I})$ and $(\overset{\xi}{K}\ 3)$, for $\xi \in \{\wedge, \vee\}$.

To prove the Normal-Form Lemma of §10.2 with **GML** replaced by **GZIML**, we proceed as in the proof in §10.2, and we use in addition the equation $f \cup g = f$ for any zero atomic component g. This equation, which is analogous to the equation $(\cup \mathbf{0} g)$ of §10.3, is derivable from $(\mathbf{0}\mathrm{I})$ and $(\cup \mathbf{0})$. We can then prove Restricted Zero-Identity Mix-Lattice Coherence as Restricted Mix-Lattice Coherence in §10.2. We use in that proof the fact that if f'' and g'' are zero atomic components of the same type, then they are equal in **GZIML** by $(\mathbf{0}\mathrm{I})$ and $(\cup \mathbf{0})$.

Restricted Zero-Identity Mix-Dicartesian Coherence is formulated analogously by replacing **ZIML** with $\mathbf{ZIML}_{\top,\bot}$ (cf. also §10.3), and is proved in the same manner. The equations

$$\overset{\wedge}{\kappa}_{\top} = \mathbf{1}_{\top} = \mathbf{0}_{\top},$$
$$\overset{\vee}{\kappa}_{\bot} = \mathbf{1}_{\bot} = \mathbf{0}_{\bot}$$

hold in $\mathbf{ZIL}_{\top,\bot}$, and hence also in $\mathbf{ZIML}_{\top,\bot}$.

Let the natural logical categories **ZIDL** and $\mathbf{ZIDL}_{\top,\bot}$ be obtained from **DL** and $\mathbf{DL}_{\top,\bot}$ respectively by adding the zero-identity arrow terms and the equation $(\mathbf{0}\mathrm{I})$. In these categories, we have an arrow of type $A \vdash B$ iff the implication $A \rightarrow B$ is a tautology.

We define $\mathbf{0}_{A,B}$ by $f \circ \mathbf{0}_A$, as before, and we can now derive $(\cup \mathbf{0})$ in the following manner. We have in **ZIDL** and $\mathbf{ZIDL}_{\top,\bot}$

$$\begin{aligned}
\mathbf{1}_A \cup \mathbf{0}_A &= \overset{\vee}{w}_A \circ (\hat{k}{}^1_{A,A} \vee \mathbf{1}_A) \circ d_{A,A,A} \circ (\mathbf{1}_A \wedge \overset{\vee}{k}{}^2_{A,A}) \circ (\mathbf{1}_A \wedge \mathbf{0}_A) \circ \hat{w}_A \\
&= \overset{\vee}{w}_A \circ (\hat{k}{}^1_{A,A} \vee \mathbf{1}_A) \circ d_{A,A,A} \circ (\mathbf{1}_A \wedge \overset{\vee}{k}{}^1_{A,A}) \circ (\mathbf{1}_A \wedge \mathbf{0}_A) \circ \hat{w}_A, \text{ with } (\mathbf{0}\mathrm{I}), \\
&= \overset{\vee}{w}_A \circ (\hat{k}{}^1_{A,A} \vee \mathbf{1}_A) \circ \overset{\vee}{k}{}^1_{A \wedge A,A} \circ (\mathbf{1}_A \wedge \mathbf{0}_A) \circ \hat{w}_A, \text{ by } (d\overset{\vee}{k}) \text{ of §11.1}, \\
&= \mathbf{1}_A,
\end{aligned}$$

by Zero-Identity Lattice Coherence, or by applying $(\overset{\xi}{k}{}^1\ nat)$ and $(\overset{\xi}{w}\overset{\xi}{k})$, for $\xi \in \{\wedge, \vee\}$ (see §9.1 and the List of Equations). From $\mathbf{1}_A \cup \mathbf{0}_A = \mathbf{1}_A$ we easily obtain $(\cup \mathbf{0})$, as we remarked above.

Then we can prove the following.

ZERO-IDENTITY DISTRIBUTIVE LATTICE COHERENCE. *The functor G from* **ZIDL** *to Rel is faithful.*

ZERO-IDENTITY DISTRIBUTIVE DICARTESIAN COHERENCE. *The functor G from* $\mathbf{ZIDL}_{\top,\bot}$ *to Rel is faithful.*

Let $\mathcal{D}$ be now the category obtained as the disjoint union of the strictified category $\mathbf{ZIDL}^{\mathbf{A}}$ and the trivial category with the single object $\emptyset$ and the single arrow $\mathbf{1}_\emptyset : \emptyset \vdash \emptyset$ (cf. §11.1). The Gentzen terms for this category $\mathcal{D}$ are defined as in §11.1 with the addition in the basis of the inductive definition that $\mathbf{0}''_p : p \vdash p$, which denotes the arrow $(\mathbf{0}_p, p, p)$ of $\mathcal{D}$, is a Gentzen term. We prove the Gentzenization Lemma of §11.1 and the Cut-Elimination Theorem of §11.2 as before. In the proof of the Gentzenization Lemma, we rely on $(\mathbf{0I}\,\xi)$. To define clusters and rank, we count among leaf formulae the occurrences of p in the type $p \vdash p$ of $\mathbf{0}''_p$. In the proof of the Cut-Elimination Theorem, we have as the only additional cases, when the complexity of our clean cut is $(0,2)$, the left-hand sides of the following equations of $\mathcal{D}$:

$$
\begin{aligned}
cut_{\emptyset,\emptyset}(\mathbf{0}''_p, \mathbf{1}''_p) &= \mathbf{0}''_p,\\
cut_{\emptyset,\emptyset}(\mathbf{1}''_p, \mathbf{0}''_p) &= \mathbf{0}''_p,\\
cut_{\emptyset,\emptyset}(\mathbf{0}''_p, \mathbf{0}''_p) &= \mathbf{0}''_p,\\
cut_{\emptyset,\emptyset}(\mathbf{0}''_p, k^L_p g) &= k^L_p g,\\
cut_{\emptyset,\emptyset}(k^R_p f, \mathbf{0}''_p) &= k^R_p f.
\end{aligned}
$$

For the first two equations we use $(cat\ 1)$, for the third $(\mathbf{00})$, for the fourth $(\hat{k}^1\ nat)$, and for the fifth $(\check{k}^2\ nat)$.

For the remainder of the proof of the Cut-Elimination Theorem, we can proceed as before, but we can also proceed differently in the cases where our clean cut is of the form $cut(f, w^L g)$ with blocked $w^L g$ tied to our clean cut. In all these cases—namely, $(w\,\gamma\,1)$, $(w\,\gamma\,2), \ldots, (w \vee 3)$—we applied an equation of $\mathcal{D}$ where on the left-hand side we have a single occurrence of $h\colon \Gamma \vdash \Delta$, while on the right-hand side we have more than one occurrence of h; this is always two occurrences of h, except in case $(w \vee 3)$, where we have three occurrences of h. Now, we can put h on the right-hand side for exactly one old occurrence of h, and replace the others by the cut-free Gentzen term h^0 obtained from the cut-free Gentzen term h by replacing every $\mathbf{1}''_p$ in h by $\mathbf{0}''_p$. It can be shown, by induction on the length of the cut-free Gentzen term h, that in $\mathcal{D}$ we have $h^0 = \mathbf{0}''_\Delta \circ h$. This new procedure would dispense us from applying the equation (wm) in the remainder of the proof. Instead, we would rely on the following equations of **ZIDL**:

$$\check{w}_A \circ (\mathbf{0}_A \vee \mathbf{1}_A) \circ m_{A,A} \circ \hat{w}_A = \mathbf{1}_A,$$
$$\check{w}_A \circ (\mathbf{1}_A \vee \mathbf{0}_A) \circ m_{A,A} \circ \hat{w}_A = \mathbf{1}_A,$$

which amount to $(\cup\mathbf{0})$.

For the remainder of the proof of Zero-Identity Distributive Lattice Coherence, we proceed as for the proof of Distributive Lattice Coherence, relying on Restricted Zero-Identity Mix-Lattice Coherence. We proceed analogously for the proof of Zero-Identity Distributive Dicartesian Coherence, relying on Restricted Zero-Identity Mix-Dicartesian Coherence.

It can be proved that **ZIDL** is maximal by imitating the proof of the maximality of **ZL** in §12.3, with the same modification we introduced above (after stating Zero-Identity Dicartesian Coherence) to establish the maximality of **ZIL**. The category $\mathbf{ZIDL}_{\top,\bot}$ is maximal in the relative sense in which $\mathbf{L}_{\top,\bot}$ is maximal (see §9.7).

Consider the natural logical categories $\mathbf{ZIDL}^-$ and $\mathbf{ZIDL}^-_{\top,\bot}$ that differ from **ZIDL** and $\mathbf{ZIDL}_{\top,\bot}$ respectively by rejecting the equation (wm) for m, or alternatively the idempotency equation ($\cup$ *idemp*) for $\cup$. In these categories hom-sets are not necessarily semilattices with unit—they must be only commutative monoids. Union of arrows becomes now disjoint union of arrows, or addition of arrows. The possibility indicated above to prove the Cut-Elimination Theorem by not relying on (wm), but on $(\cup\mathbf{0})$ instead, indicates that we could prove that there are faithful functors G from $\mathbf{ZIDL}^-$ and $\mathbf{ZIDL}^-_{\top,\bot}$ not into *Rel*, but into the category *Mat*, which is isomorphic to the skeleton of the category whose objects are finite-dimensional vector spaces over a fixed number field, and whose arrows are linear transformations. (Note that this category of vector spaces is a subcategory of the category Set_* of §9.7, where the null vector is $*$.)

More precisely, the objects of the category *Mat* are finite ordinals, i.e. natural numbers (the dimensions of our vector spaces), and an arrow of type $n \vdash m$ is an $n \times m$ matrix. Matrices that are images under the functor G will have entries that are natural numbers. Composition of arrows is matrix multiplication, and the identity arrow $\mathbf{1}_n : n \vdash n$ is the $n \times n$ identity matrix with the entries $\mathbf{1}_n(i,j) = \delta(i,j)$, where δ is the Kronecker delta.

Every $n \times m$ matrix M whose entries are only 0 and 1 may be identified with a binary relation $R^M \subseteq n \times m$ such that $M(i,j) = 1$ iff $(i,j) \in R^M$. Multiplication of such matrices is the same as composition of relations if we assume that $1+1 = 1$.

For the proof of the faithfulness of G from $\mathbf{ZIDL}^-$ and $\mathbf{ZIDL}^-_{\top,\bot}$ into *Mat* we would rely on restricted coherence results for the natural logical categories $\mathbf{ZIML}^-$ and $\mathbf{ZIML}^-_{\top,\bot}$, which are obtained from $\mathbf{ZIML}$ and $\mathbf{ZIML}_{\top,\bot}$ respectively by rejecting the equation (wm). These restricted coherence results are of the same type as those we had for $\mathbf{ML}$, $\mathbf{ML}_{\top,\bot}$, $\mathbf{ZIML}$ and $\mathbf{ZIML}_{\top,\bot}$. In producing the settled normal form, we just do not rely on the equation $(\cup\ idemp)$.

The fact that (wm) does not hold in *Mat* shows that this equation cannot be derived from the remaining equations we have used to axiomatize the equations of $\mathbf{ZIDL}$ and $\mathbf{ZIDL}_{\top,\bot}$.

We conclude our consideration of zero-identity arrows with some remarks on formulae that are isomorphic in their presence. In $\mathbf{ZIML}$ we have the isomorphism

$$\langle [\mathbf{1}_A, \mathbf{0}_A], [\mathbf{0}_A, \mathbf{1}_A] \rangle : A \vee A \vdash A \wedge A,$$

whose inverse is $m_{A,A}$. Let A and B be formulae of the language $\mathcal{L}^{\{p\}}_{\wedge,\vee}$, which is the language $\mathcal{L}_{\wedge,\vee}$ generated by $\mathcal{P} = \{p\}$. Then it is clear that there are in the category $\mathbf{L}$ arrows of type $A \vdash B$ and $B \vdash A$, and hence in $\mathbf{ZIML}$ we have the isomorphism

$$\langle [\mathbf{1}_A, \mathbf{0}_{B,A}], [\mathbf{0}_{A,B}, \mathbf{1}_B] \rangle : A \vee B \vdash A \wedge B,$$

whose inverse is $m_{A,B}$. It is the easy to conclude that in $\mathbf{ZIML}$ and $\mathbf{ZIDL}$ for every pair (C,D) of formulae of $\mathcal{L}^{\{p\}}_{\wedge,\vee}$ such that in each of C and D there are $n \geq 1$ occurrences of p we have that C and D are isomorphic. (The category $\mathbf{ZIML}$ generated by $\{p\}$ is isomorphic to the category $\mathbf{ZML}$ of §13.1 generated by $\{p\}$.) If p^n stands for any of these formulae, then the functoriality of G from **ZIML** and **ZIDL** to *Rel* implies that p^n and p^m cannot be isomorphic in **ZIML** and **ZIDL** for $n \neq m$. This characterizes completely the formulae of $\mathcal{L}_{\wedge,\vee}^{\{p\}}$ isomorphic in **ZIML** and **ZIDL**.

Let $\mathcal{L}_{\wedge,\vee,\top,\bot}^{\{p\}}$ be the language $\mathcal{L}_{\wedge,\vee,\top,\bot}$ generated by $\mathcal{P} = \{p\}$. Then every formula of $\mathcal{L}_{\wedge,\vee,\top,\bot}^{\{p\}}$ is isomorphic in $\mathbf{ZIDL}_{\top,\bot}$ to one of the form p^n for $n \geq 1$, or $p^m \wedge \bot$, or $p^m \vee \top$ for $m \geq 0$, where if $m = 0$, then $p^m \wedge \bot$ is $\bot$ and $p^m \vee \top$ is $\top$. To prove that, we use various isomorphisms of $\mathbf{ZIDL}_{\top,\bot}$, among which isomorphisms of the following types are prominent:

$$\begin{aligned}
&p^n \vee (p^m \wedge \bot) \vdash p^{n+m},\\
&p^n \wedge (p^m \vee \top) \vdash p^{n+m},\\
&(A \wedge \bot) \vee (B \wedge \bot) \vdash (A \vee B) \wedge \bot,\\
&(A \vee \top) \wedge (B \vee \top) \vdash (A \wedge B) \vee \top,\\
&(A \wedge \bot) \vee \top \vdash A \vee \top,\\
&(A \vee \top) \wedge \bot \vdash A \wedge \bot.
\end{aligned}$$

For example, the following isomorphism is of the last of these types:

$$((\check{\delta}^{\rightarrow}_{A} \circ (\mathbf{1}_A \vee \hat{\sigma}^{\rightarrow}_{\bot}) \circ d^R_{A,\top,\bot}) \wedge \mathbf{1}_\bot) \circ \hat{b}^{\rightarrow}_{A\vee\top,\bot,\bot} \circ (\mathbf{1}_{A\vee\top} \wedge \hat{w}_\bot).$$

Since in classical logic formulae in the classes p^n, $p^m \wedge \bot$ and $p^m \vee \top$ are equivalent respectively to p, $\bot$ and $\top$, formulae from distinct classes among these three cannot be isomorphic. And that, within each class, formulae with different superscripts n or m cannot be isomorphic is shown by appealing to the functoriality of G from $\mathbf{ZIDL}_{\top,\bot}$ to *Rel*.

CHAPTER 13

ZERO-MIX LATTICE CATEGORIES

Zero-mix lattice categories are categories with finite products and coproducts, with or without the terminal and initial objects, to which we add the union operation on arrows of Chapter 10 and the zero arrows of Chapter 12. This amounts to making products isomorphic to coproducts. In zero-mix lattice categories hom-sets are semilattices with unit, and these categories are related to categories whose hom-sets are commutative monoids, like linear categories, preadditive categories, additive categories and abelian categories. In zero-mix lattice categories we have dissociativity, and these categories are distributive lattice categories in the sense of Chapter 11. We prove coherence for zero-mix lattice categories with the help of composition elimination and a unique normal form inspired by linear algebra. Zero-mix lattice categories are maximal, in the sense that it is impossible to extend their axioms without collapse into preorder.

The category whose arrows are relations between finite ordinals, on which we relied throughout the book for our coherence results, is a zero-mix lattice category. This category is isomorphic to a subcategory of another zero-mix lattice category—namely, the category of semilattices with unit, which is itself a subcategory of the category Set_* of sets with a distinguished object $*$, whose arrows are $*$-preserving functions.

§13.1. Coherence of zero-mix lattice categories

To obtain the natural logical category **ZML**, we have that the logical system $\mathcal{C}(\mathbf{ZML})$ is in $\mathcal{L}_{\wedge,\vee}$, with the transformations α included in $\mathbf{1}$, b, c, w-k, m and m^{-1}. The specific equations of $\mathcal{E}(\mathbf{ZML})$ are obtained by taking the union of those of $\mathcal{E}(\mathbf{ML})$ and $\mathcal{E}(\mathbf{ZL})$ plus

$$(mm^{-1}) \quad m^{-1}_{A,B} \circ m_{A,B} = \mathbf{1}_{A\wedge B}, \qquad m_{A,B} \circ m^{-1}_{A,B} = \mathbf{1}_{A\vee B}.$$

So in **ZML** we have that $A \wedge B$ and $A \vee B$ are isomorphic. The equations $(m\ nat)$ and $(m^{-1}\ nat)$ entail each other in the presence of (mm^{-1}).

We call natural **ZML**-categories *zero-mix lattice* categories. The hom-sets in a zero-mix lattice category are semilattices with unit (see §13.3 for references concerning related kinds of categories).

According to what we had in §10.1 and §12.1, we can take for **ZML** primitives alternative to m and m^{-1}; for example, we can take $\cup$ and $\mathbf{0}$, which are defined in **ZML** as before (see §10.1 and §12.1).

In **ZML** the following equations hold:

$$(\hat{k}\,m) \quad \hat{k}^1_{A,B} = [\mathbf{1}_A, \mathbf{0}_{B,A}] \circ m_{A,B}, \qquad \hat{k}^2_{A,B} = [\mathbf{0}_{A,B}, \mathbf{1}_B] \circ m_{A,B},$$

$$(\check{k}\,m) \quad \check{k}^1_{A,B} = m_{A,B} \circ \langle \mathbf{1}_A, \mathbf{0}_{A,B}\rangle, \qquad \check{k}^2_{A,B} = m_{A,B} \circ \langle \mathbf{0}_{B,A}, \mathbf{1}_B\rangle.$$

Here is a derivation of the first $(\hat{k}\,m)$ equation:

$$\begin{aligned}\hat{k}^1_{A,B} \circ m^{-1}_{A,B} &= [\hat{k}^1_{A,B} \circ m^{-1}_{A,B} \circ \check{k}^1_{A,B}, \hat{k}^1_{A,B} \circ m^{-1}_{A,B} \circ \check{k}^2_{A,B}], \text{ by } (\vee\eta),\\ &= [\mathbf{1}_A, \mathbf{0}_{B,A}], \text{ by } (m^{-1}\,1) \text{ and } (m^{-1}\,0) \text{ of §12.1;}\end{aligned}$$

then we apply (mm^{-1}) (for $(\vee\eta)$ see the List of Equations at the end of the book). Alternatively, we can rely on the following equation of **ZL**:

$$m^{-1}_{A,B} = \langle [\mathbf{1}_A, \mathbf{0}_{B,A}], [\mathbf{0}_{A,B}, \mathbf{1}_B]\rangle.$$

We proceed analogously for the remaining three equations of $(\hat{k}\ m)$ and $(\check{k}\,m)$.

For $f\colon A \vdash B$, let us use the following abbreviations for arrow terms of $\mathcal{C}(\mathbf{ZML})$:

$$\hat{Z}^1_C f =_{df} m^{-1}_{B,C} \circ \check{k}^1_{B,C} \circ f, \qquad \check{Z}^1_C f =_{df} f \circ \hat{k}^1_{A,C} \circ m^{-1}_{A,C},$$

$$\hat{Z}^2_C f =_{df} m^{-1}_{C,B} \circ \check{k}^2_{C,B} \circ f, \qquad \check{Z}^2_C f =_{df} f \circ \hat{k}^2_{C,A} \circ m^{-1}_{C,A}.$$

With the equations $(\hat{k}\,m)$ and $(\check{k}\,m)$, it is clear that in **ZML** we have

$$\hat{Z}^1_C f = \langle f, \mathbf{0}_{A,C}\rangle \colon A \vdash B \wedge C, \qquad \check{Z}^1_C f = [f, \mathbf{0}_{C,B}] \colon A \vee C \vdash B,$$
$$\hat{Z}^2_C f = \langle \mathbf{0}_{A,C}, f\rangle \colon A \vdash C \wedge B, \qquad \check{Z}^2_C f = [\mathbf{0}_{C,B}, f] \colon C \vee A \vdash B.$$

Then we can infer that for $f_i \colon C \vdash A_i$ and $g_i \colon A_i \vdash C$, where $i \in \{1,2\}$, the following equations hold in **ZML**:

$$(\hat{Z}) \quad \langle f_1, f_2\rangle = \hat{Z}^1_{A_2} f_1 \cup \hat{Z}^2_{A_1} f_2,$$
$$(\check{Z}) \quad [g_1, g_2] = \check{Z}^1_{A_2} g_1 \cup \check{Z}^2_{A_1} g_2.$$

For $(\hat{Z})$ we have

$$\begin{aligned} \check{k}^1_{C,C} \cup \check{k}^2_{C,C} &= \check{w}_{C \vee C} \circ (\check{k}^1_{C,C} \cup \check{k}^2_{C,C}) \circ m_{C,C} \circ \hat{w}_C \\ &= m_{C,C} \circ \hat{w}_C, \end{aligned}$$

and from that, with (mm^{-1}), we obtain

$$\hat{w}_C = m^{-1}_{C,C} \circ (\check{k}^1_{C,C} \cup \check{k}^2_{C,C}),$$

which yields $(\hat{Z})$. We derive $(\check{Z})$ analogously via

$$\check{w}_C = (\hat{k}^1_{C,C} \cup \hat{k}^2_{C,C}) \circ m^{-1}_{C,C}.$$

For every arrow term $f \colon A \vdash B$ of $\mathcal{C}(\mathbf{ZML})$, in **ZML** we have also the equation

$$(\cup\mathbf{0}) \quad f \cup \mathbf{0}_{A,B} = f,$$

which we encountered already in the preceding chapter (see §12.5). Here is a derivation of this equation:

$$\begin{aligned} \mathbf{1}_A \cup \mathbf{0}_{A,A} &= \check{w}_A \circ (\mathbf{1}_A \vee \mathbf{0}_{A,A}) \circ m_{A,A} \circ \hat{w}_A \\ &= [\mathbf{1}_A, \mathbf{0}_{A,A}] \circ m_{A,A} \circ \hat{w}_A \\ &= \mathbf{1}_A, \text{ by } (\hat{k}\, m) \text{ and } (\hat{w}\hat{k}) \text{ of §9.1}; \end{aligned}$$

from that we easily obtain $(\cup\mathbf{0})$ with the help of $(\cup\circ)$.

Conversely, we can derive (mm^{-1}) from $(\cup\mathbf{0})$. Here is derivation, with subscripts omitted, of the first equation of (mm^{-1}):

$$\begin{aligned} m^{-1} \circ m &= [\langle \mathbf{1}, \mathbf{0}\rangle, \langle \mathbf{0}, \mathbf{1}\rangle] \circ ((\check{k}^1 \circ \hat{k}^1) \cup (\check{k}^2 \circ \hat{k}^2)) \\ &= ((\hat{k}^1 \wedge \mathbf{0}) \cup (\mathbf{0} \wedge \hat{k}^2)) \circ \hat{w}, \text{ with } (\cup \circ) \text{ of §10.1}, \\ &= \mathbf{1}, \text{ with } (\cup\, \xi) \text{ of §10.1 and } (\cup \mathbf{0}); \end{aligned}$$

we proceed analogously for the second one by using $m^{-1} = \langle [\mathbf{1}, \mathbf{0}], [\mathbf{0}, \mathbf{1}]\rangle$. So $(\cup \mathbf{0})$, or $\mathbf{1}_A \cup \mathbf{0}_{A,A} = \mathbf{1}_A$, could replace (mm^{-1}) for the axiomatization of $\mathcal{E}(\mathbf{ZML})$. The axiom $(\cup \mathbf{0})$ is more appropriate than (mm^{-1}) if $\cup$ and $\mathbf{0}$ are primitive instead of m and m^{-1}.

To prove the coherence of **ZML** with respect to *Rel* we introduce, in the style of the preceding chapters, a syntactical category isomorphic to **ZML** for which we can prove Composition Elimination. The syntactical system $\mathcal{C}(\mathbf{GZML})$ is formulated by combining what we had for $\mathcal{C}(\mathbf{GML})$ in §10.1 and for $\mathcal{C}(\mathbf{GZL})$ in §12.2 (which is based on §9.4), together with the primitive operations $\overset{\xi}{Z}{}^i_C$ on arrow terms, for $\xi \in \{\wedge, \vee\}$ and $i \in \{1, 2\}$. The equations of $\mathcal{E}(\mathbf{GZML})$ are obtained by assuming in addition to the equations of $\mathcal{E}(\mathbf{GML})$ and $\mathcal{E}(\mathbf{GZL})$ the equation $(\cup \mathbf{0})$ and the four equations given above immediately after the definitions of $\overset{\xi}{Z}{}^i_C$ in $\mathcal{C}(\mathbf{ZML})$. Note that in the presence of $(\cup \mathbf{0})$, which is $f \cup \mathbf{0}_{A,B} = f$, and of the analogous equation $\mathbf{0}_{A,B} \cup f = f$, we can replace ($\cup$ *assoc*) and ($\cup$ *com*) by the single equation

$$(f_1 \cup f_2) \cup (f_3 \cup f_4) = (f_1 \cup f_3) \cup (f_2 \cup f_4).$$

The syntactical category **GZML** is $\mathcal{C}(\mathbf{GZML})/\mathcal{E}(\mathbf{GZML})$, and it is isomorphic to **ZML**.

We can prove Composition Elimination for **GZML** by extending the proofs for **GML** and **GZL**, which are based on the proof of Composition Elimination for **GL** in §9.4 and §9.1.

Next we introduce some definitions analogous up to a point to those we had in Chapter 10 and §12.5. Arrow terms of $\mathcal{C}(\mathbf{GZML})$ of the form $P_1 \ldots P_n Q_1 \ldots Q_m \theta$, where $n, m \geq 0$, and θ is of the form $\mathbf{1}_p$ or $\mathbf{0}_{p,q}$, for some letters p and q, while P_i for $i \in \{1, \ldots, n\}$ is of the form $\check{K}^1_C$, or $\check{K}^2_C$, or $\hat{Z}^1_C$, or $\hat{Z}^2_C$, and Q_j for $j \in \{1, \ldots, m\}$ is of the form $\hat{K}^1_C$, or $\hat{K}^2_C$, or $\check{Z}^1_C$, or $\check{Z}^2_C$, are called *atomic terms*. We have a *zero* atomic term when θ is of the form $\mathbf{0}_{p,q}$ (where p and q can be the same letter), and a *nonzero* atomic term when θ is of the form $\mathbf{1}_p$.

Arrow terms of $\mathcal{C}(\mathbf{GZML})$ in *normal form* are defined inductively by stipulating that atomic terms are in normal form, and that if f and g are in normal form, then $f \cup g$ is in normal form.

Let f be an arrow term of $\mathcal{C}(\mathbf{GZML})$ in normal form, and let f' be a subterm of f such that f' is an atomic term, and there is no atomic subterm f'' of f with f' a proper subterm of f''. Then we say that f' is an *atomic component* of f.

Let p occur in a formula A of $\mathcal{L}_{\wedge,\vee}$ as the $x+1$-th occurrence of letter counting from the left. Then there is a unique atomic term $Q_1 \ldots Q_m \mathbf{1}_p : A \vdash p$ such that $G(Q_1 \ldots Q_m \mathbf{1}_p) = \{(x,0)\}$. We say that the word $Q_1 \ldots Q_m$ is *bound to* $(x,0)$.

Let q occur in a formula B of $\mathcal{L}_{\wedge,\vee}$ as the $y+1$-th occurrence of letter counting from the left. Then there is a unique atomic term $P_1 \ldots P_n \mathbf{1}_q : q \vdash B$ such that $G(P_1 \ldots P_n \mathbf{1}_q) = \{(0,y)\}$. We say that the word $P_1 \ldots P_n$ is *bound to* $(0,y)$.

Hence there is a unique word $P_1 \ldots P_n Q_1 \ldots Q_m$ such that $Q_1 \ldots Q_m$ is bound to $(x,0)$ and $P_1 \ldots P_n$ is bound to $(0,y)$. We say that the word $P_1 \ldots P_n Q_1 \ldots Q_m$ is *bound to* $(x,y) \in GA \times GB$.

An arrow term $f : A \vdash B$ of $\mathcal{C}(\mathbf{GZML})$ is in *settled normal form* when it is in normal form and there is a one-to-one correspondence between the set $GA \times GB$ and the set of atomic components of f, such that for every atomic component $P_1 \ldots P_n Q_1 \ldots Q_m \theta$ we have that $P_1 \ldots P_n Q_1 \ldots Q_m$ is bound to the ordered pair in $GA \times GB$ corresponding to it. To every ordered pair in $GA \times GB$ corresponds either a zero or a nonzero atomic component depending on whether θ is of the form $\mathbf{0}_{p,q}$ or $\mathbf{1}_p$. Then we can prove the following.

NORMAL-FORM LEMMA. *Every arrow term of* $\mathcal{C}(\mathbf{GZML})$ *is equal in* **GZML** *to an arrow term in settled normal form.*

PROOF. Take an arrow term $f : A \vdash B$ of $\mathcal{C}(\mathbf{GZML})$. By Composition Elimination for **GZML** there is a composition-free arrow term $f' : A \vdash B$ of $\mathcal{C}(\mathbf{GZML})$ equal to f in **GZML**.

Then we apply the equations ($\hat{K}$ 4) and ($\check{K}$ 4) (see §9.1 and §9.4), and the following equations of **GZML**:

$$\mathbf{0}_{A\wedge B,C} = \hat{K}^1_B\,\mathbf{0}_{A,C}, \qquad \mathbf{0}_{A\vee B,C} = \check{Z}^1_B\,\mathbf{0}_{A,C},$$
$$\mathbf{0}_{C,A\wedge B} = \hat{Z}^1_B\,\mathbf{0}_{C,A}, \qquad \mathbf{0}_{C,A\vee B} = \check{K}^1_B\,\mathbf{0}_{C,A},$$

in order to obtain a composition-free arrow term f'' of $\mathcal{C}(\mathbf{GZML})$, equal to f' in **GZML**, in which every **1** and and every **0** have subscripts that are letters. This procedure is arbitrary as far as zero arrow terms are concerned: we could as well base it on $\overset{\xi}{K}{}^2$ and $\overset{\xi}{Z}{}^2$ instead of $\overset{\xi}{K}{}^1$ and $\overset{\xi}{Z}{}^1$. (We could also use ($\cup\mathbf{0}$) to omit zero arrow terms, which will reappear through ($\cup\mathbf{0}$) in another garb afterwards; see below.)

Next we apply to f'' the equations ($\overset{\xi}{Z}$) and the following equations of **GZML**:

$$\overset{\xi}{X}{}^i_A(f \cup g) = \overset{\xi}{X}{}^i_A f \cup \overset{\xi}{X}{}^i_A g,$$

$$\hat{K}^i_A\check{K}^j_B f = \check{K}^j_B\hat{K}^i_A f, \qquad \check{Z}^i_A\check{K}^j_B f = \check{K}^j_B\check{Z}^i_A f,$$
$$\hat{K}^i_A\hat{Z}^j_B f = \hat{Z}^j_B\hat{K}^i_A f, \qquad \check{Z}^i_A\hat{Z}^j_B f = \hat{Z}^j_B\check{Z}^i_A f,$$

for $X \in \{K, Z\}$, $i, j \in \{1, 2\}$ and $\xi \in \{\wedge, \vee\}$, to obtain an arrow term f''' of $\mathcal{C}(\mathbf{GZML})$ in normal form equal to f'' in **GZML**.

To transform f''' into an arrow term in settled normal form, we apply the equations ($\mathbf{0}$) and ($\cup\mathbf{0}$) to put in the missing atomic components, and delete the atomic components $P_1 \ldots P_n Q_1 \ldots Q_m \mathbf{0}_{p,p}$ for which we have already the atomic components $P_1 \ldots P_n Q_1 \ldots Q_m \mathbf{1}_p$. ⊣

It is easy to establish that if $f, g\colon A \vdash B$ are arrow terms of $\mathcal{C}(\mathbf{GZML})$ in settled normal form and $Gf = Gg$, then the set of atomic components of f and the set of atomic components of g must be the same set of atomic terms. We can then easily prove the following.

ZERO-MIX LATTICE COHERENCE. *The functor G from* **ZML** *to Rel is faithful.*

PROOF. Suppose $f, g : A \vdash B$ are arrow terms of $\mathcal{C}(\mathbf{GZML})$ in settled normal form such that $Gf = Gg$. By the Normal-Form Lemma, we have the arrow terms f' and g' in settled normal form such that $f = f'$ and $g = g'$ in **GZML**. By the functoriality of G, we have $Gf = Gf'$ and $Gg = Gg'$; hence $Gf' = Gg'$. So f' and g' have equal sets of atomic components, and they must be equal in **GZML** by applying ($\cup$ *assoc*), ($\cup$ *com*) and ($\cup$ *idemp*). Therefore, $f = g$ in **GZML**. ⊣

We could have used ($\cup$ *assoc*), ($\cup$ *com*) and ($\cup$ *idemp*) to find a unique term in settled normal form equal to an arrow term of **GZML**. The advantage **ZML** has over **DL** is that, due to zero arrows, we can reach this unique composition-free normal form. For **DL**, a unique cut-free normal form was not forthcoming.

We will see in the next section that **DL** is isomorphic to a subcategory of **ZML**. For that we rely on Distributive Lattice Coherence and Zero-Mix Lattice Coherence. The unique normal form we have for **ZML** can serve as a substitute for the missing unique normal form of **DL**. For every arrow term of $\mathcal{C}(\mathbf{DL})$ we take the arrow term of $\mathcal{C}(\mathbf{GZML})$ in normal form whose image by G is the same.

§13.2. Zero-mix lattice and distributive lattice categories

To obtain the natural logical category **ZDL**, we have that the logical system $\mathcal{C}(\mathbf{ZDL})$ is in $\mathcal{L}_{\wedge,\vee}$, with the transformations α included in $\mathbf{1}$, b, c, w-k, d and m^{-1}. The specific equations of $\mathcal{E}(\mathbf{ZDL})$ are obtained by taking the union of those of $\mathcal{E}(\mathbf{DS})$ and $\mathcal{E}(\mathbf{ZL})$ plus $(d\hat{k})$ and $(d\overset{\vee}{k})$ of §11.1 and (wm) of §10.1 for $m_{A,A}$ defined by (dm) of §11.1 understood as a definition. Note that we do not assume here the equations $(m\,\hat{e})$ and $(m\,\check{e})$, which would deliver immediately the equations of $\mathcal{E}(\mathbf{DL}')$, and hence of $\mathcal{E}(\mathbf{DL})$ (see §11.1). We will see below, however, that all the equations of $\mathcal{E}(\mathbf{DL})$ are in $\mathcal{E}(\mathbf{ZDL})$.

We will show that the categories **ZML** and **ZDL** are isomorphic with the definition

$$(md) \quad d_{A,B,C} =_{df} m_{A\wedge B,C} \circ \hat{b}{}^{\rightarrow}_{A,B,C} \circ (\mathbf{1}_A \wedge m^{-1}_{B,C})$$

in $\mathcal{C}(\mathbf{ZML})$ (this definition is derived from the equation $(\hat{b}\,mL)$ of §8.3; an alternative definition can be obtained from $(\overset{\vee}{b}\,mL)$), and the definition of $m_{A,B}$ in $\mathcal{C}(\mathbf{ZDL})$ corresponding to the equation (dm) of §11.1.

It is easy to conclude from Zero-Mix Lattice Coherence of the preceding section that with (md) all the equations of $\mathcal{E}(\mathbf{ZDL})$ plus (dm) hold in **ZML**. To show that with the definition of $m_{A,B}$ corresponding to (dm) all the equations of $\mathcal{E}(\mathbf{ZML})$ hold in **ZDL**, first we derive easily $(m\ nat)$ for **ZDL**, by using naturality equations of **ZDL**. Then we infer with the help of $(\overset{\xi}{w}\overset{\xi}{k}\overset{\xi}{k})$ for $\xi \in \{\wedge, \vee\}$ (see §9.1 and the List of Equations) that in **ZDL** we have $(\check{k}^1 \circ \hat{k}^1) \cup (\check{k}^2 \circ \hat{k}^2) = m$, with subscripts omitted. Next we derive easily $(\cup \circ)$ for **ZDL** with the help of naturality and bifunctorial equations. The equation $(\cup\, \xi)$ of §10.1 is derived for **ZDL** as indicated in §10.1, and $(\cup \mathbf{0})$ is derived for **ZDL** as we derived it for **ZIDL** in §12.5. With all that, we obtain (mm^{-1}) in **ZDL** as in the preceding section. Since in **ZDL** we have all the equations of $\mathcal{E}(\mathbf{ZL})$, we have also (bm^{-1}) and (cm^{-1}) (see §12.1), which together with (mm^{-1}) yield (bm) and (cm) in **ZDL**. (The equation $(m\ nat)$ follows from $(m^{-1}\ nat)$ with the help of (mm^{-1}), but we relied on $(m\ nat)$ in the derivation of (mm^{-1}).) It remains only to derive for **ZDL** the equation obtained from (md) by replacing $m_{A\wedge B,C}$ according to (dm); namely,

$$d_{A,B,C} = (\hat{k}^1_{A\wedge B,D} \vee \mathbf{1}_C) \circ d_{A\wedge B,D,C} \circ (\mathbf{1}_{A\wedge B} \wedge \check{k}^2_{D,C}) \circ \hat{b}^{\rightarrow}_{A,B,C} \circ (\mathbf{1}_A \wedge m^{-1}_{B,C}).$$

For that it is enough to derive $(\hat{b}\ mL)$, as we did it for $\mathbf{DL}'$ in §11.1, and use moreover (mm^{-1}). We can then conclude that **ZML** and **ZDL** are isomorphic categories.

We can infer from Zero-Mix Lattice Coherence that all the equations of $\mathcal{E}(\mathbf{DL})$ are in $\mathcal{E}(\mathbf{ZDL})$. Because of the question concerning the independence of the equations $(m\ \hat{e})$ and $(m\ \check{e})$ in our axiomatization of **DL** (see §11.1), it is, however, of some interest to see how these equations are derived in $\mathcal{E}(\mathbf{ZDL})$. We derive the equations $(\hat{b}\ mL)$, $(\check{b}\ mL)$ and (cm) as we did it for $\mathbf{DL}'$ in §11.1. Then we derive the equations $(m\ \hat{c}^m)$ and $(m\ \check{c}^m)$ as we did it for **DL** in §11.1. Note that we do not need the equations $(m\ \hat{e})$ and $(m\ \check{e})$ for all these derivations. With (mm^{-1}), we easily obtain from $(m\ \hat{c}^m)$ and $(m\ \check{c}^m)$ the following equations:

$$\hat{e}'_{A,B,C,D} = m_{A\wedge C,B\wedge D} \circ \hat{c}^m_{A,B,C,D} \circ (\mathbf{1}_{A\wedge B} \wedge m^{-1}_{C,D}),$$
$$\check{e}'_{D,C,B,A} = (m^{-1}_{D,C} \vee \mathbf{1}_{B\vee A}) \circ \check{c}^m_{D,C,B,A} \circ m_{D\vee B,C\vee A}.$$

We also have in **ZDL**

$$
\begin{aligned}
& c^k_{A,C,B,D} \circ \hat{e}'_{A,B,C,D} \circ (\mathbf{1}_{A \wedge B} \wedge m_{C,D}) \\
& \quad = ((\hat{k}^1_{A,C} \vee \hat{k}^1_{B,D}) \wedge (\hat{k}^2_{A,C} \vee \hat{k}^2_{B,D})) \circ \hat{w}_{(A \wedge C) \vee (B \wedge D)} \circ m_{A \wedge C, B \wedge D} \circ \\
& \qquad\qquad \circ \hat{c}^m_{A,B,C,D}, \text{ by } (m\,\hat{c}^m), \\
& \quad = m_{A,B} \wedge m_{C,D}, \text{ by naturality equations and Lattice Coherence,}
\end{aligned}
$$

from which $(m\ \hat{e})$ follows easily with the help of (mm^{-1}). We proceed analogously for $(m\,\check{e})$ by using $(m\,\check{c}^m)$.

We have seen in §12.4 that Set_* is a zero-lattice category with $\wedge$ being $\boxtimes$ and $\vee$ being $\boxplus$. (It is also a symmetric net category with $\wedge$ being $\boxplus$ and $\vee$ being $\boxtimes$.) If Set_* with $\wedge$ being $\boxtimes$ and $\vee$ being $\boxplus$ were also a symmetric net category, and $(d\,\hat{k})$ and $(d\,\check{k})$ were moreover satisfied, then $a \boxtimes b$ and $a \boxplus b$ would be isomorphic in Set_*, which is not the case. Note that the equation (wm) played no role in inferring above that $A \wedge B$ and $A \vee B$ are isomorphic in **ZDL**; namely, in deriving the equation (mm^{-1}) for **ZDL**. Since the equations $(d\,\hat{k})$ and $(d\,\check{k})$ hold in the cartesian linearly distributive categories of [19] (cf. §11.5), this shows that no definition of $d_{a,b,c} : a \boxtimes (b \boxplus c) \rightarrow (a \boxtimes b) \boxplus c$ can support the claims made in [19] (p. 22), which we have already considered in §12.4. Since products and coproducts are unique up to isomorphism (see [94], Sections IV.1-2), there is no alternative lattice-category structure to the lattice-category structure provided by $\boxtimes$ and $\boxplus$ in Set_*. (This invalidates also Proposition 3.4, on the same page of [19].)

In **ZML**, the arrow $c^k_{A,C,B,D} \colon (A \wedge C) \vee (B \wedge D) \vdash (A \vee B) \wedge (C \vee D)$ has an inverse $c^l_{A,B,C,D} \colon (A \vee B) \wedge (C \vee D) \vdash (A \wedge C) \vee (B \wedge D)$. The natural transformation c^l could be taken as a primitive instead of m and m^{-1}, or $\cup$ and $\mathbf{0}$, because, for $f, g \colon C \vdash B$, in **ZML** we have the equations

$$
\begin{aligned}
f \cup g &= [\hat{k}^1_{B,A}, \hat{k}^2_{A,B}] \circ c^l_{B,A,A,B} \circ \langle \check{k}^1_{B,A} \circ f, \check{k}^2_{A,B} \circ g \rangle, \\
\mathbf{0}_{A,B} &= [\hat{k}^1_{B,A}, \hat{k}^2_{A,B}] \circ c^l_{B,A,A,B} \circ \langle \check{k}^2_{B,A}, \check{k}^1_{A,B} \rangle,
\end{aligned}
$$

which are easily checked with the help of Zero-Mix Lattice Coherence.

There are many ways to define $c^l_{A,B,C,D}$ in **ZML**. One way is

$$
c^l_{A,B,C,D} =_{df} \hat{e}'_{A,B,C,D} \circ (m^{-1}_{A,B} \wedge \mathbf{1}_{C \vee D}),
$$

and another

$$c^{l}_{D,B,C,A} =_{df} (\mathbf{1}_{D\wedge C} \vee m^{-1}_{B,A}) \circ \check{e}'_{D,C,B,A}.$$

These two definitions show that the equations $(m\,\hat{e})$ and $(m\,\check{e})$ of §11.1 are in **ZML** immediate consequences of

$$\begin{aligned} c^{k}_{A,C,B,D} \circ c^{l}_{A,B,C,D} &= \mathbf{1}_{(A\vee B)\wedge(C\vee D)}, \\ c^{l}_{A,C,B,D} \circ c^{k}_{A,B,C,D} &= \mathbf{1}_{(A\wedge B)\vee(C\wedge D)}. \end{aligned}$$

§13.3. Coherence of zero-mix dicartesian categories

To obtain the natural logical category $\mathbf{ZML}_{\top,\bot}$, we have that the logical system $\mathcal{C}(\mathbf{ZML}_{\top,\bot})$ is in $\mathcal{L}_{\wedge,\vee.\top,\bot}$, with the transformations α included in $\mathbf{1}$, b, c, w-k, δ-σ, m and m^{-1}. The specific equations of $\mathcal{E}(\mathbf{ZML}_{\top,\bot})$ are obtained by taking the union of those of $\mathcal{E}(\mathbf{ZML})$ and $\mathcal{E}(\mathbf{L}_{\top,\bot})$. We call natural $\mathbf{ZML}_{\top,\bot}$-categories *zero-mix dicartesian* categories.

Zero-mix dicartesian categories are *linear* categories in the sense of [89] (p. 279). The difference is that linear categories need not satisfy (wm), which amounts to $(\cup\ idemp)$. So the hom-sets of linear categories are commutative monoids, and not necessarily semilattices with unit, as in zero-mix dicartesian categories (cf. the categories $\mathbf{ZML}^{-}$ and $\mathbf{ZML}^{-}_{\top,\bot}$ below). Closely related notions are the notions of *Ab*-category (or *preadditive* category) and *additive* category, where the hom-sets are abelian groups (see [94], Sections I.8 and VIII.2, and [52], p. 60). These notions enter into the notion of *abelian* category (see [94], Section VIII.3, [52], Chapter 2, and [54], Section 1.591).

The syntactical category $\mathbf{GZML}_{\top,\bot}$ is defined by combining what we had for **GZML** and $\mathbf{GL}_{\top,\bot}$ in §9.6. We can then prove Composition Elimination for $\mathbf{GZML}_{\top,\bot}$ as for **GZML**.

We define the atomic terms of $\mathcal{C}(\mathbf{GZML}_{\top,\bot})$ as we did for $\mathcal{C}(\mathbf{GZML})$ in §13.1, save that the indices p and q of $\mathbf{0}_{p,q}$ (but not those of $\mathbf{1}_p$) can be replaced by $\top$ or $\bot$. Arrow terms of $\mathcal{C}(\mathbf{GZML}_{\top,\bot})$ in normal form, and their atomic components, are then defined analogously to what we had in §13.1. Let the settled normal form for an arrow term $f : A \vdash B$ of $\mathcal{C}(\mathbf{GZML}_{\top,\bot})$ be defined as for $\mathcal{C}(\mathbf{GZML})$ when $GA \neq \emptyset$ and $GB \neq \emptyset$. If either $GA = \emptyset$ or $GB = \emptyset$, then f is in settled normal form when it is $\mathbf{0}_{A,B}$.

We can prove as in §13.1 the Normal-Form Lemma where **GZML** is replaced by $\mathbf{GZML}_{\top,\bot}$, with the following additions. We use the following equations of $\mathbf{GZML}_{\top,\bot}$:

$$\hat{\kappa}_A = \mathbf{0}_{A,\top}, \qquad \check{\kappa}_A = \mathbf{0}_{\bot,A},$$
$$\mathbf{1}_\top = \mathbf{0}_{\top,\top}, \qquad \mathbf{1}_\bot = \mathbf{0}_{\bot,\bot},$$

together with ($\mathbf{0}$) and ($\cup\mathbf{0}$), to remove superfluous atomic components. We can then prove as before the following.

ZERO-MIX DICARTESIAN COHERENCE. *The functor G from $\mathbf{ZML}_{\top,\bot}$ to Rel is faithful.*

We prove the maximality of **ZML** and $\mathbf{ZML}_{\top,\bot}$ as we proved the maximality of **ZL** and $\mathbf{ZL}_{\top,\bot}$ in §12.3.

Let $\mathbf{ZML}^-$ and $\mathbf{ZML}^-_{\top,\bot}$ differ from **ZML** and $\mathbf{ZML}_{\top,\bot}$ by omitting (wm), or alternatively ($\cup$ *idemp*), from the specific equations. In these categories hom-sets are not necessarily semilattices with unit—they must be only commutative monoids. We can prove that there are faithful functors from $\mathbf{ZML}^-$ and $\mathbf{ZML}^-_{\top,\bot}$ into the category *Mat* of §12.5. For these proofs we proceed as for **ZML** and $\mathbf{ZML}_{\top,\bot}$. Note that we did not need (wm) for Composition Elimination in **GZML** and $\mathbf{GZML}_{\top,\bot}$. (We needed ($wm$) for the cut elimination of **DL** in Chapter 11, but not for the cut elimination of **ZIDL** in Chapter 12; see §12.5.) The settled normal form is now defined by making every ordered pair from $GA \times GB$ correspond to a nonempty set of atomic components bound to that pair; more precisely, a nonempty set of occurrences of a single arrow term bound to that pair such that each occurrence is an atomic component (cf. §13.1). This is a multiset based on a singleton. Whether zero atomic components are duplicated in this multiset is without importance, but we count the number of nonzero atomic components; this number corresponds to an entry $n \geq 1$ in the matrix. The categories $\mathbf{ZML}^-$ and $\mathbf{ZML}^-_{\top,\bot}$ are clearly not maximal, since we can add the equation (wm) without falling into preorder.

§13.4. The category $Semilat_*$

In this section we will consider as an example of a zero-mix dicartesian category the category $Semilat_*$, whose objects are semilattices with unit, and whose arrows are unit-preserving semilattice homomorphisms. This category is a subcategory of the category Set_* of §9.7. Note that $Semilat_*$ is not the category Set_*^{sl} of §10.1, which is isomorphic to the category $Semilat$ of semilattices with semilattice homomorphisms.

We want to summarize matters in this section; so we give again the following definitions from §9.7:

$$\mathrm{I} = \{*\}, \qquad a' = \{(x,*) \mid x \in a - \mathrm{I}\}, \qquad b'' = \{(*,y) \mid y \in b - \mathrm{I}\},$$

$$\begin{aligned} a \otimes b &= ((a - \mathrm{I}) \times (b - \mathrm{I})) \cup \mathrm{I}, \\ a \boxtimes b &= (a \otimes b) \cup a' \cup b'', \\ a \boxplus b &= a' \cup b'' \cup \mathrm{I}. \end{aligned}$$

If $\langle a_1, \cdot, * \rangle$ and $\langle a_2, \cdot, * \rangle$ are semilattices with unit, then we define the semilattice with unit $\langle a_1, \cdot, * \rangle \, \xi \, \langle a_2, \cdot, * \rangle$, for $\xi \in \{\wedge, \vee\}$, as $\langle a_1 \boxtimes a_2, \cdot, * \rangle$, where $\boxtimes$ corresponds to cartesian product. For $\cdot$ and $*$ we have the following clauses (taken over from §10.1):

$$\begin{aligned} &(x_1, x_2) \cdot (y_1, y_2) = (x_1 \cdot y_1, x_2 \cdot y_2), \\ &(x_1, x_2) \cdot * = * \cdot (x_1, x_2) = (x_1, x_2), \\ &* \cdot * = *. \end{aligned}$$

We have that $\top = \bot = \mathrm{I} = \{*\}$ is the trivial semilattice with unit.

The functions $\hat{k}{}^i_{a_1,a_2} \colon a_1 \boxtimes a_2 \to a_i$, for $i \in \{1,2\}$, are defined by

$$\hat{k}{}^i_{a_1,a_2}(x_1, x_2) = x_i, \qquad \hat{k}{}^i_{a_1,a_2}(*) = *;$$

for $f_i \colon c \to a_i$, the function $\langle f_1, f_2 \rangle \colon c \to a_1 \boxtimes a_2$ is defined by

$$\langle f_1, f_2 \rangle(z) = \begin{cases} (f_1(z), f_2(z)) & \text{if } f_1(z) \neq * \text{ or } f_2(z) \neq * \\ * & \text{if } f_1(z) = f_2(z) = *; \end{cases}$$

and the function $\overset{\wedge}{\kappa}_a \colon a \to \mathrm{I}$ is defined by $\overset{\wedge}{\kappa}_a(x) = *$.

The functions $\overset{\vee}{k}{}^i_{a_1,a_2} \colon a_i \to a_1 \boxtimes a_2$ are defined by

$$\begin{aligned} &\overset{\vee}{k}{}^1_{a_1,a_2}(x) = (x, *), \quad \overset{\vee}{k}{}^2_{a_1,a_2}(x) = (*, x), \quad \text{for } x \neq *, \\ &\overset{\vee}{k}{}^i_{a_1,a_2}(*) = *; \end{aligned}$$

for $g_i \colon a_i \to c$, the function $[g_1, g_2] \colon a_1 \boxtimes a_2 \to c$ is defined by

$$[g_1, g_2](x_1, x_2) = g_1(x_1) \cdot g_2(x_2),$$
$$[g_1, g_2](*) = *;$$

and the function $\check{\kappa}_a\colon \mathrm{I} \to a$ is defined by $\check{\kappa}_a\,(*) = *$. (The clauses in the definitions of $\hat{k}^i_{a_1,a_2}$, $\langle f_1, f_2\rangle$, $\hat{\kappa}_a$ and $\check{\kappa}_a$ are taken over from §9.7, where they were given for *Set$_*$*, while the clauses for $\check{k}^i_{a_1,a_2}$ and $[g_1, g_2]$ are taken over from §10.1, where they were given for Set_*^{sl}.)

For $f, g\colon a \to b$, we define the function $f \cup g\colon a \to b$ by

$$(f \cup g)(x) = f(x) \cdot g(x)$$

(as for Set_*^{sl} in §10.1), and, finally, we have the function $\mathbf{0}_{a,b}\colon a \to b$ defined by

$$\mathbf{0}_{a,b}(x) = *$$

(as for *Set$_*$* in §12.4). It is straightforward to check that with all these definitions *Semilat$_*$* is a zero-mix dicartesian category.

The category *Semilat$_*$* is a subcategory of the category *ComMon* of commutative monoids with monoid homomorphisms. By repeating what we had above, we can show that *ComMon*, with both $\wedge$ and $\vee$ being cartesian product, and both $\top$ and $\bot$ being the trivial single-element monoid, is a natural $\mathbf{ZML}^{\neg}_{\top,\bot}$-category. The category *Mat* of §12.5 is isomorphic to a subcategory of *ComMon*, which is itself a subcategory of *Set$_*$*.

Let us summarize in a table the connections between the three subcategories of *Set$_*$* that we had as examples for various kinds of lattice categories (see §9.7, §10.1, §10.3 and §12.4):

	category	$\wedge$	$\vee$	$\top$	$\bot$
$Set_*^{sl} \cong Semilat$	mix-lattice	$\otimes$	$\boxtimes$	$\mathrm{I} \cup \{x\}$	I
Set$_*$	zero-dicartesian	$\boxtimes$	$\boxplus$	I	I
Semilat$_*$	zero-mix dicartesian	$\boxtimes$	$\boxtimes$	I	I

Note that Set_*^{sl} is not a dicartesian category, but only a sesquicartesian category with $\top$ and $\bot$ as in the table. We also had in §9.7 the dicartesian category $Set_*^{\emptyset}$, where, in contradistinction to Set_*, we had that $\bot$ is $\emptyset$; but this category is not a zero-lattice category.

The category *Rel* is a zero-mix dicartesian category with both $\wedge$ and $\vee$ being $+$, and both $\top$ and $\bot$ being $\mathbf{0}$; the operation $\cup$ in *Rel* is union, and the zero arrows are empty relations.

The category *Rel* is isomorphic to a subcategory of $Semilat_*$. We define a functor F from *Rel* to $Semilat_*$ by

$$Fn = \langle \mathcal{P}n, \cup, \emptyset \rangle,$$

where $\mathcal{P}n$ is the power set of the ordinal n and $\cup$ is binary union of sets; for $R \subseteq n \times m$ and $X \in \mathcal{P}n$ we have

$$(FR)(X) = \{y \in m \mid \text{ for some } x \in X, xRy\}.$$

It is straightforward to check that F is a faithful functor from *Rel* to $Semilat_*$, which is one-one on objects. So the image under F in $Semilat_*$ is isomorphic to *Rel*. The functor F is a strong (but not strict) $\mathcal{C}(\mathbf{ZML}_{\top,\bot})$-functor (see §2.8). The semilattice with unit $\langle \mathcal{P}n, \cup, \emptyset \rangle$ is, up to isomorphism, the free semilattice with unit with n free generators.

It is known that *Rel* is isomorphic to the Kleisli category of the power-set monad (or triple) on the category of finite sets with functions (see [85], Section 0.6, p. 32). In this isomorphism, every relation $R \subseteq n \times m$ is mapped to a function $f_R \colon n \to \mathcal{P}m$ such that $y \in f_R(x)$ iff xRy, and f_R can then be extended in a unique way to an $\emptyset$-preserving semilattice homomorphism FR from $\langle \mathcal{P}n, \cup, \emptyset \rangle$ to $\langle \mathcal{P}m, \cup, \emptyset \rangle$.

CHAPTER 14

CATEGORIES WITH NEGATION

In this, final, chapter of the book we bring to completion our proposed codification of the proof theory of classical propositional logic. We first prove a general coherence result that enables us to pass from coherence proved in the absence of negation to coherence with a De Morgan negation added. De Morgan negation is involutive negation that satisfies the De Morgan laws, but does not yet amount to Boolean negation.

To obtain Boolean negation, i.e. an operation corresponding to complementation, we need extra assumptions, which, if we want coherence with respect to the category whose arrows are relations between finite ordinals, must be zero arrows. The effect of having these zero arrows, which yield the zero-identity arrows of Chapter 12, is that all theorems, i.e. all propositions proved without hypotheses, will have zero proofs. Then we prove coherence for our Boolean categories, by reducing it to a previously proved coherence result of Chapter 12.

We end this chapter with comments on alternatives to our approach in categorifying the proof theory of Boolean propositional logic. Besides the approach through bicartesian closed categories, i.e. cartesian closed categories with finite coproducts, which with natural assumptions about negation collapses into preorder, there are alternatives with relations between finite ordinals being replaced by a more complex kind of relation on the sum of the ordinals in the domain and codomain. We discuss problems that arise with these alternatives, and in particular problems we would encounter in cut elimination.

We conclude that if our codification of the general proof theory of classical propositional logic is acceptable, then this proof theory is simpler than the general proof theory of intuitionistic propositional logic, codified in bicartesian closed categories, or, equivalently, in a typed lambda calculus with product and coproduct types. In particular, equality of derivations is easily decidable for classical logic. The categorial structure of this classical proof theory is, however, quite rich. It covers all the categorial structures considered in this book, except the zero-mix lattice structure of Chapter 13, and extends them conservatively with respect to identity of proofs. It comes close to the zero-mix lattice structure, through which it is related to linear algebra.

§14.1. De Morgan Coherence

If $\mathcal{L}$ is one of the languages $\mathcal{L}_{\wedge,\vee}$ and $\mathcal{L}_{\wedge,\vee,\top,\bot}$, then let $\mathcal{L}^{\neg}$ be the language obtained by assuming in the definition of $\mathcal{L}$ that we have in addition the unary (that is, 1-ary) connective $\neg$ of *negation*. The language $\mathcal{L}^{\neg p}$ is, on the other hand, defined like $\mathcal{L}$ save that the set of letters $\mathcal{P}$ is replaced by the union of $\mathcal{P}$ and the set $\mathcal{P}^{\neg} = \{\neg p \mid p \in \mathcal{P}\}$.

The syntactical system $\mathcal{C}(\mathbf{I}^{\neg})$ is defined by taking first for its objects the formulae of $\mathcal{L}^{\neg}_{\wedge,\vee}$; next, for every $A, B \in \mathcal{L}^{\neg}_{\wedge,\vee}$ we have the primitive arrow terms

$$\mathbf{1}_A : A \vdash A,$$

$$n^{\rightarrow}_A : \neg\neg A \vdash A, \qquad n^{\leftarrow}_A : A \vdash \neg\neg A,$$

$$\hat{r}^{\rightarrow}_{A,B} : \neg(A \wedge B) \vdash \neg A \vee \neg B, \qquad \hat{r}^{\leftarrow}_{A,B} : \neg A \vee \neg B \vdash \neg(A \wedge B),$$

$$\check{r}^{\rightarrow}_{A,B} : \neg(A \vee B) \vdash \neg A \wedge \neg B, \qquad \check{r}^{\leftarrow}_{A,B} : \neg A \wedge \neg B \vdash \neg(A \vee B),$$

and as the operations on arrow terms we have composition, $\wedge$ and $\vee$. Let the family n-r be the union of the families $n^{\rightarrow}$, $n^{\leftarrow}$, $\hat{r}^{\rightarrow}$, $\hat{r}^{\leftarrow}$, $\check{r}^{\rightarrow}$ and $\check{r}^{\leftarrow}$. The equations $\mathcal{E}(\mathbf{I}^{\neg})$ are obtained by assuming the categorial equations, the bifunctorial equations for $\wedge$ and $\vee$, and the isomorphism equations

$$n^{\leftarrow}_A \circ n^{\rightarrow}_A = \mathbf{1}_{\neg\neg A}, \qquad n^{\rightarrow}_A \circ n^{\leftarrow}_A = \mathbf{1}_A,$$

$$\hat{r}^{\leftarrow}_{A,B} \circ \hat{r}^{\rightarrow}_{A,B} = \mathbf{1}_{\neg(A\wedge B)}, \qquad \hat{r}^{\rightarrow}_{A,B} \circ \hat{r}^{\leftarrow}_{A,B} = \mathbf{1}_{\neg A\vee\neg B},$$

$$\check{r}^{\leftarrow}_{A,B} \circ \check{r}^{\rightarrow}_{A,B} = \mathbf{1}_{\neg(A\vee B)}, \qquad \check{r}^{\rightarrow}_{A,B} \circ \check{r}^{\leftarrow}_{A,B} = \mathbf{1}_{\neg A\wedge\neg B}.$$

The syntactical category $\mathbf{I}^{\neg}$ is $\mathcal{C}(\mathbf{I}^{\neg})/\mathcal{E}(\mathbf{I}^{\neg})$. Due to the presence of categorial and bifunctorial equations, we can easily prove the Development Lemma (see §2.7) for $\mathbf{I}^{\neg}$ (this presupposes a definition of β-term where β can be one of the families $n^{\rightarrow}$, $n^{\leftarrow}$, $\hat{r}^{\rightarrow}$, etc.).

An arrow term of $\mathcal{C}(\mathbf{I}^{\neg})$ is called $\rightarrow$-*directed* when $\leftarrow$ does not occur as a superscript in it. The formulae of $\mathcal{L}^{\neg}_{\wedge,\vee}$ that are also formulae of $\mathcal{L}^{\neg p}_{\wedge,\vee}$ are said to be in normal form. We can then prove the Directedness Lemma (see §4.3) for $\mathbf{I}^{\neg}$. The proof of this lemma is obtained by relying on the bifunctorial equations for $\wedge$ and $\vee$. This lemma enables us to prove the following.

$\mathbf{I}^{\neg}$ COHERENCE. *The category* $\mathbf{I}^{\neg}$ *is a preorder.*

The proof is analogous to the proof of Associative Coherence in §4.3.

Consider the following definitions in $\mathcal{C}(\mathbf{I}^{\neg})$:

$$\neg\mathbf{1}_A =_{df} \mathbf{1}_{\neg A}, \qquad \neg n^{\rightarrow}_A =_{df} n^{\leftarrow}_{\neg A}, \qquad \neg n^{\leftarrow}_A =_{df} n^{\rightarrow}_{\neg A},$$

$$\neg\, \hat{r}^{\rightarrow}_{A,B} =_{df} n^{\leftarrow}_{A\wedge B} \circ (n^{\rightarrow}_A \wedge n^{\rightarrow}_B) \circ \check{r}^{\rightarrow}_{\neg A,\neg B},$$
$$\neg\, \hat{r}^{\leftarrow}_{A,B} =_{df} \check{r}^{\leftarrow}_{\neg A,\neg B} \circ (n^{\leftarrow}_A \wedge n^{\leftarrow}_B) \circ n^{\rightarrow}_{A\wedge B},$$

$$\neg\, \check{r}^{\rightarrow}_{A,B} =_{df} n^{\leftarrow}_{A\vee B} \circ (n^{\rightarrow}_A \vee n^{\rightarrow}_B) \circ \hat{r}^{\rightarrow}_{\neg A,\neg B},$$
$$\neg\, \check{r}^{\leftarrow}_{A,B} =_{df} \hat{r}^{\leftarrow}_{\neg A,\neg B} \circ (n^{\leftarrow}_A \vee n^{\leftarrow}_B) \circ n^{\rightarrow}_{A\vee B},$$

$$\neg(g \circ f) =_{df} \neg f \circ \neg g,$$

$$\neg(f \wedge g) =_{df} \hat{r}^{\leftarrow}_{A,B} \circ (\neg f \vee \neg g) \circ \hat{r}^{\rightarrow}_{D,E},$$
$$\neg(f \vee g) =_{df} \check{r}^{\leftarrow}_{A,B} \circ (\neg f \wedge \neg g) \circ \check{r}^{\rightarrow}_{D,E}.$$

It is easy to see that $\neg$ is a functor from $\mathbf{I}^{\neg}$ to $\mathbf{I}^{\neg op}$, i.e. a contravariant functor from $\mathbf{I}^{\neg}$ to $\mathbf{I}^{\neg}$. It follows easily from $\mathbf{I}^{\neg}$ Coherence that $n^{\rightarrow}$, $n^{\leftarrow}$, $\overset{\xi}{r}{}^{\rightarrow}$ and $\overset{\xi}{r}{}^{\leftarrow}$, for $\xi \in \{\wedge, \vee\}$, define natural transformations between functors defined in terms of the identity functor, the contravariant functor $\neg$ and the bifunctors ξ. Our official definition of logical category does not cover $\mathbf{I}^{\neg}$, but it is clear how we can extend this definition to cover also categories like $\mathbf{I}^{\neg}$.

The syntactical category $\mathbf{I}^{\neg}_{\top,\bot}$ is defined as $\mathbf{I}^{\neg}$ save that its objects are from $\mathcal{L}^{\neg}_{\wedge,\vee,\top,\bot}$, and we have in $\mathcal{C}(\mathbf{I}^{\neg}_{\top,\bot})$ the additional primitive arrow terms

$$\hat{\rho}^{\rightarrow}\colon \neg\top \vdash \bot, \qquad \hat{\rho}^{\leftarrow}\colon \bot \vdash \neg\top,$$

$$\check{\rho}^{\rightarrow}\colon \neg\bot \vdash \top, \qquad \check{\rho}^{\leftarrow}\colon \top \vdash \neg\bot,$$

and in $\mathcal{E}(\mathbf{I}^{\neg}_{\top,\bot})$ the additional isomorphism equations

$$\hat{\rho}^{\leftarrow} \circ \hat{\rho}^{\rightarrow} = \mathbf{1}_{\neg\top}, \qquad \hat{\rho}^{\rightarrow} \circ \hat{\rho}^{\leftarrow} = \mathbf{1}_{\bot},$$

$$\check{\rho}^{\leftarrow} \circ \check{\rho}^{\rightarrow} = \mathbf{1}_{\neg\bot}, \qquad \check{\rho}^{\rightarrow} \circ \check{\rho}^{\leftarrow} = \mathbf{1}_{\top}.$$

We call ρ the union of the families $\hat{\rho}^{\rightarrow}$, $\hat{\rho}^{\leftarrow}$, $\check{\rho}^{\rightarrow}$ and $\check{\rho}^{\leftarrow}$.

By extending the proof of $\mathbf{I}^{\neg}$ Coherence, we easily obtain the following.

$\mathbf{I}^{\neg}_{\top,\bot}$ COHERENCE. *The category $\mathbf{I}^{\neg}_{\top,\bot}$ is a preorder.*

In $\mathcal{C}(\mathbf{I}^{\neg}_{\top,\bot})$ we can introduce the definition of $\neg f$ as in $\mathcal{C}(\mathbf{I}^{\neg})$ with the following additions:

$$\neg\, \hat{\rho}^{\rightarrow} =_{df} n^{\leftarrow}_{\top} \circ \check{\rho}^{\rightarrow}, \qquad \neg\, \hat{\rho}^{\leftarrow} =_{df} \check{\rho}^{\leftarrow} \circ n^{\rightarrow}_{\top},$$

$$\neg\, \check{\rho}^{\rightarrow} =_{df} n^{\leftarrow}_{\bot} \circ \hat{\rho}^{\rightarrow}, \qquad \neg\, \check{\rho}^{\leftarrow} =_{df} \hat{\rho}^{\leftarrow} \circ n^{\rightarrow}_{\bot},$$

and obtain a functor from $\mathbf{I}^{\neg}_{\top,\bot}$ to $\mathbf{I}^{\neg op}_{\top,\bot}$.

Let $\mathcal{K}$ be a logical category in $\mathcal{L}$. Then the syntactical category $\mathcal{K}^{\neg}$, whose objects are formulae of $\mathcal{L}^{\neg}$, will be obtained from $\mathcal{K}$ as $\mathbf{I}^{\neg}$ is obtained from the variant of $\mathbf{I}$ in the language $\mathcal{L}_{\wedge,\vee}$, or as $\mathbf{I}^{\neg}_{\top,\bot}$ is obtained from the variant of $\mathbf{I}$ in the language $\mathcal{L}_{\wedge,\vee,\top,\bot}$. For example, in the syntactical system $\mathcal{C}(\mathbf{DL}^{\neg}_{\top,\bot})$, whose objects are formulae of $\mathcal{L}^{\neg}_{\wedge,\vee,\top,\bot}$, we will have besides the primitive arrow terms in the families $\mathbf{1}$, b, c, w-k, δ-σ, m and d, those in the families n-r and ρ, and the equations of $\mathcal{E}(\mathbf{DL}^{\neg}_{\top,\bot})$ will be obtained by assuming the union of those of $\mathcal{E}(\mathbf{DL}_{\top,\bot})$ and $\mathcal{E}(\mathbf{I}^{\neg}_{\top,\bot})$.

We define a functor F from $\mathcal{K}^{\neg}$ to $\mathcal{K}$ in the language $\mathcal{L}^{\neg p}$; we call the latter category $\mathcal{K}^{\neg p}$. The category $\mathcal{K}^{\neg p}$ is exactly like the old logical category $\mathcal{K}$ save that it is generated not by $\mathcal{P}$ but by $\mathcal{P} \cup \mathcal{P}^{\neg}$ (see the end of §2.7).

We define a functor F from $\mathcal{K}^{\neg}$ to $\mathcal{K}^{\neg p}$ by the following graph-morphism from $\mathcal{C}(\mathcal{K}^{\neg})$ to $\mathcal{C}(\mathcal{K}^{\neg p})$:

$Fp = p$,

$F\zeta = \zeta, \quad$ for $\zeta \in \{\top, \bot\}$,

$$
\begin{aligned}
&F(A\,\xi\,B) = FA\,\xi\,FB, \quad \text{for } \xi \in \{\wedge,\vee\},\\
&F\neg p = \neg p,\\
&F\neg\top = \bot, \qquad\qquad F\neg\bot = \top,\\
&F\neg\neg A = FA,\\
&F\neg(A\wedge B) = F\neg A \vee F\neg B, \quad F\neg(A\vee B) = F\neg A \wedge F\neg B;
\end{aligned}
$$

for $f\colon A \vdash B$ in the families n-r and ρ,

$$
\begin{aligned}
&Ff = \mathbf{1}_{FA} \quad (\text{here } FA \text{ is equal to } FB),\\
&F\alpha_{A_1,\ldots,A_n} = \alpha_{FA_1,\ldots,FA_n},\\
&F(g\circ f) = Fg\circ Ff,\\
&F(f\,\xi\,g) = Ff\,\xi\,Fg, \quad \text{for } \xi\in\{\wedge,\vee\}.
\end{aligned}
$$

It is easy to check that F is indeed a functor; namely, $f = g$ in $\mathcal{K}^{\neg}$ implies $Ff = Fg$ in $\mathcal{K}^{\neg p}$ (cf. the penultimate paragraph of §2.4).

Next, we define a functor $F^{\neg}$ from $\mathcal{K}^{\neg p}$ to $\mathcal{K}^{\neg}$ by the graph-morphism from $\mathcal{C}(\mathcal{K}^{\neg p})$ to $\mathcal{C}(\mathcal{K}^{\neg})$ for which we have $F^{\neg}A = A$, and $F^{\neg}f = f$. It is clear that F and $F^{\neg}$ are strict $\mathcal{C}(\mathcal{K})$-functors.

We define by induction on the length of $A \in \mathcal{L}^{\neg}_{\wedge,\vee}$ the arrow terms $i_A\colon A\vdash FA$ and $i_A^{-1}\colon FA\vdash A$ of $\mathcal{C}(\mathbf{I}^{\neg})$:

$$
\begin{aligned}
&i_A = i_A^{-1} = \mathbf{1}_A, \quad \text{if } A \text{ is } p \text{ or } \neg p,\\
&i_{A_1\xi A_2} = i_{A_1}\,\xi\,i_{A_2}, \quad \text{for } \xi\in\{\wedge,\vee\},\\
&i_{\neg\neg B} = i_B \circ \overrightarrow{n}_B,\\
&i_{\neg(A_1\wedge A_2)} = (i_{\neg A_1}\vee i_{\neg A_2})\circ \hat{\overrightarrow{r}}_{A_1,A_2}, \qquad i_{\neg(A_1\vee A_2)} = (i_{\neg A_1}\wedge i_{\neg A_2})\circ \check{\overrightarrow{r}}_{A_1,A_2},\\
&i^{-1}_{A_1\xi A_2} = i^{-1}_{A_1}\,\xi\,i^{-1}_{A_2}, \quad \text{for } \xi\in\{\wedge,\vee\},\\
&i^{-1}_{\neg\neg B} = \overleftarrow{n}_B \circ i^{-1}_B,\\
&i^{-1}_{\neg(A_1\wedge A_2)} = \hat{\overleftarrow{r}}_{A_1,A_2}\circ(i^{-1}_{\neg A_1}\vee i^{-1}_{\neg A_2}), \qquad i^{-1}_{\neg(A_1\vee A_2)} = \check{\overleftarrow{r}}_{A_1,A_2}\circ(i^{-1}_{\neg A_1}\wedge i^{-1}_{\neg A_2}).
\end{aligned}
$$

If $A\in\mathcal{L}^{\neg}_{\wedge,\vee,\top,\bot}$, then we define the arrow terms $i_A\colon A\vdash FA$ and $i_A^{-1}\colon FA\vdash A$ of $\mathcal{C}(\mathbf{I}^{\neg}_{\top,\bot})$ with the additional clauses:

$$
i_\zeta = \mathbf{1}_\zeta, \quad \text{for } \zeta\in\{\top,\bot\},
$$

$$i_{\neg\top} = \hat{\rho}^{\rightarrow}, \qquad i_{\neg\bot} = \check{\rho}^{\rightarrow},$$
$$i^{-1}_{\neg\top} = \hat{\rho}^{\leftarrow}, \qquad i^{-1}_{\neg\bot} = \check{\rho}^{\leftarrow}.$$

It is clear that we have $i_A^{-1} \circ i_A = \mathbf{1}_A$ and $i_A \circ i_A^{-1} = \mathbf{1}_{FA}$ in $\mathbf{I}^\neg$ or $\mathbf{I}^\neg_{\top,\bot}$. We can prove the following.

AUXILIARY LEMMA. *For every arrow term $f : A \vdash B$ of $\mathcal{C}(\mathcal{K}^\neg)$ we have $f = i_B^{-1} \circ Ff \circ i_A$ in $\mathcal{K}^\neg$.*

PROOF. We proceed by induction on the length of f.

If $f : A \vdash B$ is in the families n-r and ρ, we have that $f = i_B^{-1} \circ Ff \circ i_A$ by $\mathbf{I}^\neg$ Coherence or $\mathbf{I}^\neg_{\top,\bot}$ Coherence.

If f is $\alpha_{A_1,\ldots,A_k} : M^\mu(A_1, \ldots, A_k) \vdash N^\nu(A_1, \ldots, A_k)$, then $i_{M^\mu(A_1,\ldots,A_k)}$ is $M^\mu(i_{A_1}, \ldots, i_{A_k})$ and $i^{-1}_{N^\nu(A_1,\ldots,A_k)}$ is $N^\nu(i^{-1}_{A_1}, \ldots, i^{-1}_{A_k})$; we obtain $f = i_B^{-1} \circ Ff \circ i_A$ by using (α *nat*).

If f is $f_2 \circ f_1$, then we have

$$\begin{aligned} f_2 \circ f_1 &= i_B^{-1} \circ Ff_2 \circ i_C \circ i_C^{-1} \circ Ff_1 \circ i_A, \text{ by the induction hypothesis,} \\ &= i_B^{-1} \circ F(f_2 \circ f_1) \circ i_A. \end{aligned}$$

If f is $f_1 \xi f_2$, for $\xi \in \{\wedge, \vee\}$, then $i_{A_1 \xi A_2}$ is $i_{A_1} \xi\, i_{A_2}$ and $i^{-1}_{B_1 \xi B_2}$ is $i^{-1}_{B_1} \xi\, i^{-1}_{B_2}$; we obtain $f = i_B^{-1} \circ Ff \circ i_A$ by using bifunctorial equations. ⊣

$\mathcal{K}^\neg$-$\mathcal{K}^{\neg p}$-EQUIVALENCE. *The categories $\mathcal{K}^\neg$ and $\mathcal{K}^{\neg p}$ are equivalent via the functors F and $F^\neg$.*

PROOF. We have $FF^\neg A = A$ and $FF^\neg f = f$. That i is a natural isomorphism from the identity functor of $\mathcal{K}^\neg$ to the composite functor $F^\neg F$ is shown by the Auxiliary Lemma. ⊣

Let the functor G from $\mathcal{K}^\neg$ to *Rel* be defined by extending the definition of the functor G from $\mathcal{K}$ to *Rel* with the clauses

$$G\neg A = GA,$$
$$Gf = \mathbf{1}_{GA},$$

for every arrow term $f : A \vdash B$ in the n-r and ρ families. Here GA must be equal to GB, and $\mathbf{1}_{GA}$ is the identity relation, i.e. identity function, on GA. Then we can prove the following.

DE MORGAN COHERENCE. *If G from $\mathcal{K}$ to Rel is faithful, then G from $\mathcal{K}^{\neg}$ to Rel is faithful.*

PROOF. Suppose that for the arrow terms $f, g\colon A \vdash B$ of $\mathcal{C}(\mathcal{K}^{\neg})$ we have $Gf = Gg$. Then we have $GFf = GFg$, where F is the functor from $\mathcal{K}^{\neg}$ to $\mathcal{K}^{\neg p}$ we have defined above. Since G from $\mathcal{K}$ to *Rel* is faithful, we have that G from $\mathcal{K}^{\neg p}$ to *Rel* is faithful and hence $Ff = Fg$ in $\mathcal{K}^{\neg p}$. From $\mathcal{K}^{\neg}$-$\mathcal{K}^{\neg p}$-Equivalence we conclude that $f = g$ in $\mathcal{K}^{\neg}$. ⊣

We can define the functor $\neg$ from $\mathcal{K}^{\neg}$ to $\mathcal{K}^{\neg op}$ by extending the definitions we have for $\mathbf{I}^{\neg}$ and $\mathbf{I}^{\neg}_{\top,\bot}$ with the following clauses, provided $\mathcal{C}(\mathcal{K}^{\neg})$ has the required arrow terms on the right-hand side:

$$
\begin{aligned}
\neg\, \hat{b}^{\rightarrow}_{A,B,C} &=_{df} \hat{r}^{\leftarrow}_{A,B\wedge C} \circ (\mathbf{1}_{\neg A} \vee \hat{r}^{\leftarrow}_{B,C}) \circ \check{b}^{\leftarrow}_{\neg A,\neg B,\neg C} \circ (\hat{r}^{\rightarrow}_{A,B} \vee \mathbf{1}_{\neg C}) \circ \hat{r}^{\rightarrow}_{A\wedge B,C},\\
\neg\, \hat{b}^{\leftarrow}_{A,B,C} &=_{df} \hat{r}^{\leftarrow}_{A\wedge B,C} \circ (\hat{r}^{\leftarrow}_{A,B} \vee \mathbf{1}_{\neg C}) \circ \check{b}^{\rightarrow}_{\neg A,\neg B,\neg C} \circ (\mathbf{1}_{\neg A} \vee \hat{r}^{\rightarrow}_{B,C}) \circ \hat{r}^{\rightarrow}_{A,B\wedge C},\\
\neg\, \hat{c}_{A,B} &=_{df} \hat{r}^{\leftarrow}_{A,B} \circ \check{c}_{\neg A,\neg B} \circ \hat{r}^{\rightarrow}_{B,A},\\
\neg\, \hat{w}_{A} &=_{df} \check{w}_{\neg A} \circ \hat{r}^{\rightarrow}_{A,A},\\
\neg\, \hat{k}^{i}_{A,B} &=_{df} \hat{r}^{\leftarrow}_{A,B} \circ \check{k}^{i}_{\neg A,\neg B}, \quad \text{for } i \in \{1,2\},\\
\neg d^{L}_{A,B,C} &=_{df} \hat{r}^{\leftarrow}_{A,B\vee C} \circ (\mathbf{1}_{\neg A} \vee \check{r}^{\leftarrow}_{B,C}) \circ d^{R}_{\neg A,\neg B,\neg C} \circ (\hat{r}^{\rightarrow}_{A,B} \wedge \mathbf{1}_{\neg C}) \circ \check{r}^{\rightarrow}_{A\wedge B,C},\\
\neg d^{R}_{C,B,A} &=_{df} \hat{r}^{\leftarrow}_{C\vee B,A} \circ (\check{r}^{\leftarrow}_{C,B} \vee \mathbf{1}_{\neg A}) \circ d^{L}_{\neg C,\neg B,\neg A} \circ (\mathbf{1}_{\neg C} \wedge \hat{r}^{\rightarrow}_{B,A}) \circ \check{r}^{\rightarrow}_{C,B\wedge A},\\
\neg m_{A,B} &=_{df} \hat{r}^{\leftarrow}_{A,B} \circ m_{\neg A,\neg B} \circ \check{r}^{\rightarrow}_{A,B},\\
\neg m^{-1}_{A,B} &=_{df} \check{r}^{\leftarrow}_{A,B} \circ m^{-1}_{\neg A,\neg B} \circ \hat{r}^{\rightarrow}_{A,B},\\
\neg(f \cup g) &=_{df} \neg f \cup \neg g,\\
\neg \mathbf{0}_{A,B} &=_{df} \mathbf{0}_{\neg B,\neg A};
\end{aligned}
$$

the clauses for $\neg\, \check{b}^{\rightarrow}_{A,B,C}$ and $\neg\, \check{b}^{\leftarrow}_{A,B,C}$ are obtained from the clauses for $\neg\, \hat{b}^{\rightarrow}_{A,B,C}$ and $\neg\, \hat{b}^{\leftarrow}_{A,B,C}$ respectively by interchanging $\wedge$ and $\vee$;

$$
\begin{aligned}
\neg\, \check{c}_{A,B} &=_{df} \check{r}^{\leftarrow}_{B,A} \circ \hat{c}_{\neg A,\neg B} \circ \check{r}^{\rightarrow}_{A,B},\\
\neg\, \check{w}_{A} &=_{df} \check{r}^{\leftarrow}_{A,A} \circ \hat{w}_{\neg A},\\
\neg\, \check{k}^{i}_{A,B} &=_{df} \hat{k}^{i}_{\neg A,\neg B} \circ \check{r}^{\rightarrow}_{A,B}, \text{ for } i \in \{1,2\};
\end{aligned}
$$

$$
\begin{aligned}
\neg\, \hat{\delta}^{\rightarrow}_{A} &=_{df} \hat{r}^{\leftarrow}_{A,\top} \circ (\mathbf{1}_{\neg A} \vee \hat{\rho}^{\leftarrow}) \circ \check{\delta}^{\leftarrow}_{\neg A}, \\
\neg\, \hat{\delta}^{\leftarrow}_{A} &=_{df} \check{\delta}^{\rightarrow}_{\neg A} \circ (\mathbf{1}_{\neg A} \vee \hat{\rho}^{\rightarrow}) \circ \hat{r}^{\rightarrow}_{A,\top}, \\
\neg\, \hat{\sigma}^{\rightarrow}_{A} &=_{df} \hat{r}^{\leftarrow}_{\top,A} \circ (\hat{\rho}^{\leftarrow} \vee \mathbf{1}_{\neg A}) \circ \check{\sigma}^{\leftarrow}_{\neg A}, \\
\neg\, \hat{\sigma}^{\leftarrow}_{A} &=_{df} \check{\sigma}^{\rightarrow}_{\neg A} \circ (\hat{\rho}^{\rightarrow} \vee \mathbf{1}_{\neg A}) \circ \hat{r}^{\rightarrow}_{\top,A};
\end{aligned}
$$

the clauses for $\neg\, \check{\delta}^{\rightarrow}_{A}$, $\neg\, \check{\delta}^{\leftarrow}_{A}$, $\neg\, \check{\sigma}^{\rightarrow}_{A}$ and $\neg\, \check{\sigma}^{\leftarrow}_{A}$ are obtained from the last four clauses by interchanging $\wedge$ with $\vee$, and $\top$ with $\bot$;

$$
\neg\, \hat{\kappa}_A =_{df} \check{\kappa}_{\neg A} \circ \hat{\rho}^{\rightarrow}, \qquad\qquad \neg\, \check{\kappa}_A =_{df} \check{\rho}^{\rightarrow} \circ \hat{\kappa}_{\neg A}.
$$

The coherence of $\mathcal{K}^{\neg}$ is a sufficient (though not a necessary) condition for the correctness of these definitions. By the coherence of $\mathcal{K}^{\neg}$, we also obtain that $n^{\rightarrow}$, $n^{\leftarrow}$, $\overset{\xi}{r}{}^{\rightarrow}$ and $\overset{\xi}{r}{}^{\leftarrow}$ for $\xi \in \{\wedge, \vee\}$ define natural transformations between functors defined in terms of the identity functor, the contravariant functor $\neg$ and the bifunctor ξ.

The category $\mathbf{DL}^{\neg}$ corresponds in the following sense to the system $\mathbf{E}_{\mathrm{fde}}$ of *tautological entailments* of [1] (Section 15, and Section 18 by J.M. Dunn; see also [5]): there is an arrow of type $A \vdash B$ in $\mathbf{DL}^{\neg}$ iff $A \rightarrow B$ is a theorem of $\mathbf{E}_{\mathrm{fde}}$. The algebraic models with respect to which $\mathbf{E}_{\mathrm{fde}}$ is complete are called De Morgan lattices, distributive involution lattices or quasi-Boolean algebras (see [1], Section 18, and references therein; see also [107], Section III.3). Complementation in De Morgan lattices is not in general Boolean complementation (see the next section).

§14.2. Boolean Coherence

The syntactical system $\mathcal{C}(\mathbf{B})$ is defined by taking first for its objects the formulae of $\mathcal{L}^{\neg}_{\wedge,\vee,\top,\bot}$; next, the primitive arrow terms of $\mathcal{C}(\mathbf{B})$ are those of $\mathcal{C}(\mathbf{ZIDL}^{\neg}_{\top,\bot})$, i.e. those in the families $\mathbf{1}$, b, c, w-k, δ-σ, m, d, $\mathbf{0}$ (whose members are $\mathbf{0}_A \colon A \vdash A$), n-r and ρ, plus

$$
\begin{aligned}
\eta_A &\colon \top \vdash \neg A \vee A, \\
\varepsilon_A &\colon A \wedge \neg A \vdash \bot,
\end{aligned}
$$

for every A in $\mathcal{L}^{\neg}_{\wedge,\vee,\top,\bot}$, and the operations on arrow terms are composition, $\wedge$ and $\vee$. The equations of $\mathcal{E}(\mathbf{B})$ are obtained by assuming the union of those of $\mathcal{E}(\mathbf{ZIDL}_{\top,\bot})$, i.e. those of $\mathcal{E}(\mathbf{DL}_{\top,\bot})$ plus $(\mathbf{0}\mathrm{I})$, and those of $\mathcal{E}(\mathbf{I}^{\neg}_{\top,\bot})$. The syntactical category $\mathbf{B}$ is $\mathcal{C}(\mathbf{B})/\mathcal{E}(\mathbf{B})$. Here, $\mathbf{B}$ comes from *Boolean*.

We define $\mathbf{0}_{A,B}$ in $\mathbf{B}$ by $f \circ \mathbf{0}_A$, as we did in $\mathbf{ZIDL}_{\top,\bot}$, and we can infer that the following equation holds in $\mathbf{B}$:

$$\begin{aligned} \eta_A &= \eta_A \circ \mathbf{1}_\top, \\ &= \eta_A \circ \mathbf{0}_\top, \;\; \text{since } \mathbf{1}_\top = \hat{\kappa}_\top = \mathbf{0}_{\top,\top}, \\ &= \mathbf{0}_{\top,\neg A \vee A}. \end{aligned}$$

We derive analogously $\varepsilon_A = \mathbf{0}_{A\wedge\neg A,\bot}$. Since η_A and ε_A are zero arrows, we assume that $G\eta_A$ and $G\varepsilon_A$ are empty relations.

The following equation holds in $\mathbf{B}$:

$$(\mathbf{0}\eta\varepsilon) \quad \mathbf{0}_A = \check{\sigma}^{\rightarrow}_{A} \circ (\varepsilon_A \vee \mathbf{1}_A) \circ d_{A,\neg A,A} \circ (\mathbf{1}_A \wedge \eta_A) \circ \hat{\delta}^{\leftarrow}_{A},$$

since

$$\begin{aligned} \varepsilon_A &= \varepsilon_A \circ \mathbf{0}_{A\wedge\neg A} \\ &= \varepsilon_A \circ (\mathbf{0}_A \wedge \mathbf{0}_{\neg A}), \;\; \text{by } (\mathbf{0}\mathrm{I}\;\xi) \text{ of §12.5}, \end{aligned}$$

which with naturality and bifunctorial equations yields that the right-hand side *RHS* of $(\mathbf{0}\eta\varepsilon)$ is equal to $\mathit{RHS} \circ \mathbf{0}_A$. The equation $(\mathbf{0}\eta\varepsilon)$ shows that we need not take $\mathbf{0}_A$ as a primitive arrow term: we can take it as defined in terms of η_A and ε_A. We could then conceive of $\mathbf{B}$ as obtained by extending $\mathbf{DL}^{\neg}_{\top,\bot}$ with the arrows η_A and ε_A and the equations $(\mathbf{0}\mathrm{I})$ for defined $\mathbf{0}_A$.

The following equations too hold in $\mathbf{B}$:

$$\begin{aligned} &(\eta\wedge) && \eta_{B\wedge A} = (\hat{r}^{\leftarrow}_{B,A} \vee \mathbf{1}_{B\wedge A}) \circ \check{c}_{\neg B\vee\neg A,B\wedge A} \circ \check{e}'_{B,A,\neg B,\neg A} \circ \\ &&& \qquad\qquad \circ (\check{c}_{B,\neg B} \wedge \check{c}_{A,\neg A}) \circ (\eta_B \wedge \eta_A) \circ \hat{\delta}^{\leftarrow}_{\top}, \\ &(\eta\vee) && \eta_{B\vee A} = (\check{r}^{\leftarrow}_{B,A} \vee \mathbf{1}_{B\vee A}) \circ \check{e}'_{\neg B,\neg A,B,A} \circ (\eta_B \wedge \eta_A) \circ \hat{\delta}^{\leftarrow}_{\top}, \\ &(\varepsilon\wedge) && \varepsilon_{A\wedge B} = \check{\delta}^{\rightarrow}_{\bot} \circ (\varepsilon_A \vee \varepsilon_B) \circ \hat{e}'_{A,B,\neg A,\neg B} \circ (\mathbf{1}_{A\wedge B} \wedge \hat{r}^{\rightarrow}_{A,B}), \\ &(\varepsilon\vee) && \varepsilon_{A\vee B} = \check{\delta}^{\rightarrow}_{\bot} \circ (\varepsilon_A \vee \varepsilon_B) \circ (\hat{c}_{\neg A,A} \vee \hat{c}_{\neg B,B}) \circ \hat{e}'_{\neg A,\neg B,A,B} \circ \\ &&& \qquad\qquad \circ \hat{c}_{A\vee B,\neg A\wedge\neg B} \circ (\mathbf{1}_{A\vee B} \wedge \check{r}^{\rightarrow}_{A,B}), \end{aligned}$$

$$\begin{aligned} &(\eta\top) && \eta_\top = (\hat{\rho}^{\leftarrow} \vee \mathbf{1}_\top) \circ \check{\sigma}^{\leftarrow}_{\top}, && (\varepsilon\top) && \varepsilon_\top = \hat{\sigma}^{\rightarrow}_{\bot} \circ (\mathbf{1}_\top \wedge \hat{\rho}^{\rightarrow}), \\ &(\eta\bot) && \eta_\bot = (\check{\rho}^{\leftarrow} \vee \mathbf{1}_\bot) \circ \check{\delta}^{\leftarrow}_{\top}, && (\varepsilon\bot) && \varepsilon_\bot = \hat{\delta}^{\rightarrow}_{\bot} \circ (\mathbf{1}_\bot \wedge \check{\rho}^{\rightarrow}). \end{aligned}$$

The syntactical system $\mathcal{C}(\mathbf{C})$ is defined by taking first for its objects the formulae of $\mathcal{L}^{\neg p}_{\wedge,\vee,\top,\bot}$, namely $\mathcal{L}_{\wedge,\vee,\top,\bot}$ generated by $\mathcal{P}\cup\mathcal{P}^{\neg}$; next, the primitive arrow terms of $\mathcal{C}(\mathbf{C})$ are those of $\mathcal{C}(\mathbf{ZIDL}_{\top,\bot})$, i.e. those in the families $\mathbf{1}$, b, c, w-k, m, d, δ-σ and $\mathbf{0}$ (whose members are $\mathbf{0}_A\colon A\vdash A$) plus

$$\eta_p\colon \top\vdash\neg p\vee p,$$
$$\varepsilon_p\colon p\wedge\neg p\vdash\bot,$$

for every $p\in\mathcal{P}$, and the operations on arrow terms are composition, $\wedge$ and $\vee$. As equations of $\mathcal{E}(\mathbf{C})$ we assume those of $\mathcal{E}(\mathbf{ZIDL}_{\top,\bot})$, i.e. those of $\mathcal{E}(\mathbf{DL}_{\top,\bot})$ plus $(\mathbf{0}\mathrm{I})$. The syntactical category $\mathbf{C}$ is $\mathcal{C}(\mathbf{C})/\mathcal{E}(\mathbf{C})$.

We define $\mathbf{0}_{A,B}$ in $\mathbf{C}$ by $f\circ\mathbf{0}_A$, as we did in $\mathbf{ZIDL}_{\top,\bot}$ and $\mathbf{B}$, and as in $\mathbf{B}$ we infer $\eta_p=\mathbf{0}_{\top,\neg p\vee p}$, $\varepsilon_p=\mathbf{0}_{p\wedge\neg p,\bot}$ and $(\mathbf{0}\eta\varepsilon)$ with A replaced by p.

The category $\mathbf{C}^{\neg}$ is obtained from $\mathbf{C}$ as $\mathbf{I}^{\neg}_{\top,\bot}$ is obtained from $\mathbf{I}_{\top,\bot}$; namely, we have as objects the formulae of $\mathcal{L}^{\neg}_{\wedge,\vee,\top,\bot}$, we add to $\mathbf{C}$ the arrows n-r and ρ, and we assume in addition the equations of $\mathcal{E}(\mathbf{I}^{\neg}_{\top,\bot})$.

The category $\mathbf{C}^{\neg}$ is isomorphic to the category $\mathbf{B}$. We define η_A and ε_A in $\mathbf{C}^{\neg}$ inductively with the help of $(\eta\,\xi)$ and $(\varepsilon\,\xi)$ for $\xi\in\{\wedge,\vee\}$, and $(\eta\zeta)$ and $(\varepsilon\zeta)$ for $\zeta\in\{\top,\bot\}$. Our purpose now is to show that the functor G from $\mathbf{C}$ to *Rel* is faithful, and use this coherence result, together with results analogous to $\mathcal{K}^{\neg}$-$\mathcal{K}^{\neg p}$-Equivalence and De Morgan Coherence, to infer that G from $\mathbf{B}$ to *Rel* is faithful, i.e. that $\mathbf{B}$ is coherent.

Let $A[\top]$ be a formula of $\mathcal{L}^{\neg p}_{\wedge,\vee,\top,\bot}$ with a particular occurrence of $\top$, and let $A[B]$ be obtained from $A[\top]$ by replacing this particular occurrence of $\top$ by the formula B of $\mathcal{L}^{\neg p}_{\wedge,\vee,\top,\bot}$. Then it is clear that there is an η-term $A[\eta_p]\colon A[\top]\vdash A[\neg p\vee p]$.

We define the arrow term $\hat{g}_{A[B]}\colon A[\top]\wedge B\vdash A[B]$ of $\mathcal{C}(\mathbf{DL}_{\top,\bot})$ by induction on the length of $A[\top]$:

$$\hat{g}_B=\hat{\sigma}^{\rightarrow}_B\colon \top\wedge B\vdash B,$$

$$\hat{g}_{C\wedge A[B]}=(\mathbf{1}_C\wedge\hat{g}_{A[B]})\circ\hat{b}^{\leftarrow}_{C,A[\top],B}\colon (C\wedge A[\top])\wedge B\vdash C\wedge A[B],$$

$$\hat{g}_{A[B]\wedge C}=(\hat{g}_{A[B]}\wedge\mathbf{1}_C)\circ(\hat{c}_{B,A[\top]}\wedge\mathbf{1}_C)\circ\hat{b}^{\rightarrow}_{B,A[\top],C}\circ\hat{c}_{A[\top]\wedge C,B}\colon$$
$$(A[\top]\wedge C)\wedge B\vdash A[B]\wedge C,$$

$$\hat{g}_{C\vee A[B]}=(\mathbf{1}_C\vee\hat{g}_{A[B]})\circ d^R_{C,A[\top],B}\colon (C\vee A[\top])\wedge B\vdash C\vee A[B],$$

$$\hat{g}_{A[B]\vee C} = (\hat{g}_{A[B]} \vee \mathbf{1}_C)\circ(\hat{c}_{B,A[\top]} \vee \mathbf{1}_C)\circ d^L_{B,A[\top],C}\circ \hat{c}_{A[\top]\vee C,B}\colon$$
$$(A[\top]\vee C)\wedge B \vdash A[B]\vee C.$$

Then we can prove that the following equations hold in **C**:

$$(g\eta)\quad A[\eta_p] = \hat{g}_{A[\neg p\vee p]}\circ(\mathbf{1}_{A[\top]}\wedge \mathbf{0}_{\neg p\vee p})\circ(\mathbf{1}_{A[\top]}\wedge\eta_p)\circ\hat{\delta}^{\leftarrow}_{A[\top]}.$$

We proceed by induction on the length of $A[\top]$. If A is $\top$, then

$$\hat{\sigma}^{\rightarrow}_{\neg p\vee p}\circ(\mathbf{1}_\top\wedge\mathbf{0}_{\neg p\vee p})\circ(\mathbf{1}_\top\wedge\eta_p)\circ\hat{\delta}^{\leftarrow}_\top$$
$$= \mathbf{0}_{\neg p\vee p}\circ\eta_p, \text{ by } (\hat{\delta}\hat{\sigma}) \text{ and naturality equations,}$$
$$= \eta_p, \text{ by } (\mathbf{0}\mathrm{I}) \text{ and } \mathbf{0}_\top = \hat{\kappa}_\top = \mathbf{1}_\top.$$

If A is $C\wedge D[\top]$, then

$$(\mathbf{1}_C\wedge\hat{g}_{D[B]})\circ\hat{b}^{\leftarrow}_{C,D[\top],B}\circ(\mathbf{1}_{C\wedge D[\top]}\wedge\mathbf{0}_{\neg p\vee p})\circ(\mathbf{1}_{C\wedge D[\top]}\wedge\eta_p)\circ\hat{\delta}^{\leftarrow}_{C\wedge D[\top]}$$
$$= (\mathbf{1}_C\wedge\hat{g}_{D[B]})\circ(\mathbf{1}_C\wedge(\mathbf{1}_{D[\top]}\wedge\mathbf{0}_{\neg p\vee p}))\circ(\mathbf{1}_C\wedge(\mathbf{1}_{D[\top]}\wedge\eta_p))\circ$$
$$\circ(\mathbf{1}_C\wedge\hat{\delta}^{\leftarrow}_{D[\top]}), \text{ by } (\hat{b}^{\leftarrow}\ nat),\ (\hat{b}\hat{\delta}) \text{ and isomorphisms,}$$
$$= \mathbf{1}_C\wedge D[\eta_p], \text{ by the induction hypothesis.}$$

We proceed analogously when A is $D[\top]\wedge C$, $C\vee D[\top]$ and $D[\top]\vee C$, by applying naturality equations and Distributive Dicartesian Coherence.

For $A[\bot]$ a formula of $\mathcal{L}^{\neg p}_{\wedge,\vee,\top,\bot}$ with a particular occurrence of $\bot$, and $A[B]$ obtained from $A[\bot]$ by replacing this particular occurrence of $\bot$ by B, we have an ε-term $A[\varepsilon_q]\colon A[q\wedge\neg q]\vdash A[\bot]$.

We define the arrow term $\check{g}_{A[B]}\colon A[B]\vdash A[\bot]\vee B$ of $\mathcal{C}(\mathbf{DL}_{\top,\bot})$ by

$$\check{g}_B = \check{\sigma}^{\leftarrow}_B\colon B\vdash\top\vee B,$$
$$\check{g}_{C\vee A[B]} = \check{b}^{\rightarrow}_{C,A[\bot],B}\circ(\mathbf{1}_C\vee\check{g}_{A[B]})\colon C\vee A[B]\vdash(C\vee A[\bot])\vee B,$$
$$\check{g}_{A[B]\vee C} = \check{c}_{A[\bot]\vee C,B}\circ\check{b}^{\leftarrow}_{B,A[\bot],C}\circ(\check{c}_{B,A[\bot]}\vee\mathbf{1}_C)\circ(\check{g}_{A[B]}\vee\mathbf{1}_C)\colon$$
$$A[B]\vee C\vdash(A[\bot]\vee C)\vee B,$$
$$\check{g}_{C\wedge A[B]} = d^L_{C,A[\bot],B}\circ(\mathbf{1}_C\wedge\check{g}_{A[B]})\colon C\wedge A[B]\vdash(C\wedge A[\bot])\vee B,$$
$$\check{g}_{A[B]\wedge C} = \check{c}_{A[\bot]\wedge C,B}\circ d^R_{B,A[\bot],C}\circ(\check{c}_{B,A[\bot]}\wedge\mathbf{1}_C)\circ(\check{g}_{A[B]}\wedge\mathbf{1}_C)\colon$$
$$A[B]\wedge C\vdash(A[\bot]\wedge C)\vee B.$$

Then, in a dual manner, we can prove by induction on the length of $A[\bot]$ that the following equations, analogous to $(g\eta)$, hold in C:

$$(g\varepsilon)\quad A[\varepsilon_q] = \check{\delta}^{\rightarrow}_{A[\bot]} \circ (\mathbf{1}_{A[\bot]} \vee \varepsilon_q) \circ (\mathbf{1}_{A[\bot]} \vee \mathbf{0}_{q\wedge\neg q}) \circ \check{g}_{A[q\wedge\neg q]}.$$

From $(g\eta)$ and $(g\varepsilon)$, we easily infer with naturality and bifunctorial equations the following equations of $\mathbf{C}$:

for $f\colon B \vdash A[\top]$,

$$(g\eta f)\quad A[\eta_p] \circ f = \hat{g}_{A[\neg p\vee p]} \circ (f \wedge \mathbf{0}_{\neg p\vee p}) \circ (\mathbf{1}_B \wedge \eta_p) \circ \hat{\delta}^{\leftarrow}_{B},$$

for $f\colon A[\bot] \vdash C$,

$$(g\varepsilon f)\quad f \circ A[\varepsilon_q] = \check{\delta}^{\rightarrow}_{C} \circ (\mathbf{1}_C \vee \varepsilon_q) \circ (f \vee \mathbf{0}_{q\wedge\neg q}) \circ \check{g}_{A[q\wedge\neg q]}.$$

Note that in the relations $G(A[\eta_p] \circ f)$ and $G(\hat{g}_{A[\neg p\vee p]} \circ (f \wedge \mathbf{0}_{\neg p\vee p}))$ we have the same sets of ordered pairs, and in the relations $G(f \circ A[\varepsilon_q])$ and $G((f \vee \mathbf{0}_{q\wedge\neg q}) \circ \check{g}_{A[q\wedge\neg q]})$ we also have the same sets of ordered pairs.

Suppose now that we have two arrow terms $f_1, f_2\colon B \vdash C$ of $\mathcal{C}(\mathbf{C})$ such that $Gf_1 = Gf_2$. Let $p_1, \ldots, p_n$, with $n \geq 0$, be the set of all occurrences of letters that are subscripts of subterms of f_1 and f_2 of the form η_{p_i}, where $i \in \{1, \ldots, n\}$, and let $q_1, \ldots, q_m$, with $m \geq 0$, be the set of all occurrences of letters that are subscripts of subterms of f_1 and f_2 of the form ε_{q_j}, where $j \in \{1, \ldots, m\}$ (the same letter may be repeated in $p_1, \ldots, p_n$, or $q_1, \ldots, q_m$).

We introduce the following abbreviations by induction:

$\eta^0 = \mathbf{1}_B$,	$\varepsilon^0 = \mathbf{1}_C$,
$\eta^{k+1} = \eta^k \wedge \eta_{p_{k+1}}$,	$\varepsilon^{k+1} = \varepsilon^k \vee \varepsilon_{q_{k+1}}$,
$\hat{\mathbf{0}}^0 = \mathbf{1}_B$,	$\check{\mathbf{0}}^0 = \mathbf{1}_C$,
$\hat{\mathbf{0}}^{k+1} = \hat{\mathbf{0}}^k \wedge \mathbf{0}_{\neg p_{k+1}\vee p_{k+1}}$,	$\check{\mathbf{0}}^{k+1} = \check{\mathbf{0}}^k \vee \mathbf{0}_{q_{k+1}\wedge\neg q_{k+1}}$,
$\hat{B}^0_\top = B$,	$\check{C}^0_\bot = C$,
$\hat{B}^{k+1}_\top = \hat{B}^k_\top \wedge \top$,	$\check{C}^{k+1}_\bot = \check{C}^k_\bot \vee \bot$,
$\hat{\delta}^0 = \mathbf{1}_B$,	$\check{\delta}^0 = \mathbf{1}_C$,
$\hat{\delta}^{k+1} = \hat{\delta}^{\leftarrow}_{\hat{B}^k_\top} \circ \hat{\delta}^k$,	$\check{\delta}^{k+1} = \check{\delta}^k \circ \check{\delta}^{\rightarrow}_{\check{C}^k_\bot}$,

$$\begin{array}{llll}
\hat{B}{}_p^0 & = B, & \overset{\vee}{C}{}_q^0 & = C, \\
\hat{B}{}_p^{k+1} & = \hat{B}{}_p^k \wedge(\neg p_{k+1} \vee p_{k+1}), & \overset{\vee}{C}{}_q^{k+1} & = \overset{\vee}{C}{}_q^k \vee(q_{k+1} \wedge \neg q_{k+1}), \\
\hat{h}{}_f^0 & = \mathbf{1}_B, & \overset{\vee}{h}{}_f^0 & = \mathbf{1}_C,
\end{array}$$

$$\hat{h}{}_f^{k+1} = \begin{cases} \hat{h}{}_f^k \wedge \mathbf{1}_{\neg p_{k+1} \vee p_{k+1}} & \text{if } \eta_{p_{k+1}} \text{ is in } f \\ \hat{h}{}_f^k \circ \hat{k}{}^1_{\hat{B}{}_p^k, \neg p_{k+1} \vee p_{k+1}} & \text{if } \eta_{p_{k+1}} \text{ is not in } f, \end{cases}$$

$$\overset{\vee}{h}{}_f^{k+1} = \begin{cases} \overset{\vee}{h}{}_f^k \vee \mathbf{1}_{q_{k+1} \vee \neg q_{k+1}} & \text{if } \varepsilon_{q_{k+1}} \text{ is in } f \\ \overset{\vee}{k}{}^1_{\overset{\vee}{C}{}_q^k, q_{k+1} \wedge \neg q_{k+1}} \circ \overset{\vee}{h}{}_f^k & \text{if } \varepsilon_{q_{k+1}} \text{ is not in } f. \end{cases}$$

Then, by relying on $(g\eta f)$ and $(g\varepsilon f)$, for $i \in \{1,2\}$ we obtain in $\mathbf{C}$ the equations

$$f_i = \overset{\vee}{\delta}{}^m \circ \varepsilon^m \circ \overset{\vee}{\mathbf{0}}{}^m \circ \overset{\vee}{h}{}^m_{f_i} \circ f'_i \circ \hat{h}{}^n_{f_i} \circ \overset{\wedge}{\mathbf{0}}{}^n \circ \eta^n \circ \overset{\wedge}{\delta}{}^n$$

where f'_i is an arrow term of $\mathcal{C}(\mathbf{ZIDL}^{\neg p}_{\top,\bot})$, and for f''_i being

$$\overset{\vee}{\mathbf{0}}{}^m \circ \overset{\vee}{h}{}^m_{f_i} \circ f'_i \circ \hat{h}{}^n_{f_i} \circ \overset{\wedge}{\mathbf{0}}{}^n$$

we have $Gf''_1 = Gf''_2$. Since f''_1 and f''_2 are also arrow terms of $\mathcal{C}(\mathbf{ZIDL}^{\neg p}_{\top,\bot})$, by Zero-Identity Distributive Dicartesian Coherence of §12.5 we conclude that $f''_1 = f''_2$ in $\mathbf{ZIDL}^{\neg p}_{\top,\bot}$, and hence $f_1 = f_2$ in $\mathbf{C}$. This establishes that the functor G from $\mathbf{C}$ to *Rel* is faithful.

We prove as in the preceding section $\mathcal{K}^{\neg}$-$\mathcal{K}^{\neg p}$ Equivalence for $\mathcal{K}^{\neg p}$ being $\mathbf{C}$ and $\mathcal{K}^{\neg}$ being $\mathbf{C}^{\neg}$. (We stipulate that $F\beta_p = \beta_p$ and $F^{\neg}\beta_p = \beta_p$ for $\beta \in \{\eta, \varepsilon\}$.) Then, as in the proof of De Morgan Coherence, we use the faithfulness of G from $\mathbf{C}$ to *Rel* and $\mathcal{K}^{\neg}$-$\mathcal{K}^{\neg p}$ Equivalence to establish that G from $\mathbf{C}^{\neg}$ to *Rel* is faithful. The isomorphism of the categories $\mathbf{C}^{\neg}$ and $\mathbf{B}$ then yields the following.

BOOLEAN COHERENCE. *The functor G from $\mathbf{B}$ to Rel is faithful.*

We can define the functor $\neg$ from $\mathbf{B}$ to $\mathbf{B}^{op}$ by extending the definitions in the preceding section with

$$\neg\eta_A =_{df} \hat{\rho}^{\leftarrow} \circ \varepsilon_A \circ (n^{\rightarrow}_{A} \wedge \mathbf{1}_{\neg A}) \circ \check{r}^{\rightarrow}_{\neg A,A},$$
$$\neg\varepsilon_A =_{df} \hat{r}^{\leftarrow}_{A,\neg A} \circ (\mathbf{1}_{\neg A} \vee n^{\leftarrow}_{A}) \circ \eta_A \circ \check{\rho}^{\rightarrow}.$$

The transformations η and ε are dinatural transformations (see [94], Section IX.4), which means that for $f\colon A \vdash B$ we have

$$(\mathbf{1}_{\neg A} \vee f) \circ \eta_A = (\neg f \vee \mathbf{1}_B) \circ \eta_B,$$
$$\varepsilon_A \circ (\mathbf{1}_A \wedge \neg f) = \varepsilon_B \circ (f \wedge \mathbf{1}_{\neg B}).$$

These equations are satisfied trivially in **B**, because η_A, η_B, ε_A and ε_B are zero arrows.

We leave open the question of maximality for the category **B**. In the light of the results of §9.7, it seems natural to conjecture that this category is not maximal. Note, however, that the category **ZIDL**, which is isomorphic to a subcategory of **B** (see Chart 3), and covers the conjunction-disjunction fragment of classical propositional logic, is maximal (see §12.5). For the conjunction-disjunction fragment of classical propositional logic with the constants $\top$ and $\bot$, we have the category $\mathbf{ZIDL}_{\top,\bot}$, which is also isomorphic to a subcategory of **B**, and is maximal in the relative sense in which $\mathbf{L}_{\top,\bot}$ is maximal (see §9.7 and §12.5). The technique of §9.7 suggests how to prove some sort of relative maximality also for **B**.

§14.3. Boolean categories

A distributive dicartesian category $\mathcal{A}$ for which we have a functor $\neg$ from $\mathcal{A}$ to $\mathcal{A}^{op}$, natural isomorphisms like those in the families n-r and ρ, and dinatural transformations ε and η will be called a *Boolean* category.

A Boolean category is called a *zero-identity* Boolean category when for every $\mathbf{0}_a$ defined by $(\mathbf{0}\eta\varepsilon)$ of the preceding section, where A is replaced by the object a, we have for every $f, g\colon a \vdash b$ the equation $(\mathbf{0}\mathrm{I})$, namely $f \circ \mathbf{0}_a = \mathbf{0}_b \circ g$. The category **B** is the zero-identity Boolean category generated by $\mathcal{P}$.

The connection of Boolean categories with Boolean algebras is the following. We have an arrow of type $A \vdash B$ in **B** iff $A \rightarrow B$ is a tautology of propositional logic. This is how **B** is connected to classical propositional logic. A partially ordered Boolean category, which must be a zero-identity

Boolean category, is a Boolean algebra (in which top and bottom are not necessarily distinct).

Note that the equations of **B** cover cut elimination—i.e. they enable us to prove a cut-elimination theorem, such as we had in §11.2. They cover first the cut elimination of $\mathbf{ZIDL}_{\top,\bot}$. As far as negation is concerned, the key equation is $(\mathbf{0}\eta\varepsilon)$, whose right-hand side corresponds to the cut

$$\frac{f\colon (\emptyset,\wedge) \vdash \neg A^{\vee} A \qquad g\colon A^{\wedge}\neg A \vdash (\emptyset,\vee)}{cut_{(\emptyset,\wedge),(\emptyset,\vee)}(f,g)\colon A \vdash A}$$

With $(\mathbf{0}\eta\varepsilon)$, we have that $cut_{(\emptyset,\wedge),(\emptyset,\vee)}(f,g)$ is not equal to $\mathbf{1}_A\colon A \vdash A$, but to $\mathbf{0}_A\colon A \vdash A$.

To prove a cut-elimination theorem, we can rely on Gentzen terms like those in §11.1, to which we would add dual primitive Gentzen terms and Gentzen operations where $\neg A \wedge \neg B$, $\neg A \vee \neg B$, $\top$ and $\bot$ are replaced respectively by $\neg(A \vee B)$, $\neg(A \wedge B)$, $\neg\bot$ and $\neg\top$. For example, we would have the operation

$$\frac{f\colon \Gamma^{\wedge}\neg A^{\wedge}\neg B \vdash \Delta}{\neg\vee^{L} f =_{dn} f \circ (\Gamma^{\wedge}\mathbf{1}^{e}_{A^{\wedge}B}) \circ (\Gamma^{\wedge}\check{r}^{\rightarrow\prime\prime}_{A,B})\colon \Gamma^{\wedge}\neg(A \vee B) \vdash \Delta}$$

Our strictification result should be adjusted to support such operations. As additional primitive Gentzen terms, we would have $\mathbf{1}''_{\neg p}\colon \neg p \vdash \neg p$, $\mathbf{0}''_{p}\colon p \vdash p$, $\mathbf{0}''_{\neg p}\colon \neg p \vdash \neg p$, $\eta''_{p}\colon (\emptyset,\wedge) \vdash \neg p^{\vee} p$ and $\varepsilon''_{p}\colon p^{\wedge}\neg p \vdash (\emptyset,\vee)$, and we would have the additional Gentzen operations

$$\frac{f\colon \Gamma^{\wedge} A \vdash \Delta}{\neg\neg^{L} f =_{dn} f \circ (\Gamma^{\wedge} n^{\rightarrow\prime\prime}_{A})\colon \Gamma^{\wedge}\neg\neg A \vdash \Delta}$$

$$\frac{f\colon \Gamma \vdash A^{\vee}\Delta}{\neg\neg^{R} f =_{dn} (n^{\leftarrow\prime\prime}_{A}{}^{\vee}\Delta) \circ f\colon \Gamma \vdash \neg\neg A^{\vee}\Delta}$$

A similar idea underlies a sequent system in [1] (Section 17) for tautological entailments, which correspond to De Morgan lattices.

The usual introduction and elimination rules for negation:

$$\frac{f : \Gamma \vdash A^{\vee}\Delta}{\neg^L f : \Gamma^{\wedge}\neg A \vdash \Delta} \qquad \frac{f : \Gamma^{\wedge} A \vdash \Delta}{\neg^R f : \Gamma \vdash \neg A^{\vee}\Delta}$$

would be *admissible* in the cut-free system; i.e., we can find cut-free Gentzen terms that define $\neg^L f$ and $\neg^R f$. This does not mean that the operations $\neg^L$ and $\neg^R$ are defined in terms of the postulated Gentzen operations; in such a case, we would speak of *derivable* rules. The arrow terms $\neg^L f$ and $\neg^R f$ involve zero arrow terms.

Zero-identity arrows make equal in **B** many arrow terms of the same type involving negation. In particular, all arrow terms of the type $\top \vdash A$ where A is a tautology are equal. However, **B** is far from being a preorder.

There is an argument from which it is usually concluded that it is hopeless to try to find a categorification of Boolean algebras. All plausible candidates seem to be categories that are preorders. To present this argument, we rely on notions defined in [85]. The argument is based on the fact that in every bicartesian closed category (i.e. cartesian closed category with finite coproducts), for every object a there is at most one arrow of type $a \vdash \bot$, for $\bot$ an initial object. In [85] the discovery of that fact is credited to Joyal (p. 116), and the fact is established (on p. 67, Proposition 8.3) by relying on a proposition of Freyd (see [53], p. 7, Proposition 1.12) to the effect that if in a cartesian closed category the hom-set $Hom(a, \bot)$ is not empty, then $a \cong \bot$; that is, a is isomorphic to $\bot$. Here is a simpler proof of the same fact (from [37], Section 5).

PROPOSITION 1. *In every cartesian closed category with an initial object $\bot$ we have that $Hom(a, \bot)$ is either empty or a singleton.*

PROOF. In every cartesian closed category with $\bot$ we have $\hat{k}^1_{\bot,\bot} = \hat{k}^2_{\bot,\bot}$: $\bot \wedge \bot \vdash \bot$, because $Hom(\bot \wedge \bot, \bot) \cong Hom(\bot, \bot^{\bot})$. Then for $f, g : a \vdash \bot$ we have $\hat{k}^1_{\bot,\bot} \circ \langle f, g\rangle = \hat{k}^2_{\bot,\bot} \circ \langle f, g\rangle$, and so $f = g$. ⊣

In [85] (p. 67) it is concluded from Proposition 1 that if in a bicartesian closed category for every object a we have $a \cong \neg\neg a$, where the negation $\neg b$ is $\bot^b$ (which corresponds to $b \to \bot$), then this category is a preorder.

If the requirement $a \cong \neg\neg a$ is deemed too strong, here are other similar propositions (taken over from [37], Section 5), in which preordering is inferred from other natural requirements.

PROPOSITION 2. *Every cartesian closed category with an initial object* $\bot$ *in which we have a natural transformation whose members are* $n_a^{\rightarrow} : \neg\neg a \vdash a$ *is a preorder.*

PROOF. Take $f, g\colon \neg\neg a \vdash b$, and take the canonical arrow $n_a^{\leftarrow} : a \vdash \neg\neg a$, which we have by the cartesian closed structure of our category. Then we have $\neg\neg(f \circ n_a^{\leftarrow}) = \neg\neg(g \circ n_a^{\leftarrow})$ by Proposition 1, and from

$$n_b^{\rightarrow} \circ \neg\neg(f \circ n_a^{\leftarrow}) = n_b^{\rightarrow} \circ \neg\neg(g \circ n_a^{\leftarrow}),$$

by the naturality of $n^{\rightarrow}$, we infer

$$f \circ n_a^{\leftarrow} \circ n_a^{\rightarrow} = g \circ n_a^{\leftarrow} \circ n_a^{\rightarrow}.$$

Since $n_a^{\leftarrow} \circ n_a^{\rightarrow} = \mathbf{1}_{\neg\neg a}$ by Proposition 1, we have $f = g$.

Then, for $\top$ terminal, we have

$$\begin{aligned} Hom(c, d) &\cong Hom(\top, d^C) \\ &\cong Hom(\neg\neg\top, d^C), \end{aligned}$$

since $\top \cong \neg\neg\top$, and $Hom(\neg\neg\top, d^C)$ is at most a singleton, as we have shown above. ⊣

PROPOSITION 3. *Every bicartesian closed category in which we have a dinatural transformation whose members are* $\eta_a\colon \top \vdash \neg a \vee a$ *is a preorder.*

PROOF. Take $f, g\colon \top \vdash a$. Then $\neg f = \neg g$ by Proposition 1, and from

$$(\neg f \vee \mathbf{1}_a) \circ \eta_a = (\neg g \vee \mathbf{1}_a) \circ \eta_a,$$

by the dinaturality of η we infer

$$(\mathbf{1}_{\neg\top} \vee f) \circ \eta_\top = (\mathbf{1}_{\neg\top} \vee g) \circ \eta_\top.$$

Since $\eta_\top : \top \vdash \neg\top \vee \top$ is an isomorphism, we obtain $\mathbf{1}_{\neg\top} \vee f = \mathbf{1}_{\neg\top} \vee g$, from which $f = g$ follows with the help of $\check{\sigma}^{\leftarrow}$. ⊣

(For an argument along similar lines, based on De Morgan isomorphisms, see [118].)

With Boolean categories, and with **B** in particular, we have, however, that A is isomorphic to $\neg\neg A$, and we have also that $n^{\rightarrow}$ is a natural transformation and η a dinatural transformation, without falling into preorder. We believe that our notion of Boolean category, which does not imply preorder, gives a reasonable categorification of the concept of Boolean algebra. Moreover, this notion delivers cut elimination, as we have indicated above.

That η_A and ε_A ended up by being zero arrows in **B** is dictated by *Rel*, since we have no other choice in *Rel* for $G\eta_A$ and $G\varepsilon_A$ save the empty relation. With another category, replacing *Rel* for coherence results, we need not take η_A and ε_A as zero arrows.

A category that could replace *Rel* is the category whose objects are finite ordinals and whose arrows are *split equivalence relations* (see [46] and [47]). These are equivalence relations defined on the sum of the ordinals in the source and target. For that category, $G\eta_p$ and $G\varepsilon_p$ would correspond to the diagrams

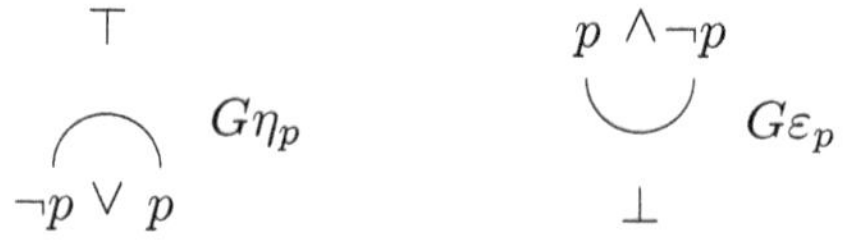

In that context, the left-hand side $\mathbf{0}_A$ of $(0\eta\varepsilon)$ would be replaced by $\mathbf{1}_A$, and that equation would become a triangular equation of adjunction (see [94], Section IV.1). But in this direction there is a heavy price to pay. The transformations in the families w-k, δ-σ and κ cannot remain natural if we want coherence. (Lack of naturality for these transformations jeopardizes cut elimination.) For example, for the following instance of $(\hat{w}\ nat)$:

$$\hat{w}_{\neg p \vee p} \circ \eta_p = (\eta_p \wedge \eta_p) \circ \hat{w}_{\top}$$

we would not have that $G(\hat{w}_{\neg p \vee p} \circ \eta_p)$ is equal to $G((\eta_p \wedge \eta_{\neg p}) \circ \hat{w}_{\top})$, as can be seen from the diagrams

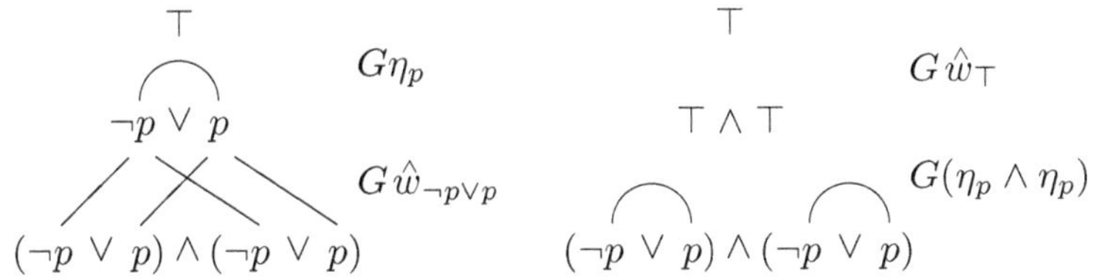

One may perhaps envisage a categorification of Boolean algebras where these transformations are not always natural, as they are in our Boolean categories. (In [62], Section 5.4, rejecting bifunctoriality is envisaged for the same purpose; we have no reason to reject bifunctoriality here.) Problems would, however, not cease once naturality is rejected for w-k, δ-σ and κ in the presence of negation.

The question is should $G\hat{w}_p$ be the relation in the left one or in the right one of the following two diagrams:

The second option, induced by dealing with equivalence relations, or by connecting all letters that must remain the same in generalizing proofs, would lead to abolishing the naturality of $\hat{w}$ even in the absence of negation. For example, in the following instance of ($\hat{w}$ *nat*):

$$\hat{w}_p \circ \check{\kappa}_p = (\check{\kappa}_p \wedge \check{\kappa}_p) \circ \hat{w}_\bot$$

we do not have that $G(\hat{w}_p \circ \check{\kappa}_p)$ is equal to $G((\check{\kappa}_p \wedge \check{\kappa}_p) \circ \hat{w}_\bot)$:

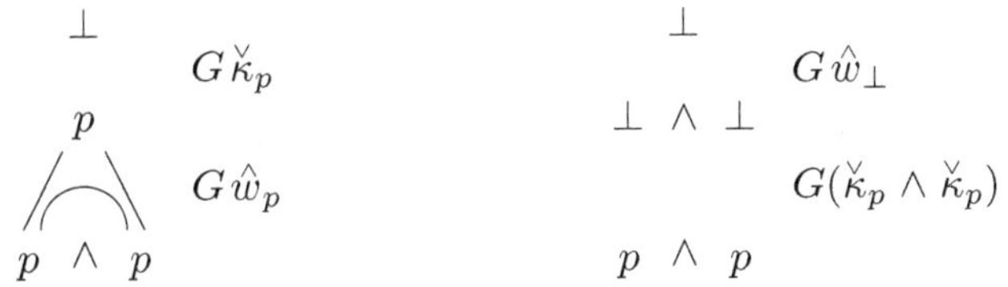

We obtain similarly that $\check{\kappa}$ cannot be natural.

If, on the other hand, we keep for $G\hat{w}_p$ the relation in the left diagram—the same we had in *Rel*—there would still be problems. One problem is with cut elimination. If f is the arrow term

$$c^k_{\top,\neg p,p,\top} \circ (\overset{\wedge}{\sigma}{}^{\leftarrow}_{\neg p} \vee \overset{\wedge}{\delta}{}^{\leftarrow}_{p}) \circ \eta_p,$$

then we would have that Gf is

$$\top$$

$$(\top \vee p) \wedge (\neg p \vee \top)$$

and no cut-free term whose main operation is conjunction introduction can yield such a graph. Although in the introduction we have motivated the category *Rel* by the Generality Conjecture, it should be stressed that this category does not always correspond to the intuitive idea of generality. This is so even if we do not consider the split equivalence relations of [46] and [47], but stay within *Rel*.

For arrows of *Rel* capturing generality it is natural to assume that they are *difunctional* in the sense of [108] (Section 7), a binary relation R being difunctional when $R \circ R^{-1} \circ R \subseteq R$ (in other words, if xRz, yRz and yRu, then xRu). It is easy to see that the image under G of an arrow of **DL** is not necessarily difunctional. For example, $G(m_{p,p} \cup (\overset{\vee}{k}{}^1_{p,p} \circ \hat{k}{}^2_{p,p}))$ is not a difunctional relation. The claim made in [37] (Section 4) that the image under G of any arrow of **L** is difunctional is not correct. A counterexample is provided by $G\langle[\hat{k}{}^1_{p,p}, \mathbf{1}_p], [\hat{k}{}^2_{p,p}, \mathbf{1}_p]\rangle$, which is not difunctional.

The category *Rel* captures, however, the intuitive idea of generality for all categories in Chart 1 except for **L** and the three categories above **L**. It captures this idea for the category **MDS**, too, and for all categories below **MDS** in Chart 2.

§14.4. Concluding remarks

Our coherence results show that a number of logical categories that we have investigated here are isomorphic to subcategories of the Boolean category **B**, and **B** is isomorphic to a subcategory of $\mathbf{ZML}^{\neg}_{\top,\bot}$. We record all these results about subcategories in Charts 1-3. Such results about subcategories are sometimes taken for granted, and, indeed, they are not surprising, but it is not trivial to establish them. One means of proving them is via coherence, which, as we have seen, is often established with considerable effort. (Another means can be via maximality.)

In general, we have the following situation. Suppose a syntactical system $\mathcal{S}'$ is a subsystem of a syntactical system $\mathcal{S}$. Suppose also that we have the syntactical categories $\mathcal{S}'/\mathcal{E}'$ and $\mathcal{S}/\mathcal{E}$ such that the set of equations $\mathcal{E}'$ is a subset of $\mathcal{E}$, the functor G from $\mathcal{S}'/\mathcal{E}'$ to *Rel* is faithful, and there is a functor from $\mathcal{S}/\mathcal{E}$ to *Rel* that extends G. Then $\mathcal{S}'/\mathcal{E}'$ is isomorphic to a subcategory of $\mathcal{S}/\mathcal{E}$.

To show that, it is enough to show that the identity maps on the objects and arrow terms of $\mathcal{S}'$ induce a functor from $\mathcal{S}'/\mathcal{E}'$ to $\mathcal{S}/\mathcal{E}$ that is inclusion on objects and one-one on arrows (see the penultimate paragraph of §2.4). This amounts to showing that for f and g arrow terms of $\mathcal{S}'$ of the same type we have $f = g$ in $\mathcal{S}'/\mathcal{E}'$ iff $f = g$ in $\mathcal{S}/\mathcal{E}$. It is clear that $f = g$ in $\mathcal{S}'/\mathcal{E}'$ implies $f = g$ in $\mathcal{S}/\mathcal{E}$, since $\mathcal{E}'$ is included in $\mathcal{E}$. For the converse, from $f = g$ in $\mathcal{S}/\mathcal{E}$ we infer $Gf = Gg$, and then, by the faithfulness of G assumed above, we obtain that $f = g$ in $\mathcal{S}'/\mathcal{E}'$ (cf. the end of §4.3). Note that $\mathcal{S}'/\mathcal{E}'$ is only isomorphic to a subcategory of $\mathcal{S}/\mathcal{E}$, and is not actually a subcategory of $\mathcal{S}/\mathcal{E}$, because an arrow term f of $\mathcal{S}'$ stands in $\mathcal{S}'/\mathcal{E}'$ for an equivalence class of arrow terms (this is an arrow of $\mathcal{S}'/\mathcal{E}'$) that is a subset, maybe proper, of the equivalence class for which f stands in $\mathcal{S}/\mathcal{E}$ (see §2.3).

Note that if for $\mathcal{S}'$ a subsystem of $\mathcal{S}$ and $\mathcal{E}'$ a subset of $\mathcal{E}$ we have that $\mathcal{S}'/\mathcal{E}'$ is a preorder, then we can ascertain that $\mathcal{S}'/\mathcal{E}'$ is isomorphic to a subcategory of $\mathcal{S}/\mathcal{E}$ without appealing to the functor G and *Rel*. We have such a situation with many of our categories where coherence amounts to preorder, but we also have it where preorder does not amount to coherence, as with the categories $\mathbf{S}'$ and $\hat{\mathbf{S}}'$ of §6.5.

There are coherence results with respect to *Rel*, related to the coherence results of this book, about categories that have arrows of the w kind, but not those of the k kind, and vice versa, about categories that have arrows of the k kind, but not those of the w kind. These categories are tied to substructural logics: the former to relevant logic, and the latter to affine logic. These coherences are proved in [102] for logical categories in the language $\mathcal{L}_{\wedge,\top}$ in between $\hat{\mathbf{S}}_\top$ and $\hat{\mathbf{L}}_\top$.

Speaking of categories tied to relevant logic, there is in the neighbourhood a sort of category interesting for strictification. If to the relevant natural logical category in $\mathcal{L}_\wedge$, which is like $\hat{\mathbf{L}}$ save that it lacks the natural transformations $\hat{k}^i$, we add the natural isomorphism $\hat{w}^{-1}$, whose members are $\hat{w}_A^{-1}\colon A \wedge A \vdash A$, and whose inverse is $\hat{w}$, then we obtain a groupoid that is a preorder. (The logical principle standing behind $\hat{w}^{-1}$ is called *mingle* in relevant logic; see [1], Section 8.15.) If this groupoid, which is a preorder, happens to be the category $\mathcal{G}$ involved in the strictification of some category, the equivalence classes introduced by strictification will correspond to finite nonempty *sets*. When $\mathcal{G}$ is $\hat{\mathbf{S}}$ plus $\hat{c}_{A,A} = \mathbf{1}_{A\wedge A}$, then we have *multisets* instead of sets (cf. §7.7), and with $\hat{\mathbf{A}}$ we have *sequences* (see §4.5).

We need not, however, assume that $\hat{w}^{-1}$ is an isomorphism. We can keep just $\hat{w}_A^{-1} \circ \hat{w}_A = \mathbf{1}_A$, reject $\hat{w}_A \circ \hat{w}_A^{-1} = \mathbf{1}_{A\wedge A}$, and have only equations that will yield coherence with respect to *Rel* with the assumption that $G\hat{w}_A^{-1}$ is equal to $G(\hat{k}^1_{A,A} \cup \hat{k}^2_{A,A})$. Defining $\hat{w}_A^{-1}$ by $\hat{k}^1_{A,A} \cup \hat{k}^2_{A,A}$ we have all these equations in **DL**, and hence also in **B**. (Another possibility is to define $\hat{w}_A^{-1}$ as just $\hat{k}^1_{A,A}$, or just $\hat{k}^2_{A,A}$.)

If we are right that **B** provides a reasonable notion of identity of proofs in classical propositional logic, and if bicartesian closed categories provide the notion of identity of proofs in intuitionistic propositional logic, we can conclude that the general proof theory of the former logic is simpler than that of the latter. Equality of derivations in classical propositional logic, i.e. equality of arrow terms in **B**, would be decided via *Rel*, in an elementary way. It was assumed before that classical general proof theory should be simpler, because it was assumed that all derivations with the same premise and conclusion are equal in classical logic. In other words, it was assumed that for a given premise and conclusion we cannot have more than one proof. We do not agree with that, and though we provide with **B** a relatively simple codification of that proof theory, it is not that simple.

It is true that all theorems, i.e. all propositions proved without hypotheses, will have zero proofs, which is not the case in the standard formulations of intuitionistic general proof theory. If we are right, with the theorems of classical logic we do not find a record of the deductive metatheory. But this metatheory of proofs from hypotheses exists, and it is not trivial. Our charts (see Charts 1-3) give an idea of the number of important mathematical structures that enter into the notion of Boolean category. We can also note in Chart 3 how with **B** we have come close to $\mathbf{ZML}^{\neg}_{\top,\bot}$, which is related to linear algebra.

Problems Left Open

1) How to axiomatize the equations $\mathcal{E}$ such that $\mathcal{C}(\mathbf{DA}_{\top,\bot})/\mathcal{E}$ is a preorder (see §7.9)?

2) Let $\mathcal{C}(\mathbf{DS}_{\top,\bot})$ be like $\mathcal{C}(\mathbf{DA}_{\top,\bot})$ with the transformation c added. How to axiomatize the equations $\mathcal{E}$ such that $\mathcal{C}(\mathbf{DS}_{\top,\bot})/\mathcal{E}$ is coherent with respect to *Rel* (see §7.9)?

3) How to axiomatize equations for mix-bimonoidal categories, symmetric or not symmetric, dissociative or not dissociative, for which one could prove coherence with respect to *Rel* (see Chapter 8)?

4) Can one prove coherence with respect to *Rel* for $\mathbf{ML}$, $\mathbf{ML}_{\top,\bot}$, $\mathbf{ZIML}$ and $\mathbf{ZIML}_{\top,\bot}$ (see §10.2, §10.3 and §12.5)? If not, what extended axiomatization delivers coherence?

5) For $\mathcal{K}$ being $\mathbf{ML}$, $\mathbf{ML}_{\top,\bot}$, $\mathbf{DL}$ or $\mathbf{DL}_{\top,\bot}$, prove that one could properly extend $\mathcal{E}(\mathcal{K})$ without falling into preorder (see §10.3 and §11.5). Classify the equations that give such extensions.

6) Can one derive the equations $(m\ \hat{e})$ and $(m\ \check{e})$ from the remaining axioms of $\mathcal{E}(\mathbf{DL})$ (see §11.1)?

7) Find concrete examples, distinct from $\mathbf{DL}$, $\mathbf{DL}_{\top,\bot}$ and $\mathbf{B}$, of distributive lattice, distributive dicartesian and Boolean categories in which $\wedge$ and $\vee$ are not isomorphic (see Chapters 11 and 14).

8) Consider the maximality question for the category $\mathbf{B}$ (see the end of §14.2).

List of Equations

We list here the equations assumed as axioms for the logical categories in our book. Besides that, we list prominent equations and definitions that were used in derivations or alternative axiomatizations. We mention in parentheses the sections where the equations were first introduced. A number of equations for $\vee$ were not stated explicitly in the main text, but appear here for the first time. We put the equations for $\vee$ immediately below the dual equations for $\wedge$. Otherwise, the list follows the order in which the equations appear in the book.

Categorial equations:

$(cat\ 1)\quad f \circ \mathbf{1}_a = \mathbf{1}_b \circ f = f : a \vdash b$ (§2.2)

$(cat\ 2)\quad h \circ (g \circ f) = (h \circ g) \circ f$ (§2.2)

Bifunctorial equations:

$(\wedge 1)\quad \mathbf{1}_A \wedge \mathbf{1}_B = \mathbf{1}_{A \wedge B}$ (§2.7)

$(\vee 1)\quad \mathbf{1}_A \vee \mathbf{1}_B = \mathbf{1}_{A \vee B}$ (§2.7)

$(\wedge 2)\quad (g_1 \circ f_1) \wedge (g_2 \circ f_2) = (g_1 \wedge g_2) \circ (f_1 \wedge f_2)$ (§2.7)

$(\vee 2)\quad (g_1 \circ f_1) \vee (g_2 \circ f_2) = (g_1 \vee g_2) \circ (f_1 \vee f_2)$ (§2.7)

Naturality equations (for $f : A \vdash D$, $g : B \vdash E$ and $h : C \vdash F$):

$(\hat{b}^{\rightarrow}\ nat)\quad ((f \wedge g) \wedge h) \circ \hat{b}^{\rightarrow}_{A,B,C} = \hat{b}^{\rightarrow}_{D,E,F} \circ (f \wedge (g \wedge h))$ (§2.7)

$(\check{b}^{\rightarrow}\ nat)\quad ((f \vee g) \vee h) \circ \check{b}^{\rightarrow}_{A,B,C} = \check{b}^{\rightarrow}_{D,E,F} \circ (f \vee (g \vee h))$ (§2.7)

$(\hat{b}^{\leftarrow}\ nat)\quad (f \wedge (g \wedge h)) \circ \hat{b}^{\leftarrow}_{A,B,C} = \hat{b}^{\leftarrow}_{D,E,F} \circ ((f \wedge g) \wedge h)$ (§2.7)

$(\check{b}^{\leftarrow}\ nat)\quad (f \vee (g \vee h)) \circ \check{b}^{\leftarrow}_{A,B,C} = \check{b}^{\leftarrow}_{D,E,F} \circ ((f \vee g) \vee h)$ (§2.7)

$(\hat{\delta}^{\rightarrow}\ nat)\quad f \circ \hat{\delta}^{\rightarrow}_{A} = \hat{\delta}^{\rightarrow}_{D} \circ (f \wedge \mathbf{1}_{\top})$ (§2.7)

$(\check{\delta}^{\rightarrow}\ nat)\quad f \circ \check{\delta}^{\rightarrow}_{A} = \check{\delta}^{\rightarrow}_{D} \circ (f \vee \mathbf{1}_{\bot})$ (§2.7)

$(\hat{\delta}^{\leftarrow}\ nat)\quad (f \wedge \mathbf{1}_{\top}) \circ \hat{\delta}^{\leftarrow}_{A} = \hat{\delta}^{\leftarrow}_{D} \circ f$ (§2.7)

$(\check{\delta}^{\leftarrow}\ nat)\quad (f \vee \mathbf{1}_{\bot}) \circ \check{\delta}^{\leftarrow}_{A} = \check{\delta}^{\leftarrow}_{D} \circ f$ (§2.7)

$(\hat{\sigma}\!\rightarrow\ nat)$	$f \circ \hat{\sigma}^{\rightarrow}_{A} = \hat{\sigma}^{\rightarrow}_{D} \circ (\mathbf{1}_{\top} \wedge f)$	(§2.7)
$(\check{\sigma}\!\rightarrow\ nat)$	$f \circ \check{\sigma}^{\rightarrow}_{A} = \check{\sigma}^{\rightarrow}_{D} \circ (\mathbf{1}_{\bot} \vee f)$	(§2.7)
$(\hat{\sigma}\!\leftarrow\ nat)$	$(\mathbf{1}_{\top} \wedge f) \circ \hat{\sigma}^{\leftarrow}_{A} = \hat{\sigma}^{\leftarrow}_{D} \circ f$	(§2.7)
$(\check{\sigma}\!\leftarrow\ nat)$	$(\mathbf{1}_{\bot} \vee f) \circ \check{\sigma}^{\leftarrow}_{A} = \check{\sigma}^{\leftarrow}_{D} \circ f$	(§2.7)
$(\hat{c}\ nat)$	$(g \wedge f) \circ \hat{c}_{A,B} = \hat{c}_{D,E} \circ (f \wedge g)$	(§2.7)
$(\check{c}\ nat)$	$(g \vee f) \circ \check{c}_{B,A} = \check{c}_{E,D} \circ (f \vee g)$	(§2.7)
$(\hat{w}\ nat)$	$(f \wedge f) \circ \hat{w}_{A} = \hat{w}_{D} \circ f$	(§2.7)
$(\check{w}\ nat)$	$f \circ \check{w}_{A} = \check{w}_{D} \circ (f \vee f)$	(§2.7)
$(\hat{k}^1\ nat)$	$f \circ \hat{k}^1_{A,B} = \hat{k}^1_{D,E} \circ (f \wedge g)$	(§2.7)
$(\check{k}^1\ nat)$	$(g \vee f) \circ \check{k}^1_{B,A} = \check{k}^1_{E,D} \circ g$	(§2.7)
$(\hat{k}^2\ nat)$	$g \circ \hat{k}^2_{A,B} = \hat{k}^2_{D,E} \circ (f \wedge g)$	(§2.7)
$(\check{k}^2\ nat)$	$(g \vee f) \circ \check{k}^2_{B,A} = \check{k}^2_{E,D} \circ f$	(§2.7)
$(\hat{\kappa}\ nat)$	$\mathbf{1}_{\top} \circ \hat{\kappa}_{A} = \hat{\kappa}_{D} \circ f$	(§2.7)
$(\check{\kappa}\ nat)$	$f \circ \check{\kappa}_{A} = \check{\kappa}_{D} \circ \mathbf{1}_{\bot}$	(§2.7)
$(d^L\ nat)$	$((f \wedge g) \vee h) \circ d^L_{A,B,C} = d^L_{D,E,F} \circ (f \wedge (g \vee h))$	(§2.7)
$(d^R\ nat)$	$(h \vee (g \wedge f)) \circ d^R_{C,B,A} = d^R_{F,E,D} \circ ((h \vee g) \wedge f)$	(§2.7)
$(m\ nat)$	$(f \vee g) \circ m_{A,B} = m_{D,E} \circ (f \wedge g)$	(§2.7)
$(m^{-1}\ nat)$	$(f \wedge g) \circ m^{-1}_{A,B} = m^{-1}_{D,E} \circ (f \vee g)$	

Specific and other equations:

$(\hat{b}\,5)$	$\hat{b}^{\rightarrow}_{A\wedge B,C,D} \circ \hat{b}^{\rightarrow}_{A,B,C\wedge D} = (\hat{b}^{\rightarrow}_{A,B,C} \wedge \mathbf{1}_D) \circ \hat{b}^{\rightarrow}_{A,B\wedge C,D} \circ (\mathbf{1}_A \wedge \hat{b}^{\rightarrow}_{B,C,D})$	(§4.2)
$(\check{b}\,5)$	$\check{b}^{\rightarrow}_{A\vee B,C,D} \circ \check{b}^{\rightarrow}_{A,B,C\vee D} = (\check{b}^{\rightarrow}_{A,B,C} \vee \mathbf{1}_D) \circ \check{b}^{\rightarrow}_{A,B\vee C,D} \circ (\mathbf{1}_A \vee \check{b}^{\rightarrow}_{B,C,D})$	
$(\hat{b}\hat{b})$	$\hat{b}^{\leftarrow}_{A,B,C} \circ \hat{b}^{\rightarrow}_{A,B,C} = \mathbf{1}_{A\wedge(B\wedge C)},\quad \hat{b}^{\rightarrow}_{A,B,C} \circ \hat{b}^{\leftarrow}_{A,B,C} = \mathbf{1}_{(A\wedge B)\wedge C}$	(§4.3)
$(\check{b}\check{b})$	$\check{b}^{\leftarrow}_{A,B,C} \circ \check{b}^{\rightarrow}_{A,B,C} = \mathbf{1}_{A\vee(B\vee C)},\quad \check{b}^{\rightarrow}_{A,B,C} \circ \check{b}^{\leftarrow}_{A,B,C} = \mathbf{1}_{(A\vee B)\vee C}$	
$(\hat{\delta}\hat{\delta})$	$\hat{\delta}^{\rightarrow}_{A} \circ \hat{\delta}^{\leftarrow}_{A} = \mathbf{1}_A,\quad \hat{\delta}^{\leftarrow}_{A} \circ \hat{\delta}^{\rightarrow}_{A} = \mathbf{1}_{A\wedge\top}$	(§4.6)
$(\check{\delta}\check{\delta})$	$\check{\delta}^{\rightarrow}_{A} \circ \check{\delta}^{\leftarrow}_{A} = \mathbf{1}_A,\quad \check{\delta}^{\leftarrow}_{A} \circ \check{\delta}^{\rightarrow}_{A} = \mathbf{1}_{A\vee\bot}$	

$(\hat{\sigma}\,\hat{\sigma})\quad \hat{\sigma}^{\rightarrow}_{A}\circ\hat{\sigma}^{\leftarrow}_{A}=\mathbf{1}_A,\quad \hat{\sigma}^{\leftarrow}_{A}\circ\hat{\sigma}^{\rightarrow}_{A}=\mathbf{1}_{\top\wedge A}$ (§4.6)

$(\check{\sigma}\,\check{\sigma})\quad \check{\sigma}^{\rightarrow}_{A}\circ\check{\sigma}^{\leftarrow}_{A}=\mathbf{1}_A,\quad \check{\sigma}^{\leftarrow}_{A}\circ\check{\sigma}^{\rightarrow}_{A}=\mathbf{1}_{\bot\vee A}$

$(\hat{b}\,\hat{\delta}\,\hat{\sigma})\quad \hat{b}^{\rightarrow}_{A,\top,C}=(\hat{\delta}^{\leftarrow}_{A}\wedge\mathbf{1}_C)\circ(\mathbf{1}_A\wedge\hat{\sigma}^{\rightarrow}_{C})$ (§4.6)

$(\check{b}\,\check{\delta}\,\check{\sigma})\quad \check{b}^{\rightarrow}_{A,\bot,C}=(\check{\delta}^{\leftarrow}_{A}\vee\mathbf{1}_C)\circ(\mathbf{1}_A\vee\check{\sigma}^{\rightarrow}_{C})$

$(\hat{b}\,\hat{\delta})\quad \hat{b}^{\rightarrow}_{A,B,\top}=\hat{\delta}^{\leftarrow}_{A\wedge B}\circ(\mathbf{1}_A\wedge\hat{\delta}^{\rightarrow}_{B})$ (§4.6)

$(\check{b}\,\check{\delta})\quad \check{b}^{\rightarrow}_{A,B,\bot}=\check{\delta}^{\leftarrow}_{A\vee B}\circ(\mathbf{1}_A\vee\check{\delta}^{\rightarrow}_{B})$

$(\hat{b}\,\hat{\sigma})\quad \hat{b}^{\rightarrow}_{\top,B,C}=(\hat{\sigma}^{\leftarrow}_{B}\wedge\mathbf{1}_C)\circ\hat{\sigma}^{\rightarrow}_{B\wedge C}$ (§4.6)

$(\check{b}\,\check{\sigma})\quad \check{b}^{\rightarrow}_{\bot,B,C}=(\check{\sigma}^{\leftarrow}_{B}\vee\mathbf{1}_C)\circ\check{\sigma}^{\rightarrow}_{B\vee C}$

$(\hat{\delta}\,\hat{\sigma})\quad \hat{\delta}^{\rightarrow}_{\top}=\hat{\sigma}^{\rightarrow}_{\top}$ (§4.6)

$(\check{\delta}\,\check{\sigma})\quad \check{\delta}^{\rightarrow}_{\bot}=\check{\sigma}^{\rightarrow}_{\bot}$

$(\hat{c}\,\hat{c})\quad \hat{c}_{B,A}\circ\hat{c}_{A,B}=\mathbf{1}_{A\wedge B}$ (§5.1)

$(\check{c}\,\check{c})\quad \check{c}_{A,B}\circ\check{c}_{B,A}=\mathbf{1}_{A\vee B}$

$(\hat{b}\,\hat{c})\quad \hat{c}_{A,B\wedge C}=\hat{b}^{\rightarrow}_{B,C,A}\circ(\mathbf{1}_B\wedge\hat{c}_{A,C})\circ\hat{b}^{\leftarrow}_{B,A,C}\circ(\hat{c}_{A,B}\wedge\mathbf{1}_C)\circ\hat{b}^{\rightarrow}_{A,B,C}$ (§5.1)

$(\check{b}\,\check{c})\quad \check{c}_{B\vee C,A}=\check{b}^{\rightarrow}_{B,C,A}\circ(\mathbf{1}_B\vee\check{c}_{C,A})\circ\check{b}^{\leftarrow}_{B,A,C}\circ(\check{c}_{B,A}\vee\mathbf{1}_C)\circ\check{b}^{\rightarrow}_{A,B,C}$

$(\hat{c}\,\hat{\delta}\,\hat{\sigma})\quad \hat{c}_{A,\top}=\hat{\sigma}^{\leftarrow}_{A}\circ\hat{\delta}^{\rightarrow}_{A}$ (§5.3)

$(\check{c}\,\check{\delta}\,\check{\sigma})\quad \check{c}_{\bot,A}=\check{\sigma}^{\leftarrow}_{A}\circ\check{\delta}^{\rightarrow}_{A}$

$(\hat{c}\,\bot)\quad \hat{c}_{C,C}=\mathbf{1}_{C\wedge C},$ for letterless C (§6.4)

$(\check{c}\,\top)\quad \check{c}_{C,C}=\mathbf{1}_{C\vee C},$ for letterless C (§6.4)

$(\hat{c}\,\mathbf{1})\quad \hat{c}_{A,A}=\mathbf{1}_{A\wedge A}$ (§6.5)

$(\check{c}\,\mathbf{1})\quad \check{c}_{A,A}=\mathbf{1}_{A\vee A}$ (§6.5)

$(d^L\wedge)\quad d^L_{A\wedge B,C,D}=(\hat{b}^{\rightarrow}_{A,B,C}\vee\mathbf{1}_D)\circ d^L_{A,B\wedge C,D}\circ(\mathbf{1}_A\wedge d^L_{B,C,D})\circ\hat{b}^{\leftarrow}_{A,B,C\vee D}$ (§7.2)

$(d^L\vee)\quad d^L_{D,C,B\vee A}=\check{b}^{\leftarrow}_{D\wedge C,B,A}\circ(d^L_{D,C,B}\vee\mathbf{1}_A)\circ d^L_{D,C\vee B,A}\circ(\mathbf{1}_D\wedge\check{b}^{\rightarrow}_{C,B,A})$ (§7.2)

$(d^R\wedge)$ $d^R_{D,C,B\wedge A} = (\mathbf{1}_D \vee \hat{b}^{\leftarrow}_{C,B,A}) \circ d^R_{D,C\wedge B,A} \circ (d^R_{D,C,B} \wedge \mathbf{1}_A) \circ \hat{b}^{\rightarrow}_{D\vee C,B,A}$ (§7.2)

$(d^R\vee)$ $d^R_{A\vee B,C,D} = \check{b}^{\rightarrow}_{A,B,C\wedge D} \circ (\mathbf{1}_A \vee d^R_{B,C,D}) \circ d^R_{A,B\vee C,D} \circ (\check{b}^{\leftarrow}_{A,B,C} \wedge \mathbf{1}_D)$ (§7.2)

$(d\hat{b})$ $d^R_{A\wedge B,C,D} \circ (d^L_{A,B,C} \wedge \mathbf{1}_D) = d^L_{A,B,C\wedge D} \circ (\mathbf{1}_A \wedge d^R_{B,C,D}) \circ \hat{b}^{\leftarrow}_{A,B\vee C,D}$ (§7.2)

$(d\check{b})$ $(d^R_{A,B,C} \vee \mathbf{1}_D) \circ d^L_{A\vee B,C,D} = \check{b}^{\rightarrow}_{A,B\wedge C,D} \circ (\mathbf{1}_A \vee d^L_{B,C,D}) \circ d^R_{A,B,C\vee D}$ (§7.2)

(d^Rc) $d^R_{C,B,A} = \check{c}_{C,B\wedge A} \circ (\hat{c}_{A,B} \vee \mathbf{1}_C) \circ d^L_{A,B,C} \circ (\mathbf{1}_A \wedge \check{c}_{B,C}) \circ \hat{c}_{C\vee B,A}$ (§7.6)

$\hat{e}_{A,B,C,D} =_{df} d_{A,D,B\wedge C} \circ (\mathbf{1}_A \wedge \check{c}_{D,B\wedge C}) \circ (\mathbf{1}_A \wedge d_{B,C,D}) \circ \hat{b}^{\leftarrow}_{A,B,C\vee D}$:
$(A\wedge B)\wedge(C\vee D) \vdash (A\wedge D)\vee(B\wedge C)$ (§7.6)

$\hat{e}'_{A,B,C,D} =_{df} \hat{e}_{A,B,D,C} \circ (\mathbf{1}_{A\wedge B} \wedge \check{c}_{D,C})$:
$(A\wedge B)\wedge(C\vee D) \vdash (A\wedge C)\vee(B\wedge D)$ (§7.6)

$(\hat{e})$ $\check{c}_{B\wedge C,A\wedge D} \circ \hat{e}_{A,B,C,D} = \hat{e}'_{B,A,C,D} \circ (\hat{c}_{A,B} \wedge \mathbf{1}_{C\vee D})$ (§7.6)

$\check{e}_{D,C,B,A} =_{df} \check{b}^{\leftarrow}_{D\wedge C,B,A} \circ (d_{D,C,B} \vee \mathbf{1}_A) \circ (\hat{c}_{C\vee B,D} \vee \mathbf{1}_A) \circ d_{C\vee B,D,A}$:
$(C\vee B)\wedge(D\vee A) \vdash (D\wedge C)\vee(B\vee A)$ (§7.6)

$\check{e}'_{D,C,B,A} =_{df} (\hat{c}_{C,D} \vee \mathbf{1}_{B\vee A}) \circ \check{e}_{C,D,B,A}$:
$(D\vee B)\wedge(C\vee A) \vdash (D\wedge C)\vee(B\vee A)$ (§7.6)

$(\check{e})$ $(\mathbf{1}_{D\wedge C} \vee \check{c}_{B,A}) \circ \check{e}'_{D,C,A,B} = \check{e}_{D,C,B,A} \circ \hat{c}_{D\vee A,C\vee B}$ (§7.6)

$(\hat{\sigma}\, d^L)$ $d^L_{\top,B,C} = (\hat{\sigma}^{\leftarrow}_B \vee \mathbf{1}_C) \circ \hat{\sigma}^{\rightarrow}_{B\vee C}$ (§7.9)

$(\check{\delta}\, d^L)$ $d^L_{A,B,\bot} = \check{\delta}^{\leftarrow}_{A\wedge B} \circ (\mathbf{1}_A \wedge \check{\delta}^{\rightarrow}_B)$ (§7.9)

$(\hat{\delta}\, d^R)$ $d^R_{C,B,\top} = (\mathbf{1}_C \vee \hat{\delta}^{\leftarrow}_B) \circ \hat{\delta}^{\rightarrow}_{C\vee B}$ (§7.9)

$(\check{\sigma}\, d^R)$ $d^R_{\bot,B,A} = \check{\sigma}^{\leftarrow}_{B\wedge A} \circ (\check{\sigma}^{\rightarrow}_B \wedge \mathbf{1}_A)$ (§7.9)

$f \diamond g =_{df} (f \vee g) \circ m_{A,B}$, for $f\colon A \vdash D$ and $g\colon B \vdash E$ (§8.1)

$(\diamond)$ $(g_1 \diamond g_2) \circ (f_1 \wedge f_2) = (g_1 \vee g_2) \circ (f_1 \diamond f_2) = (g_1 \circ f_1) \diamond (g_2 \circ f_2)$ (§8.1)

$m_{A,B} =_{df} \mathbf{1}_A \diamond \mathbf{1}_B$ (§8.1)

(bm) $(m_{A,B} \vee \mathbf{1}_C) \circ m_{A\wedge B,C} \circ \hat{b}^{\rightarrow}_{A,B,C} = \check{b}^{\rightarrow}_{A,B,C} \circ m_{A,B\vee C} \circ (\mathbf{1}_A \wedge m_{B,C})$ (§8.2)

$((f \diamond g) \diamond h) \circ \hat{b}^{\rightarrow}_{A,B,C} = \check{b}^{\rightarrow}_{D,E,F} \circ (f \diamond (g \diamond h))$ (§8.2)

$(\hat{b}\, mL)$ $m_{A\wedge B,C} \circ \hat{b}^{\rightarrow}_{A,B,C} = d^L_{A,B,C} \circ (\mathbf{1}_A \wedge m_{B,C})$ (§8.3)

$(\check{b}\, mL)$ $\check{b}^{\rightarrow}_{A,B,C} \circ m_{A,B\vee C} = (m_{A,B} \vee \mathbf{1}_C) \circ d^L_{A,B,C}$ (§8.3)

$(\hat{b}\, mR)$ $m_{C,B\wedge A} \circ \hat{b}^{\leftarrow}_{C,B,A} = d^R_{C,B,A} \circ (m_{C,B} \wedge \mathbf{1}_A)$ (§8.3)

$(\check{b}\, mR)$ $\check{b}^{\leftarrow}_{C,B,A} \circ m_{C\vee B,A} = (\mathbf{1}_C \vee m_{B,A}) \circ d^R_{C,B,A}$ (§8.3)

(cm) $m_{B,A} \circ \hat{c}_{A,B} = \check{c}_{B,A} \circ m_{A,B}$ (§8.4)

$(g \diamond f) \circ \hat{c}_{A,B} = \check{c}_{E,D} \circ (f \diamond g)$ (§8.4)

$(\hat{b}\hat{w})$ $\hat{b}^{\rightarrow}_{A,A,A} \circ (\mathbf{1}_A \wedge \hat{w}_A) \circ \hat{w}_A = (\hat{w}_A \wedge \mathbf{1}_A) \circ \hat{w}_A$ (§9.1)

$(\check{b}\check{w})$ $\check{w}_A \circ (\mathbf{1}_A \vee \check{w}_A) \circ \check{b}^{\leftarrow}_{A,A,A} = \check{w}_A \circ (\check{w}_A \wedge \mathbf{1}_A)$

$(\hat{c}\hat{w})$ $\hat{c}_{A,A} \circ \hat{w}_A = \hat{w}_A$ (§9.1)

$(\check{c}\check{w})$ $\check{w}_A \circ \check{c}_{A,A} = \check{w}_A$

$\hat{c}^m_{A,B,C,D} =_{df} \hat{b}^{\rightarrow}_{A,C,B\wedge D} \circ (\mathbf{1}_A \wedge (\hat{b}^{\leftarrow}_{C,B,D} \circ (\hat{c}_{B,C} \wedge \mathbf{1}_D) \circ \hat{b}^{\rightarrow}_{B,C,D})) \circ \hat{b}^{\leftarrow}_{A,B,C\wedge D}\colon$
$(A \wedge B) \wedge (C \wedge D) \vdash (A \wedge C) \wedge (B \wedge D)$ (§9.1)

$(\hat{b}\,\hat{c}\,\hat{w})$ $\hat{w}_{A\wedge B} = \hat{c}^m_{A,A,B,B} \circ (\hat{w}_A \wedge \hat{w}_B)$ (§9.1)

$\check{c}^m_{A,B,C,D} =_{df} \check{b}^{\rightarrow}_{A,B,C\vee D} \circ (\mathbf{1}_A \vee (\check{b}^{\leftarrow}_{B,C,D} \circ (\check{c}_{B,C} \vee \mathbf{1}_D) \circ \check{b}^{\rightarrow}_{C,B,D})) \circ \check{b}^{\leftarrow}_{A,C,B\vee D}\colon$
$(A \vee C) \vee (B \vee D) \vdash (A \vee B) \vee (C \vee D)$

$(\check{b}\,\check{c}\,\check{w})$ $\check{w}_{A\vee B} = (\check{w}_A \vee \check{w}_B) \circ \check{c}^m_{A,A,B,B}$

$(\hat{b}\hat{k})$ $(\hat{k}^1_{A,B} \wedge \mathbf{1}_C) \circ \hat{b}^{\rightarrow}_{A,B,C} = \mathbf{1}_A \wedge \hat{k}^2_{B,C}$ (§9.1)

$(\check{b}\check{k})$ $\check{b}^{\leftarrow}_{A,B,C} \circ (\check{k}^1_{A,B} \vee \mathbf{1}_C) = \mathbf{1}_A \vee \check{k}^2_{B,C}$

$(\hat{c}\hat{k})$ $\hat{k}^2_{A,B} = \hat{k}^1_{B,A} \circ \hat{c}_{A,B}$ (§9.1)

$(\check{c}\check{k})$ $\check{k}^2_{A,B} = \check{c}_{A,B} \circ \check{k}^1_{B,A}$

$(\hat{w}\hat{k})$ $\hat{k}^i_{A,A} \circ \hat{w}_A = \mathbf{1}_A,$ for $i \in \{1,2\}$ (§9.1)

$(\check{w}\check{k})$ $\check{w}_A \circ \check{k}^i_{A,A} = \mathbf{1}_A,$ for $i \in \{1,2\}$

$(\hat{b}\hat{k}1)$ $\hat{k}^1_{A\wedge B,C} = (\mathbf{1}_A \wedge \hat{k}^1_{B,C}) \circ \hat{b}^{\leftarrow}_{A,B,C}$ (§9.1)

$(\check{b}\check{k}1)$ $\check{k}^1_{A\vee B,C} = \check{b}^{\rightarrow}_{A,B,C} \circ (\mathbf{1}_A \vee \check{k}^1_{B,C})$

$(\hat{b}\hat{k}2)$ $\hat{k}^2_{C,B\wedge A} = (\hat{k}^2_{C,B} \wedge \mathbf{1}_A) \circ \hat{b}^{\rightarrow}_{C,B,A}$ (§9.1)

$(\check{b}\check{k}2)$ $\check{k}^2_{C,B\vee A} = \check{b}^{\leftarrow}_{C,B,A} \circ (\check{k}^2_{C,B} \vee \mathbf{1}_A)$

$(\hat{w}\hat{k}\hat{k})$ $(\hat{k}^1_{A,A} \wedge \hat{k}^2_{A,A}) \circ \hat{w}_{A\wedge A} = \mathbf{1}_{A\wedge A}$ (§9.1)

$(\check{w}\check{k}\check{k})$ $\check{w}_{A\vee A} \circ (\check{k}^1_{A,A} \vee \check{k}^2_{A,A}) = \mathbf{1}_{A\vee A}$

$(\langle -,- \rangle)$ $\langle f_1, f_2\rangle =_{df} (f_1 \wedge f_2) \circ \hat{w}_C$, for $f_1 : C \vdash A_1$ and $f_2 : C \vdash A_2$ (§9.1)

$[g_1, g_2] =_{df} \check{w}_C \circ (g_1 \vee g_2)$, for $g_1 : A_1 \vdash C$ and $g_2 : A_2 \vdash C$ (§9.4)

$(\wedge)$ $f \wedge g = \langle f \circ \hat{k}^1_{A,B}, g \circ \hat{k}^2_{A,B}\rangle$, for $f : A \vdash D$ and $g : B \vdash E$ (§9.1)

$(\vee)$ $f \vee g = [\check{k}^1_{D,E} \circ f, \check{k}^2_{D,E} \circ g]$, for $f : A \vdash D$ and $g : B \vdash E$

$(\hat{b}^{\rightarrow})$ $\hat{b}^{\rightarrow}_{A,B,C} = \langle \mathbf{1}_A \wedge \hat{k}^1_{B,C}, \hat{k}^2_{B,C} \circ \hat{k}^2_{A,B\wedge C}\rangle$ (§9.1)

$(\check{b}^{\leftarrow})$ $\check{b}^{\leftarrow}_{A,B,C} = [\mathbf{1}_A \vee \check{k}^1_{B,C}, \check{k}^2_{A,B\vee C} \circ \check{k}^2_{B,C}]$

$(\hat{b}^{\leftarrow})$ $\hat{b}^{\leftarrow}_{C,B,A} = \langle \hat{k}^1_{C,B} \circ \hat{k}^1_{C\wedge B,A}, \hat{k}^2_{C,B} \wedge \mathbf{1}_A\rangle$ (§9.1)

$(\check{b}^{\rightarrow})$ $\check{b}^{\rightarrow}_{C,B,A} = [\check{k}^1_{C\vee B,A} \circ \check{k}^1_{C,B}, \check{k}^2_{C,B} \vee \mathbf{1}_A]$

$(\hat{c})$ $\hat{c}_{A,B} = \langle \hat{k}^2_{A,B}, \hat{k}^1_{A,B}\rangle$ (§9.1)

$(\check{c})$ $\check{c}_{A,B} = [\check{k}^2_{A,B}, \check{k}^1_{A,B}]$

$(\hat{w})$ $\hat{w}_A = \langle \mathbf{1}_A, \mathbf{1}_A\rangle$ (§9.1)

$(\check{w})$ $\check{w}_A = [\mathbf{1}_A, \mathbf{1}_A]$

$(\wedge\beta)$ $\hat{k}^i_{A_1,A_2} \circ \langle f_1, f_2\rangle = f_i$, for $f_i : C \vdash A_i$ and $i \in \{1, 2\}$ (§9.1)

$(\vee\beta)$ $[g_1, g_2] \circ \check{k}^i_{A_1,A_2} = g_i$, for $g_i : A_i \vdash C$ and $i \in \{1, 2\}$

$(\wedge\eta)$ $\langle \hat{k}^1_{A_1,A_2} \circ h, \hat{k}^2_{A_1,A_2} \circ h\rangle = h$, for $h : C \vdash A_1 \wedge A_2$ (§9.1)

$(\vee\eta)$ $[h \circ \check{k}^1_{A_1,A_2}, h \circ \check{k}^2_{A_1,A_2}] = h$, for $h : A_1 \vee A_2 \vdash C$

$\hat{K}^1_{A_2} g_1 =_{df} g_1 \circ \hat{k}^1_{A_1,A_2}$, for $g_1 : A_1 \vdash C$ (§9.1)

$\check{K}^1_{A_2} f_1 =_{df} \check{k}^1_{A_1,A_2} \circ f_1$, for $f_1 : C \vdash A_1$ (§9.4)

$\hat{K}^2_{A_1}\, g_2 =_{df} g_2 \circ \hat{k}^2_{A_1,A_2}$, for $g_2\colon A_2 \vdash C$ (§9.1)

$\check{K}^2_{A_1}\, f_2 =_{df} \check{k}^2_{A_1,A_2} \circ f_2$, for $f_2\colon C \vdash A_2$ (§9.4)

for $i \in \{1,2\}$, and f, g, f_i and g_i of appropriate types,

$(\hat{K}1)$ $g \circ \hat{K}^i_A f = \hat{K}^i_A (g \circ f)$ (§9.1)

$(\check{K}1)$ $\check{K}^i_A g \circ f = \check{K}^i_A (g \circ f)$ (§9.4)

$(\hat{K}2)$ $\hat{K}^i_A g \circ \langle f_1, f_2\rangle = g \circ f_i$ (§9.1)

$(\check{K}2)$ $[g_1, g_2] \circ \check{K}^i_A f = g_i \circ f$ (§9.4)

$(\hat{K}3)$ $\langle g_1, g_2\rangle \circ f = \langle g_1 \circ f, g_2 \circ f\rangle$ (§9.1)

$(\check{K}3)$ $g \circ [f_1, f_2] = [g \circ f_1, g \circ f_2]$ (§9.4)

$(\hat{K}4)$ $\mathbf{1}_{A\wedge B} = \langle \hat{K}^1_B \mathbf{1}_A, \hat{K}^2_A \mathbf{1}_B\rangle$ (§9.1)

$(\check{K}4)$ $\mathbf{1}_{A\vee B} = [\check{K}^1_B \mathbf{1}_A, \check{K}^2_A \mathbf{1}_B]$ (§9.4)

$(\hat{K}5)$ $\hat{K}^i_A \langle f, g\rangle = \langle \hat{K}^i_A f, \hat{K}^i_A g\rangle$ (§9.1)

$(\check{K}5)$ $\check{K}^i_A [f, g] = [\check{K}^i_A f, \check{K}^i_A g]$

$\hat{k}^1_{A_1,A_2} =_{df} \hat{K}^1_{A_2} \mathbf{1}_{A_1}$ (§9.1)

$\check{k}^1_{A_1,A_2} =_{df} \check{K}^1_{A_2} \mathbf{1}_{A_1}$

$\hat{k}^2_{A_1,A_2} =_{df} \hat{K}^2_{A_1} \mathbf{1}_{A_2}$ (§9.1)

$\check{k}^2_{A_1,A_2} =_{df} \check{K}^2_{A_1} \mathbf{1}_{A_2}$

$(\hat{k}\hat{\delta})$ $\hat{k}^1_{A,\top} = \hat{\delta}^{\rightarrow}_A$ (§9.2)

$(\check{k}\check{\delta})$ $\check{k}^1_{A,\bot} = \check{\delta}^{\leftarrow}_A$

$(\hat{w}\hat{\delta})$ $\hat{w}_\top = \hat{\delta}^{\leftarrow}_\top$ (§9.2)

$(\check{w}\check{\delta})$ $\check{w}_\bot = \check{\delta}^{\rightarrow}_\bot$

$\hat{\kappa}_A =_{df} \hat{k}^2_{A,\top} \circ \hat{\delta}^{\leftarrow}_A$ (§9.2)

$\check{\kappa}_A =_{df} \check{\delta}^{\rightarrow}_A \circ \hat{k}^2_{A,\bot}$ (§9.6)

$(\hat{\kappa}\mathbf{1})$ $\hat{\kappa}_\top = \mathbf{1}_\top$ (§9.2)

$(\check{\kappa}\mathbf{1})$ $\check{\kappa}_\bot = \mathbf{1}_\bot$ (§9.6)

$(\hat{\kappa})$ $\hat{\kappa}_A = f, \quad \text{for } f\colon A \vdash \top$ (§9.2)

$(\check{\kappa})$ $\check{\kappa}_A = f, \quad \text{for } f\colon \bot \vdash A$ (§9.6)

$\hat{\delta}^{\leftarrow}_{A} = \langle \mathbf{1}_A, \hat{\kappa}_A \rangle$ (§9.2)

$\hat{\sigma}^{\leftarrow}_{A} = \langle \hat{\kappa}_A, \mathbf{1}_A \rangle$ (§9.2)

$\hat{k}^1_{A,B} = \hat{\delta}^{\rightarrow}_{A} \circ (\mathbf{1}_A \wedge \hat{\kappa}_B)$ (§9.2)

$\hat{k}^2_{A,B} = \hat{\sigma}^{\rightarrow}_{B} \circ (\hat{\kappa}_A \wedge \mathbf{1}_B)$ (§9.2)

$(\hat{k}\hat{k})$ $\hat{k}^1_{p,p} = \hat{k}^2_{p,p}$ (§9.3)

$(\check{k}\check{k})$ $\check{k}^1_{p,p} = \check{k}^2_{p,p}$ (§9.5)

(*in-out*) $\langle [f,g],[h,j] \rangle = [\langle f,h \rangle, \langle g,j \rangle]$ (§9.4)

$c^k_{A,B,C,D} =_{df} \langle \hat{k}^1_{A,B} \vee \hat{k}^1_{C,D}, \hat{k}^2_{A,B} \vee \hat{k}^2_{C,D} \rangle\colon$
$(A \wedge B) \vee (C \wedge D) \vdash (A \vee C) \wedge (B \vee D)$ (§9.4)

$c^k_{A,B,C,D} = [\check{k}^1_{A,C} \wedge \check{k}^1_{B,D}, \check{k}^2_{A,C} \wedge \check{k}^2_{B,D}]$ (§9.4)

$\hat{w}_{A \vee B} = c^k_{A,A,B,B} \circ (\hat{w}_A \vee \hat{w}_B)$ (§9.4)

$\check{w}_{A \wedge B} = (\check{w}_A \wedge \check{w}_B) \circ c^k_{A,B,A,B}$ (§9.4)

$\hat{c}^m_{A,B,C,D} = \langle \hat{k}^1_{A,B} \wedge \hat{k}^1_{C,D}, \hat{k}^2_{A,B} \wedge \hat{k}^2_{C,D} \rangle$ (§9.4)

$\check{c}^m_{D,C,B,A} = [\check{k}^1_{D,C} \vee \check{k}^1_{B,A}, \check{k}^2_{D,C} \vee \check{k}^2_{B,A}]$ (§9.4)

$(\hat{\bot})$ $\hat{c}_{\bot,\bot} = \mathbf{1}_{\bot \wedge \bot}$ (§9.6)

$(\check{\top})$ $\check{c}_{\top,\top} = \mathbf{1}_{\top \vee \top}$ (§9.6)

$(\hat{k}\,\bot)$ $\hat{k}^1_{\bot,\bot} = \hat{k}^2_{\bot,\bot}$ (§9.6)

$(\check{k}\,\top)$ $\check{k}^1_{\top,\top} = \check{k}^2_{\top,\top}$ (§9.6)

$(\hat{K}\bot)$ $\hat{K}^1_{\bot,\bot}\,\mathbf{1}_\bot = \hat{K}^2_{\bot,\bot}\,\mathbf{1}_\bot$ (§9.6)

$(\check{K}\top)$ $\check{K}^1_{\top,\top}\,\mathbf{1}_\top = \check{K}^2_{\top,\top}\,\mathbf{1}_\top$ (§9.6)

$(\hat{k}\check{k})$ $\check{k}^1_{p,\top} \circ \hat{k}^1_{p,\bot} = \check{k}^2_{p,\top} \circ \mathbf{0}_{\bot,\top} \circ \hat{k}^2_{p,\bot}, \quad \text{for } \mathbf{0}_{\bot,\top} = \hat{\kappa}_\bot = \check{\kappa}_\top$ (§9.7)

$(\hat{k}\check{\kappa})$ $\hat{k}^1_{p,\bot} = \check{\kappa}_p \circ \hat{k}^2_{p,\bot}$ (§9.7)

$(\check{k}\hat{\kappa})$ $\check{k}^1_{p,\top} = \check{k}^2_{p,\top} \circ \hat{\kappa}_p$ (§9.7)

$(\hat{k}\check{k}\ fg)$ $\check{k}^1_{b,\top}\circ f\circ\hat{k}^1_{a,\bot} = \check{k}^1_{b,\top}\circ g\circ\hat{k}^1_{a,\bot}$, for $f,g\colon a\vdash b$ (§9.7)

(wm) $\check{w}_A\circ m_{A,A}\circ\hat{w}_A = \mathbf{1}_A$ (§10.1)

$f\cup g =_{df} \check{w}_B\circ(f\Diamond g)\circ\hat{w}_A$, for $f,g\colon A\vdash B$ (§10.1)

$m_{A,B} =_{df} \check{K}^1_B\hat{K}^1_B\mathbf{1}_A\cup\check{K}^2_A\hat{K}^2_A\mathbf{1}_B$ (§10.1)

$(\cup\circ)$ $(f\cup g)\circ h = (f\circ h)\cup(g\circ h),\quad h\circ(f\cup g) = (h\circ f)\cup(h\circ g)$ (§10.1)

$(\cup\ assoc)$ $f\cup(g\cup h) = (f\cup g)\cup h$ (§10.1)

$(\cup\ com)$ $f\cup g = g\cup f$ (§10.1)

$(\cup\ idemp)$ $f\cup f = f$ (§10.1)

$(\cup\wedge)$ $(f_1\cup f_2)\wedge(g_1\cup g_2) = (f_1\wedge g_1)\cup(f_2\wedge g_2)$ (§10.1)

$(\cup\vee)$ $(f_1\cup f_2)\vee(g_1\cup g_2) = (f_1\vee g_1)\cup(f_2\vee g_2)$ (§10.1)

$c^k_{A,C,B,D}\circ m_{A\wedge C,B\wedge D}\circ\hat{c}^m_{A,B,C,D} = m_{A,B}\wedge m_{C,D}$ (§10.1)

$\check{c}^m_{A,B,C,D}\circ m_{A\vee C,B\vee D}\circ c^k_{A,B,C,D} = m_{A,B}\vee m_{C,D}$ (§10.1)

$(m\top)$ $m_{A,\top} = \check{k}^1_{A,\top}\circ\hat{k}^1_{A,\top}$ (§10.3)

$(m\bot)$ $m_{A,\bot} = \check{k}^1_{A,\bot}\circ\hat{k}^1_{A,\bot}$ (§10.3)

$m_{A,C} = \check{k}^1_{A,C}\circ\hat{k}^1_{A,C}$, for letterless C (§10.3)

$m_{C,A} = \check{k}^2_{C,A}\circ\hat{k}^2_{C,A}$, for letterless C (§10.3)

$(\cup\mathbf{0}\top)$ $\mathbf{1}_{A\vee\top}\cup\check{K}^2_A\hat{\kappa}_{A\vee\top} = \mathbf{1}_{A\vee\top}$ (§10.3)

$(\cup\mathbf{0}\bot)$ $\mathbf{1}_{A\wedge\bot}\cup\hat{K}^2_A\check{\kappa}_{A\wedge\bot} = \mathbf{1}_{A\wedge\bot}$ (§10.3)

$(\cup\mathbf{0}g)$ $f\cup g = f$, for a null term g (§10.3)

$(d\hat{k})$ $\hat{k}^2_{A,B\vee C} = (\hat{k}^2_{A,B}\vee\mathbf{1}_C)\circ d_{A,B,C}$ (§11.1)

$(d\check{k})$ $\check{k}^1_{C\wedge B,A} = d_{C,B,A}\circ(\mathbf{1}_C\wedge\check{k}^1_{B,A})$ (§11.1)

(dm) $m_{A,C} = (\hat{k}^1_{A,B}\vee\mathbf{1}_C)\circ d_{A,B,C}\circ(\mathbf{1}_A\wedge\check{k}^2_{B,C})$ (§11.1)

$m_{A,C} = (\mathbf{1}_A\vee\hat{k}^2_{B,C})\circ d^R_{A,B,C}\circ(\check{k}^1_{A,B}\wedge\mathbf{1}_C)$ (§11.1)

$(m\,\hat{e})$ $c^k_{A,C,B,D}\circ\hat{e}'_{A,B,C,D} = m_{A,B}\wedge\mathbf{1}_{C\vee D}$ (§11.1)

$(m\,\check{e})$ $\check{e}'_{D,C,B,A}\circ c^k_{D,C,B,A} = \mathbf{1}_{D\wedge C}\vee m_{B,A}$ (§11.1)

$$(\hat{k}^1_{A,C} \vee \hat{k}^1_{B,D}) \circ \hat{e}'_{A,B,C,D} = m_{A,B} \circ \hat{k}^1_{A\wedge B,C\vee D} \qquad (\S 11.1)$$

$$\check{e}'_{D,C,B,A} \circ (\check{k}^2_{D,B} \wedge \check{k}^2_{C,A}) = \check{k}^2_{D\wedge C,B\vee A} \circ m_{B,A} \qquad (\S 11.1)$$

$$(m\,\hat{c}^m) \quad m_{A\wedge C,B\wedge D} \circ \hat{c}^m_{A,B,C,D} = \hat{e}'_{A,B,C,D} \circ (\mathbf{1}_{A\wedge B} \wedge m_{C,D}) \qquad (\S 11.1)$$

$$(m\,\check{c}^m) \quad \check{c}^m_{D,C,B,A} \circ m_{D\vee B,C\vee A} = (m_{D,C} \vee \mathbf{1}_{B\vee A}) \circ \check{e}'_{D,C,B,A} \qquad (\S 11.1)$$

$$\hat{s}_{A,C,D} =_{df} \hat{e}'_{A,A,C,D} \circ (\hat{w}_A \wedge \mathbf{1}_{C\vee D}) \colon\ A \wedge (C \vee D) \vdash (A \wedge C) \vee (A \wedge D) \qquad (\S 11.3)$$

$$\check{s}_{D,C,A} =_{df} (\mathbf{1}_{D\wedge C} \vee \check{w}_A) \circ \check{e}'_{D,C,A,A} \colon\ (D \vee A) \wedge (C \vee A) \vdash (D \wedge C) \vee A \qquad (\S 11.3)$$

$$\hat{t}_{A,C,D} =_{df} (\check{w}_A \wedge \mathbf{1}_{A\vee D}) \circ c^k_{A,C,A,D} \colon\ (A \wedge C) \vee (A \wedge D) \vdash A \wedge (C \vee D) \qquad (\S 11.3)$$

$$\check{t}_{D,C,A} =_{df} c^k_{D,C,A,A} \circ (\mathbf{1}_{D\wedge C} \vee \hat{w}_A) \colon\ (D \wedge C) \vee A \vdash (D \vee A) \wedge (C \vee A) \qquad (\S 11.3)$$

$$\hat{t}_{A,C,D} \circ \hat{s}_{A,C,D} = \mathbf{1}_{A\wedge(C\vee D)} \qquad (\S 11.3)$$

$$\check{s}_{D,C,A} \circ \check{t}_{D,C,A} = \mathbf{1}_{(D\wedge C)\vee A} \qquad (\S 11.3)$$

$$d_{A,B,C} = (\mathbf{1}_{A\wedge B} \vee \hat{k}^2_{A,C}) \circ \hat{e}'_{A,A,B,C} \circ (\hat{w}_A \wedge \mathbf{1}_{B\vee C}) \qquad (\S 11.3)$$

$$d_{C,B,A} = (\mathbf{1}_{C\wedge B} \vee \check{w}_A) \circ \check{e}'_{C,B,A,A} \circ (\check{k}^1_{C,A} \wedge \mathbf{1}_{B\vee A}) \qquad (\S 11.3)$$

$$(d\top\top) \quad d_{A,\top,\top} = \check{k}^1_{A\wedge\top,\top} \circ (\mathbf{1}_A \wedge \hat{\kappa}_{\top\vee\top}) \qquad (\S 11.3)$$

$$(d\bot\bot) \quad d_{\bot,\bot,C} = (\check{\kappa}_{\bot\wedge\bot} \vee \mathbf{1}_C) \circ \hat{k}^2_{\bot,\bot\vee C} \qquad (\S 11.3)$$

$$(m^{-1}0) \quad \hat{k}^2_{A,B} \circ m^{-1}_{A,B} \circ \check{k}^1_{A,B} = \hat{k}^1_{B,A} \circ m^{-1}_{B,A} \circ \check{k}^2_{B,A} \qquad (\S 12.1)$$

$$(m^{-1}1) \quad \hat{k}^1_{A,B} \circ m^{-1}_{A,B} \circ \check{k}^1_{A,B} = \hat{k}^2_{B,A} \circ m^{-1}_{B,A} \circ \check{k}^2_{B,A} = \mathbf{1}_A \qquad (\S 12.1)$$

$$(bm^{-1}) \quad (m^{-1}_{A,B} \wedge \mathbf{1}_C) \circ m^{-1}_{A\vee B,C} \circ \check{b}^{\rightarrow}_{A,B,C} = \hat{b}^{\rightarrow}_{A,B,C} \circ m^{-1}_{A,B\wedge C} \circ (\mathbf{1}_A \vee m^{-1}_{B,C}) \qquad (\S 12.1)$$

$$(cm^{-1}) \quad m^{-1}_{B,A} \circ \check{c}_{B,A} = \hat{c}_{A,B} \circ m^{-1}_{A,B} \qquad (\S 12.1)$$

$$\mathbf{0}_{A,B} =_{df} \hat{k}^2_{A,B} \circ m^{-1}_{A,B} \circ \check{k}^1_{A,B} = \hat{k}^1_{B,A} \circ m^{-1}_{B,A} \circ \check{k}^2_{B,A} \qquad (\S 12.1)$$

$$m^{-1}_{A,B} =_{df} \langle [\mathbf{1}_A, \mathbf{0}_{B,A}], [\mathbf{0}_{A,B}, \mathbf{1}_B] \rangle = [\langle \mathbf{1}_A, \mathbf{0}_{A,B} \rangle, \langle \mathbf{0}_{B,A}, \mathbf{1}_B \rangle] \qquad (\S 12.1)$$

$$f \circ \mathbf{0}_{A,A} = \mathbf{0}_{B,B} \circ f = \mathbf{0}_{A,B}, \quad \text{for } f \colon A \vdash B \qquad (\S 12.1)$$

(**0**) $f \circ \mathbf{0}_{C,A} = \mathbf{0}_{C,B}, \quad \mathbf{0}_{B,C} \circ f = \mathbf{0}_{A,C}, \quad \text{for } f : A \vdash B$ (§12.1)

(**0** *nat*) $f \circ \mathbf{0}_{A,A} = \mathbf{0}_{B,B} \circ f, \quad \text{for } f : A \vdash B$ (§12.1)

(**0**$\wedge$) $\mathbf{0}_{A,C} \wedge \mathbf{0}_{B,D} = \mathbf{0}_{A \wedge B, C \wedge D}$ (§12.1)

(**0**$\vee$) $\mathbf{0}_{A,C} \vee \mathbf{0}_{B,D} = \mathbf{0}_{A \vee B, C \vee D}$ (§12.1)

$$d^{-1}_{A,B,C} =_{df} \langle \mathbf{1}_A \vee \hat{k}^1_{B,C}, [\mathbf{0}_{A,C}, \hat{k}^2_{B,C}] \rangle = [\langle \check{k}^1_{A,B}, \mathbf{0}_{A,C} \rangle, \check{k}^2_{A,B} \wedge \mathbf{1}_C] : A \vee (B \wedge C) \vdash (A \vee B) \wedge C$$ (§12.1)

$$\mathbf{0}_{A,C} =_{df} \hat{k}^2_{A \vee B, C} \circ d^{-1}_{A,B,C} \circ \check{k}^1_{A, B \wedge C}$$ (§12.1)

(d^{-1} 1) $\hat{k}^1_{A \vee B, C} \circ d^{-1}_{A,B,C} \circ \check{k}^1_{A, B \wedge C} = \check{k}^1_{A,B}$ (§12.1)

(d^{-1} 2) $\hat{k}^2_{A \vee B, C} \circ d^{-1}_{A,B,C} \circ \check{k}^2_{A, B \wedge C} = \hat{k}^2_{B,C}$ (§12.1)

(d^{-1} 3) $\hat{k}^1_{A \vee B, C} \circ d^{-1}_{A,B,C} \circ \check{k}^2_{A, B \wedge C} = \check{k}^2_{A,B} \circ \hat{k}^1_{B,C}$ (§12.1)

$\hat{\kappa}_A = f = \mathbf{0}_{A,\top}, \quad \text{for } f : A \vdash \top$ (§12.1)

$\check{\kappa}_A = f = \mathbf{0}_{\bot,A}, \quad \text{for } f : \bot \vdash A$ (§12.1)

$\hat{\kappa}_\bot = \check{\kappa}_\top = \mathbf{0}_{\bot,\top}$ (§12.1)

$\hat{\kappa}_\top = \mathbf{1}_\top = \mathbf{0}_{\top,\top}$ (§12.1)

$\check{\kappa}_\bot = \mathbf{1}_\bot = \mathbf{0}_{\bot,\bot}$ (§12.1)

$\hat{k}^1_{\bot,\bot} = \hat{k}^2_{\bot,\bot} = \mathbf{0}_{\bot \wedge \bot, \bot}$ (§12.1)

$\check{k}^1_{\top,\top} = \check{k}^2_{\top,\top} = \mathbf{0}_{\top, \top \vee \top}$ (§12.1)

(**0**$\top\bot$) $\mathbf{0}_{A,B} = \check{\kappa}_B \circ \mathbf{0}_{\top,\bot} \circ \hat{\kappa}_A$ (§12.1)

(**0**I) $f \circ \mathbf{0}_A = \mathbf{0}_B \circ g, \quad \text{for } f, g : A \vdash B$ (§12.5)

$f \circ \mathbf{0}_A = g \circ \mathbf{0}_A, \quad \text{for } f, g : A \vdash B$ (§12.5)

$\mathbf{0}_B \circ f = \mathbf{0}_B \circ g, \quad \text{for } f, g : A \vdash B$ (§12.5)

(**00**) $\mathbf{0}_A \circ \mathbf{0}_A = \mathbf{0}_A$ (§12.5)

$\mathbf{0}_{A,B} =_{df} f \circ \mathbf{0}_A, \quad \text{for } f : A \vdash B$ (§12.5)

(**0**I$\wedge$) $\mathbf{0}_A \wedge \mathbf{0}_B = \mathbf{0}_{A \wedge B}$ (§12.5)

(**0**I$\vee$) $\mathbf{0}_A \vee \mathbf{0}_B = \mathbf{0}_{A \vee B}$ (§12.5)

($\cup$**0**) $f \cup \mathbf{0}_{A,B} = f, \quad \text{for } f : A \vdash B$ (§12.5)

$$\mathbf{1}_A \cup \mathbf{0}_A = \mathbf{1}_A \qquad (\S 12.5)$$

(mm^{-1}) $$m^{-1}_{A,B} \circ m_{A,B} = \mathbf{1}_{A\wedge B}, \quad m_{A,B} \circ m^{-1}_{A,B} = \mathbf{1}_{A\vee B} \qquad (\S 13.1)$$

$(\hat{k}\,m)$ $$\hat{k}^1_{A,B} = [\mathbf{1}_A, \mathbf{0}_{B,A}] \circ m_{A,B}, \quad \hat{k}^2_{A,B} = [\mathbf{0}_{A,B}, \mathbf{1}_B] \circ m_{A,B} \qquad (\S 13.1)$$

$(\check{k}\,m)$ $$\check{k}^1_{A,B} = m_{A,B} \circ \langle \mathbf{1}_A, \mathbf{0}_{A,B}\rangle, \quad \check{k}^2_{A,B} = m_{A,B} \circ \langle \mathbf{0}_{B,A}, \mathbf{1}_B\rangle \qquad (\S 13.1)$$

for $f\colon A \vdash B$,

$$\hat{Z}^1_C f =_{df} m^{-1}_{B,C} \circ \check{k}^1_{B,C} \circ f = \langle f, \mathbf{0}_{A,C}\rangle : A \vdash B \wedge C \qquad (\S 13.1)$$

$$\check{Z}^1_C f =_{df} f \circ \hat{k}^1_{A,C} \circ m^{-1}_{A,C} = [f, \mathbf{0}_{C,B}] : A \vee C \vdash B \qquad (\S 13.1)$$

$$\hat{Z}^2_C f =_{df} m^{-1}_{C,B} \circ \check{k}^2_{C,B} \circ f = \langle \mathbf{0}_{A,C}, f\rangle : A \vdash C \wedge B \qquad (\S 13.1)$$

$$\check{Z}^2_C f =_{df} f \circ \hat{k}^2_{C,A} \circ m^{-1}_{C,A} = [\mathbf{0}_{C,B}, f] : C \vee A \vdash B \qquad (\S 13.1)$$

$(\hat{Z})$ $$\langle f_1, f_2\rangle = \hat{Z}^1_{A_2} f_1 \cup \hat{Z}^2_{A_1} f_2, \text{ for } f_1\colon C \vdash A_1 \text{ and } f_2\colon C \vdash A_2 \qquad (\S 13.1)$$

$(\check{Z})$ $$[g_1, g_2] = \check{Z}^1_{A_2} g_1 \cup \check{Z}^2_{A_1} g_2, \text{ for } g_1\colon A_1 \vdash C \text{ and } g_2\colon A_2 \vdash C \qquad (\S 13.1)$$

$$\hat{w}_C = m^{-1}_{C,C} \circ (\check{k}^1_{C,C} \cup \check{k}^2_{C,C}) \qquad (\S 13.1)$$

$$\check{w}_C = (\hat{k}^1_{C,C} \cup \hat{k}^2_{C,C}) \circ m^{-1}_{C,C} \qquad (\S 13.1)$$

$$(f_1 \cup f_2) \cup (f_3 \cup f_4) = (f_1 \cup f_3) \cup (f_2 \cup f_4) \qquad (\S 13.1)$$

(md) $$d_{A,B,C} =_{df} m_{A\wedge B,C} \circ \hat{b}^{\rightarrow}_{A,B,C} \circ (\mathbf{1}_A \wedge m^{-1}_{B,C}) \qquad (\S 13.2)$$

$$\hat{e}'_{A,B,C,D} = m_{A\wedge C,B\wedge D} \circ \hat{c}^m_{A,B,C,D} \circ (\mathbf{1}_{A\wedge B} \wedge m^{-1}_{C,D}) \qquad (\S 13.2)$$

$$\check{e}'_{D,C,B,A} = (m^{-1}_{D,C} \vee \mathbf{1}_{B\vee A}) \circ \check{c}^m_{D,C,B,A} \circ m_{D\vee B,C\vee A} \qquad (\S 13.2)$$

$$c^l_{A,B,C,D} =_{df} \hat{e}'_{A,B,C,D} \circ (m^{-1}_{A,B} \wedge \mathbf{1}_{C\vee D}) : (A \vee B) \wedge (C \vee D) \vdash (A \wedge C) \vee (B \wedge D) \qquad (\S 13.2)$$

$$c^l_{D,B,C,A} =_{df} (\mathbf{1}_{D\wedge C} \vee m^{-1}_{B,A}) \circ \check{e}'_{D,C,B,A} \qquad (\S 13.2)$$

$$f \cup g = [\hat{k}^1_{B,A}, \hat{k}^2_{A,B}] \circ c^l_{B,A,A,B} \circ \langle \check{k}^1_{B,A} \circ f, \check{k}^2_{A,B} \circ g\rangle \qquad (\S 13.2)$$

$$\mathbf{0}_{A,B} = [\hat{k}^1_{B,A}, \hat{k}^2_{A,B}] \circ c^l_{B,A,A,B} \circ \langle \check{k}^2_{B,A}, \check{k}^1_{A,B}\rangle \qquad (\S 13.2)$$

$$c^k_{A,C,B,D} \circ c^l_{A,B,C,D} = \mathbf{1}_{(A\vee B)\wedge(C\vee D)} \qquad (\S 13.2)$$

$$c^l_{A,C,B,D} \circ c^k_{A,B,C,D} = \mathbf{1}_{(A\wedge B)\vee(C\wedge D)} \qquad (\S 13.2)$$

$$n^{\leftarrow}_A \circ n^{\rightarrow}_A = \mathbf{1}_{\neg\neg A} \qquad (\S 14.1)$$

$$n^{\rightarrow}_A \circ n^{\leftarrow}_A = \mathbf{1}_A \qquad (\S 14.1)$$

$$\hat{r}^{\leftarrow}_{A,B} \circ \hat{r}^{\rightarrow}_{A,B} = \mathbf{1}_{\neg(A \wedge B)} \qquad (\S 14.1)$$

$$\hat{r}^{\rightarrow}_{A,B} \circ \hat{r}^{\leftarrow}_{A,B} = \mathbf{1}_{\neg A \vee \neg B} \qquad (\S 14.1)$$

$$\check{r}^{\leftarrow}_{A,B} \circ \check{r}^{\rightarrow}_{A,B} = \mathbf{1}_{\neg(A \vee B)} \qquad (\S 14.1)$$

$$\check{r}^{\rightarrow}_{A,B} \circ \check{r}^{\leftarrow}_{A,B} = \mathbf{1}_{\neg A \wedge \neg B} \qquad (\S 14.1)$$

$$\hat{\rho}^{\leftarrow} \circ \hat{\rho}^{\rightarrow} = \mathbf{1}_{\neg\top} \qquad (\S 14.1)$$

$$\hat{\rho}^{\rightarrow} \circ \hat{\rho}^{\leftarrow} = \mathbf{1}_{\bot} \qquad (\S 14.1)$$

$$\check{\rho}^{\leftarrow} \circ \check{\rho}^{\rightarrow} = \mathbf{1}_{\neg\bot} \qquad (\S 14.1)$$

$$\check{\rho}^{\rightarrow} \circ \check{\rho}^{\leftarrow} = \mathbf{1}_{\top} \qquad (\S 14.1)$$

$(\mathbf{0}\eta\varepsilon)$ $$\mathbf{0}_A = \check{\sigma}^{\rightarrow}_A \circ (\varepsilon_A \vee \mathbf{1}_A) \circ d_{A,\neg A,A} \circ (\mathbf{1}_A \wedge \eta_A) \circ \hat{\delta}^{\leftarrow}_A \qquad (\S 14.2)$$

$(\eta\wedge)$ $$\eta_{B \wedge A} = (\hat{r}^{\leftarrow}_{B,A} \vee \mathbf{1}_{B \wedge A}) \circ \check{c}_{\neg B \vee \neg A, B \wedge A} \circ \check{e}'_{B,A,\neg B,\neg A} \circ \\ \circ (\check{c}_{B,\neg B} \wedge \check{c}_{A,\neg A}) \circ (\eta_B \wedge \eta_A) \circ \hat{\delta}^{\leftarrow}_{\top} \qquad (\S 14.2)$$

$(\eta\vee)$ $$\eta_{B \vee A} = (\check{r}^{\leftarrow}_{B,A} \vee \mathbf{1}_{B \vee A}) \circ \check{e}'_{\neg B,\neg A,B,A} \circ (\eta_B \wedge \eta_A) \circ \hat{\delta}^{\leftarrow}_{\top} \qquad (\S 14.2)$$

$(\varepsilon\wedge)$ $$\varepsilon_{A \wedge B} = \check{\delta}^{\rightarrow}_{\bot} \circ (\varepsilon_A \vee \varepsilon_B) \circ \hat{e}'_{A,B,\neg A,\neg B} \circ (\mathbf{1}_{A \wedge B} \wedge \hat{r}^{\rightarrow}_{A,B}) \qquad (\S 14.2)$$

$(\varepsilon\vee)$ $$\varepsilon_{A \vee B} = \check{\delta}^{\rightarrow}_{\bot} \circ (\varepsilon_A \vee \varepsilon_B) \circ (\hat{c}_{\neg A,A} \vee \hat{c}_{\neg B,B}) \circ \hat{e}'_{\neg A,\neg B,A,B} \circ \\ \circ \hat{c}_{A \vee B,\neg A \wedge \neg B} \circ (\mathbf{1}_{A \vee B} \wedge \check{r}^{\rightarrow}_{A,B}) \qquad (\S 14.2)$$

$(\eta\top)$ $$\eta_{\top} = (\hat{\rho}^{\leftarrow} \vee \mathbf{1}_{\top}) \circ \check{\sigma}^{\leftarrow}_{\top} \qquad (\S 14.2)$$

$(\eta\bot)$ $$\eta_{\bot} = (\check{\rho}^{\leftarrow} \vee \mathbf{1}_{\bot}) \circ \check{\delta}^{\leftarrow}_{\top} \qquad (\S 14.2)$$

$(\varepsilon\top)$ $$\varepsilon_{\top} = \hat{\sigma}^{\rightarrow}_{\bot} \circ (\mathbf{1}_{\top} \wedge \hat{\rho}^{\rightarrow}) \qquad (\S 14.2)$$

$(\varepsilon\bot)$ $$\varepsilon_{\bot} = \hat{\delta}^{\rightarrow}_{\bot} \circ (\mathbf{1}_{\bot} \wedge \check{\rho}^{\rightarrow}) \qquad (\S 14.2)$$

$$(\mathbf{1}_{\neg A} \vee f) \circ \eta_A = (\neg f \vee \mathbf{1}_B) \circ \eta_B, \quad \text{for } f \colon A \vdash B \qquad (\S 14.2)$$

$$\varepsilon_A \circ (\mathbf{1}_A \wedge \neg f) = \varepsilon_B \circ (f \wedge \mathbf{1}_{\neg B}), \quad \text{for } f \colon A \vdash B \qquad (\S 14.2)$$

List of Categories

We list in the table below the logical categories and some other related categories we deal with in the book. We present the categories involving $\vee$ immediately below the dual categories involving $\wedge$. Otherwise, we follow the ascending order in which the categories appear in the Charts that follow, which is close to the order in which they appear in the book.

category	language	families	specific equations	section
$\hat{\mathbf{I}}$	$\mathcal{L}_\wedge$	$\mathbf{1}$		§4.1
$\check{\mathbf{I}}$	$\mathcal{L}_\vee$	$\mathbf{1}$		
$\mathbf{I}$	$\mathcal{L}_{\wedge,\vee}$	$\mathbf{1}$		§4.1
$\hat{\mathbf{A}}^{\to}$	$\mathcal{L}_\wedge$	$\mathbf{1}$, $\hat{b}^{\to}$	$(\hat{b}\,5)$	§4.2
$\hat{\mathbf{A}}$	$\mathcal{L}_\wedge$	$\mathbf{1}$, $\hat{b}$	$(\hat{b}\,5)$, $(\hat{b}\hat{b})$	§4.3
$\check{\mathbf{A}}$	$\mathcal{L}_\vee$	$\mathbf{1}$, $\check{b}$	$(\check{b}\,5)$, $(\check{b}\check{b})$	§6.1
$\hat{\mathbf{A}}_\top$	$\mathcal{L}_{\wedge,\top}$	$\mathbf{1}$, $\hat{b}$, $\hat{\delta}$-$\hat{\sigma}$	$(\hat{b}\,5)$, $(\hat{b}\hat{b})$, $(\hat{\delta}\,\hat{\delta})$, $(\hat{\sigma}\hat{\sigma})$, $(\hat{b}\,\hat{\delta}\,\hat{\sigma})$	§4.6
$\check{\mathbf{A}}_\bot$	$\mathcal{L}_{\vee,\bot}$	$\mathbf{1}$, $\check{b}$, $\check{\delta}$-$\check{\sigma}$	$(\check{b}\,5)$, $(\check{b}\check{b})$, $(\check{\delta}\,\check{\delta})$, $(\check{\sigma}\check{\sigma})$, $(\check{b}\,\check{\delta}\,\check{\sigma})$	§6.1
$\mathbf{A}$	$\mathcal{L}_{\wedge,\vee}$	$\mathbf{1}$, b	$(\hat{b}\,5)$, $(\hat{b}\hat{b})$, $(\check{b}\,5)$, $(\check{b}\check{b})$	§6.1
$\mathbf{A}_{\top,\bot}$	$\mathcal{L}_{\wedge,\vee,\top,\bot}$	$\mathbf{1}$, b, δ-σ	$(\hat{b}\,5)$, $(\hat{b}\hat{b})$, $(\check{b}\,5)$, $(\check{b}\check{b})$, $(\hat{\delta}\,\hat{\delta})$, $(\hat{\sigma}\hat{\sigma})$, $(\hat{b}\,\hat{\delta}\,\hat{\sigma})$, $(\check{\delta}\,\check{\delta})$, $(\check{\sigma}\check{\sigma})$, $(\check{b}\,\check{\delta}\,\check{\sigma})$	§6.1
$\hat{\mathbf{S}}$	$\mathcal{L}_\wedge$	$\mathbf{1}$, $\hat{b}$, $\hat{c}$	$(\hat{b}\,5)$, $(\hat{b}\hat{b})$, $(\hat{c}\hat{c})$, $(\hat{b}\hat{c})$	§5.1
$\hat{\mathbf{S}}'$	$\mathcal{L}_\wedge$	$\mathbf{1}$, $\hat{b}$, $\hat{c}$	$(\hat{b}\,5)$, $(\hat{b}\hat{b})$, $(\hat{c}\hat{c})$, $(\hat{b}\hat{c})$, $(\hat{c}\,\mathbf{1})$	§6.5
$\check{\mathbf{S}}$	$\mathcal{L}_\vee$	$\mathbf{1}$, $\check{b}$, $\check{c}$	$(\check{b}\,5)$, $(\check{b}\check{b})$, $(\check{c}\check{c})$, $(\check{b}\check{c})$	§6.3
$\hat{\mathbf{S}}_\top$	$\mathcal{L}_{\wedge,\top}$	$\mathbf{1}$, $\hat{b}$, $\hat{c}$, $\hat{\delta}$-$\hat{\sigma}$	$(\hat{b}\,5)$, $(\hat{b}\hat{b})$, $(\hat{c}\hat{c})$, $(\hat{b}\hat{c})$, $(\hat{\delta}\,\hat{\delta})$, $(\hat{\sigma}\hat{\sigma})$, $(\hat{b}\,\hat{\delta}\,\hat{\sigma})$	§5.3
$\check{\mathbf{S}}_\bot$	$\mathcal{L}_{\vee,\bot}$	$\mathbf{1}$, $\check{b}$, $\check{c}$, $\check{\delta}$-$\check{\sigma}$	$(\check{b}\,5)$, $(\check{b}\check{b})$, $(\check{c}\check{c})$, $(\check{b}\check{c})$, $(\check{\delta}\,\check{\delta})$, $(\check{\sigma}\check{\sigma})$, $(\check{b}\,\check{\delta}\,\check{\sigma})$	§6.4

category	language	families	specific equations	section
$\mathbf{S}$	$\mathcal{L}_{\wedge,\vee}$	$\mathbf{1}$, b, c	$(\hat{b}\,5)$, $(\hat{b}\hat{b})$, $(\hat{c}\hat{c})$, $(\hat{b}\hat{c})$, $(\check{b}\,5)$, $(\check{b}\check{b})$, $(\check{c}\check{c})$, $(\check{b}\check{c})$	§6.3
$\mathbf{S}'$	$\mathcal{L}_{\wedge,\vee}$	$\mathbf{1}$, b, c	$(\hat{b}\,5)$, $(\hat{b}\hat{b})$, $(\hat{c}\hat{c})$, $(\hat{b}\hat{c})$, $(\check{b}\,5)$, $(\check{b}\check{b})$, $(\check{c}\check{c})$, $(\check{b}\check{c})$, $(\hat{c}\,\mathbf{1})$, $(\check{c}\,\mathbf{1})$	§6.5
$\mathbf{S}_{\top,\bot}$	$\mathcal{L}_{\wedge,\vee,\top,\bot}$	$\mathbf{1}$, b, c, δ-σ	$(\hat{b}\,5)$, $(\hat{b}\hat{b})$, $(\hat{c}\hat{c})$, $(\hat{b}\hat{c})$, $(\check{b}\,5)$, $(\check{b}\check{b})$, $(\check{c}\check{c})$, $(\check{b}\check{c})$, $(\hat{\delta}\hat{\delta})$, $(\hat{\sigma}\hat{\sigma})$, $(\hat{b}\hat{\delta}\hat{\sigma})$, $(\check{\delta}\check{\delta})$, $(\check{\sigma}\check{\sigma})$, $(\check{b}\check{\delta}\check{\sigma})$, $(\hat{c}\,\bot)$, $(\check{c}\,\top)$	§6.4
$\hat{\mathbf{L}}$	$\mathcal{L}_{\wedge}$	$\mathbf{1}$, $\hat{b}$, $\hat{c}$, $\hat{w}$-$\hat{k}$	$(\hat{b}\,5)$, $(\hat{b}\hat{b})$, $(\hat{c}\hat{c})$, $(\hat{b}\hat{c})$, $(\hat{b}\hat{w})$, $(\hat{c}\hat{w})$, $(\hat{b}\hat{c}\hat{w})$, $(\hat{b}\hat{k})$, $(\hat{c}\hat{k})$, $(\hat{w}\hat{k})$	§9.1
$\check{\mathbf{L}}$	$\mathcal{L}_{\vee}$	$\mathbf{1}$, $\check{b}$, $\check{c}$, $\check{w}$-$\check{k}$	$(\check{b}\,5)$, $(\check{b}\check{b})$, $(\check{c}\check{c})$, $(\check{b}\check{c})$, $(\check{b}\check{w})$, $(\check{c}\check{w})$, $(\check{b}\check{c}\check{w})$, $(\check{b}\check{k})$, $(\check{c}\check{k})$, $(\check{w}\check{k})$	§9.4
$\hat{\mathbf{L}}_{\top}$	$\mathcal{L}_{\wedge,\top}$	$\mathbf{1}$, $\hat{b}$, $\hat{c}$, $\hat{w}$-$\hat{k}$, $\hat{\delta}$-$\hat{\sigma}$	$(\hat{b}\,5)$, $(\hat{b}\hat{b})$, $(\hat{c}\hat{c})$, $(\hat{b}\hat{c})$, $(\hat{b}\hat{w})$, $(\hat{c}\hat{w})$, $(\hat{b}\hat{c}\hat{w})$, $(\hat{b}\hat{k})$, $(\hat{c}\hat{k})$, $(\hat{w}\hat{k})$, $(\hat{\delta}\hat{\delta})$, $(\hat{\sigma}\hat{\sigma})$, $(\hat{b}\hat{\delta}\hat{\sigma})$, $(\hat{k}\hat{\delta})$	§9.2
$\check{\mathbf{L}}_{\bot}$	$\mathcal{L}_{\vee,\bot}$	$\mathbf{1}$, $\check{b}$, $\check{c}$, $\check{w}$-$\check{k}$, $\check{\delta}$-$\check{\sigma}$	$(\check{b}\,5)$, $(\check{b}\check{b})$, $(\check{c}\check{c})$, $(\check{b}\check{c})$, $(\check{b}\check{w})$, $(\check{c}\check{w})$, $(\check{b}\check{c}\check{w})$, $(\check{b}\check{k})$, $(\check{c}\check{k})$, $(\check{w}\check{k})$, $(\check{\delta}\check{\delta})$, $(\check{\sigma}\check{\sigma})$, $(\check{b}\check{\delta}\check{\sigma})$, $(\check{k}\check{\delta})$	§9.6
$\hat{\mathbf{L}}_{\vee}$	$\mathcal{L}_{\wedge,\vee}$	$\mathbf{1}$, $\hat{b}$, $\hat{c}$, $\hat{w}$-$\hat{k}$	$(\hat{b}\,5)$, $(\hat{b}\hat{b})$, $(\hat{c}\hat{c})$, $(\hat{b}\hat{c})$, $(\hat{b}\hat{w})$, $(\hat{c}\hat{w})$, $(\hat{b}\hat{c}\hat{w})$, $(\hat{b}\hat{k})$, $(\hat{c}\hat{k})$, $(\hat{w}\hat{k})$	§9.4
$\check{\mathbf{L}}_{\wedge}$	$\mathcal{L}_{\wedge,\vee}$	$\mathbf{1}$, $\check{b}$, $\check{c}$, $\check{w}$-$\check{k}$	$(\check{b}\,5)$, $(\check{b}\check{b})$, $(\check{c}\check{c})$, $(\check{b}\check{c})$, $(\check{b}\check{w})$, $(\check{c}\check{w})$, $(\check{b}\check{c}\check{w})$, $(\check{b}\check{k})$, $(\check{c}\check{k})$, $(\check{w}\check{k})$	§9.4

category	language	families	specific equations	section
L	$\mathcal{L}_{\wedge,\vee}$	$\mathbf{1}$, b, c, w-k	$(\hat{b}5)$, $(\hat{b}\hat{b})$, $(\hat{c}\hat{c})$, $(\hat{b}\hat{c})$, $(\check{b}5)$, $(\check{b}\check{b})$, $(\check{c}\check{c})$, $(\check{b}\check{c})$, $(\hat{b}\hat{w})$, $(\hat{c}\hat{w})$, $(\hat{b}\hat{c}\hat{w})$, $(\check{b}\check{w})$, $(\check{c}\check{w})$, $(\check{b}\check{c}\check{w})$, $(\hat{b}\hat{k})$, $(\hat{c}\hat{k})$, $(\hat{w}\hat{k})$, $(\check{b}\check{k})$, $(\check{c}\check{k})$, $(\check{w}\check{k})$	§9.4
$\mathbf{L}_{\top}$	$\mathcal{L}_{\wedge,\vee,\top}$	$\mathbf{1}$, b, c, w-k, $\hat{\delta}$-$\hat{\sigma}$	$(\hat{b}5)$, $(\hat{b}\hat{b})$, $(\hat{c}\hat{c})$, $(\hat{b}\hat{c})$, $(\check{b}5)$, $(\check{b}\check{b})$, $(\check{c}\check{c})$, $(\check{b}\check{c})$, $(\hat{b}\hat{w})$, $(\hat{c}\hat{w})$, $(\hat{b}\hat{c}\hat{w})$, $(\check{b}\check{w})$, $(\check{c}\check{w})$, $(\check{b}\check{c}\check{w})$, $(\hat{b}\hat{k})$, $(\hat{c}\hat{k})$, $(\hat{w}\hat{k})$, $(\check{b}\check{k})$, $(\check{c}\check{k})$, $(\check{w}\check{k})$, $(\hat{\delta}\hat{\delta})$, $(\hat{\sigma}\hat{\sigma})$, $(\hat{b}\hat{\delta}\hat{\sigma})$, $(\hat{k}\hat{\delta})$, $(\check{c}\top)$	§9.6
$\mathbf{L}_{\bot}$	$\mathcal{L}_{\wedge,\vee,\bot}$	$\mathbf{1}$, b, c, w-k, $\check{\delta}$-$\check{\sigma}$	$(\hat{b}5)$, $(\hat{b}\hat{b})$, $(\hat{c}\hat{c})$, $(\hat{b}\hat{c})$, $(\check{b}5)$, $(\check{b}\check{b})$, $(\check{c}\check{c})$, $(\check{b}\check{c})$, $(\hat{b}\hat{w})$, $(\hat{c}\hat{w})$, $(\hat{b}\hat{c}\hat{w})$, $(\check{b}\check{w})$, $(\check{c}\check{w})$, $(\check{b}\check{c}\check{w})$, $(\hat{b}\hat{k})$, $(\hat{c}\hat{k})$, $(\hat{w}\hat{k})$, $(\check{b}\check{k})$, $(\check{c}\check{k})$, $(\check{w}\check{k})$, $(\check{\delta}\check{\delta})$, $(\check{\sigma}\check{\sigma})$, $(\check{b}\check{\delta}\check{\sigma})$, $(\check{k}\check{\delta})$, $(\hat{c}\bot)$	§9.6
$\mathbf{L}_{\top,\bot}$	$\mathcal{L}_{\wedge,\vee,\top,\bot}$	$\mathbf{1}$, b, c, w-k, δ-σ	$(\hat{b}5)$, $(\hat{b}\hat{b})$, $(\hat{c}\hat{c})$, $(\hat{b}\hat{c})$, $(\check{b}5)$, $(\check{b}\check{b})$, $(\check{c}\check{c})$, $(\check{b}\check{c})$, $(\hat{b}\hat{w})$, $(\hat{c}\hat{w})$, $(\hat{b}\hat{c}\hat{w})$, $(\check{b}\check{w})$, $(\check{c}\check{w})$, $(\check{b}\check{c}\check{w})$, $(\hat{b}\hat{k})$, $(\hat{c}\hat{k})$, $(\hat{w}\hat{k})$, $(\check{b}\check{k})$, $(\check{c}\check{k})$, $(\check{w}\check{k})$, $(\hat{\delta}\hat{\delta})$, $(\hat{\sigma}\hat{\sigma})$, $(\hat{b}\hat{\delta}\hat{\sigma})$, $(\check{\delta}\check{\delta})$, $(\check{\sigma}\check{\sigma})$, $(\check{b}\check{\delta}\check{\sigma})$, $(\hat{k}\hat{\delta})$, $(\check{k}\check{\delta})$, $(\hat{c}\bot)$, $(\check{c}\top)$	§9.6
DI	$\mathcal{L}_{\wedge,\vee}$	$\mathbf{1}$, d		§7.1
$\mathbf{D}^L\mathbf{A}$	$\mathcal{L}_{\wedge,\vee}$	$\mathbf{1}$, b, d^L	$(\hat{b}5)$, $(\hat{b}\hat{b})$, $(d^L\wedge)$, $(\check{b}5)$, $(\check{b}\check{b})$, $(d^L\vee)$	§7.5

category	language	families	specific equations	section
DA	$\mathcal{L}_{\wedge,\vee}$	$\mathbf{1}$, b, d	$(\hat{b}\,5)$, $(\hat{b}\hat{b})$, $(d^L\wedge)$, $(d^R\wedge)$, $(\check{b}\,5)$, $(\check{b}\check{b})$, $(d^L\vee)$, $(d^R\vee)$, $(d\hat{b})$, $(d\check{b})$	§7.2
DA$_{\top,\bot}$	$\mathcal{L}_{\wedge,\vee,\top,\bot}$	$\mathbf{1}$, b, δ-σ, d	$(\hat{b}\,5)$, $(\hat{b}\hat{b})$, $(d^L\wedge)$, $(d^R\wedge)$, $(\check{b}\,5)$, $(\check{b}\check{b})$, $(d^L\vee)$, $(d^R\vee)$, $(d\hat{b})$, $(d\check{b})$, $(\hat{\delta}\,\hat{\delta})$, $(\hat{\sigma}\hat{\sigma})$, $(\hat{b}\hat{\delta}\hat{\sigma})$, $(\check{\delta}\,\check{\delta})$, $(\check{\sigma}\check{\sigma})$, $(\check{b}\check{\delta}\check{\sigma})$, $(\hat{\sigma}\,d^L)$, $(\check{\delta}\,d^L)$, $(\hat{\delta}\,d^R)$, $(\check{\sigma}\,d^R)$	§7.9
DS	$\mathcal{L}_{\wedge,\vee}$	$\mathbf{1}$, b, c, d	$(\hat{b}\,5)$, $(\hat{b}\hat{b})$, $(\hat{c}\hat{c})$, $(\hat{b}\hat{c})$, $(\check{b}\,5)$, $(\check{b}\check{b})$, $(\check{c}\check{c})$, $(\check{b}\check{c})$, $(d^L\wedge)$, $(d^L\vee)$, $(d^R\wedge)$, $(d^R\vee)$, $(d\hat{b})$, $(d\check{b})$, (d^Rc)	§7.6
MI	$\mathcal{L}_{\wedge,\vee}$	$\mathbf{1}$, m		§8.1
MA	$\mathcal{L}_{\wedge,\vee}$	$\mathbf{1}$, b, m	$(\hat{b}\,5)$, $(\hat{b}\hat{b})$, $(\check{b}\,5)$, $(\check{b}\check{b})$, (bm)	§8.2
MS	$\mathcal{L}_{\wedge,\vee}$	$\mathbf{1}$, b, c, m	$(\hat{b}\,5)$, $(\hat{b}\hat{b})$, $(\hat{c}\hat{c})$, $(\hat{b}\hat{c})$, $(\check{b}\,5)$, $(\check{b}\check{b})$, $(\check{c}\check{c})$, $(\check{b}\check{c})$, (bm), (cm)	§8.5
ML	$\mathcal{L}_{\wedge,\vee}$	$\mathbf{1}$, b, c, w-k, m	$(\hat{b}\,5)$, $(\hat{b}\hat{b})$, $(\hat{c}\hat{c})$, $(\hat{b}\hat{c})$, $(\check{b}\,5)$, $(\check{b}\check{b})$, $(\check{c}\check{c})$, $(\check{b}\check{c})$, $(\hat{b}\hat{w})$, $(\hat{c}\hat{w})$, $(\hat{b}\hat{c}\hat{w})$, $(\check{b}\check{w})$, $(\check{c}\check{w})$, $(\check{b}\check{c}\check{w})$, $(\hat{b}\hat{k})$, $(\hat{c}\hat{k})$, $(\hat{w}\hat{k})$, $(\check{b}\check{k})$, $(\check{c}\check{k})$, $(\check{w}\check{k})$, (bm), (cm), (wm)	§10.1
ML$_{\top,\bot}$	$\mathcal{L}_{\wedge,\vee,\top,\bot}$	$\mathbf{1}$, b, c, w-k, m, δ-σ	$(\hat{b}\,5)$, $(\hat{b}\hat{b})$, $(\hat{c}\hat{c})$, $(\hat{b}\hat{c})$, $(\check{b}\,5)$, $(\check{b}\check{b})$, $(\check{c}\check{c})$, $(\check{b}\check{c})$, $(\hat{b}\hat{w})$, $(\hat{c}\hat{w})$, $(\hat{b}\hat{c}\hat{w})$, $(\check{b}\check{w})$, $(\check{c}\check{w})$, $(\check{b}\check{c}\check{w})$, $(\hat{b}\hat{k})$, $(\hat{c}\hat{k})$, $(\hat{w}\hat{k})$, $(\check{b}\check{k})$, $(\check{c}\check{k})$, $(\check{w}\check{k})$, (bm), (cm), (wm), $(\hat{\delta}\,\hat{\delta})$, $(\hat{\sigma}\hat{\sigma})$, $(\hat{b}\hat{\delta}\hat{\sigma})$, $(\check{\delta}\,\check{\delta})$, $(\check{\sigma}\check{\sigma})$, $(\check{b}\check{\delta}\check{\sigma})$, $(\hat{k}\,\hat{\delta})$, $(\check{k}\,\check{\delta})$, $(\hat{c}\,\bot)$, $(\check{c}\,\top)$, $(m\top)$, $(m\bot)$	§10.3

category	language	families	specific equations	section
MDI	$\mathcal{L}_{\wedge,\vee}$	**1**, d, m		§8.1
MDA	$\mathcal{L}_{\wedge,\vee}$	**1**, b, d, m	$(\hat{b}5)$, $(\hat{b}\hat{b})$, $(d^L\wedge)$, $(d^R\wedge)$, $(\check{b}5)$, $(\check{b}\check{b})$, $(d^L\vee)$, $(d^R\vee)$, $(d\hat{b})$, $(\hat{b}mL)$, $(\hat{b}mR)$, $(d\check{b})$, $(\check{b}mL)$, $(\check{b}mR)$	§8.3
MDS	$\mathcal{L}_{\wedge,\vee}$	**1**, b, c, d, m	$(\hat{b}5)$, $(\hat{b}\hat{b})$, $(\hat{c}\hat{c})$, $(\hat{b}\hat{c})$, $(\check{b}5)$, $(\check{b}\check{b})$, $(\check{c}\check{c})$, $(\check{b}\check{c})$, $(d^L\wedge)$, $(d^L\vee)$, $(d^R\wedge)$, $(d^R\vee)$, $(d\hat{b})$, $(\hat{b}mL)$, $(\hat{b}mR)$, $(d\check{b})$, $(\check{b}mR)$, $(\check{b}mL)$, (d^Rc), (cm)	§8.4
DL	$\mathcal{L}_{\wedge,\vee}$	**1**, b, c, w-k, m, d	$(\hat{b}5)$, $(\hat{b}\hat{b})$, $(\hat{c}\hat{c})$, $(\hat{b}\hat{c})$, $(\check{b}5)$, $(\check{b}\check{b})$, $(\check{c}\check{c})$, $(\check{b}\check{c})$, $(\hat{b}\hat{w})$, $(\hat{c}\hat{w})$, $(\hat{b}\hat{c}\hat{w})$, $(\check{b}\check{w})$, $(\check{c}\check{w})$, $(\check{b}\check{c}\check{w})$, $(\hat{b}\hat{k})$, $(\hat{c}\hat{k})$, $(\hat{w}\hat{k})$, $(\check{b}\check{k})$, $(\check{c}\check{k})$, $(\check{w}\check{k})$, (bm), (cm), (wm), (dm), $(d^L\wedge)$, $(d^L\vee)$, $(d^R\wedge)$, $(d^R\vee)$, $(d\hat{b})$, $(d\hat{k})$, $(m\hat{e})$, $(d\check{b})$, $(d\check{k})$, $(m\check{e})$, (d^Rc)	§11.1
$\mathbf{DL}_{\top,\bot}$	$\mathcal{L}_{\wedge,\vee,\top,\bot}$	**1**, b, c, w-k, m, d, δ-σ	$(\hat{b}5)$, $(\hat{b}\hat{b})$, $(\hat{c}\hat{c})$, $(\hat{b}\hat{c})$, $(\check{b}5)$, $(\check{b}\check{b})$, $(\check{c}\check{c})$, $(\check{b}\check{c})$, $(\hat{b}\hat{w})$, $(\hat{c}\hat{w})$, $(\hat{b}\hat{c}\hat{w})$, $(\check{b}\check{w})$, $(\check{c}\check{w})$, $(\check{b}\check{c}\check{w})$, $(\hat{b}\hat{k})$, $(\hat{c}\hat{k})$, $(\hat{w}\hat{k})$, $(\check{b}\check{k})$, $(\check{c}\check{k})$, $(\check{w}\check{k})$, (bm), (cm), (wm), (dm), $(d^L\wedge)$, $(d^L\vee)$, $(d^R\wedge)$, $(d^R\vee)$, $(d\hat{b})$, $(d\hat{k})$, $(m\hat{e})$, $(d\check{b})$, $(d\check{k})$, $(m\check{e})$, (d^Rc), $(\hat{\delta}\hat{\delta})$, $(\hat{\sigma}\hat{\sigma})$, $(\hat{b}\hat{\delta}\hat{\sigma})$, $(\check{\delta}\check{\delta})$, $(\check{\sigma}\check{\sigma})$, $(\check{b}\check{\delta}\check{\sigma})$, $(\hat{k}\hat{\delta})$, $(\check{k}\check{\delta})$, $(\hat{c}\bot)$, $(\check{c}\top)$	§11.5

category	language	families	specific equations	section
ZIL	$\mathcal{L}_{\wedge,\vee}$	$\mathbf{1}$, b, c, w-k, $\mathbf{0}_A$	$(\hat{b}\,5)$, $(\hat{b}\hat{b})$, $(\hat{c}\hat{c})$, $(\hat{b}\hat{c})$, $(\check{b}\,5)$, $(\check{b}\check{b})$, $(\check{c}\check{c})$, $(\check{b}\check{c})$, $(\hat{b}\hat{w})$, $(\hat{c}\hat{w})$, $(\hat{b}\hat{c}\hat{w})$, $(\check{b}\check{w})$, $(\check{c}\check{w})$, $(\check{b}\check{c}\check{w})$, $(\hat{b}\hat{k})$, $(\hat{c}\hat{k})$, $(\hat{w}\hat{k})$, $(\check{b}\check{k})$, $(\check{c}\check{k})$, $(\check{w}\check{k})$, $(\mathbf{0}\mathrm{I})$	§12.5
ZL	$\mathcal{L}_{\wedge,\vee}$	$\mathbf{1}$, b, c, w-k, m^{-1}	$(\hat{b}\,5)$, $(\hat{b}\hat{b})$, $(\hat{c}\hat{c})$, $(\hat{b}\hat{c})$, $(\check{b}\,5)$, $(\check{b}\check{b})$, $(\check{c}\check{c})$, $(\check{b}\check{c})$, $(\hat{b}\hat{w})$, $(\hat{c}\hat{w})$, $(\hat{b}\hat{c}\hat{w})$, $(\check{b}\check{w})$, $(\check{c}\check{w})$, $(\check{b}\check{c}\check{w})$, $(\hat{b}\hat{k})$, $(\hat{c}\hat{k})$, $(\hat{w}\hat{k})$, $(\check{b}\check{k})$, $(\check{c}\check{k})$, $(\check{w}\check{k})$, $(m^{-1}\,0)$, $(m^{-1}\,1)$	§12.1
ZIL$_{\top,\bot}$	$\mathcal{L}_{\wedge,\vee,\top,\bot}$	$\mathbf{1}$, b, c, w-k, δ-σ, $\mathbf{0}_A$	$(\hat{b}\,5)$, $(\hat{b}\hat{b})$, $(\hat{c}\hat{c})$, $(\hat{b}\hat{c})$, $(\check{b}\,5)$, $(\check{b}\check{b})$, $(\check{c}\check{c})$, $(\check{b}\check{c})$, $(\hat{b}\hat{w})$, $(\hat{c}\hat{w})$, $(\hat{b}\hat{c}\hat{w})$, $(\check{b}\check{w})$, $(\check{c}\check{w})$, $(\check{b}\check{c}\check{w})$, $(\hat{b}\hat{k})$, $(\hat{c}\hat{k})$, $(\hat{w}\hat{k})$, $(\check{b}\check{k})$, $(\check{c}\check{k})$, $(\check{w}\check{k})$, $(\mathbf{0}\mathrm{I})$ $(\hat{\delta}\,\hat{\delta})$, $(\hat{\sigma}\hat{\sigma})$, $(\hat{b}\,\hat{\delta}\,\hat{\sigma})$, $(\check{\delta}\,\check{\delta})$, $(\check{\sigma}\check{\sigma})$, $(\check{b}\,\check{\delta}\,\check{\sigma})$, $(\hat{k}\,\hat{\delta})$, $(\check{k}\,\check{\delta})$, $(\hat{c}\,\bot)$, $(\check{c}\,\top)$,	§12.5
ZL$_{\top,\bot}$	$\mathcal{L}_{\wedge,\vee,\top,\bot}$	$\mathbf{1}$, b, c, w-k, δ-σ, m^{-1}	$(\hat{b}\,5)$, $(\hat{b}\hat{b})$, $(\hat{c}\hat{c})$, $(\hat{b}\hat{c})$, $(\check{b}\,5)$, $(\check{b}\check{b})$, $(\check{c}\check{c})$, $(\check{b}\check{c})$, $(\hat{b}\hat{w})$, $(\hat{c}\hat{w})$, $(\hat{b}\hat{c}\hat{w})$, $(\check{b}\check{w})$, $(\check{c}\check{w})$, $(\check{b}\check{c}\check{w})$, $(\hat{b}\hat{k})$, $(\hat{c}\hat{k})$, $(\hat{w}\hat{k})$, $(\check{b}\check{k})$, $(\check{c}\check{k})$, $(\check{w}\check{k})$, $(m^{-1}\,0)$, $(m^{-1}\,1)$ $(\hat{\delta}\,\hat{\delta})$, $(\hat{\sigma}\hat{\sigma})$, $(\hat{b}\,\hat{\delta}\,\hat{\sigma})$, $(\check{\delta}\,\check{\delta})$, $(\check{\sigma}\check{\sigma})$, $(\check{b}\,\check{\delta}\,\check{\sigma})$, $(\hat{k}\,\hat{\delta})$, $(\check{k}\,\check{\delta})$, $(\hat{c}\,\bot)$, $(\check{c}\,\top)$,	§12.1

category	language	families	specific equations	section
ZIML	$\mathcal{L}_{\wedge,\vee}$	$\mathbf{1}$, b, c, w-k, m, $\mathbf{0}_A$	$(\hat{b}5)$, $(\hat{b}\hat{b})$, $(\hat{c}\hat{c})$, $(\hat{b}\hat{c})$, $(\check{b}5)$, $(\check{b}\check{b})$, $(\check{c}\check{c})$, $(\check{b}\check{c})$, $(\hat{b}\hat{w})$, $(\hat{c}\hat{w})$, $(\hat{b}\hat{c}\hat{w})$, $(\check{b}\check{w})$, $(\check{c}\check{w})$, $(\check{b}\check{c}\check{w})$, $(\hat{b}\hat{k})$, $(\hat{c}\hat{k})$, $(\hat{w}\hat{k})$, $(\check{b}\check{k})$, $(\check{c}\check{k})$, $(\check{w}\check{k})$, (bm), (cm), (wm), $(\mathbf{0I})$, $(\cup\mathbf{0})$	§12.5
ZIML$_{\top,\bot}$	$\mathcal{L}_{\wedge,\vee,\top,\bot}$	$\mathbf{1}$, b, c, w-k, m, δ-σ, $\mathbf{0}_A$	$(\hat{b}5)$, $(\hat{b}\hat{b})$, $(\hat{c}\hat{c})$, $(\hat{b}\hat{c})$, $(\check{b}5)$, $(\check{b}\check{b})$, $(\check{c}\check{c})$, $(\check{b}\check{c})$, $(\hat{b}\hat{w})$, $(\hat{c}\hat{w})$, $(\hat{b}\hat{c}\hat{w})$, $(\check{b}\check{w})$, $(\check{c}\check{w})$, $(\check{b}\check{c}\check{w})$, $(\hat{b}\hat{k})$, $(\hat{c}\hat{k})$, $(\hat{w}\hat{k})$, $(\check{b}\check{k})$, $(\check{c}\check{k})$, $(\check{w}\check{k})$, (bm), (cm), (wm), $(\hat{\delta}\hat{\delta})$, $(\hat{\sigma}\hat{\sigma})$, $(\hat{b}\hat{\delta}\hat{\sigma})$, $(\check{\delta}\check{\delta})$, $(\check{\sigma}\check{\sigma})$, $(\check{b}\check{\delta}\check{\sigma})$, $(\hat{k}\hat{\delta})$, $(\check{k}\check{\delta})$, $(\hat{c}\bot)$, $(\check{c}\top)$, $(m\top)$, $(m\bot)$, $(\mathbf{0I})$, $(\cup\mathbf{0})$	§12.5
ZIDL	$\mathcal{L}_{\wedge,\vee}$	$\mathbf{1}$, b, c, w-k, m, d, $\mathbf{0}_A$	$(\hat{b}5)$, $(\hat{b}\hat{b})$, $(\hat{c}\hat{c})$, $(\hat{b}\hat{c})$, $(\check{b}5)$, $(\check{b}\check{b})$, $(\check{c}\check{c})$, $(\check{b}\check{c})$, $(\hat{b}\hat{w})$, $(\hat{c}\hat{w})$, $(\hat{b}\hat{c}\hat{w})$, $(\check{b}\check{w})$, $(\check{c}\check{w})$, $(\check{b}\check{c}\check{w})$, $(\hat{b}\hat{k})$, $(\hat{c}\hat{k})$, $(\hat{w}\hat{k})$, $(\check{b}\check{k})$, $(\check{c}\check{k})$, $(\check{w}\check{k})$, (bm), (cm), (wm), (dm), $(d^L\wedge)$, $(d^L\vee)$, $(d^R\wedge)$, $(d^R\vee)$, $(d\hat{b})$, $(d\hat{k})$, $(m\hat{e})$, $(d\check{b})$, $(d\check{k})$, $(m\check{e})$, (d^Rc), $(\mathbf{0I})$	§12.5

category	language	families	specific equations	section
$\mathbf{ZIDL}_{\top,\bot}$	$\mathcal{L}_{\wedge,\vee,\top,\bot}$	$\mathbf{1}$, b, c, w-k, m, d, δ-σ, $\mathbf{0}_A$	$(\hat{b}\,5)$, $(\hat{b}\hat{b})$, $(\hat{c}\,\hat{c})$, $(\hat{b}\,\hat{c})$, $(\check{b}\,5)$, $(\check{b}\check{b})$, $(\check{c}\,\check{c})$, $(\check{b}\,\check{c})$, $(\hat{b}\hat{w})$, $(\hat{c}\hat{w})$, $(\hat{b}\,\hat{c}\hat{w})$, $(\check{b}\check{w})$, $(\check{c}\check{w})$, $(\check{b}\,\check{c}\check{w})$, $(\hat{b}\hat{k})$, $(\hat{c}\hat{k})$, $(\hat{w}\hat{k})$, $(\check{b}\check{k})$, $(\check{c}\check{k})$, $(\check{w}\check{k})$, (bm), (cm), (wm), (dm), $(d^L\wedge)$, $(d^L\vee)$, $(d^R\wedge)$, $(d^R\vee)$, $(d\hat{b})$, $(d\hat{k})$, $(m\,\hat{e})$, $(d\check{b})$, $(d\check{k})$, $(m\,\check{e})$, (d^Rc), $(\mathbf{0}\mathrm{I})$, $(\hat{\delta}\,\hat{\delta})$, $(\hat{\sigma}\hat{\sigma})$, $(\hat{b}\,\hat{\delta}\,\hat{\sigma})$, $(\check{\delta}\,\check{\delta})$, $(\check{\sigma}\check{\sigma})$, $(\check{b}\,\check{\delta}\,\check{\sigma})$, $(\hat{k}\,\hat{\delta})$, $(\check{k}\,\check{\delta})$, $(\hat{c}\,\bot)$, $(\check{c}\,\top)$	§12.5
$\mathbf{B}$	$\mathcal{L}^{\neg}_{\wedge,\vee,\top,\bot}$	$\mathbf{1}$, b, c, w-k, m, d, δ-σ, $\mathbf{0}_A$, n-r, ρ, η, ε	$(\hat{b}\,5)$, $(\hat{b}\hat{b})$, $(\hat{c}\,\hat{c})$, $(\hat{b}\,\hat{c})$, $(\check{b}\,5)$, $(\check{b}\check{b})$, $(\check{c}\,\check{c})$, $(\check{b}\,\check{c})$, $(\hat{b}\hat{w})$, $(\hat{c}\hat{w})$, $(\hat{b}\,\hat{c}\hat{w})$, $(\check{b}\check{w})$, $(\check{c}\check{w})$, $(\check{b}\,\check{c}\check{w})$, $(\hat{b}\hat{k})$, $(\hat{c}\hat{k})$, $(\hat{w}\hat{k})$, $(\check{b}\check{k})$, $(\check{c}\check{k})$, $(\check{w}\check{k})$, (bm), (cm), (wm), (dm), $(d^L\wedge)$, $(d^L\vee)$, $(d^R\wedge)$, $(d^R\vee)$, $(d\hat{b})$, $(d\hat{k})$, $(m\,\hat{e})$, $(d\check{b})$, $(d\check{k})$, $(m\,\check{e})$, (d^Rc), $(\mathbf{0}\mathrm{I})$, $(\hat{\delta}\,\hat{\delta})$, $(\hat{\sigma}\hat{\sigma})$, $(\hat{b}\,\hat{\delta}\,\hat{\sigma})$, $(\check{\delta}\,\check{\delta})$, $(\check{\sigma}\check{\sigma})$, $(\check{b}\,\check{\delta}\,\check{\sigma})$, $(\hat{k}\,\hat{\delta})$, $(\check{k}\,\check{\delta})$, $(\hat{c}\,\bot)$, $(\check{c}\,\top)$, n-r and ρ isomorphisms	§14.2

category	language	families	specific equations	section
ZML	$\mathcal{L}_{\wedge,\vee}$	**1**, b, c, w-k, m, m^{-1}	$(\hat{b}\,5)$, $(\hat{b}\hat{b})$, $(\hat{c}\hat{c})$, $(\hat{b}\hat{c})$, $(\check{b}\,5)$, $(\check{b}\check{b})$, $(\check{c}\check{c})$, $(\check{b}\check{c})$, $(\hat{b}\hat{w})$, $(\hat{c}\hat{w})$, $(\hat{b}\hat{c}\hat{w})$, $(\check{b}\check{w})$, $(\check{c}\check{w})$, $(\check{b}\check{c}\check{w})$, $(\hat{b}\hat{k})$, $(\hat{c}\hat{k})$, $(\hat{w}\hat{k})$, $(\check{b}\check{k})$, $(\check{c}\check{k})$, $(\check{w}\check{k})$, (bm), (cm), (wm), $(m^{-1}\,0)$, $(m^{-1}\,1)$, (mm^{-1})	§13.1
$\mathbf{ZML}_{\top,\bot}$	$\mathcal{L}_{\wedge,\vee,\top,\bot}$	**1**, b, c, w-k, δ-σ, m, m^{-1}	$(\hat{b}\,5)$, $(\hat{b}\hat{b})$, $(\hat{c}\hat{c})$, $(\hat{b}\hat{c})$, $(\check{b}\,5)$, $(\check{b}\check{b})$, $(\check{c}\check{c})$, $(\check{b}\check{c})$, $(\hat{b}\hat{w})$, $(\hat{c}\hat{w})$, $(\hat{b}\hat{c}\hat{w})$, $(\check{b}\check{w})$, $(\check{c}\check{w})$, $(\check{b}\check{c}\check{w})$, $(\hat{b}\hat{k})$, $(\hat{c}\hat{k})$, $(\hat{w}\hat{k})$, $(\check{b}\check{k})$, $(\check{c}\check{k})$, $(\check{w}\check{k})$, (bm), (cm), (wm), $(m^{-1}\,0)$, $(m^{-1}\,1)$, (mm^{-1}), $(\hat{\delta}\,\hat{\delta})$, $(\hat{\sigma}\hat{\sigma})$, $(\hat{b}\hat{\delta}\hat{\sigma})$, $(\check{\delta}\,\check{\delta})$, $(\check{\sigma}\check{\sigma})$, $(\check{b}\check{\delta}\check{\sigma})$, $(\hat{k}\hat{\delta})$, $(\check{k}\check{\delta})$, $(\hat{c}\,\bot)$, $(\check{c}\,\top)$,	§13.3
$\mathbf{ZML}^{\neg}_{\top,\bot}$	$\mathcal{L}^{\neg}_{\wedge,\vee,\top,\bot}$	**1**, b, c, w-k, δ-σ, m, m^{-1}, n-r, ρ	$(\hat{b}\,5)$, $(\hat{b}\hat{b})$, $(\hat{c}\hat{c})$, $(\hat{b}\hat{c})$, $(\check{b}\,5)$, $(\check{b}\check{b})$, $(\check{c}\check{c})$, $(\check{b}\check{c})$, $(\hat{b}\hat{w})$, $(\hat{c}\hat{w})$, $(\hat{b}\hat{c}\hat{w})$, $(\check{b}\check{w})$, $(\check{c}\check{w})$, $(\check{b}\check{c}\check{w})$, $(\hat{b}\hat{k})$, $(\hat{c}\hat{k})$, $(\hat{w}\hat{k})$, $(\check{b}\check{k})$, $(\check{c}\check{k})$, $(\check{w}\check{k})$, (bm), (cm), (wm), $(m^{-1}\,0)$, $(m^{-1}\,1)$, (mm^{-1}), $(\hat{\delta}\,\hat{\delta})$, $(\hat{\sigma}\hat{\sigma})$, $(\hat{b}\hat{\delta}\hat{\sigma})$, $(\check{\delta}\,\check{\delta})$, $(\check{\sigma}\check{\sigma})$, $(\check{b}\check{\delta}\check{\sigma})$, $(\hat{k}\hat{\delta})$, $(\check{k}\check{\delta})$, $(\hat{c}\,\bot)$, $(\check{c}\,\top)$, n-r and ρ isomorphisms	§14.1

Charts

The charts we present on the next three pages are to be read as follows. When the category $\mathcal{A}$ is joined by an upward-going line to the category $\mathcal{B}$, this means that $\mathcal{A}$ is isomorphic to a subcategory of $\mathcal{B}$. The assertion that $\mathcal{A}$ is isomorphic to a subcategory of $\mathcal{B}$ is established by appealing to the fact that $\mathcal{A}$ is a preorder or that $\mathcal{A}$ is coherent with respect to *Rel*, as explained in §14.4. The three charts could be combined into a single chart by pasting them together in growing order over the parts in which they overlap. For practical, and aesthetical, reasons we have preferred not to make this pasting, and have three separate charts.

We have established coherence with respect to *Rel* for all categories in the charts except $\hat{\mathbf{S}}'$, $\mathbf{S}'$, $\mathbf{DA}_{\top,\bot}$, **ML**, $\mathbf{ML}_{\top,\bot}$, **ZIML** and $\mathbf{ZIML}_{\top,\bot}$. For $\hat{\mathbf{S}}'$ and $\mathbf{S}'$ we have that they are preorders, though they are not coherent with respect to *Rel* (see §6.5). The category $\mathbf{DA}_{\top,\bot}$ was considered in §7.9. For **ML**, $\mathbf{ML}_{\top,\bot}$, **ZIML** and $\mathbf{ZIML}_{\top,\bot}$ we have proved only a restricted form of coherence (see §§10.2-3 and §12.5). This explains the absence of some lines in Charts 2 and 3.

Of the categories with negation, we have mentioned only two, **B** and $\mathbf{ZML}^{\neg}_{\top,\bot}$, at the top of Chart 3, which is also the top of all the charts pasted together. We have, however, coherence for categories with negation where we have coherence without negation (see §14.1), and there are replicas of Charts 2 and 3 involving such categories.

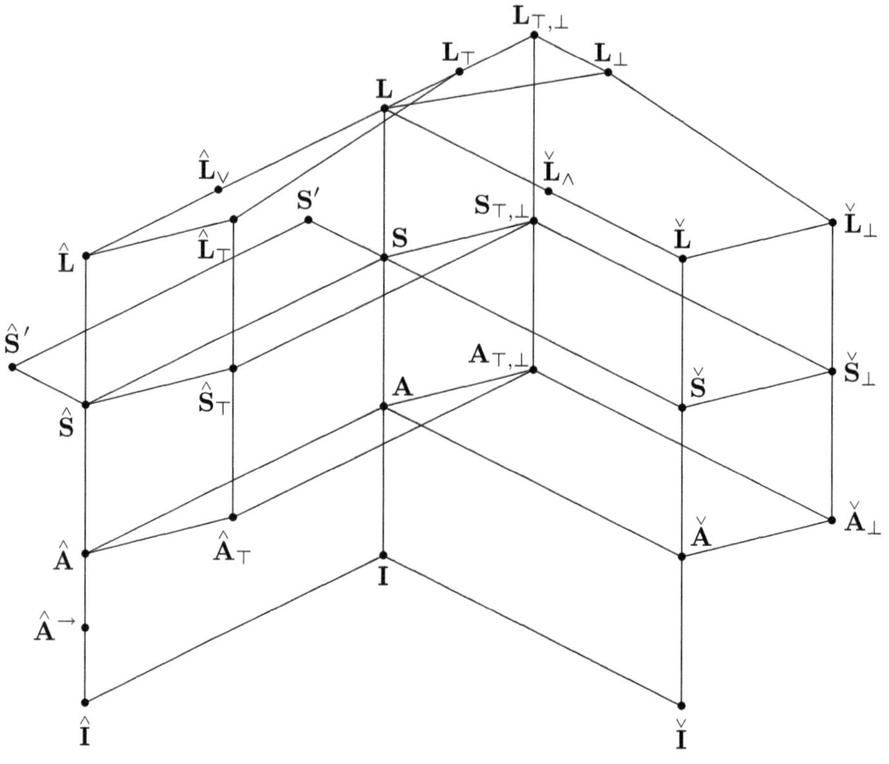
$\mathbf{L}_{\top,\bot}$
$\mathbf{L}_{\top}$
$\mathbf{L}_{\bot}$
$\mathbf{L}$
$\hat{\mathbf{L}}_{\vee}$
$\check{\mathbf{L}}_{\wedge}$
$\mathbf{S}'$
$\mathbf{S}_{\top,\bot}$
$\hat{\mathbf{L}}$
$\hat{\mathbf{L}}_{\top}$
$\mathbf{S}$
$\check{\mathbf{L}}$
$\check{\mathbf{L}}_{\bot}$
$\hat{\mathbf{S}}'$
$\mathbf{A}_{\top,\bot}$
$\check{\mathbf{S}}_{\bot}$
$\mathbf{A}$
$\hat{\mathbf{S}}_{\top}$
$\check{\mathbf{S}}$
$\hat{\mathbf{S}}$
$\check{\mathbf{A}}_{\bot}$
$\check{\mathbf{A}}$
$\hat{\mathbf{A}}$
$\hat{\mathbf{A}}_{\top}$
$\mathbf{I}$
$\hat{\mathbf{A}}^{\rightarrow}$
$\hat{\mathbf{I}}$
$\check{\mathbf{I}}$

CHART 1

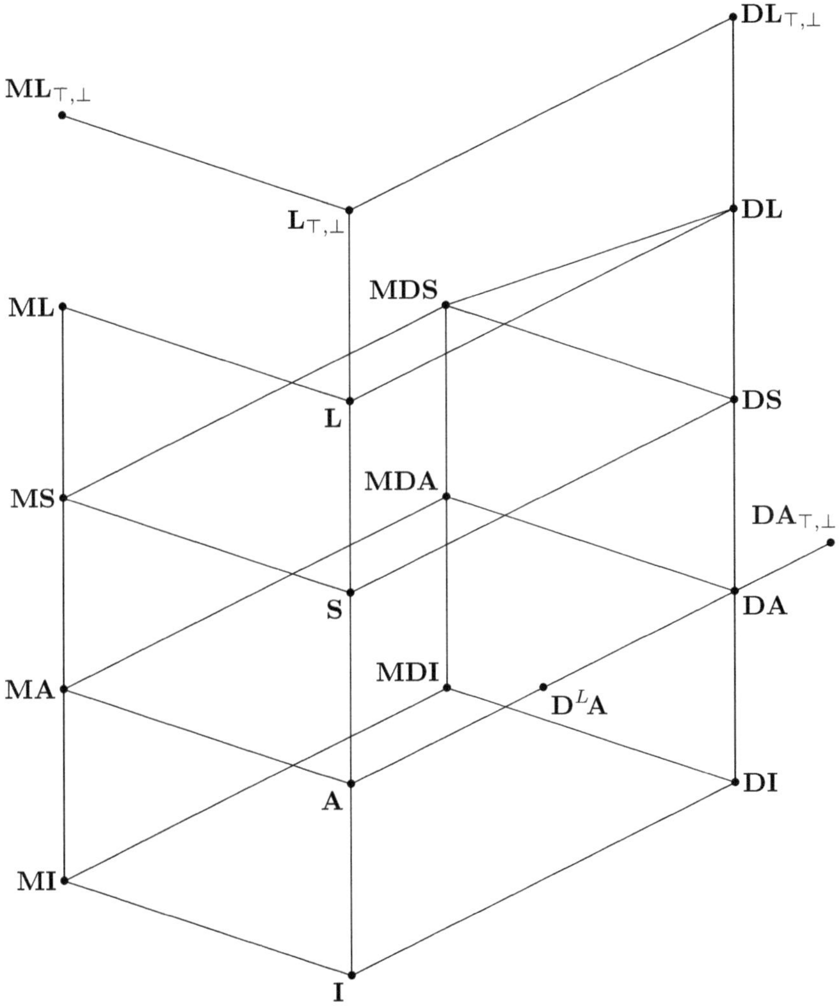
DL$_{\top,\bot}$
ML$_{\top,\bot}$
L$_{\top,\bot}$
DL
MDS
ML
DS
L
MDA
MS
DA$_{\top,\bot}$
S
DA
MDI
MA
D^{L}A
DI
A
MI
I

CHART 2

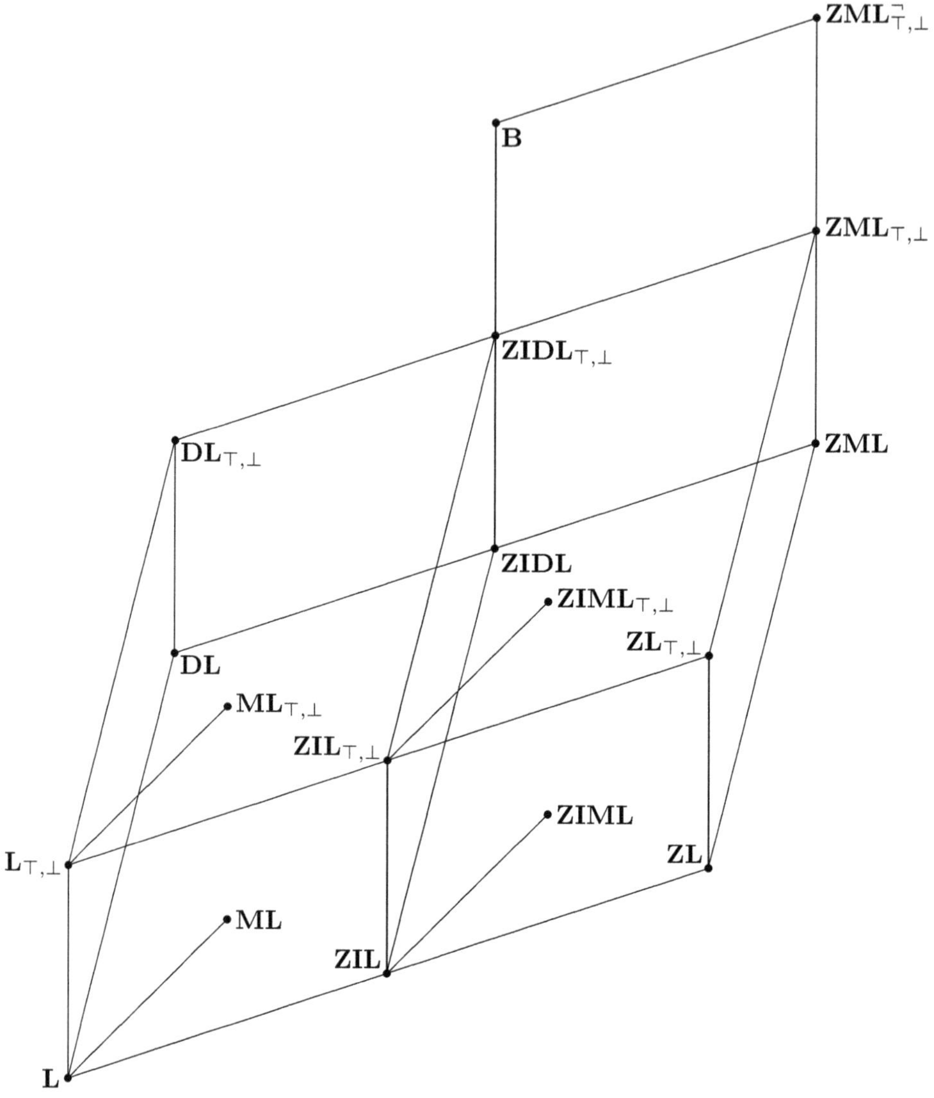
$\mathbf{ZML}^{\neg}_{\top,\bot}$
B
$\mathbf{ZML}_{\top,\bot}$
$\mathbf{ZIDL}_{\top,\bot}$
$\mathbf{DL}_{\top,\bot}$
ZML
ZIDL
$\mathbf{ZIML}_{\top,\bot}$
$\mathbf{ZL}_{\top,\bot}$
DL
$\mathbf{ML}_{\top,\bot}$
$\mathbf{ZIL}_{\top,\bot}$
ZIML
$\mathbf{L}_{\top,\bot}$
ZL
ML
ZIL
L

CHART 3

BIBLIOGRAPHY

[§...] The sections where a reference is mentioned are listed at the end in square brackets.

[1] A.R. ANDERSON and N.D. BELNAP, ***Entailment: The Logic of Relevance and Necessity, Vol. I***, Princeton University Press, Princeton, 1975 [§7.1, §14.1, §§14.3-4]

[2] J.C. BAEZ and J. DOLAN, *Categorification*, ***Higher Category Theory*** (E. Getzler et al., editors), Contemporary Mathematics, vol. 230, American Mathematical Society, Providence, 1998, pp. 1-36 [§1.2]

[3] C. BALTEANU, Z. FIEDOROWICZ, R. SCHWÄNZL and R. VOGT, *Iterated monoidal categories*, ***Advances in Mathematics***, vol. 176 (2003), pp. 277-349 [§9.4]

[4] H.P. BARENDREGT, ***The Lambda Calculus: Its Syntax and Semantics***, North-Holland, Amsterdam, 1981 [§4.2, §9.3]

[5] N.D. BELNAP, *A useful four-valued logic*, ***Modern Uses of Multiple-Valued Logic*** (J.M. Dunn and G. Epstein, editors), Reidel, Dordrecht, 1977, pp. 8-37 [§14.1]

[6] G. BIRKHOFF, ***Lattice Theory***, third edition, American Mathematical Society, Providence, 1967 (first edition, 1940) [§7.1, §12.1]

[7] R. BLUTE, *Linear logic, coherence and dinaturality*, ***Theoretical Computer Science***, vol. 115 (1993), pp. 3-41 [§8.1]

[8] R. BLUTE, J.R.B. COCKETT, R.A.G. SEELY and T.H. TRIMBLE, *Natural deduction and coherence for weakly distributive categories*, ***Journal of Pure and Applied Algebra***, vol. 113 (1996), pp. 229-296 [§7.9]

[9] M. BORISAVLJEVIĆ, *A cut-elimination proof in intuitionistic predicate logic*, ***Annals of Pure and Applied Logic***, vol. 99 (1999), pp. 105-136 [Preface, §11.1]

[10] ——, *Two measures for proving Gentzen's Hauptsatz without mix*, ***Archive for Mathematical Logic***, vol. 42 (2003), pp. 371-387 [§11.1]

[11] M. BORISAVLJEVIĆ, K. DOŠEN and Z. PETRIĆ, *On permuting cut with contraction*, ***Mathematical Structures in Computer Science***, vol. 10 (2000), pp. 99-136 [Preface, §§11.1-2]

[12] R. BRAUER, *On algebras which are connected with the semisimple continuous groups*, ***Annals of Mathematics***, vol. 38 (1937), pp. 857-872 [§1.4]

[13] K. BRÜNNLER and A.F. TIU, *A local system for classical logic*, ***Logic for Programming, Artificial Intelligence and Reasoning***, Lecture Notes in Computer Science, vol. 2250, Springer, Berlin, 2001, pp. 347-361 [§7.1, §11.1]

[14] W. BURNSIDE, ***Theory of Groups of Finite Order***, second edition, Cambridge University Press, Cambridge, 1911 (reprint, Dover, New York, 1955) [§5.2]

[15] S.R. BUSS, *The undecidability of k-provability*, ***Annals of Pure and Applied Logic***, vol. 53 (1991), pp. 75-102 [§1.4]

[16] A. CARBONE, *Interpolants, cut elimination and flow graphs for the propositional calculus*, ***Annals of Pure and Applied Logic***, vol. 83 (1997), pp. 249-299 [§1.4]

[17] J.R.B. COCKETT, *Introduction to distributive categories*, ***Mathematical Structures in Computer Science***, vol. 3 (1993), pp. 277-307 [§1.2, §11.1]

[18] J.R.B. COCKETT and R.A.G. SEELY, *Weakly distributive categories*, ***Applications of Categories in Computer Science*** (M.P. Fourman et al., editors), Cambridge University Press, Cambridge, 1992, pp. 45-65 [§§7.1-2]

[19] ——, *Weakly distributive categories*, ***Journal of Pure and Applied Algebra***, vol. 114 (1997), pp. 133-173 (updated version available at: http://www.math.mcgill.ca/rags) [Preface, §1.2, Chapter 7, §§7.1-2, §7.9, §11.1, §11.3, §12.4, §13.2]

[20] ——, *Proof theory for full intuitionistic linear logic, bilinear logic and mix categories*, ***Theory and Application of Categories***, vol. 3 (1997), pp. 85-131 [§7.9, §8.1, §8.5, §11.1]

[21] ——, *Finite sum-product logic*, ***Theory and Application of Categories***, vol. 8 (2001), pp. 63-99 [§9.6]

[22] J.R.B. Cockett, J. Koslowski and R.A.G. Seely, *Introduction to linear bicategories*, ***Mathematical Structures in Computer Science***, vol. 10 (2000), pp. 165-203 [§1.2, §7.2, §7.9]

[23] H.S.M. Coxeter and W.O.J. Moser, ***Generators and Relations for Discrete Groups***, Springer, Berlin, 1957 [§5.1]

[24] L. Crane and D.N. Yetter, *Examples of categorification*, ***Cahiers de Topologie et Géométrie Différentielle Catégoriques***, vol. 39 (1998), pp. 3-25 [§1.2]

[25] Dj. Čubrić, *Interpolation property for bicartesian closed categories*, ***Archive for Mathematical Logic***, vol. 33 (1994), pp. 291-319 [Preface]

[26] ——, *On the semantics of the universal quantifier*, ***Annals of Pure and Applied Logic***, vol. 87 (1997), pp. 209-239 [Preface]

[27] ——, *Embedding of a free cartesian closed category into the category of sets*, ***Journal of Pure and Applied Algebra***, vol. 126 (1998), pp. 121-147 [Preface, §9.3]

[28] H.B. Curry, *A note on the reduction of Gentzen's calculus LJ**, ***Bulletin of the American Mathematical Society***, vol. 45 (1939), pp. 288-293 [§1.3]

[29] ——, ***Foundations of Mathematical Logic***, McGraw-Hill, New York, 1963 (second edition, Dover, New York, 1977) [§1.3]

[30] V. Danos and L. Regnier, *The structure of multiplicatives*, ***Archive for Mathematical Logic***, vol. 28 (1989), pp. 181-203 [§7.2, §7.8, §8.1]

[31] N. Dershowitz and Z. Manna, *Proving termination with multiset orderings*, ***Communications of the Association for Computing Machinery***, vol. 22 (1979), pp. 465-476 [§11.2]

[32] K. Došen, *Modal translations in substructural logics*, ***Journal of Philosophical Logic***, vol. 21 (1992), pp. 283-336 [§11.2]

[33] ——, *Deductive completeness*, ***The Bulletin of Symbolic Logic***, vol. 2 (1996), pp. 243-283, 523 (for corrections see [34], Section 5.1.7, and [36]) [§1.3]

[34] ——, ***Cut Elimination in Categories***, Kluwer, Dordrecht, 1999 [Preface, §§1.3-5, §9.3]

[35] ——, *Models of deduction*, ***Proof-Theoretic Semantics*** (R. Kahle and P. Schroeder-Heister, editors), to appear as a special issue of ***Synthese***, 1999 [§1.3]

[36] ——, *Abstraction and application in adjunction*, ***Proceedings of the Tenth Congress of Yugoslav Mathematicians*** (Z. Kadelburg, editor), Faculty of Mathematics, University of Belgrade, Belgrade, 2001, pp. 33-46 (available at: http://arXiv.org/math.CT/0111061) [§1.3]

[37] ——, *Identity of proofs based on normalization and generality*, ***The Bulletin of Symbolic Logic***, vol. 9 (2003), pp. 477-503 (version with corrected remark on difunctionality available at: http://arXiv.org/math.LO/0208094) [Preface, Chapter 1, §1.3, §14.3]

[38] ——, *Simplicial endomorphisms*, preprint (available at: http://arXiv.org/math.GT/0301302), 2003 [§1.4]

[39] K. Došen and Z. Petrić, *Modal functional completeness*, ***Proof Theory of Modal Logic*** (H. Wansing, editor), Kluwer, Dordrecht, 1996, pp. 167-211 [§1.3, §9.1, §11.1]

[40] ——, *Cartesian isomorphisms are symmetric monoidal: A justification of linear logic*, ***The Journal of Symbolic Logic***, vol. 64 (1999), pp. 227-242 [§9.1, §11.2, §11.5]

[41] ——, *The maximality of the typed lambda calculus and of cartesian closed categories*, ***Publications de l'Institut Mathématique (N.S.)***, vol. 68(82) (2000), pp. 1-19 (available at: http://arXiv.org/math.CT/9911073) [§1.5, §9.3]

[42] ——, *The maximality of cartesian categories*, ***Mathematical Logic Quarterly***, vol. 47 (2001), pp. 137-144 (available at: http://arXiv.org/math.CT/911059) [§§1.4-5, Chapter 9, §§9.2-3]

[43] ——, *Coherent bicartesian and sesquicartesian categories*, ***Proof Theory in Computer Science*** (R. Kahle et al., editors), Lecture Notes in Computer Science, vol. 2183, Springer, Berlin, 2001, pp. 78-92 (available at: http://arXiv.org/math.CT/0006091) [§1.4, Chapter 9, §9.6]

[44] ———, *Bicartesian coherence*, ***Studia Logica***, vol. 71 (2002), pp. 331-353 (version with some corrections in the proof of maximality available at: http://arXiv.org/math.CT/0006052) [§§1.4-5, Chapter 9, §§9.4-6, §11.1]

[45] ———, *Self-adjunctions and matrices*, ***Journal of Pure and Applied Algebra***, vol. 184 (2003), pp. 7-39 (unabridged version available at: http://arXiv.org/math.GT/0111058) [§1.4]

[46] ———, *Generality of proofs and its Brauerian representation*, ***The Journal of Symbolic Logic***, vol. 68 (2003), pp. 740-750 (available at: http://arXiv.org/math.LO/0211090) [§1.4, §14.3]

[47] ———, *A Brauerian representation of split preorders*, ***Mathematical Logic Quarterly***, vol. 49 (2003), pp. 579-586 (available at: http://arXiv.org/math.LO/0211277) [§1.4, §14.3]

[48] B. ECKMANN and P.J. HILTON, *Group-like structures in general categories I: Multiplications and comultiplications*, ***Mathematische Annalen***, vol. 145 (1962), pp. 227-255 [§9.1]

[49] S. EILENBERG and G.M. KELLY, *A generalization of the functorial calculus*, ***Journal of Algebra***, vol. 3 (1966), pp. 366-375 [Preface, §1.4]

[50] ———, *Closed categories*, ***Proceedings of the Conference on Categorical Algebra, La Jolla 1965*** (S. Eilenberg et al., editors), Springer, Berlin, 1966, pp. 421-562 [§9.1]

[51] A. FLEURY and C. RETORÉ, *The mix rule*, ***Mathematical Structures in Computer Science***, vol. 4 (1994), pp. 273-285 [§8.1]

[52] P.J. FREYD, ***Abelian Categories: An Introduction to the Theory of Functors***, Harper & Row, New York, 1964 [§13.3]

[53] ———, *Aspects of topoi*, ***Bulletin of the Australian Mathematical Society***, vol. 7 (1972), pp. 1-76 [§14.3]

[54] P.J. FREYD and A. SCEDROV, ***Categories, Allegories***, North-Holland, Amsterdam, 1990 [§13.3]

[55] G. GENTZEN, *Untersuchungen über das logische Schließen*, ***Mathematische Zeitschrift***, vol. 39 (1935), pp. 176-210, 405-431 (English translation: *Investigations into logical deduction*, in [57], pp. 68-131) [Chapter 1, §1.3, §1.6, §7.6, §8.1, §§11.1-2]

[56] ———, *Neue Fassung des Widerspruchsfreiheitsbeweises für die reine Zahlentheorie*, ***Forschungen zur Logik und zur Grundlegung der exakten Wissenschaften (N.S.)***, vol. 4 (1938), pp. 19-44 (English translation: *New version of the consistency proof for elementary number theory*, in [57], pp. 252-286) [§11.2]

[57] ———, ***The Collected Papers of Gerhard Gentzen*** (M.E. Szabo, editor), North-Holland, Amsterdam, 1969

[58] J.-Y. GIRARD, *Linear logic*, ***Theoretical Computer Science***, vol. 50 (1987), pp. 1-101 [§7.2, §7.8, §8.1]

[59] J.-Y. GIRARD, P. TAYLOR and Y. LAFONT, ***Proofs and Types***, Cambridge University Press, Cambridge, 1989 [§1.6, §11.1]

[60] V.N. GRISHIN, *A generalization of the Ajdukiewicz-Lambek system* (in Russian), ***Investigations in Nonclassical Logics and Formal Systems*** (in Russian), Nauka, Moscow, 1983, pp. 315-334 (*Mathematical Reviews* 85f:03068) [§7.1]

[61] D. HILBERT and W. ACKERMANN, ***Grundzüge der theoretischen Logik***, Springer, Berlin, 1928 (English translation of the second edition from 1938, ***Principles of Mathematical Logic***, Chelsea, New York, 1950) [§1.5, §9.3]

[62] J.M.E. HYLAND, *Proof theory in the abstract*, ***Annals of Pure and Applied Logic***, vol. 114 (2002), pp. 43-78 [§1.6, §11.1, §14.3]

[63] J.M.E. HYLAND and V. DE PAIVA, *Full intuitionistic linear logic (extended abstract)*, ***Annals of Pure and Applied Logic***, vol. 64 (1993), pp. 273-291 [§7.1]

[64] C.B. JAY, *Coherence in category theory and the Church-Rosser property*, ***Notre Dame Journal of Formal Logic***, vol. 33 (1992), pp. 140-143 [Preface]

[65] V.F.R. JONES, *A quotient of the affine Hecke algebra in the Brauer algebra*, ***L'Enseignement Mathématique (2)***, vol. 40 (1994), pp. 313-344 [§1.4]

[66] A. JOYAL and R. STREET, *The geometry of tensor calculus, I*, ***Advances in Mathematics***, vol. 88 (1991), pp. 55-112 [§4.6]
[67] ——, *Braided tensor categories*, ***Advances in Mathematics***, vol. 102 (1993), pp. 20-78 [Preface, §1.7, §3.1, §§4.5-7, §9.4]
[68] C. KASSEL, ***Quantum Groups***, Springer, Berlin, 1995 [§1.4]
[69] L.H. KAUFFMAN and S.L. LINS, ***Temperley-Lieb Recoupling Theory and Invariants of 3-Manifolds***, Annals of Mathematical Studies, vol. 134, Princeton University Press, Princeton, 1994 [§1.4]
[70] G.M. KELLY, *On Mac Lane's conditions for coherence of natural associativities, commutativities, etc.*, ***Journal of Algebra***, vol. 1 (1964), pp. 397-402 [§4.6, §5.3]
[71] ——, *Many-variable functorial calculus, I*, in [74], pp. 66-105 [§11.1]
[72] ——, *An abstract approach to coherence*, in [74], pp. 106-147 [Preface, §9.2]
[73] ——, *A cut-elimination theorem*, in [74], pp. 196-213 [Preface]
[74] G.M. KELLY et al., editors, ***Coherence in Categories***, Lecture Notes in Mathematics, vol. 281, Springer, Berlin, 1972 [§1.1]
[75] G.M. KELLY and M.L. LAPLAZA, *Coherence for compact closed categories*, ***Journal of Pure and Applied Algebra***, vol. 19 (1980), pp. 193-213 [Preface, §1.1]
[76] G.M. KELLY and S. MAC LANE, *Coherence in closed categories*, ***Journal of Pure and Applied Algebra***, vol. 1 (1971), pp. 97-140, 219 [Preface, §1.1, §1.4]
[77] S.C. KLEENE, ***Introduction to Metamathematics***, North-Holland, Amsterdam, 1952 [Chapter 1]
[78] G. KREISEL, *A survey of proof theory II*, ***Proceedings of the Second Scandinavian Logic Symposium*** (J.E. Fenstad, editor), North-Holland, Amsterdam, 1971, pp. 109-170 [§1.5]
[79] J. LAMBEK, *Deductive systems and categories I: Syntactic calculus and residuated categories*, ***Mathematical Systems Theory***, vol. 2 (1968), pp. 287-318 [Preface, §1.1, §1.4]
[80] ——, *Deductive systems and categories II: Standard constructions and closed categories*, ***Category Theory, Homology Theory and their Applications I***, Lecture Notes in Mathematics, vol. 86, Springer, Berlin, 1969, pp. 76-122 [Preface, §1.1, §1.4, §7.8]
[81] ——, *Deductive systems and categories III: Cartesian closed categories, intuitionist propositional calculus, and combinatory logic*, ***Toposes, Algebraic Geometry and Logic*** (F.W. Lawvere, editor), Lecture Notes in Mathematics, vol. 274, Springer, Berlin, 1972, pp. 57-82 [Preface, §1.4]
[82] ——, *Functional completeness of cartesian categories*, ***Annals of Mathematical Logic***, vol. 6 (1974), pp. 259-292 [Preface, §§1.3-4]
[83] ——, *Multicategories revisited*, ***Categories in Computer Science and Logic*** (J.W. Gray and A. Scedrov, editors), American Mathematical Society, Providence, 1989, pp. 217-239 [§7.8]
[84] ——, *From categorial grammar to bilinear logic*, ***Substructural Logics*** (K. Došen and P. Schroeder-Heister, editors), Oxford University Press, Oxford, 1993, pp. 207-237 [§7.1]
[85] J. LAMBEK and P.J. SCOTT, ***Introduction to Higher Order Categorical Logic***, Cambridge University Press, Cambridge, 1986 [Preface, Chapter 1, §§1.1-3, §2.2, §9.1, §9.6, §13.4, §14.3]
[86] M.L. LAPLAZA, *Coherence for distributivity*, in [74], pp. 29-65 [§11.3]
[87] ——, *A new result of coherence for distributivity*, in [74], pp. 214-235 [§11.3]
[88] F.W. LAWVERE, *Adjointness in foundations*, ***Dialectica***, vol. 23 (1969), pp. 281-296 [Preface, §1.3]
[89] F.W. LAWVERE and S.H. SCHANUEL, ***Conceptual Mathematics: A First Introduction to Categories***, Cambridge University Press, Cambridge, 1997 [§1.2, §11.1, §12.1, §13.3]
[90] T. LEINSTER, *A survey of definitions of n-category*, ***Theory and Application of Categories***, vol. 10 (2002), pp. 1-70 [§1.1, §1.4]

[91] W.B.R. LICKORISH, ***An Introduction to Knot Theory***, Springer, Berlin, 1997 [§1.1, §1.4]

[92] J.-L. LODAY et al., editors, ***Operads: Proceedings of Renaissance Conferences***, Contemporary Mathematics, vol. 202, American Mathematical Society, Providence, 1997

[93] S. MAC LANE, *Natural associativity and commutativity*, ***Rice University Studies, Papers in Mathematics***, vol. 49 (1963), pp. 28-46 [Preface, §1.1, §1.4, §2.9, §§3.1-2, Chapter 4, §§4.2-3, §4.6, Chapter 5, §5.1, §5.3]

[94] ——, ***Categories for the Working Mathematician***, Springer, Berlin, 1971 (incorporated in the expanded edition [96]) [Chapter 1, §1.7, §§3.1-2, Chapter 4, §§4.2-3, §4.6, §5.1, §7.9, §11.3, §12.1, §§13.2-3, §§14.2-3]

[95] ——, *Why commutative diagrams coincide with equivalent proofs*, ***Algebraist's Homage: Papers in Ring Theory and Related Topics*** (S.A. Amitsur et al., editors), Contemporary Mathematics, vol. 13, American Mathematical Society, Providence, 1982, pp. 387-401 [Preface]

[96] ——, ***Categories for the Working Mathematician***, second edition, Springer, Berlin, 1998 [§1.7, §2.8, §§3.1-2, §4.5, §4.7, Chapter 5, §5.1]

[97] S. MAEHARA, *Eine Darstellung der intuitionistischen Logik in der klassischen*, ***Nagoya Mathematical Journal***, vol. 7 (1954), pp. 45-64 [§1.3, §11.2]

[98] M. MAKKAI and G.E. REYES, ***First Order Categorical Logic***, Lecture Notes in Mathematics, vol. 611, Springer, Berlin, 1977 [Chapter 2]

[99] C.R. MANN, *The connection between equivalence of proofs and cartesian closed categories*, ***Proceedings of the London Mathematical Society*** (3), vol. 31 (1975), pp. 289-310 [§1.4]

[100] G.E. MINTS, *Category theory and proof theory* (in Russian), ***Aktual'nye voprosy logiki i metodologii nauki***, Naukova Dumka, Kiev, 1980, pp. 252-278 (English translation, with permuted title, in: G. E. Mints, *Selected Papers in Proof Theory*, Bibliopolis, Naples, 1992) [§9.2]

[101] Y.N. MOSCHOVAKIS, *What is an algorithm?*, ***Mathematics Unlimited—2001 and Beyond*** (B. Engquist and W. Schmid, editors), Springer, Berlin, 2001, pp. 919-936 [§1.3]

[102] Z. PETRIĆ, *Coherence in substructural categories*, ***Studia Logica***, vol. 70 (2002), pp. 271-296 (available at: http://arXiv.org/math.CT/0006061) [§1.4, §9.2, §11.1, §14.4]

[103] ——, *G-Dinaturality*, ***Annals of Pure and Applied Logic***, vol. 122 (2003), pp. 131-173 (available at: http://arXiv.org/math.CT/0012019) [§1.1, §1.3, §11.1]

[104] V.V. PRASOLOV and A.B. SOSINSKIĬ, ***Knots, Links, Braids and 3-Manifolds*** (in Russian), MCNMO, Moscow, 1997 [§1.4]

[105] D. PRAWITZ, ***Natural Deduction: A Proof-Theoretical Study***, Almqvist & Wiksell, Stockholm, 1965 [Chapter 1, §1.3]

[106] ——, *Ideas and results in proof theory*, ***Proceedings of the Second Scandinavian Logic Symposium*** (J.E. Fenstad, editor), North-Holland, Amsterdam, 1971, pp. 235-307 [§1.3, §1.5]

[107] H. RASIOWA ***An Algebraic Approach to Non-Classical Logics***, North-Holland, Amsterdam, 1974 [§14.1]

[108] J. RIGUET, *Relations binaires, fermetures, correspondances de Galois*, ***Bulletin de la Société mathématique de France***, vol. 76 (1948), pp. 114-155 [§1.4, §14.3]

[109] R.R. SCHNECK, *Natural deduction and coherence for non-symmetric linearly distributive categories*, ***Theory and Applications of Categories***, vol. 6 (1999), pp. 105-146 [§7.9]

[110] M.C. SHUM, *Tortile tensor categories*, ***Journal of Pure and Applied Algebra***, vol. 93 (1994), pp. 57-110 [§3.2]

[111] A.K. SIMPSON, *Categorical completeness results for the simply-typed lambda-calculus*, ***Typed Lambda Calculi and Applications*** (M. Dezani-Ciancaglini and G. Plotkin, editors), Lecture Notes in Computer Science, vol. 902, Springer, Berlin, 1995, pp. 414-427 [Preface, §1.5, §9.3]

[112] S.V. SOLOVIEV, *The category of finite sets and cartesian closed categories* (in Russian), ***Zapiski Nauchnykh Seminarov (LOMI)***, vol. 105 (1981), pp. 174-194 (English translation in ***Journal of Soviet Mathematics***, vol. 22 (1983), pp. 1387-1400) [Preface]

[113] ——, *On the conditions of full coherence in closed categories*, ***Journal of Pure and Applied Algebra***, vol. 69 (1990), pp. 301-329 (Russian version in ***Matematicheskie metody postroeniya i analiza algoritmov***, A.O. Slisenko and S.V. Soloviev, editors, Nauka, Leningrad, 1990, pp. 163-189) [Preface, §7.9]

[114] ——, *Proof of a conjecture of S. Mac Lane*, ***Annals of Pure and Applied Logic***, vol. 90 (1997), pp. 101-162 [Preface]

[115] R. STATMAN, *λ-definable functionals and $\beta\eta$-conversion*, ***Archiv für mathematische Logik und Grundlagenforschung***, vol. 23 (1983), pp. 21-26 [§1.5]

[116] J.D. STASHEFF, *The pre-history of operads*, in [92], pp. 9-14 [§1.1]

[117] ——, *From operads to 'physically' inspired theories*, in [92], pp. 53-81 [§1.1]

[118] M.E. SZABO, *A categorical characterization of Boolean algebras*, ***Algebra Universalis***, vol. 4 (1974), pp. 192-194 [§14.3]

[119] ——, *A counter-example to coherence in cartesian closed categories*, ***Canadian Mathematical Bulletin***, vol. 18 (1975), pp. 111-114 [§1.4]

[120] ——, *Polycategories*, ***Communications in Algebra***, vol. 3(8) (1975), pp. 663-689 [§7.9]

[121] ——, ***Algebra of Proofs***, North-Holland, Amsterdam, 1978 [Preface, §1.6]

[122] A.S. TROELSTRA and H. SCHWICHTENBERG, ***Basic Proof Theory***, Cambridge University Press, Cambridge, 1996 (second edition, 2000) [Preface, Chapter 1, §9.2]

[123] V.G. TURAEV, *Operator invariants of tangles and R-matrices* (in Russian), ***Izvestiya Akademii Nauk SSSR, Seriya Matematicheskaya***, vol. 53 (1989), pp. 1073-1107 (English translation in ***Mathematics of the USSR-Izvestiya***, vol. 35, 1990, pp. 411-444) [§1.4]

[124] J. VON PLATO, *A proof of Gentzen's* Hauptsatz *without multicut*, ***Archive for Mathematical Logic***, vol. 40 (2001), pp. 9-18 [§11.1]

[125] R. VOREADOU, ***Coherence and Non-Commutative Diagrams in Closed Categories***, Memoirs of the American Mathematical Society, no 182, American Mathematical Society, Providence, 1977 [§1.1]

[126] H. WENZL, *On the structure of Brauer's centralizer algebras*, ***Annals of Mathematics***, vol. 128 (1988), pp. 173-193 [§1.4]

[127] F. WIDEBÄCK, ***Identity of Proofs***, doctoral dissertation, University of Stockholm, Almqvist & Wiksell, Stockholm, 2001 [§1.5]

[128] D.N. YETTER, *Markov algebras*, ***Braids*** (J.S. Birman and A. Libgober, editors), Contemporary Mathematics, vol. 78, American Mathematical Society, Providence, 1988, pp. 705-730 [§1.4]

[129] R. ZACH, *Completeness before Post: Bernays, Hilbert, and the development of propositional logic*, ***The Bulletin of Symbolic Logic***, vol. 5 (1999), pp. 331-366 [§1.5, §9.3]

INDEX

www.ingramcontent.com/pod-product-compliance
Ingram Content Group UK Ltd.
Pitfield, Milton Keynes, MK11 3LW, UK
UKHW021316180426
11947UKWH00015B/1265

9 781904 987062